Corrosion Control in Petroleum Production

3rd Edition

Robert J. Franco and Timothy Bieri

Cover photographs provided courtesy of BP PLC

NACE International
The Worldwide Corrosion Authority

©2020 by NACE International
All Rights Reserved
Printed in the United States of America

ISBN: 978-1-57590-389-7

Reproduction of the contents in whole or part or transfer into electronic or photographic storage
without permission of the copyright owner is strictly forbidden.

Neither NACE International, its officers, directors, nor members thereof accept any responsibility for the use of the methods and materials discussed herein. No authorization is implied concerning the use of patented or copyrighted material. The information is advisory only and the use of the materials and methods is solely at the risk of the user.

NACE International
The Worldwide Corrosion Authority
15835 Park Ten Place
Houston, TX 77084
nace.org

Contents

Foreword xvii
In Memoriam xix
About the Authors xxi
Reviewers and Contributors xxiii

CHAPTER 1: OVERVIEW 1
  1.1 Definition and Importance 1
  1.2 The Corrosion Cell 1
      1.2.1 Controlling Corrosion 4
  1.3 Corrosive Environments 4
      1.3.1 External Environments 5
            1.3.1.1 Atmospheric Corrosion 5
            1.3.1.2 Corrosion Under Insulation (CUI) 5
            1.3.1.3 Soil Corrosion 5
            1.3.1.4 Submerged Corrosion 6
  1.4 Internal Environments 6
  1.5 Equipment, Systems, and Facilities 10
      1.5.1 Wells 11
            1.5.1.1 Artificial Lift Wells 12
            1.5.1.2 Flowing Wells 14
            1.5.1.3 Casing Corrosion 15
      1.5.2 Surface Equipment 17
            1.5.2.1 Wellheads and Christmas Trees 17
            1.5.2.2 Flow Lines 17
      1.5.3 Processing Equipment 18
            1.5.3.1 Separators 18
            1.5.3.2 Gas Sweetening 19
            1.5.3.3 Gas Dehydration 19

    1.5.4 Storage Tanks 19
    1.5.5 Piping 20
        1.5.5.1 Flow Velocity 21
        1.5.5.2 Turbulence Areas... 22
        1.5.5.3 Stagnant Zones and Dead Legs 23
        1.5.5.4 Flanged/Bolted Equipment 25
    1.5.6 Water Systems.. 27
1.6 Drilling Operations 28
1.7 Enhanced Oil Recovery Projects.. 28
    1.7.1 WAG Injection 28
    1.7.2 In Situ Combustion 29
1.8 Carbon Capture and Sequestration... 29
    1.8.1 Coal Seam Gas... 30
1.9 Offshore Operations 30
    1.9.1 Reservoir Souring.. 30
    1.9.2 Corrosion Control 31
        1.9.2.1 Control of External Corrosion 31
1.10 Corrosion Under Insulation.. 33
1.11 Summary... 35
References... 35
Bibliography... 36
    APPENDIX 1.A: Internal Corrosion and Corrosion Control in Oil and Gas Production ... 37

CHAPTER 2: CORROSION BASICS 43
2.1 Introduction 43
2.2 Why Iron and Steels Corrode... 43
2.3 Corrosion Reactions... 44
2.4 Corrosion Variables.. 46
    2.4.1 Conductivity... 46
    2.4.2 pH.. 46
    2.4.3 Dissolved Gases... 47
        2.4.3.1 Dissolved Oxygen (DO) 48
        2.4.3.2 Dissolved Carbon Dioxide (CO_2) 50
        2.4.3.3 Other Factors Affecting CO_2 Corrosion 54
        2.4.3.4 Steel Chemistry and Microstructure 54
        2.4.3.5 Dissolved Hydrogen Sulfide (H_2S)... 55
2.5 Microbiological Influences 59
    2.5.1 SRBs... 59
    2.5.2 APBs 60
    2.5.3 MOBs... 60
    2.5.4 Field Monitoring.. 60
    2.5.5 Rapidly Evolving Understanding.. 61
2.6 Physical Variables 61
    2.6.1 Temperature 61
    2.6.2 Pressure 62
    2.6.3 Velocity... 62
2.7 Corrosion Product and Scale Analysis... 62
    2.7.1 Elemental Analysis.. 63
    2.7.2 X-Ray Diffraction... 63

  2.7.3 Original Manufacturing Defects 63
  2.7.4 Chemical Analysis Inside Cracks 64
  2.7.5 Quick Field Analysis Methods 64
References 65
Bibliography 67
Appendix 2.A: Solubility of Oxygen in Water 68
Appendix 2.B: Minimum Partial Pressure of H_2S to Form Mackinawite 69

CHAPTER 3: FORMS (OR TYPES) OF CORROSION 71
 3.1 **Introduction** 71
 3.2 **Definitions and Classifications** 72
  3.2.1 Mechanisms Involving Wall-Loss Damage 73
  3.2.2 Mechanisms Not Necessarily Involving Significant Wall-Loss Damage 74
  3.2.3 Alternative Definitions API 579-1/ASME FFS-1, "Fitness-For-Service" 74
  3.2.4 Applications of Definitions 75
 3.3 **General Corrosion** 76
 3.4 **Localized Corrosion** 79
  3.4.1 Pitting Corrosion 79
 3.5 **Concentration Cells** 85
  3.5.1 Crevice Corrosion 86
  3.5.2 Oxygen Tubercles 88
  3.5.3 Differential Aeration Cells 88
  3.5.4 Scale and Deposits 90
    3.4.5.1 MIC 92
  3.5.5 Localized Corrosion and Pitting of Metals 94
    3.5.5.1 Carbon and Low-Alloy Steel 94
    3.5.5.2 Corrosion-Resistant Alloys (CRAs) 95
    3.5.5.3 Other Alloys 96
 3.6 **Galvanic Corrosion** 97
  3.6.1 Area Principle 98
  3.6.2 Galvanic Series 99
  3.6.3 Oxygen and Depolarizers 100
  3.6.4 Polarity Reversal 100
  3.6.5 Mill Scale 100
  3.6.6 Ringworm Corrosion and Preferential Weld Corrosion (PWC) 101
    3.6.6.1 Ringworm Corrosion 101
    3.6.6.2 PWC of Steel 102
    3.6.6.3 PWC of Austenitic Stainless Steel 103
    3.6.6.4 PWC Caused by MIC 105
 3.7 **Erosion-Corrosion, Impingement, and Wear** 106
  3.7.1 Erosion-Corrosion 106
  3.7.2 Wear Corrosion 108
  3.7.3 Impingement Corrosion 109
  3.7.4 Erosion 111
 3.8 **Cavitation** 112
 3.9 **Intergranular Corrosion** 113
  3.9.1 Sensitization of Austenitic Stainless Steel 113
  3.9.2 Preventing Intergranular Corrosion 114
 3.10 **Hydrogen-Induced Failures** 115

- 3.11 Hydrogen Blistering and Hydrogen-Induced Cracking 115
  - 3.11.1 Intrusive Hydrogen Probes 116
  - 3.11.2 Patch (Nonintrusive) Hydrogen Probes 116
- 3.12 Hydrogen Embrittlement 118
- 3.13 SSC 119
  - 3.13.1 SSC Susceptibility and Steel Strength 119
  - 3.13.2 SSC Is Typically Intergranular 120
  - 3.13.3 Conditions Required for SSC 121
  - 3.13.4 Factors Affecting SSC Susceptibility 122
- 3.14 High Temperature Attack 122
- 3.15 Fatigue of Metals in Air 123
- 3.16 Corrosion Fatigue 125
- 3.17 Stress Corrosion Cracking (SCC) 129
  - 3.17.1 Carbon and Low-Alloy Steels 130
    - 3.17.1.1 Alkaline SCC of Carbon and Low-Alloy Steels 130
    - 3.17.1.2 External SCC of Carbon and Low-Alloy Steel Underground Pipelines ... 131
    - 3.17.1.3 Other SCC Environments Affecting Carbon and Low-Alloy Steels 132
  - 3.17.2 High Strength Steels 132
  - 3.17.3 Austenitic Alloys 133
    - 3.17.3.1 Temperature 133
    - 3.17.3.2 pH 133
    - 3.17.3.3 Oxygen 134
    - 3.17.3.4 Strain Hardening 135
    - 3.17.3.5 Thermal Insulation 135
    - 3.17.3.6 Sensitized Microstructure 135
    - 3.17.3.7 Machining Grades 136
  - 3.17.4 Copper-Based Alloys 136
  - 3.17.5 Aluminum Alloys 137
  - 3.17.6 Nickel and Nickel-Based Alloys 137
  - 3.17.7 Titanium and Titanium Alloys 137
- 3.18 Liquid Metal Embrittlement 138
- References 139
- Bibliography 141

CHAPTER 4: METALLIC MATERIALS SELECTION 143
- 4.1 Introduction 143
- 4.2 Basic Metallurgy 145
  - 4.2.1 Crystalline Structure 145
    - 4.2.1.1 Grain Boundaries 146
    - 4.2.1.2 Isotropic and Anisotropic Polycrystalline Metals 146
    - 4.2.1.3 Effect of Grain Size on Mechanical Properties 147
  - 4.2.2 Alloys 149
- 4.3 Steels 150
  - 4.3.1 Structure of Steel 150
  - 4.3.2 Names for Steels 151
- 4.4 Heat Treatment of Steels 152
  - 4.4.1 Annealing 153
  - 4.4.2 Normalizing 153
  - 4.4.3 Spheroidizing 153

    4.4.4 Quenching and Tempering 153
        4.4.4.1 Carbide Distribution and Corrosion Resistance 154
    4.4.5 Localized Heat Treatments.. 155
        4.4.5.1 Ringworm Corrosion 155
        4.4.5.2 Weld Line Corrosion 155
4.5 Alloys.. 156
    4.5.1 Homogeneous and Heterogeneous Alloys 156
    4.5.2 Ferrous and Nonferrous Alloys 156
    4.5.3 Corrosion Resistant Alloys (CRAs) 157
    4.5.4 Heat-Treated Alloys 157
4.6 Ferrous Alloys 158
    4.6.1 Carbon and Alloy Steels... 158
    4.6.2 Chromium-Containing Steels 158
        4.6.2.1 Chromium-Molybdenum Steels 159
    4.6.3 Nickel-Containing Steels 159
    4.6.4 Stainless Steels 160
        4.6.4.1 Martensitic Stainless Steels 160
        4.6.4.2 Ferritic Stainless Steels 162
        4.6.4.3 Austenitic Stainless Steels 163
        4.6.4.4 Precipitation Hardening (PH) Stainless Steels 165
        4.6.4.5 Duplex Stainless Steels 167
        4.6.4.6 Highly Alloyed Austenitic Stainless Steels 170
    4.6.5 Cast Irons 171
4.7 Nonferrous Alloys... 171
    4.7.1 Solid-Solution Nickel-Based Alloys 171
    4.7.2 Precipitation-Hardened Nickel-Based Alloys 173
    4.7.3 Cobalt-Based Alloys 174
    4.7.4 Titanium Alloys 174
    4.7.5 Copper-Based Alloys 175
    4.7.6 Aluminum and Aluminum Alloys 176
    4.7.7 Tungsten Alloys 176
4.8 Specifications, Standards, and Codes 176
References... 178
Bibliography... 180

CHAPTER 5: NONMETALLIC MATERIALS SELECTION 181
5.1 Introduction 181
5.2 Cement and Concrete 181
5.3 Polymers 183
5.4 Thermoplastics 183
    5.4.1 Amorphous Thermoplastics 184
    5.4.2 Semicrystalline Thermoplastics 184
    5.4.3 Service Considerations for Thermoplastic Materials...................................... 184
5.5 Thermosetting.. 187
5.6 Nonmetallic Pipe 190
    5.6.1 Advantages and Disadvantages of Nonmetallic Pipe..................................... 190
        5.6.1.1 Advantages 190
        5.6.1.2 Disadvantages 190
        5.6.1.3 Disadvantages Unique to FRP.. 191

- 5.6.2 Factors to Consider Before Selecting Nonmetallic Pipe 191
  - 5.6.2.1 Chemical Resistance 191
  - 5.6.2.2 Mechanical Properties 192
  - 5.6.2.3 Fire Performance 192
  - 5.6.2.4 Economics 192
- 5.6.3 Joining Methods for Pipe 193
  - 5.6.3.1 Butt Fusion (Heat Welding) 193
  - 5.6.3.2 Solvent Welding 194
  - 5.6.3.3 Threaded Connections 194
  - 5.6.3.4 Flange Connections 195
  - 5.6.3.5 Eliminate the Connection 195
- 5.6.4 Typical Applications of Pipe in Oil and Gas Production 197
  - 5.6.4.1 Thermoplastics 197
  - 5.6.4.2 FRP (Thermosetting) 198
  - 5.6.4.3 Spoolable Pipe 199
  - 5.6.4.4 Flexible Pipe 199
- 5.6.5 Internal Polymer Liners for Pipes and Tanks 200
  - 5.6.5.1 Polyethylene Liner 200
  - 5.6.5.2 Other Polymer Liner Materials for Pipelines 203
  - 5.6.5.3 Tank and Slurry Pipeline Liners 203

5.7 FRP Pipe for Oil and Gas Production 204
- 5.7.1 FRP Pipe Manufacturing 204
- 5.7.2 Designing an FRP Pipe System 205
- 5.7.3 FRP Pipe for Flowlines, Well Lines, Manifolds 206

5.8 FRP Tanks 207
5.9 FRP Sucker Rods 209
5.10 FRP Production and Injection Tubing 209
5.11 FRP Offshore Firewater Pipe 210
5.12 Composite Repair Systems for Corroded Steel 211
- 5.12.1 Cold Weld Compounds 212

5.13 Elastomers 212
- 5.13.1 Seals and Packing 212
  - 5.13.1.1 O-Rings 213
  - 5.13.1.2 Gaskets 213
  - 5.13.1.3 Other Seals 213
  - 5.13.1.4 Flange Isolation 213
- 5.13.2 Classification of Elastomers 214
  - 5.13.2.1 Classification of Commonly Used Seal Materials 214
- 5.13.3 Chemical Resistance of Elastomers and Thermoplastic Seals 215
  - 5.13.3.1 Explosive Decomposition and Chemical Degradation 215
  - 5.13.3.2 Maximum and Minimum Service Temperature 216

5.14 Offshore Riser Splash Zone Protection 216
References 217
Appendix 5.A: Compilation of Standards Concerning Nonmetallic
Materials Used in Petroleum Production 220

CHAPTER 6: PROTECTIVE COATINGS 227
6.1 Introduction 227

- 6.2 **Coating Formulation** 230
  - 6.2.1 Resin or Binder 231
  - 6.2.2 Pigment 231
  - 6.2.3 Vehicle 231
  - 6.2.4 Additives 232
- 6.3 **Coating Types** 232
  - 6.3.1 Coating Classification or Type 232
  - 6.3.2 Coating Thickness 232
  - 6.3.3 Chemical Composition 233
  - 6.3.4 Curing Methods 233
  - 6.3.5 Reinforced Coatings 234
- 6.4 **Coating Systems** 234
  - 6.4.1 Primers 235
  - 6.4.2 Intermediate Coats 236
  - 6.4.3 Topcoats 236
- 6.5 **Coating Selection** 237
  - 6.5.1 Service Environment 237
  - 6.5.2 Functional Requirements 238
  - 6.5.3 Application Considerations 238
- 6.6 **Coating Application** 241
  - 6.6.1 Surface Preparation 241
  - 6.6.2 Coating the Structure 244
  - 6.6.3 Inspection 245
    - 6.6.3.1 Inspection Timing 246
    - 6.6.3.2 Inspection Tools 246
- 6.7 **Coating of Production Facilities and Equipment** 250
  - 6.7.1 Atmospheric Protection 250
    - 6.7.1.1 Offshore Platforms 251
    - 6.7.1.2 Corrosion Under Insulation (CUI) 254
  - 6.7.2 Vessels and Tanks 254
  - 6.7.3 Tubular Goods 255
  - 6.7.4 Flowlines, Gathering Systems, Injection Lines, and Piping 258
  - 6.7.5 External Pipeline Coatings 259
    - 6.7.5.1 Subsea 263
  - 6.7.6 Misconceptions 264
- **References** 264

CHAPTER 7: CATHODIC PROTECTION 267
- 7.1 **Introduction** 267
- 7.2 **Cathodic Protection Principles** 267
- 7.3 **Cathodic Protection Systems** 268
  - 7.3.1 Galvanic Anode Cathodic Protection Systems 269
    - 7.3.1.1 Galvanic Anodes 269
    - 7.3.1.2 Design and Installation of Sacrificial Anode Systems 271
  - 7.3.2 Impressed Current Cathodic Protection Systems 271
    - 7.3.2.1 Impressed Current Anodes 273
    - 7.3.2.2 Impressed Current Cathodic Protection Power Sources 276
    - 7.3.2.3 Cathodic Protection Interference 276

- 7.4 **Criteria for Cathodic Protection** 278
  - 7.4.1 Know Your Reference 278
  - 7.4.2 Important Terms 279
  - 7.4.3 Criteria 280
- 7.5 **Survey and Test Methods in Cathodic Protection** 280
  - 7.5.1 Structure-to-Electrolyte Potential Measurements 281
  - 7.5.2 Current Flow (IR Drop) 283
  - 7.5.3 Soil (or Water) Resistivity Measurements 285
  - 7.5.4 Current Requirement Surveys 286
- 7.6 **Application of Cathodic Protection** 287
  - 7.6.1 Cathodic Protection Design 287
  - 7.6.2 Well Casings 289
  - 7.6.3 Flowlines, Gathering Lines, Distribution Lines, Injection Lines, and Pipelines ..290
  - 7.6.4 Surface Equipment and Vessels 290
  - 7.6.5 Storage Tanks 293
    - 7.6.5.1 Internal Protection 293
    - 7.6.5.2 External Protection 294
  - 7.6.6 Offshore Structures 295
    - 7.6.6.1 Platform Design 295
    - 7.6.6.2 Criteria for Protection 295
    - 7.6.6.3 Galvanic Anodes 296
    - 7.6.6.4 Impressed Current Systems 297
    - 7.6.6.5 Subsea Structures 297
- 7.7 **Role of Protective Coatings** 298
- **References** 299

CHAPTER 8: CHEMICAL TREATMENT 301
- 8.1 **Introduction** 301
- 8.2 **Fundamentals of Corrosion Inhibitors** 302
- 8.3 **Inorganic Corrosion Inhibitors** 303
  - 8.3.1 Anodic Inorganic Inhibitors 303
  - 8.3.2 Cathodic Inorganic Inhibitors 303
- 8.4 **Organic Corrosion Inhibitors** 304
- 8.5 **Corrosion Inhibitor Formulations** 305
  - 8.5.1 Introductory Chemistry of Inhibitor Molecules 305
  - 8.5.2 Solvents and Co-solvents 306
  - 8.5.3 Additives 307
- 8.6 **Corrosion Inhibitor Properties** 307
  - 8.6.1 Inhibits Corrosion in the Well or System 308
  - 8.6.2 Film Persistency 309
  - 8.6.3 Filming Efficiency (Percent Protection) 309
  - 8.6.4 Filming Time 309
  - 8.6.5 Solubility or Dispersibility 310
    - 8.6.5.1 Effect of Temperature 311
    - 8.6.5.2 Solubility or Dispersibility in the Carrier Fluid 311
  - 8.6.6 Partitioning Between Oil and Water 311
  - 8.6.7 Compatibility of Corrosion Inhibitors with Other Chemicals 312
  - 8.6.8 Emulsification Properties 313
  - 8.6.9 Pour Point 313

8.6.10 Freeze Point	313
8.6.11 Freeze-Thaw Stability	313
8.6.12 Thermal Stability	313
8.6.12.1 Phase Behavior	314
8.6.13 Materials Compatibility	314
8.6.14 Mobility	314
8.6.15 Foaming Properties	315
8.6.16 Compatibility with Downstream Processes	315
8.6.17 Environmental Concerns	315
8.7 Selecting the Inhibitor	316
8.7.1 Laboratory Inhibitor Testing	317
8.8 Inhibitor Application	319
8.9 Inhibition of Producing Wells	319
8.9.1 Overview of Treatment Methods	319
8.9.2 Tubing Displacement Treatment	320
8.9.3 Nitrogen Squeeze and Nitrogen Displacement Treatments	321
8.9.4 Batch and Fall	321
8.9.5 Partial Tubing Displacement and Yo-Yo Treatments	322
8.9.6 Treating Strings	323
8.9.7 Batch Treating Frequency for Gas Wells	323
8.9.8 Gas-Lift Oil Wells	323
8.9.9 Hydraulic Pump Oil Wells	324
8.9.10 Electrical Submersible Pump (ESP) Oil Wells	325
8.9.11 Formation Squeeze Treatment	325
8.9.12 Continuous Treatment	326
8.9.13 Treatments Down the Tubing-Casing Annulus for Wells with an Open Annulus (Without Packers)	326
8.9.13.1 Fluid Level	326
8.9.13.2 Batch-and-Flush Treatment	327
8.9.13.3 Extended Batch Treatment	327
8.9.13.4 Automated (Semibatch) Treatment	328
8.9.13.5 Batch-and-Circulate Treatment	328
8.9.13.6 Batch Treating Frequency for Open-Annulus Oil Wells	328
8.9.13.7 Continuous Treatment	328
8.9.13.8 Infrequently Used Downhole Treating Methods	328
8.10 Inhibition of Surface Facilities	329
8.10.1 Flowlines and Gathering Systems	329
8.10.1.1 Top of the Line Corrosion	330
8.10.1.2 Batch Treating Lines Using Pigs	331
8.10.1.3 Batching Pigs	332
8.10.2 Production and Plant Facilities	332
8.10.2.1 Vessels and Tanks	332
8.10.2.2 Piping Systems	333
8.10.2.3 Gas Handling Systems and Plants	333
8.10.2.4 Gas Compression Systems	334
8.10.2.5 Gas Sweetening Systems	335
8.10.2.6 Cooling Water Systems	336
8.10.2.7 Heating and Cooling Media	337

8.11 Water Injection Systems 338
8.12 Gas Injection Wells 339
8.13 Other Types of Inhibitors 339
 8.13.1 Well Acidizing Inhibitors 340
 8.13.2 Volatile Corrosion Inhibitors 341
8.14 Horizontal Wells 341
 8.14.1 Definition of Horizontal Well 342
 8.14.2 Unique Properties of Shale Oil 342
 8.14.3 Materials and Corrosion Concerns 342
 8.14.4 Corrosion Inhibition 343
  8.14.4.1 Inhibiting Shale Oil Wells 343
  8.14.4.2 Inhibiting Shale Gas Wells 344
8.15 Chemical Delivery and Reliability 344
 8.15.1 Injection Fittings 345
8.16 Comparison with Other Corrosion Control Options 345
8.17 Summary of Corrosion Inhibition..346
References 347
Bibliography..348
Appendix 8.A: Downhole Corrosion Inhibitor Application Techniques for
 Oil and Gas Wells 349
 8.A.1 Tubing Displacement 349
 8.A.2 Nitrogen Formation Squeeze or Tubing Displacement 351
 8.A.3 Batch and Fall, Partial Tubing Displacement, and Yo-Yo Treatments..............352
 8.A.4 Treating String...354
 8.A.5 Formation Squeeze 355
 8.A.6 Weighted Liquids...357
 8.A.7 Dump Bailer 359
 8.A.8 Wash Bailer 360
 8.A.9 Inhibitor Sticks 362
 8.A.10 Chemical Injector Valve 363
 8.A.11 Gas-Lift Gas..365
 8.A.12 Batch and Flush/Circulate Treatment Down Annulus.....................................367
Appendix 8.B: Circulation Time Required for Rod Pumped Well Batch
 and Circulate Treatment 369

CHAPTER 9: INTERNAL CORROSIVE ENVIRONMENT CONTROL..371
9.1 Introduction 371
9.2 Designing to Avoid Corrosion and Designing for Corrosion Control............................372
 9.2.1 Process Layout 372
 9.2.2 Design for Coating 373
 9.2.3 Design for Inspection 374
 9.2.4 Design for Monitoring 375
 9.2.5 Design for Chemical Treatment 375
 9.2.6 Design for Pigging 375
 9.2.7 Mixing Waters from Various Sources 376
9.3 Piping Design 376
 9.3.1 Water Traps 377
 9.3.2 Dead Legs...377
 9.3.3 Velocity Effects 378

- 9.3.4 Sand Effects 378
- 9.3.5 Pipe Support Design 379
- 9.3.6 Thermally Insulated Piping 380
- **9.4 Electrical Isolation** 380
  - 9.4.1 Insulating Flange Kits and Monolithic Isolating Joints 381
  - 9.4.2 Avoiding Galvanic Couples 383
- **9.5 Oxygen Exclusion** 385
  - 9.5.1 Sources of Oxygen in Producing Wells 385
  - 9.5.2 Sources of Oxygen in Water Systems 385
    - 9.5.2.1 Design of Gas Blanket Systems 386
    - 9.5.2.2 Other Sources of Oxygen Entry in Water Systems 388
  - 9.5.3 Sources of Oxygen in Gas Plants 389
- **9.6 Oxygen Removal** 390
  - 9.6.1 Gas Stripping Towers 391
  - 9.6.2 Vacuum Stripping Towers 392
  - 9.6.3 Chemical Scavenging 393
    - 9.6.3.1 Commonly Used Scavengers 393
    - 9.6.3.2 Interference with Oil Field Biocides 395
    - 9.6.3.3 Calculating Dosage of an Oxygen Scavenger 395
    - 9.6.3.4 Injection Point Corrosion 395
    - 9.6.3.5 Reaction (Holding) Time and Need for Catalyst 395
    - 9.6.3.6 Calcium Sulfate Precipitation 396
    - 9.6.3.7 Alternative Scavengers 397
- **9.7 Dehydration of Gas** 397
  - 9.7.1 Water Dew Point 397
  - 9.7.2 Dehydration Versus Dehumidification 398
  - 9.7.3 Hydrate Control Is Not Corrosion Control 398
  - 9.7.4 Gas Dehydration Using Glycol 398
    - 9.7.4.1 pH Control 398
    - 9.7.4.2 Filtration 399
    - 9.7.4.3 Reflux Tower Corrosion 399
  - 9.7.5 Gas Dehydration Using Molecular Sieve 399
- **9.8 Control of Deposits (Cleanliness)** 399
  - 9.8.1 Pigging 400
    - 9.8.1.1 Production Pipelines Benefiting Most from Maintenance Pigging 400
    - 9.8.1.2 Maintenance Pig Selection 401
    - 9.8.1.3 Progressive Pigging 402
    - 9.8.1.4 Chemical Batching Pigs 402
    - 9.8.1.5 Maintenance Pigging 402
  - 9.8.2 Flushing 404
  - 9.8.3 Filtration 404
- **9.9 Control of Microbiologically Influenced Corrosion (MIC)** 405
  - 9.9.1 Water Injection Systems 406
  - 9.9.2 Reservoir Souring 408
  - 9.9.3 Hydrotest Waters 409
    - 9.9.3.1 Chemical Treatment of Hydrotest Water 409
- **9.10 Other Methods and Environments** 411
  - 9.10.1 Mothballing, Layup, and Decommissioning 411
    - 9.10.1.1 Decommissioning 412

   9.10.1.2 Mothballing 412
   9.10.1.3 Dry Layup 412
   9.10.1.4 Wet Layup 412
  9.10.2 pH Control 413
   9.10.2.1 pH Control in Wet Acid Gas 413
   9.10.2.2 pH Stabilization in Wet Acid Gas Pipelines 413
   9.10.2.3 pH Control in Open and Closed Cooling or Heating Water Systems 414
References 414
Bibliography 415
Appendix 9.A: Checklist for Troubleshooting Oxygen-Free Water Injection Systems 417

CHAPTER 10: CORROSION MONITORING AND INSPECTION 419
10.1 Introduction 419
10.2 Location 420
10.3 Indirect Techniques 421
 10.3.1 Online Techniques 421
  10.3.1.1 Process Parameters 422
  10.3.1.2 Fluid Detection 422
 10.3.2 Offline Techniques 423
  10.3.2.1 Chemical Methods 423
 10.3.3 Gas Analysis 431
 10.3.4 Microbiological Detection and Monitoring 431
10.4 Direct Techniques 431
 10.4.1 Intrusive Techniques 432
  10.4.1.1 Access Fittings 432
  10.4.1.2 Mass Loss Coupons 434
  10.4.1.3 Extended-Analysis (EA) Coupon 444
  10.4.1.4 Electrical Resistance Probes 444
  10.4.1.5 Test Spools 448
  10.4.1.6 Electrochemical Methods 449
 10.4.2 Nonintrusive Techniques (Inspection) 451
  10.4.2.1 Visual Testing (VT) 452
  10.4.2.2 Ultrasonic Testing (UT) 453
  10.4.2.3 Radiographic Inspection (RT) 456
  10.4.2.4 Magnetic Particle Inspection (MPI) 458
  10.4.2.5 Penetrant Testing (PT) 460
  10.4.2.6 Eddy Current Testing (ET) 460
10.5 Intrusive Inspection 461
  10.5.1.1 Downhole Inspection 462
  10.5.1.2 Pipeline Inline Inspection 464
  10.5.1.3 Tank Bottom Inspection 466
10.6 Records and Failure Reports 467
10.7 Summary 467
References 470

CHAPTER 11: CORROSION MANAGEMENT 473
11.1 Introduction 473
11.2 Risk 474

    11.2.1 Risk Assessment..474
        11.2.1.1 Setting the Stage 475
        11.2.1.2 Segments or Circuits...475
        11.2.1.3 Threat Assessment 475
        11.2.1.4 Barriers 477
        11.2.1.5 Probability of Failure (POF)..479
        11.2.1.6 Consequence of Failure (COF) 479
        11.2.1.7 Risk Matrix 479
11.3 Corrosion Control Programs 481
    11.3.1 Unsustainable Cycle..482
    11.3.2 Sustainable Cycle: Plan, Do, Check, Act 483
        11.3.2.1 Plan 483
        11.3.2.2 Do 483
        11.3.2.3 Check 484
        11.3.2.4 Act..484
    11.3.3 Program Integration 485
        11.3.3.1 Ask Yourself 485
        11.3.3.2 Consider the Unexpected—Look at the Big Picture 485
11.4 Economics of Corrosion Management 486
    11.4.1 Consider It an Investment 487
    11.4.2 Comparative Economic Evaluations 487
    11.4.3 Another Way: The Break-Even Figure 488
11.5 Summary 489
References...489

Foreword

The objective of this revision of *Corrosion Control in Petroleum Production* is to provide an expanded and updated introduction, overview, and reference on the subject. It is written for people who are not corrosion specialists, but who have responsibility for corrosion control or mechanical integrity in oil and gas production operations. A goal is to provide the information and terminology necessary for the reader to be able to understand the subject, know the challenges to be addressed, and communicate with corrosion specialists, consultants, suppliers, and vendors to ensure the success of corrosion control programs. It is hoped that the addition of such items as checklists, charts, summary tables, guidelines, and field examples will enhance its value as a reference. The book covers the basic causes and types of oil field corrosion, methods of control, and related material. While some theory is presented, the practical aspects are emphasized. References are cited for each chapter, as well as some bibliographies.

Although this edition is listed as the "Third Edition," it is actually the fourth writing. The original book, published in 1958 as *Corrosion of Oil- and Gas-Well Equipment*, was a joint effort of the American Petroleum Institute (API) and NACE (now NACE International). In 1979, NACE published an expanded version as "Technical Practices Committee (TPC) Publication 5" with the present title. A second API edition, published in 1990, is still available from API. However, this "Third Edition" of the NACE International book is so much more updated than the API version.

Those readers who are familiar with previous editions will recognize that most of this current book is either completely new material, or older material that has been significantly reworked to modernize the information. Some of the text, however, has been taken almost directly from the second edition with appropriate revisions and editing made to improve the readability, and for consistency with the rest of the book. Likewise, new illustrations and photographs abound throughout, but a number of illustrations and photographs of corrosion failures are the same in both editions. This just proves that—when it comes to corrosion—some things never change! The fundamentals remain the same, but the applications widen into more challenging producing environments and geographic locations.

Regarding the figures, in this edition numerous black and white figures and photographs are included that were originally printed in the first or second editions. Although they are presented herein without attribution, they have been used with permission from NACE International. Please note that although every effort has been made to faithfully duplicate them (some of them being completely redrawn), both photography and graphics creation from the late 1950s (1st edition) and even the 1990s (2nd edition) is not what it is today, hence the evident quality issues.

This third edition greatly expands the global nature of our industry, and adds examples of offshore applications for fixed platform and floating production, arctic- and other cold-climate services, pipelines, and large onshore facilities. This third edition updates technology in many areas such as laboratory evaluation testing of corrosion inhibitors, and metallic and nonmetallic materials selections. NACE International standards are updated and referenced throughout the book, as are other national and international standards.

The authors of the third edition would like to acknowledge the work of Harry Byars for his extensive revision to the second edition in 1999. It is always easier to revise than to begin with a blank sheet of paper. Harry did a thorough job addressing a wide range of topics. We have maintained his emphasis on addressing field applications.

The authors would also like to acknowledge the editorial contributions of Elaine R. Firestone in improving the organization and clarity of the book's content and improving the quality of several of the figures. Her thorough work made for a more reader-friendly end product.

Lastly, while many of the fundamental corrosion challenges remain unchanged, there have been significant changes in the oil and gas industry. It is difficult to predict the future, but we are on the edge of some exciting advances and new challenges. Perhaps future editions of this book will be titled *Corrosion Control in Energy Production* to reflect the changing emphasis from fossil fuels to noncarbon-based energy resources. Hopefully, today's reader will become tomorrow's editor.

Brand names of products or services mentioned in this book are for informational purposes only and do not constitute an endorsement of these products or services by either the authors or NACE International.

In Memoriam

Harry G. Byars worked for ARCO Oil and Gas Co. for 38 years where he was primarily involved with internal corrosion control by chemical and environmental means, and corrosion monitoring. Harry was a very active member of NACE from the time he first joined. He was the vice chair of Technical Committee T-1 on Corrosion of Oil and Gas Well Equipment from 1965–1967, and the chair of T-1 from 1967–1972. He was the chair of the Technical Practices Committee (TPC) Operations Committee from 1974–1992. He continued as a member of the Operations Committee until his death, and was always respected as a trusted adviser. He was also the liaison between the TPC/Technical Coordination Committee (TCC) oil and gas production committees and similar API committees, attending meetings of each organization and reporting back to the other, and served on Specific Technology Groups (STGs) 11, 31, 32, 33, 35, 60, and 62.

In addition to technical committees, Harry served on administrative committees that included the Publications Activities Committee, Internal Corrosion Committee, and Course Quality Committee. He was a NACE instructor and NACE-certified Corrosion Specialist. A NACE Fellow, Harry received the R.A. Brannon Award in 1982 and the T.J. Hull Award in 1999. He received his 60-year pin at CORROSION 2012 in Salt Lake City, Utah. NACE International recognizes the contribution of Harry Byars annually by awarding the Harry Byars Memorial Award, which is presented for Recognition of Continuous Service to Technical Committee Work.

About the Authors

Robert J. Franco is the owner of Franco Corrosion Consulting, LLC, which specializes in consulting to oil and gas producers and pipeline companies. Bob is a 50-year Life Member of NACE International who retired from ExxonMobil Production Co. in December 2012 after over 43 years in materials engineering and corrosion control with various Exxon and ExxonMobil companies in both downstream and upstream businesses. At retirement, Bob was the Senior Technical Materials & Corrosion Consultant—the highest technical position for this engineering discipline. His duties at ExxonMobil Production Co. included responsibility for improving and stewarding global corrosion control and mechanical integrity programs for upstream operations, and maintaining and growing the company's global corrosion network. His current consulting business includes internal corrosion control in upstream and midstream oil and gas transmission pipelines, development of risk-based inspection strategies for sour upstream operations, and instructing in-house corrosion engineering courses. Bob has published articles and conference papers on corrosion and failure analysis. He is a registered Professional Engineer in New Jersey and California.

Timothy Bieri is a Principal Corrosion Management Engineer with BP America's Upstream Engineering Center. He has been with BP for 16 years and currently provides technical support to global projects and operations. In addition, Tim also contributes to the health of the corrosion discipline, standards and practice, and integrated corrosion management. Prior to joining BP, Tim worked for the consulting engineering firms CC Technologies and Coffman Engineers for a combined 13 years. He has been an active member of NACE International for more than 25 years and has participated on several committees, as well as served on the Board of Directors. In addition, he is a NACE certified CP Specialist and Internal Corrosion Specialist. He is also a registered Professional Engineer in Alaska.

Reviewers and Contributors

A revision of this magnitude required contributions by many individuals in the corrosion community. The authors would like to thank the peer reviewers for their many valuable comments and updates to the technology in their respective areas.

Peer Reviewer	Corporate Affiliation
Dennis Enning	ExxonMobil
Andre Gokool	BP
Weiji Huang	ExxonMobil
Bryan Hutton	Lubrizol
Danny Keck	KCS Enterprises
Ardjan Kopliku	BP
Jim Mason	Mason Materials Development
Matt Matlas	CI Corrosion Solutions
Doug Moore	Carboline
Greg Ruschau	ExxonMobil
Stephen Smith	ExxonMobil, retired
Sai Venkateswaran	BP
Sol Zittrer	Georg Fischer Piping Systems

The authors would like to thank the contributors to this book. The subjects covered in this book are very visual in nature and having the proper images helps illustrate the topics. Many of the photographs in this book are courtesy of Nalco Champion, an Ecolab company. Most of the photographs of oil field failures come from years of failure analysis work performed by the late Bill Tillis who worked for NL Industries and Nalco-Exxon Energy Chemicals. Bill's efforts were published in a booklet titled, "Corrosion in the Petroleum Industry," by Ondeo Nalco in 2002. Bill's catalog of corrosion failures is a valuable resource to any corrosion engineer working in oil and gas production. Additional contributors include:

Additional Contributor	Corporate Affiliation
Lance Awalt	NOV Fiberglass Systems
Mike Beutler	SeaCorp Grp GF Harvel LLC
Dale Belmont	LF Manufacturing
Andy Bodington	BP
Jim Britton	Deepwater Corrosion Services
John Broomfield	Broomfield Consultants
Jane Brown	Brown Corrosion Services
Julia Burova	Belzona, Inc.
Mark Byerley	Tinker-Rasor
Heath Casteel	Performance Pipe
Bill Corbet	KTA-Tator
Jim Dillon	Nalco-Champion, an Ecolab Company
Rick Eckert	DNV
Rusty Fontenot	BP
Peter Harkins	Tinker-Rasor
Bill Hedges	BP
Danny Keck	KCS Enterprises
Kevin Miller	BP
Sam Mishael	Chevron
David Morris	Aegion
Crystal Myers	GPA Midstream Association
Richard Northrop	DeFelsko
Chris Reiter	Polyflow LLC
Kathleen Rosario	Cosasco
Kip Sprague	BP

Overview

Robert J. Franco

1.1 Definition and Importance

NACE defines corrosion as "the deterioration of a material, usually a metal, by a chemical or electrochemical reaction with its environment."[1] The most common metal in oil production operations is steel. The external environments that reside outside pipe and petroleum production facilities are air, water, and soils, and the internal environments are produced fluids (oil, gas, and water) and other substances (such as chemicals, glycols, coolants, etc.).

Corrosion is a major contributor to loss of pressure containment in downhole- and surface facility pressure equipment in oil and gas production. Its consequences include leaks and ruptures that result in pollution, fire, injury to facility personnel and neighbors, and financial loss to the owner from equipment down time, replacement, and loss of production.

Controlling corrosion does not mean eliminating it, but instead, developing a technically sound approach to selecting the most economical materials and corrosion control methods, monitoring equipment health, developing and implementing inspection plans, replacing or repairing equipment before it fails, and using the monitoring and inspection results to improve the corrosion control program.

1.2 The Corrosion Cell

Corrosion is an electrochemical cell that derives its electromotive force (emf) from a chemical process. The basic corrosion cell (Figure 1.1) has four components: an anode and a cathode connected by a metallic path in contact with an electrolyte. Referring to Figure 1.1, as metal atoms in the crystal structure dissolve in the electrolyte (typically a water phase), they enter the electrolyte as positively

---

1. NACE Corrosion Glossary, https://www.nace.org/glossary/corrosion-terminology-c, Accessed 27 April 2019. Accessible to NACE International members only.

charged metal ions, such as Fe^{+2}. The corresponding numbers of electrons (2) travel through the metallic path to the area known as the cathode. Using the notations of conventional electrical current, current flows from + to −, hence from the anode to cathode. Using this convention, current is discharged at the anode into the electrolyte and returns through the metallic path. However, in electron theory, the electron flow is from anode to cathode in the metallic path, and positively charged ions that are dissolved in the electrolyte migrate to the cathode to maintain charge neutrality. Corrosion engineers use conventional current flow notation (current flows from anode-to-cathode in the electrolyte, + to -). They say that current is discharged at the anode. Corrosion scientists, however, use electron theory (current flows by electron migration from anode-to-cathode in the metallic path, + to -). Electrons are discharged from the anode. The most important fact the reader should remember is that corrosion occurs at the anode.

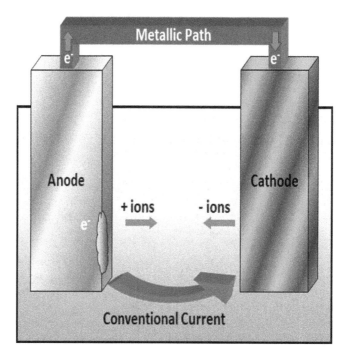

Figure 1.1 Schematic of a basic corrosion cell (electrochemical cell), illustrating an anodic area (the anode) and the cathodic area (the cathode) on the surface of steel immersed in an electrolyte (in blue). Corrosion occurs at the anode. Conventional current flow occurs from the anode to the cathode in the electrolyte.

Figure 1.1 depicts the corrosion cell as if two different metals are involved in creating the anode and cathode. Chapter 3, Section 3.6 describes galvanic corrosion, in which two different metals are connected in an electrolyte. Unlike what is depicted in Figure 1.1, the anode and cathode may be different areas of the same metal surface, as shown in Chapter 7, Figure 7.1A. The reason local anodes and cathodes are created is because of differences in chemical composition within the metal, the presence of local inhomogeneities within the metal (e.g., inclusions, grain boundaries, precipitates), and differences in chemical composition in the electrolyte, especially where the different electrolyte compositions are in contact with the metal surface. The presence of deposits on part of the metal surface can also create local anodes underneath deposits. These reasons for having local anodes and cathodes are explored further in Chapter 3, Section 3.5. Corrosion that occurs in oil and gas production is predominated by the creation of local anodes and cathodes in the same metal.

After iron ions, Fe^{+2}, enter the electrolyte, they can react with hydroxyl ions in the electrolyte to form a ferrous hydroxide corrosion product: $Fe(OH)_2$.

$$Fe^{+2} + 2OH^- = Fe(OH)_2$$

Other corrosion products may form depending on what other ions are present, such as sulfides, carbonates, and dissolved oxygen.

The formation of ferrous hydroxide corrosion product removes some of the hydroxy ions in solution, leaving H^+ ions (water is H_2O, or H^+ and OH^-, so removing OH^- by forming the ferrous hydroxide corrosion product leaves an excess of H^+). H^+ ions migrate to the cathode where they encounter electrons that have traveled through the metallic path and they form hydrogen atoms:

$$H^+ + 1e^- = H$$

Atomic hydrogen atoms react with each other to form hydrogen molecules:

$$H + H = H_2$$

In the presence of "poisons" such as sulfide ions or arsenic ions, formation of molecular hydrogen may be partially prevented, leaving atomic hydrogen on the metal surface. Atomic hydrogen is a small atom that easily diffuses into the iron crystal structure of steel, resulting in a variety of hydrogen damage mechanisms, which are discussed in Chapter 3, Sections 3.10 through 3.13, inclusively. If molecular hydrogen forms, hydrogen molecules can escape into the electrolyte, or they may accumulate around cathodic sites as gas bubbles attached to the cathode. Factors such as a low velocity electrolyte allow the gas bubbles to remain in place on the surface. A covering of gas bubbles can slow the reaction on the cathode surface, thereby slowing the overall corrosion reaction in a process called "cathodic polarization."

There are a number of important points to remember about the corrosion cell:

- The electrolyte is water in some form, and it does not take a lot—a thin film of liquid will do. (Chapter 10, Figure 10.14 shows a thin film of electrolyte in a wet gas piping system).
- Corrosion occurs at the anode.
- If any one or more of the four items of the corrosion cell is missing—anode, cathode, metallic path, or electrolyte—corrosion will not occur.
- If the circuit is interrupted in any way, then corrosion may be controlled (corrosion rate reduced).

The subsequent chapters of this book will describe how to remove one or more components of the corrosion cell and how to interrupt the circuit to control corrosion. The next section provides an introduction.

1.2.1 Controlling Corrosion

Many methods of breaking up the corrosion cell are available for use in the oil and gas industry. The most common methods of corrosion control are listed in Table 1.1, as well as where discussions of each method can be found in this book.

(Note: the causes of corrosion, corrosion mechanisms, and types of corrosion attack in oil and gas production are discussed in greater detail in Chapters 2 and 3.) Oil and gas production equipment and terminology are found in the American Petroleum Institute (API) book, *Introduction to Oil and Gas Production*, cited in this chapter's bibliography.

Table 1.1 Corrosion control methods used in oil and gas production.

Corrosion Control Method	Chapter

1. Cathodic Protection
    a. Externally to protect buried or submerged piping and equipment
    b. Internally in water portions of tanks and vessels
	7

2. Coating
    a. Externally as atmospheric coatings and pipeline coatings
    b. Internally in tubing, piping, and vessels
	6

3. Corrosion Inhibitors and Chemical treatment—Internally in wells, facility piping, and water systems	8
4. Corrosion-Resistant Materials—Both metals (such as stainless steels) and nonmetals (such as plastic and fiberglass piping and vessels)	4 and 5
5. Control of the Corrosive Environment—Internally: dehydration, deaeration, solids and scale control, oxygen-free operation, microbiological control, design to avoid building in corrosion problems, etc.	9

The selection of the corrosion control technique for a specific application is a technical and economic decision. That is, the selection is based on determining the techniques that are technically appropriate for the application, plus an economic comparison to select the most profitable solution for the life of the project. Often, two or more techniques in combination often are the most profitable (e.g., the use of an external pipe coating supplemented with cathodic protection of buried pipelines).

1.3 Corrosive Environments

Table 1.1 mentions both internal and external environments. When discussing corrosive environments, it is very important to remember that these are separate. What is inside a pipe does not typically affect the corrosion on the outside of the pipe and vice versa, unless a leak occurs allowing the inside environment to leak onto the exterior of the pipe or a neighboring pipe. Of course, the internal or external temperature does affect the temperature on the other side of the pipe, which may affect corrosion.

1.3.1 External Environments

External environments include a wide range of corrosive conditions. Atmospheric corrosion includes aggressively corrosive marine and industrial atmospheres. Corrosion under (thermal) insulation, or CUI, is a localized environment against the metal surface. Equipment buried underground is subjected to soil-side corrosion and localized corrosion at the air-soil interface. Lastly, submerged corrosion includes a wide range of different water compositions—the most aggressive being aerated seawater.

1.3.1.1 Atmospheric Corrosion

As far as atmospheric corrosion is concerned, the severity of corrosion depends on the moisture, pollutants, salts, and temperature of the air. Consequently, the severity of atmospheric corrosion is quite dependent on geographical location and the degree of industrial activity. Table 1.2 ranks the severity of different atmospheric conditions. Thus, atmospheric corrosion severity varies from its worst at an industrial area on the coast, to the least at a rural location in the desert. A quantitative approach for estimating the corrosion rate of steel and other alloys in various atmospheres is provided in the International Standards Organization (ISO) 9223 "Classification of the Corrosivity of Atmospheres."[1] Calculation methods in the standard take into account humidity, temperature, pollution (chlorides and sulfur dioxide), and airborne contaminants such as particulate deposition rate.

Table 1.2 Severity of the atmosphere.

Category	Rating	Pollution Level and Locations
C5	Very High Corrosivity	Very high pollution and/or strong effects of chlorides (example marine and/or seacoast with sea spray)
C4	High Corrosivity	Polluted urban areas
C3	Medium Corrosivity	Medium pollution (e.g., urban areas)
C2	Low Corrosivity	Low pollution (e.g., rural areas and small towns)
C1	Very Low Corrosivity	Very low pollution and time of wetness (e.g., deserts)

1.3.1.2 Corrosion Under Insulation (CUI)

CUI is a pernicious external corrosion problem that plagues thermally insulated equipment (vessels and piping) in producing, gas processing, and dehydration facilities.[2] CUI is discussed further in Section 1.10 (this chapter), Chapter 3, Section 3.3, and Chapter 9, Section 9.3.6.

1.3.1.3 Soil Corrosion

The severity of corrosion for buried structures or facilities sitting directly on soil is largely dependent on the composition and condition of the soil around and/or under the structure (Table 1.3).

Table 1.3 Factors affecting the corrosivity of soils.

Factor	Descriptive Information
Moisture	How wet is the soil, swamp or dry sand dunes? Wet soil is usually more corrosive.
Salts that may be present	Some happen naturally, such as in coastal areas, others happen because of brine leaks from previous corrosion or other failures. Higher salt content increases the soil conductivity and increases corrosion.
Temperature of the soil at the pipe surface	A hot line can have a surface temperature hotter than the average soil temperature. Higher temperature areas tend to become more corroded.
Oxygen availability	Corrosion occurs in oxygen-deficient areas if the buried structure has oxygen available to it elsewhere.

Regarding the role of oxygen, the amount of oxygen in the soil is a function of depth. Soil on the surface may have a relatively high oxygen content. The oxygen level will decrease with depth from the surface and with the difference in concentration. Oxygen concentration cells can occur, particularly with large diameter lines. The top of the pipe is nearer to the surface; therefore, the oxygen content is higher than under the bottom of the pipe, and thus a corrosion cell is set up. Oxygen concentration is also higher in disturbed soil, such as from pipeline trenching, versus compacted soil nearby.

1.3.1.4 Submerged Corrosion

Like soils, the severity of corrosion for facilities in water is largely dependent on the composition and conditions of the water (Table 1.4).

Table 1.4 Factors affecting the corrosivity of waters.

Factors	Water Corrosivity Concerns
Dissolved salts	The amount and composition are important. Is it seawater, brackish wetlands water, or freshwater from a river or lake?
Dissolved oxygen	The dissolved oxygen content will vary with the salt content of the water and the temperature, as well as the water's depth. The saltier the water, the less oxygen it will hold; and the lower the temperature, the more oxygen will be dissolved. The oxygen content at the surface will approach saturation, and it decreases with depth.
Velocity	Velocity plays an important role particularly if the water has a high velocity and contains suspended solids, such as the tidal currents in Cook Inlet, Alaska.
Marine growth	This can set up differential concentration cells, which will accelerate corrosion.

1.4 Internal Environments

Corrosion inside pipes and equipment involves the numerous factors that make up the internal environments (Table 1.5). Many of these factors are further examined in Chapter 2. These environments not only involve chemical and physical properties of metals and fluids, they include the physical characteristics of the system (i.e., temperature, pressure, flow rates, gas-oil and water-oil ratios, pipe sizes, schedules, and exposure times). In produced crude, crude oil that wets the steel surface usually protects against corrosion. However, if the surface is water-wetted, then corrosion continues. Wettability is a property that can be measured in the laboratory. In addition, water can exist as an emulsion where water and crude droplets are dispersed and not segregated into two layers. If water droplets are surrounded by oil as the matrix, then corrosion rates will be minimized. However, corrosion is more significant in emulsions that have water as the matrix that surrounds oil droplets.

Table 1.5 Factors affecting corrosion rates in production environments.

Factors	Descriptive Information	Refer to
Properties of the fluid (compositions, ratios of gas, oil and water phases, oil wetability, etc.)	Higher concentrations (partial pressures) of H_2S and CO_2 in the gas increase corrosion. Higher water content increases corrosion.	Section 2.4
Physical conditions in the system (such as temperatures, pressures, and flow rates)	Higher temperatures and flow rates typically increase corrosion. Higher pressures increase partial pressure.	Section 2.4 and 2.6
Temperature and temperature differentials	Higher temperatures and larger temperature differentials typically increase corrosion. Refer to Section 2.6	Section 2.6
Potential differences (bimetallic couples)	Larger potential differences between anode and cathode tend to increase corrosion.	Section 3.6
Heat treatment of the metal	Differences in microstructure caused by localized heating can result in ringworm corrosion, weld line corrosion	Section 4.4
Surface conditions (cleanliness, dirt, debris, deposits, scales, corrosion products)	Dirty surface promote corrosion, typically under-deposit corrosion and microbiologically-influence corrosion. Corrosion products may slow down corrosion. Mill scale increases corrosion	Section 2.5 and Section 3.5, Section 3.6.5, and Section 9.8
Velocity (erosion effects of high velocities)	Higher velocity tends to increase corrosion. Higher velocity with suspended sand in fluid increases erosion.	Section 1.5.5
Stagnant conditions ("dead" areas)	Stagnant conditions result in deposition of suspended materials, leading to under-deposit corrosion or microbiologically influenced corrosion.	Section 1.5.5 and Section 2.5
Impurities	Impurities in the produced fluid, such as dissolved oxygen, can greatly increase corrosion.	Section 2.4.3
Time (exposure or contact)	For mechanisms that involve weight (mass) loss, a longer time exposure increases corrosion. Example: acidizing jobs.	Section 3.4.1
Stresses (imposed stresses and built in stresses from fabrication)	Corrosion increases at tool marks on downhole tubing; Tensile stress may lead to stress corrosion cracking or other cracking mechanisms.	Section 3.13 and Section 3.17
Differential aeration	Differences in oxygen concentration can increase corrosion in areas that are depleted in oxygen. Crevices may create differential aeration conditions.	Section 3.5
Differential concentration	Differences in the concentration of solutes in the electrolyte can increase corrosion. Crevices may create differential concentration conditions.	Section 3.5
Biological and microbiological	Marine growth can create differential aeration cells. Microbiologically Influenced Corrosion can cause damage to carbon steel and corrosion resistant alloys.	Section 2.5, Section 3.5.4, and Section 9.9
Crude oil wettability properties	Crude oils that wet the metal surface and repel water slow down corrosion. Crude oil can be incorporated into the corrosion inhibitor "tail" to reduce corrosion.	Section 8.4 and Section 8.14.2

The single most key factor in most oil and gas system corrosion is the presence of free water on the metal surface. It does not take much water; a thin film on the metal surface is sufficient to create a problem. This water may contain dissolved salts, acids, acid gases (carbon dioxide [CO_2] and/or hydrogen sulfide [H_2S]), and oxygen (O_2), which may affect corrosion rates. Internal corrosion can be expected to occur anywhere water can condense, flow, or collect in wells, lines, vessels, or equipment. The severity of the corrosion (i.e., how soon it will create problems and cause failures) is influenced by many factors and will vary from system to system.

The corrosive environments can be changed by something as simple as changes in times and schedules. (See the sidebar on Changing Times.)

These internal and external environments will affect all types of production equipment, systems, and facilities, both onshore and offshore:

- Drilling wells
- Producing oil wells
- Gas wells
- Water supply wells
- Injection wells (i.e., water flood, water disposal, or gas injection and storage)
- Piping, flow lines, gathering lines, and injection lines
- Vessels
- Tankage
- Heat Exchangers (shell and tube or air-cooled)
- Auxiliary equipment (i.e., furnaces, heaters, pumps, dehydration, sweetening, and other plant type equipment)

Each field project is not only unique, but each type of equipment and system has multiple corrosion environments. For example, the environment in the tubing-casing annulus of a well is different from the environment inside the tubing; conditions inside the tubing vary from bottom to top in the well. Downhole conditions are different from surface conditions due to differences in pressures, temperatures, and compositions. The conditions before a choke are different from those after a choke. The conditions before heaters are different from those after heaters, etc.

Changing Times—Are Environments Really What They Seem?

Sometimes apparently unrelated events can change the corrosive environment, as cited in the following two cases.

The first case involves sucker rod pumping oil wells in the East Texas Field. The crude produced in the east Texas field has unusual oil wetting properties. Downhole failure studies in the late 1950s revealed that the first corrosion failures (rod or tubing) occurred after a well's water cut reached about 85% (an unusually high water cut, indicating good crude oil wettability). Above that water percentage the corrosion was very aggressive, but the wells responded unusually well to corrosion inhibition. Routine inhibitor treatments virtually eliminated downhole corrosion failures. Wishing to avoid the first failure, the operator initiated the practice of adding the well to the inhibition program when a well test indicated 75% water. This practice essentially eliminated downhole failures for quite some time. Then, failures started on wells that were not on the inhibition schedule—wells making considerably less than 75% water. Furthermore, these failures were occurring within a few joints of the pump. What had happened?

This was at a time of severe proration. Full allowable wells could be produced only a few days each month. This was also during one of the oil and gas industry's "downsizing" cycles. To operate with fewer field personnel, pumper assignments were adjusted so that each person had two assignments (two leases or two sets of wells). The operator would alternate the wells producing, that is, operate one set of wells during the last half of a month and the first half of the next month. The operator would then shut those wells in and move to the other assignment and produce those wells during the last half of that month, the first half of the second month. He would shut the second wells in and return to his previous assignment, and the cycle would be repeated.

With this approach, a given well would be shut-in for at least 30 days each cycle. Thirty days was long enough for the crude and water to separate in the tubing and for the water to settle. Thus, the rods and tubing above the pump were exposed to essentially 100% water. The shut-in time also was long enough for the oil to desorb off the rods and tubing, leaving them water-wet and corroding.

Something as seemingly unrelated to corrosion as the pumper's work schedule could change a well's status from "noncorrosive" to "corrosive." The problem was handled by adding all water-producing oil wells to the inhibition schedule.

The second case involves a small water flood. The source water for the flood was fresh water from a shallow aquifer. The purchased water was untreated and contained dissolved oxygen (DO). Sulfur dioxide and a catalyst were injected into the water stream entering the source water tank to remove oxygen. DO measurement indicated the scavenger was doing its job and reducing the oxygen to less than 50 parts/billion (ppb). Corrosion monitoring, however, showed that corrosion was not under control. Indications were that oxygen and higher-than-design velocities were both involved.

A review of the operation revealed that the injection plant was being operated only during the operator's daily 8-hour shift. The required daily injection volumes were injected—but in less than 8 hours—not over the 24 hours that the reported 8,000 barrels per day (1272 m^3/d) would imply. The water velocities were more than three times the velocities that would be assumed. Furthermore, during the day, at the >24,000 barrels per day rate (>3816 m^3/d), the water channeled through the water tank so fast that the oxygen scavenger and the oxygen did not have time to react. Thus, near the plant, oxygenated water was corroding the pipe, and the higher velocities kept a ready supply of oxygen available. During the off hours, when the pumps were not running, the oxygen scavenger and oxygen had time to fully react. When DO tests were run in the morning, the oxygen was gone, and specification water was injected. This is another case where the operating schedule had a significant effect on the corrosion control effort.

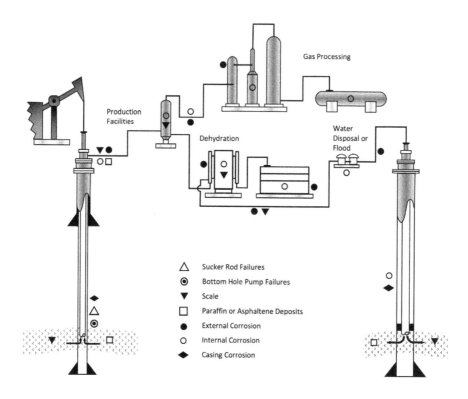

Figure 1.2 Locations and types of petroleum production problems.

1.5 Equipment, Systems, and Facilities

Figure 1.2 is a schematic of an oil field that shows locations and types of corrosion and other production problems. As noted, internal corrosion can occur anywhere in a production system where water (produced or condensed) is present, from the bottom of the well, to the equipment used to process the oil and gas for sale, and the water for injection. Areas most susceptible to internal corrosion are stagnant flow, turbulent flow (tees, elbows, etc.) gas breakout from water solutions due to change in pressure and temperature, under deposits (sand, scales, corrosion products), and near points of oxygen entry.

Appendix 1.A lists internal corrosion and corrosion control for various types of wells, facilities, and equipment. The usual problem areas, evaluation methods (detection and monitoring), and the most common corrosion control method are listed for each item.

1.5.1 Wells

All wells have five basic environment areas for potential corrosion problems (Figures 1.3 and 1.4):

- External casing
- Tubing-casing annuli (external tubing/internal casing)
- Internal casing below the bottom of the packer in packered wells, or below the bottom of the tubing in wells without packers (see Chapter 8 for an explanation of packers and packered wells)
- Internal tubing
- Internal wellhead and Christmas tree

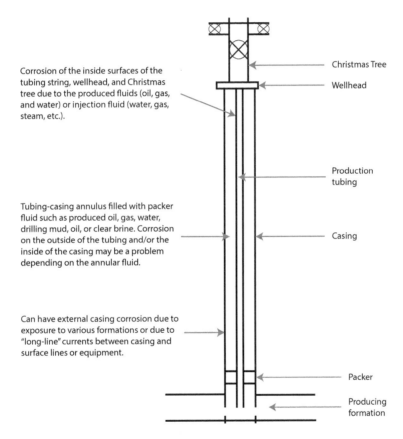

Figure 1.3 Schematic of a producing or injection well completed with a packer. Note the comments on areas of possible corrosion problems.

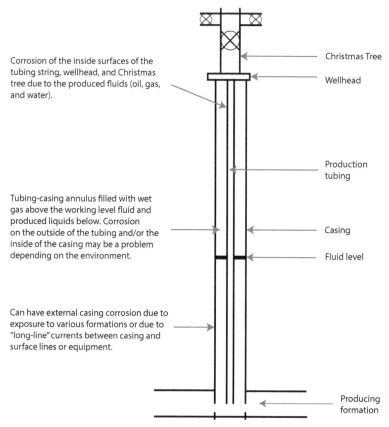

Figure 1.4 Schematic of a producing well completed with an open annulus (without a packer). Note the comments on areas of possible corrosion problems.

1.5.1.1 Artificial Lift Wells

Most oil wells throughout the world are produced with some form of artificial lift, although naturally flowing oil wells can be found in prolific fields. Artificial lift includes pumping (sucker rod pumping, electrical submersible [ESP], hydraulic pumps and progressive cavity pumps), and gas lift. The selection of the artificial lift method is outside the scope of this book. Sucker rod pumping and ESP are common pumped wells. The other type of artificial lift is "gas lift," in which gas injected down the annulus or a second tubing string is used to lighten the hydraulic load (oil or oil/water column) to lift the fluids to the surface. Gas lift wells are capable of higher production rates than pumped wells at greater lift depths. In general, gas lift wells are completed with a packer, while pumping wells are not (the annuli are open). Corrosion inhibition practices for artificial lift wells are discussed in Chapter 8.

Most of today's operators attempt to keep their pumped wells "pumped off," e.g., maintain a minimum fluid level over the pump intake. Thus, when pumping, there is very little liquid in the tubing-casing annulus, and what there is usually will be oil. The gases in the annular space are saturated with water at bottomhole conditions. The water will condense where the temperature decreases. Any CO_2 or H_2S will dissolve in the water, lower the pH, and increase the potential for corrosion. Corrosion often occurs on the internal walls of the casing where water has condensed because of cooler zones outside the casing.

Sucker Rod Pumping Wells

Corrosion-related problems involve rods, tubing, and pumps. The most common problems have been rod breaks (parts) caused by corrosion fatigue and tubing leaks caused by corrosion/wear. Photographic examples of these are provided in Chapter 3. Both failure mechanisms can be profitably controlled by proper design of the rod string and corrosion inhibition programs. The third category—corrosion and failure of the sucker rod pumps—is controlled by selecting the proper pump design and the proper combination of pump part materials. The selection of materials for bottomhole pumps is of utmost importance. Pumps must be able to resist the corrosive nature of the produced fluids and at the same time pump abrasives such as sand without experiencing failures. Many times, pump alloys can be selected to virtually eliminate downhole pump problems.[3-4] Other types of failures in rod pumped wells (including failures of sucker rod boxes, rod pin ends, and rod body problems) have been drastically reduced or eliminated by the correct use of improved makeup, correct handling techniques, and improved pumping practices. These are spelled out in detail in API RP 11BR.[5]

ESP Pumped Wells

Corrosion problems in ESP wells may occur in the tubing-casing annulus, to the ESP components themselves, and inside the tubing. Because the ESP is basically a high-speed multistage centrifugal pump, the usual method of corrosion control is selection of materials to match well conditions (materials selected for oil wells will not necessarily be the same as those selected for water flood water supply wells).

In corrosive situations, the tubing may be internally coated and corrosion inhibitors may be used. Because the ESP's intake is normally above the motor and motor protector, those parts are often protected with coating systems. The power cable run in the annulus is quite often armored cable. The cable sheath is a corrosion-resistant alloy (CRA)—often a high nickel alloy. Galvanized cable sheaths are susceptible to corrosion.

Hydraulic Pumped Wells

Corrosion problems in these wells usually occur inside the production tubing. If the power fluid tubing is installed as a concentric string, its exterior will be exposed to the same produced fluids as the tubing interior. The power fluid side usually does not present corrosion problems unless the power fluid is recycled produced water and the operation allows oxygen (air) entry. Historically, corrosion inhibitors introduced with the power fluid have been used to control corrosion of the produced fluids.

Progressive Cavity Pumped Wells

From the corrosion and corrosion control standpoints, progressive cavity pumped wells are like sucker rod pumps. Other than the design of the pump itself, the main operating difference is the rotating drive shaft of the progressive cavity pump rather than the reciprocating sucker rod string. Rod-on-tubing corrosion/wear can be a significant problem, so inhibition and materials selection should be the choices for corrosion control. Refer to Chapter 8 for methods to inhibit these wells.

Gas-lifted oil wells usually contain packers and the lift gas is dry; thus, annular corrosion is not a severe problem. When the lift gas is wet or there is no packer, annular corrosion can be a problem if the lift gas is high in acid gas (or if the gas in question is gasoline plant residue gas that contains oxygen). Oxygen is not unusual if the gasoline plant has a vacuum gathering system. In most gas lift wells, internal tubing corrosion may be controlled with specialized corrosion inhibition programs, internal plastic coatings, and CRA tubing strings.

1.5.1.2 Flowing Wells

The term "flowing wells" is used to describe both oil wells and gas production wells. Corrosion control of water injection wells is discussed in Chapter 9, Sections 9.5 and 9.9.

Oil Wells

Oil wells will usually flow when initially completed and continue to produce in this manner until the bottomhole pressure declines to a level too low to lift the fluids to the surface. Deep, high-pressure oil wells (and those offshore) routinely are completed with packers (Figure 1.3). Low-pressure or shallow oil wells have been completed with an open annulus (e.g., without a packer [Figure 1.4]). From the corrosion and corrosion control standpoints, the main difference is the corrosive environment in the tubing-casing annulus—produced fluids in the open annulus completion—packer fluids in wells with packers.

Most naturally flowing oil wells do not have internal corrosion problems. The oil tends to oil-wet the steel surfaces. Later in the well's life, as water percentages and water volumes increase, the tendency is to water-wet the steel. Water-wet steel will corrode, and corrosion control will be required.

Gas Wells

Gas wells, on the other hand, almost always are completed with a packer. Gas wells are usually the most corrosive during their early life when the pressures are the highest. Many gas wells are referred to as "gas condensate wells." In the classic gas condensate well, the fluids are all in the vapor state in the reservoir; as the gas travels up the well, hydrocarbons and water condense. Corrosion occurs where water condenses and wets the steel. Because the gas and liquid phases are a function of pressure, temperature, and gas composition, the water-wet area may "move." In such cases, corrosion will occur in a "zone" that can move during the life of the well. In many cases, corrosion will be a problem in the upper portion of a well during its early life (when the amount of condensed water may exceed the amount of produced water) and will move downhole as the well is depleted (when most of the water is formation water). Inhibitors, coatings, and CRA are used alone or together, depending on the problems in a well and the economics of the well.

Of special interest to some operators has been the deep (>3,000 m [>10,000 ft]), high- bottomhole pressure (>60 MPa [>10,000 psi]), hot (>120°C [>250°F]) gas wells with high mole percentages and resulting high partial pressures of CO_2 and/or H_2S. These deep, hot, acid gas wells represent special cases. The technology and economics for control of corrosion is an evolving and changing situation; therefore, each well should be approached with the latest developments. Depending on workover costs, these wells are potential candidates for CRA tubulars.

Auxiliary Equipment

Auxiliary equipment that may be used downhole includes such items as packers, tail pipe, safety valves, etc. Sometimes, these are referred to as well "jewelry." These are especially critical in high-pressure gas wells and often are made from selected corrosion-resistant materials. It is possible to eliminate many high-pressure well failures by proper design and materials of construction. Caution is required even when using CRA because well work operations (wirelining, acidizing, etc.) may damage even CRA materials.

1.5.1.3 Casing Corrosion

The casing in producing wells may vary from a limited amount of surface pipe and a single production string in shallow low-pressure wells, to several strings of concentric casing in deep, high-pressure wells.

External Casing Corrosion

Wells are potentially susceptible to external corrosion. External casing corrosion can occur because of exposure to various subsurface formations or to corrosion currents traveling up the casing, discharging from the casing—causing external casing corrosion—into the soil, and onto surface lines or equipment. Casing strings exposed to the drilled hole are typically protected by cement, but sometimes the cement job is inadequate, allowing a corrosive external soil or water aquifer to contact the outermost casing string. In addition, if the flowline from that well, but not the casing, is protected by an impressed current cathodic protection (CP) system, the applied current could travel to the casing string and jump off the casing to return to the flowline to complete the electrical circuit. Deep pits are located where the current is discharged from the casing. This is a form of CP interference corrosion and can be mitigated by bonding the casing and flowline with a resistance bond. Refer to Chapter 7, Section 7.6.2 for further information.

Local anodes can form on the external surface of casing when corrosion cells are created between different geological zones, when the casing is exposed to a particularly aggressive water zone, or when the casing is anodic to surface lines, tank bottoms, or other wells that are connected. External casing corrosion has not been a problem in all fields; however, it is severe enough in some fields to cause casing failures. In many cases, CP has been found to be an adequate remedy for external casing corrosion. However, interference from impressed current CP may also be the cause of external casing corrosion. If the flowline from that well, but not the casing, is protected by an impressed current CP system, the applied current could travel in soil or water to the casing string, travel along the external surface of the casing, then jump off the casing in soil or water to return to the flowline to complete the electrical circuit. Deep pits are located where the current discharged from the casing. This is a form of CP interference corrosion and can be mitigated by bonding the casing and the exterior of casing with a resistance bond. Refer to Chapter 7, Sections 7.6.2 and 7.6.3 for additional information.

In the drilling and completion of a well, drilling fluid left between the drilled hole and the external surface of the casing should be considered. Such fluids can contribute to external corrosion of the casing. Some operators attempt to cement the entire production casing (at least from the casing shoe up to the inside of the intermediate string of surface pipe). A good cement sheath will reduce the amount of bare casing exposed to corrosion, as well as reduce the amount of CP required. Sometimes the cement job is inadequate, allowing a corrosive external soil or water aquifer zone to contact the outermost casing string, requiring CP. Although it is not typical to apply impressed current to casing, operators can be made aware of the need by determining if CP is used in nearby fields or by running downhole CP potential logs, which identify anodic areas and can prove the effectiveness of CP.

Surface Casing Corrosion

Surface casing can also corrode externally to an extent where corrosion threatens the integrity of the well. This has occurred in areas where external water (rain, seawater, firewater, deluge, etc.) can accumulate against the surface casing (e.g., in well cellars) above the cement line below the wellhead. The wellhead can subside and eventually collapse if this problem is not mitigated. Mitigation may require sleeves to repair structural damage, and external immersion-service coatings to mitigate further corrosion.

Internal Casing Corrosion

Internal casing can occur below the bottom of the packer in packered wells where the internal surface of the casing is exposed to corrosive produced fluids. In an open annulus well, internal casing corrosion can occur below the working fluid level. Condensation of wet gas above the fluid level may also result in corrosion.

Internal casing corrosion can occur in the tubing-casing annulus. Corrosion is largely dependent on the composition of the annular fluids or packer fluids. Low-pressure and artificial lift wells are usually completed without a packer (Figure 1.4). Thus, the annular space is exposed to wet gas above the annular fluid level and the produced liquids below the fluid level. The water vapor in the gas space may condense where the inner casing wall is cooled by a geothermal temperature gradient, and acid gases may dissolve in the water to cause internal casing corrosion. The liquid below the fluid level will usually be oil even in high water cut wells, and corrosion in that area may not be a problem. Packered annuli are standard in gas wells, high-pressure oil wells and, water injection wells. The fluid in the annulus may be produced, drilling, or completion fluids that were in the well bore when the packer was set (Figure 1.3). More and more "packer fluids" specifically designed to meet requirements of a well are being used. Many contain a corrosion inhibitor and biocide.

Packer Fluids

Packer fluids are generally three types: (1) oil-based (weighted or unweighted), (2) water-based weighted with solids, or (3) water-based with no undissolved solids (the "clear brines"). From the corrosion and corrosion control standpoints, the clean oil or oil-based fluids are most desirable because they wet the surfaces of the casing and tubing with noncorrosive oil. Corrosion can be controlled in clear brine packer fluids using an oxygen scavenger, a water-soluble corrosion inhibitor, and a biocide (water-soluble quaternary amines function as both biocide and inhibitor).

Water-based fluids weighted with solids (including drilling mud) are the least desirable. Water-based muds are corrosive and can become more corrosive with time because of the breakdown of certain components. Inhibition is not practical because of the high, undissolved solids content of such muds, which adsorb corrosion inhibitors and prevent them reaching the steel surface. The very heavy zinc salt brines present special problems because of their native corrosiveness. Additional information on the chemical treatment of packer fluids is found in Chapter 8, Section 8.6.5.

1.5.2 Surface Equipment

1.5.2.1 Wellheads and Christmas Trees

Wellhead and Christmas tree equipment can be severely corroded, especially on high-pressure gas wells. High velocities and turbulence prevent the formation of protective films, and the wellhead and tree components and associated valves are vulnerable to erosion-corrosion as protective corrosion products are removed and fresh steel surfaces are exposed to further corrosion. Flanges present an area for crevice corrosion to occur because of shielding by the gasket material. In addition, wells that produce a high-mole percentage of acid gases usually subject the wellhead to attack. The use of corrosion-resistant metals is the usual solution. API Specification 6A discusses material requirements for various types of service and various pressure levels.[6] As defined by API "a wellhead is all permanent equipment between the uppermost portion of the surface casing and the tubing head adapter connection." API defines a Christmas Tree as "an assembly of valves and fittings attached to the uppermost connection of the tubing head, used to control well production."

1.5.2.2 Flow Lines

Flow lines (sometimes called well lines) can range from new pipe to downgraded used tubing, depending on the specific situation. Accumulation of deposits and other debris in the lines can promote corrosion and pitting, resulting in leaks. Water will tend to settle out and flow along the bottom of the line. In those cases, the bottom of the line will be exposed to water only, the metal will be water-wetted, and corrosion can occur. Optimum sizing of equipment, depending on anticipated velocities, is important. Internal corrosion of flow lines can be mitigated by inhibitors, and to a lesser degree by coating. Internal erosion or erosion-corrosion problems occur in flow lines that transport solids (e.g., sand), particularly downstream of the wellhead choke at the first few elbows and pipe bends. Mitigation includes installing sand control methods downhole (e.g., gravel packing and sand screens), as well as controlling the production rate.

In many older producing areas, flow lines were laid on top of the ground. In such cases, external corrosion was due to atmospheric conditions on the top side, and soil conditions on the bottom. Thus, the severity of corrosion was largely dependent on the location and moisture available (both in the atmosphere and soil). When lines laid on the ground experienced corrosion failures, the usual approach was to repair the leak with a pipe clamp or to rotate the pipe to move the thicker steel to the soil side. As fields were unitized and/or higher pressures were carried into central locations, the lines were buried, and cathodic protection and external pipe coatings came into widespread use to control external corrosion.

1.5.3 Processing Equipment

Processing equipment includes separators and gas processing.

1.5.3.1 Separators

Gas-oil separators can present a serious corrosion problem in areas where free water and a high partial pressure of acid gas(es) exist. A separator reduces the velocity of the produced fluids, thereby creating quiescent zones along the bottom where deposits of sand and/or corrosion product settle and cause under-deposit corrosion. Mineral scales can be detrimental to sustained production and may cause hot spots and premature failure of fired heater tubes. Corrosion is sometimes associated with the scale, especially downstream of regulators and dehydration equipment. Internal corrosion in processing equipment is controlled according to the type of process. In separators, corrosion inhibitors from downhole and flowline inhibition are carried into the vessel and help control corrosion of carbon steel. However, sand deposition may require on-stream jetting or periodic cleanout to allow proper oil-water separation and, secondarily to control corrosion. Separators are rarely cleaned and internally inspected for corrosion purposes unless the vessel must be opened for other reasons. If the downhole completion and flow lines use CRA construction, separators may also be constructed of CRA (solid or clad). The selection of the CRA alloy should consider under-deposit corrosion resistance.

As their name implies, separators physically separate one or more phases of produced oil and gas. They include two- and three phase separators:

- Two-phase separators include liquids separation (oil plus water from gas), emulsion breakers (a water-crude emulsion into two phases), water removal from crude oil (crude dehydration), water removal from gas (gas dehydration), and gas removal from crude oil (crude stabilization).
- Three phase separators separate crude from gas from water.

Equipment used for crude oil dehydration (such as free-water knockouts, gun-barrels and emulsion treaters [heater-treaters]) is subject to scale and corrosion problems similar to three phase separators. Elevated temperatures increase the corrosion and scaling tendency in heater-treaters and similar vessels used to heat incoming liquids to break emulsions. Inhibitors that might perform well in a hydrocarbon-water mixture in the flow line may not protect the water section of a heater. Therefore, heater-treaters and similar vessels often are internally coated and/or have CP in the free-water sections. Inhibitor effectiveness can be approximated in laboratory tests.

Gas plant processing equipment is subject to corrosion, especially if oxygen enters with the incoming gas, or with steam used to aid fractionation or with water used to remove salt from liquid hydrocarbon streams. Internal corrosion probably will occur where acid gases are present, and water is condensed from the hydrocarbon stream in a fractionation process. Gas plant equipment is susceptible to external CUI, which normally constitutes the largest corrosion threat to vessels and equipment, especially those that do not operate with free water present internally.

1.5.3.2 Gas Sweetening

Gas sweetening is the removal of CO_2 and/or H_2S from produced gas. It is usually accomplished by contacting the produced gas with alkaline amine solutions. Amine sweetening equipment corrosion is caused by improper acid gas loading (the quantity of the acid gases CO_2 and H_2S per quantity of the amine), oxygen entry, or amine degradation that creates corrosive heat-stable salts. Corrosion is controlled by operating the system at the proper loading and maintaining amine quality, including maintaining an upper limit on the concentration of heat-stable salts. In certain instances, operators have found it more economical to upgrade to CRA (typically austenitic stainless steel) in certain parts of the plant, such as reboiler tubes, or to cladding-selected areas of the absorber (contactor) and regenerator (stripper) towers rather than perform all necessary system controls. Refer to Chapter 8, Section 8.10.2.5 for additional information on gas sweetening systems that utilize amines.

1.5.3.3 Gas Dehydration

Gas dehydration is removal of liquid or vapor water in gas to obtain a desired dew point, which is a measure of dryness. The dew point is the temperature-pressure relationship that exists when free water begins to condense out of a gas. The colder the dew point, the drier the gas. Gas dehydration is accomplished in three ways:

1. Methanol (methyl alcohol) injection into gas flowlines to prevent freezing and hydrate (methane-ice) formation. It is primarily used on an intermittent basis when the normal hydrate prevention system has failed. It is not normally associated with corrosion problems.
2. Contacting the gas with a glycol solution, either ethylene glycol or tri-ethylene glycol. This operation is continuous and incurs the potential for corrosion in the water-rich glycol solution. Corrosion control requires pH adjustment.
3. Contacting the wet gas with a bed of pellets of aluminum silicate, called mol sieve dehydration. It is not normally associated with corrosion problems.

Refer to Chapter 9, Section 9.7 for a more detailed description of gas dehydration.

1.5.4 Storage Tanks

Lease storage tanks are typically fixed roof tanks; they are subject to both internal and external corrosion. Internal corrosion is a problem on the underside of the roof when water droplets condense in the presence of acid gas and/or oxygen, and where water and/or solids settle on the bottom. Oxygen should not be a problem today because most air quality regulations do not allow tank vents open to the atmosphere. However, some vapor recovery systems may draw sufficient vacuum on the tank to cause the vacuum relief to open and allow air entry. Additionally, the inert gas blanket system must be maintained at a slight positive pressure. There are occasions when it gets turned off during maintenance work and not turned back on.

The internal attack on the bottom of storage tanks is a result of the water layer that is generally found on the bottom of all tanks. Under-deposit (concentration cell) corrosion will occur on the bottom under sludge deposits and under corrosion product. Corrosion of the oil tank's shell (walls) normally is negligible. However, severe attack has been noted on the portion below the oil-water interface. Sometimes, severe hydrogen blistering is a major storage tank problem. (A photograph of blistering and hydrogen induced cracking is shown in Chapter 3, Figure 3.53.) Figure 1.5 shows the corrosion zones inside a sour crude tank. Coatings are often used to mitigate roof and tank bottom internal corrosion. Tank bottoms and up to 1 m (3 ft) up the sides of the wall are usually coated. The inside surfaces of fixed-roof roofs often contain a lot of structural angle iron that must be prepared for proper coating. NACE SP0178 should be consulted for surface preparation.[7]

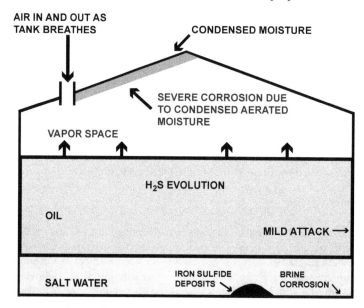

Figure 1.5 Corrosive zones and causes in sour crude storage tanks. In addition to deposits of iron sulfide, there are crude tank deposits (wax, paraffins, asphaltene), sand, and microbiological sludge. The latter two deposits and iron sulfide contribute to under-deposit and/or microbiologically influenced corrosion (MIC).

External surfaces (foundation-side) of tank bottoms are subject to corrosion where water or water wet-soil is present. Foundation design, cathodic protection, and coatings are used to control those problems (refer to Chapter 7, Section 7.6.5). Regulations concerning leak prevention and corrosion control are continually changing for aboveground storage tanks (ASTs), and the current local regulations must be considered in the planning and execution of a corrosion control program.

1.5.5 Piping

Piping is the primary means of transporting produced fluids for further processing or to sales. Corrosion considerations in piping systems fall into three categories, discussed below:

- Flow velocity
- Turbulence areas
- Stagnant zones and dead legs

1.5.5.1 Flow Velocity

API RP 14E, "Recommended Practice for Design and Installation of Offshore Production Platform Piping Systems" provides guidance regarding how to determine the fluid erosional (or critical) velocity in solids-free, multiphase fluids above which corrosion is expected to accelerate.[8] Corrosion is expected to be controlled if the fluid velocity remains below the critical velocity.

$$V_e = \frac{C}{\sqrt{\rho_m}} \tag{1.1}$$

where

V_e = fluid erosional velocity, ft/s (m/s),

C = empirical constant (guidance for values to use is provided in RP 14E, repeated below), with English units herein, and

ρ_m = gas/liquid mixture density at flowing pressure and temperature, lb/ft³ (kg/m³).

The reader is referred to API RP 14E for applications expressed in metric units.[8]

Industry experience to date indicates that for solids-free fluids, "C-Factors" (i.e., the empirical constant) of 100 for continuous service, and 125 for intermittent service are conservative for carbon steel. For solids-free fluids where corrosion is not anticipated, or when corrosion is controlled by inhibition or by employing corrosion resistant alloys, a C-Factor of 150–200 may be used for continuous service; values up to 250 have been used successfully for intermittent service. For CRA piping, higher values of the C-Factor have been used without problems.

Since the inception of API RP 14E, the erosional velocity equation has been used by many operators to estimate safe production velocities in erosive-corrosive service. The widespread use of the API RP 14E equation is a result of it being simple to apply and requiring little in the way of inputs. According to RP API 14E "The following procedure for establishing an 'erosional velocity' can be used where no specific information as to the erosive/corrosive properties of the fluid is available." Erosion and erosion-corrosion are expected to be controlled if the fluid velocity remains below the critical velocity.

If solids production, such as erosive sand is anticipated, fluid velocities should be significantly reduced below the erosional velocity for carbon steel and CRA, and more complex methods for determining acceptable velocity should be used (refer to Chapter 9, Section 9.3.4). However, there are currently no simple or readily available alternative formulas for calculating the erosional velocity in fluids that carry solids so in many cases, users must resort to a semiempirical approach that includes operational experience.

The API RP 14E equation has many limitations:[9]

- There is very little scientific backing for this equation. In addition, API RP 14E fails to provide an adequate empirical foundation that correlates erosion and erosion-corrosion characteristics to C-Factor in oil and gas systems.

- It should not be used for calculating threshold flow velocity and sizing pipes containing solid particles in a flowing fluid. A body of research related to erosion and erosion-corrosion has shown that this equation neglects the effect of flow regime-dependent wear characteristics and solids loading.
- It should be limited to multiphase flow (liquid-gas).
- It provides no quantification when solids or corrosion is expected.
- It does not apply to single-phase flow (e.g., liquid or gas), because all data indicates that in the absence of solids there is no erosion or the need to limit velocity.
- It is often quoted to be overly conservative and to unjustifiably restrict the production rate or overestimate required pipe sizes.

There are indications that higher velocities are tolerable in many situations in the absence of sand production. The user may apply different values of the C-Factor where specific application studies have shown them to be appropriate or the user may apply more complex computational approaches. The designer can adjust the empirical constant (i.e., the C-Factor) upward to allow for higher velocity if there is evidence from that field or facility that higher velocity can be tolerated safely.

Corrosion resistant alloys can operate at much higher C-Factors than carbon steel. At least one major oil company does not use the API RP 14E equation with corrosion resistant alloys. As velocity increases, pressure drop increases, and this becomes the practical limiting factor for design in the absence of a C-Factor.

1.5.5.2 Turbulence Areas

Certain corrosive environments, such as wet CO_2, are sensitive to turbulent flow and high wall-shear stress. These areas include elbows, tees, and other changes in fluid momentum. Where solids and/or corrosive contaminants are present, or where C-Factor values higher than 100 for continuous service are used, periodic surveys to assess pipe wall thickness should be considered. The design of any piping system where solids are anticipated should consider the installation of sand probes, cushion flow tees, and a minimum of 1 m (3 ft) of straight piping downstream of choke outlets.

Figure 1.6 shows the installation of cushion flow tees, which consist of sacrificial heavy wall pipe installed in areas of high turbulence between the choke and flowline. Solids entrained in the fluid flow impact the cushion tee and lose momentum before the flow changes direction.

Figure 1.6 Cushion tee installation (circled) installed between the choke and flowline.

1.5.5.3 Stagnant Zones and Dead Legs

Stagnant zones and dead legs are sections of piping with little or no flow. A dead leg is a section of pipework or instrumentation bridles on pressure vessels and tanks, which contains hydrocarbon fluids and/or water under the following stagnant conditions:

- No measurable flow (i.e., stagnant condition)
- Low flow (<1 m/s [3 ft/s]) or intermittent flow due to operational and/or production changes, or plant design and/or construction.

A dead leg over time can accumulate corrosive materials (e.g., sludge, sediments, biofilms, water), resulting in internal corrosion. When closure of a valve or installation of a blind flange creates such locations, they are often referred to as "process dead legs" or "stagnant zones." Examples of stagnant zones include control valve bypasses and piping used only for startup and shutdown. Locations of piping configurations that contain permanent areas with little or no flow are typically referred to as "dead legs" (e.g., straight runs of pipe at the end of a header). If water and oil are well mixed in the main fluid flow, their separation into two distinct layers is possible in stagnant zones and dead legs. Free water then contacts steel surfaces and promotes corrosion. In addition, suspended solids may settle under low- or no-flow conditions, which interferes with corrosion inhibition and promotes MIC and/or concentration cell corrosion. Refer to Chapter 3, Section 3.5 for further information on these corrosion mechanisms.

Categories

Dead legs can be divided into three main categories:

1. Permanent or physical dead legs: Sections of piping or pipelines subject to long-term stagnation, which have been built into a facility or have arisen from modifications over the course of the life of the facility.
2. Operational dead legs: Sections of piping or pipelines that are stagnant because of temporary changes, repairs, or other operational reasons (e.g., loading or unloading lines, lines only used for start-up or shutdown). Typically, dead legs are considered operational dead legs when they are not under flowing conditions or in intermittent use for short periods of time.
3. Mothballed equipment, or equipment temporarily removed from service, is susceptible to dead leg corrosion, especially in the absence of a preventative measure to mitigate the corrosion threat (e.g., chemical treatment).

Typical Dead Leg Locations

The following are typical dead leg locations:

- Blanked branches (flow prevented with a blind flange or closed valve)
- Lines with normally closed block valves
- Lines with one end blanked
- Pressurized dummy support legs
- Stagnant control valve bypass piping
- Spare pump piping

- Level bridles
- Relief valve inlet and outlet header piping
- Pump trim bypass lines
- High-point vents
- Sample points
- Drains and pump-out lines
- Bleeders
- Steam out connections
- Instrument tapping and/or connections
- Nozzles for thermowells on channel heads and piping
- Equipment bypass lines
- Startup and shutdown lines
- Blanked off small bore connections
- Lines in intermittent service

These locations are shown in Figure 1.7. In Chapter 9, Figure 9.1 shows a photograph of dead legs in a plant facility.

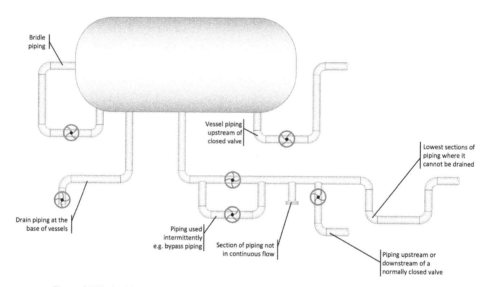

Figure 1.7 Typical locations where dead legs are formed in piping and pressure vessels.

There are systems that contain dead legs by design. Examples include firewater, sand jetting, drain, flare, and pig launcher/receiver barrels. For example, pressurized, fully water-filled firewater lines are shut-in and remain stagnant except during periodic testing or use during a fire. Internal corrosion may be a concern in firewater systems constructed from bare or internally coated carbon steel or corrosion resistant alloys with inadequate resistance to pitting corrosion, especially when using seawater. Flare piping can accumulate liquid, or it may be sloped by design to avoid water accumulation and prevent dead legs from filling with corrosive liquid.

Dead legs are not always caused by low spots in the line or by blocked flow. In some instances, a dead leg can form in pipework that does not extend below the pipe plane, such as in wet gas lines (geometries above the horizontal plane) because of water condensation, at the ends of manifolds, or due to shut-in of a duplicate system (e.g., the use of only one export pump creates a dead leg in the piping of the idle pump).

Lines may contain one or more dead legs. It is not normally practical to control internal corrosion in dead legs, and many operators depend on eliminating dead legs by redesigning the piping configuration, changing the operating procedures to periodically flow through or flush the line, or by performing periodic inspection of those dead legs that remain in service. In some cases, lines can be drained intermittently, or injected with corrosion inhibitor. However, corrosion inhibitors require flow for optimum effectiveness, so inhibition may not be effective, especially under deposits. Integrity management of piping systems containing one-off dead legs starts with an inventory of where dead legs exist, and which dead legs require replacement or periodic inspection to mitigate the risk of a leak or rupture. Systems with dead legs by design are generally managed from a whole-system perspective rather than as a one-off dead leg, with consideration given to materials of construction and whether internal corrosion can be detected during operation and function testing, e.g., by opening the pig trap door to inspect the launcher or receiver barrel.

The best way to avoid stagnant zones is to eliminate them in the design phase. However, if stagnant zones and dead legs are identified in an operating facility, their detrimental effect should be assessed, and the most practical means of remediation should be recommended to operations personnel. Refer to Chapter 9, Section 9.3.2 for additional information.

1.5.5.4 Flanged/Bolted Equipment

Flanged joints are used in most surface equipment from the wellhead through the choke, flowlines, and all piping systems. The reliability of flanged equipment depends on the integrity of bolted joints; therefore, bolts are an integral component that requires consideration for corrosion control from the design, through the operation and maintenance phases of the equipment life. In most situations, bolts are made of low-alloy steel, and are heat treated for strength and toughness where required. Low-alloy steel bolts are susceptible to atmospheric corrosion at the location of the facility, which means that offshore and coastal facilities experience the most bolt corrosion problems. Coatings have had varying degrees of success because sharp angles at the threads and thread tolerances prevent good surface preparation, and they require coating thickness to be kept to a minimum, respectively. Torque tool marks make coating damage inevitable, especially at the corners of nuts and hex heads. Access for inspecting for bolt corrosion is often limited to close visual inspection, but the area between the flange faces, the threads covered by the nuts, and the bolt shaft inside the flanges are practically impossible to thoroughly visually inspect. Special inspection devices for cross-sectional area loss, such as phased array ultrasonic inspection, can be considered if breaking the connection is not permitted. However, valuable information can be gleaned from visual inspection.

Figure 1.8 shows low-alloy steel bolts in varying degrees of corrosion deterioration in an offshore environment, as they appear to the eye. These bolts were originally coated with a fluorocarbon polytetrafluoroethylene (PTFE) coating without any prime coat. Remediation is required before Condition (D) is reached.

Figure 1.8 Progressive corrosion of PTFE-coated bolts in an offshore marine atmosphere: (A) recently installed—minimal corrosion initiated at coating breakdown caused by tool damage; (B) mild general corrosion; (C) evidence of significant corrosion on bolt and nut surfaces; and (D) heavy corrosion and metal loss evident—consider replacement.

Bolt Failures Due to Exposure to H_2S

In addition to external corrosion, carbon and low-alloy steel bolts exposed to high-pressure sour gas that contains H_2S can experience sulfide stress cracking (SSC) if the bolts were not manufactured to meet the modified hardness requirements in accordance with NACE MR0175/ISO 15156.[4] Refer to Chapter 3, Section 3.13 for additional information concerning this failure mechanism. Designers may incorrectly believe that when a high-pressure gas leaks to atmospheric pressure, it is incapable of causing this form of cracking because the H_2S partial pressure is lower than the reported threshold in this standard (0.05 psia [0.3 kPa]).[2] However, bolts not manufactured to NACE MR0175 modified hardness may have very high yield strength and high hardness (exceedance of hardness HRC 40 has been reported), which makes them susceptible to SSC failure at a lower H_2S partial pressure.

---

2. The partial pressure of a gas in a gas mixture is calculated and written in this book in terms of absolute pressure. Absolute Pressure = Gauge Pressure + Atmospheric Pressure. However, all other quantitative pressure terms other than partial pressure are written in *gauge* pressure.

Bolt Failures Subsea

This subsection is grouped with other bolt-related topics. Although not surface equipment, there is concern for critical subsea bolting and fasteners, and their risk of failure as observed in recent near-miss incidents in sea-bottom drill-through equipment. Evidence suggests that these events were associated with environmental cracking in high strength (high hardness) materials by hydrogen embrittlement because of either (a) internal hydrogen from manufacturing or electrolytic application of protective coatings, and/or (b) external hydrogen from in-service CP from galvanic and/or impressed current systems. Bolt metallurgy is an important variable in distinguishing between high-strength low-alloy steel, precipitation-hardened nickel based alloys, and other CRA bolt failures. Other environmental cracking mechanisms (e.g., chloride stress corrosion cracking, sulfide stress cracking, liquid metal embrittlement) can also induce embrittlement in certain high-strength bolting materials under subsea conditions. API and NACE are working to understand the failure mechanisms and develop standards in this area.[3]

Stainless Steel Bolt Failures in Coastal and Marine Environments

Another potential integrity problem has occurred with the failure of stainless steel (SS) bolts in marine environments located on topsides offshore and coastal facilities. These failures have been prevalent in instrumentation (level bridles, transmitters, etc.) and 3-piece ball valves with small bolt diameters (≤22 mm [0.875 in]). In most of the failures, the stud bolts were identified as ASTM A320 Gr B8M Class 2 strain hardened Type 316 austenitic stainless steel bolts.[10] Strain hardening and the relative ease in over-torqueing the small diameter bolts resulted in susceptibility to chloride stress corrosion cracking failures at ambient temperature conditions of the marine (offshore) environment in the Gulf of Mexico, Southeast Asia, Australia, and other seasonally warm climates. The internal production environment has not played a significant role in the external failures; however, it becomes a consideration in assessing the "consequence of failure" of existing equipment because bolt failures have resulted in the release of hydrocarbon gas. Refer to Chapter 3, Section 3.17.3 for additional information on chloride stress corrosion cracking.

1.5.6 Water Systems

Water flood or saltwater disposal (SWD) systems are subject to corrosion, primarily because of the entry of air into the system, poor water quality, microbiological activity, or acid gases in the water.[11] Small amounts of corrosion products can cause plugging of both equipment and injection wells. Oxygen can also cause injection well plugging if the injected water contains dissolved iron or manganese compounds that react with the oxygen to form insoluble products (solids). Well plugging can become a severe problem in systems handling produced water because the insoluble products become oil coated and tend to agglomerate and can plug even relatively porous formations. Water source wells, gathering lines and storage tanks, injection pumps and lines, and injection wells all have potential corrosion problems. Control of corrosion in water systems requires many of the following mitigation steps:

- Oxygen exclusion using good equipment design and operating practices (Chapter 9, Section 9.5)
- Dissolved gas removal (Chapter 9, Section 9.6)

---

3. For a discussion of hydrogen embrittlement, refer to Section 1.2, and Chapter 3, Sections 3.10–3.12, inclusive. For chloride stress corrosion cracking, sulfide stress cracking, and liquid metal embrittlement, see the respective sections in Chapter 3.

- Inhibitor and/or oxygen scavenger injection (Chapter 8, Section 8.11, and Chapter 9, Section 9.6.3)
- Microbiological control (Chapter 9, Section 9.9)
- Application of internal coatings and linings in various portions of oil field water systems (Chapter 6, Section 6.7.4)
- Installation of nonmetallic pipe (such as fiber reinforced plastic [FRP] piping) for surface and downhole applications, particularly in SWD applications (Chapter 5, Sections 5.6.5 and 5.7.3)

1.6 Drilling Operations

Oxygen and alternating stresses (fatigue) are considered the principal cause of failure in drill pipe. Pits caused by oxygen corrosion internal to the drill pipe, slip marks, and mechanical scratches serve as stress raisers, concentrating and increasing local stresses in drill pipe. These stress raisers enable fatigue cracks to start under fluctuating or reversing loads (refer to Chapter 3, Sections 3.15 and 3.16 for a discussion of fatigue and corrosion fatigue, respectively). During the drilling operation, the presence of oxygen in the drilling mud aggravates the fatigue action. As a corrosion pit deepens, corrosion fatigue cracking proceeds at an accelerated rate. Each of these actions aggravates the other, and a vicious cycle is created. The failure occurs when a crack or pit progresses all the way through the pipe wall and fluid is forced out. Fluid cutting rapidly enlarges even a tiny perforation into a sizable hole. This action is sometimes referred to as "drill pipe wash out." The pipe, thus weakened, can be easily twisted off. (A picture of a fatigue failure of a heavy weight drill pipe connector is shown in Chapter 3, Figure 3.62).

External corrosion of drill pipes does not usually form pits. Because of continual rubbing on the sides of the hole or intermediate casing strings, external wear is more likely. Also, oxygen content often is reduced because of the reaction with the mud components by the time the mud starts its return trip up the outside of the drill pipe. Internal corrosion pits are more prevalent, and for this reason, fatigue failures usually start on the inside of the pipe. Internal pipe coatings have been used to minimize this corrosion. The most common corrosion control method in drilling is to control the mud pH and maintain a basic condition (pH above 7).

1.7 Enhanced Oil Recovery Projects

Injection systems for enhanced oil recovery (EOR) projects can present unique corrosion and control problems.

1.7.1 WAG Injection

Projects that alternately inject water and gas (often referred to as "water alternating gas systems" [WAG]) have injection wells that must be designed to handle the alternate environments. Often, the water and gas are brought to the well through separate distribution lines, which are designed to handle corrosion due to the gas being injected or designed to meet the needs of the injection water. In cases when the gas is very dry, bare carbon steel lines are quite satisfactory. Bare carbon steel also may be used successfully with many waters if air-free (oxygen-free) conditions are maintained.

However, the tree, wellhead, tubing, and downhole equipment must be designed to withstand both injection environments. The period during the switch from one injection media to the other is the most challenging, particularly when the injection gas contains acid gases (CO_2 or H_2S).

Composite lined steel tubulars have been successfully used in water injection or water alternating gas (WAG) service. They consist of a fiber reinforced polymer (FRP) liner with a cement grouting that attaches the liner to a structural steel base pipe. Pressure containment is provided by the steel base pipe, while the liner protects the steel pipe against internal corrosion by providing a barrier between the metal and corrosive oil field fluids and gases. Care is required to maintain the corrosion barrier through the connection by properly installing a special barrier ring. Failure to install the corrosion barrier ring properly may allow corrosive fluids to access the coupling surface, which would result in corrosion of the steel substrate and loss of containment. The combination of FRP and steel provides the corrosion resistance of FRP, but with the strength of standard tubulars. The temperature and pressure limits provided by the manufacturers should be observed because of the use of nonmetallic materials.

Composite liners are superior to thin film polymer coatings, which may contain application defects or become damaged during transit, make-up, or well interventions (e.g., wireline and coil tubing). Testing and field history have shown composite-lined tubulars can tolerate repeated wireline runs without compromising the integrity of the system. Thus, composite-lined tubulars should not be confused with coated tubulars.

1.7.2 In Situ Combustion

In situ combustion projects (also called fire floods) involve injecting air or oxygen gas into a reservoir to sustain a fire that has been ignited in the reservoir. The objective is to heat the oil, and thus increase its ability to flow through the formation. Corrosion on the injection side may be controlled by injecting only dry air or dry oxygen. Corrosion at the producing well will tend to increase as combustion products break through (CO, CO_2, and NO_x), particularly if there is excess oxygen, and as the temperature increases as the fire front nears the well. Until oxygen breaks through or extreme temperatures occur, conventional corrosion inhibitor programs may be successful, although the treating frequencies and inhibitor dosages may need to be increased. When the fire front reaches the producing well bore, wells may be shut in unless special high temperature metals have been used.

1.8 Carbon Capture and Sequestration

With today's emphasis on reducing greenhouse gases (principally CO_2 and methane), carbon capture projects are being considered. Projects that capture and sequester CO_2 may involve injecting the captured gas downhole to enhance oil recovery. Similar to the corrosion rates of materials in an aqueous CO_2 environment, the corrosion rates of materials in CO_2 sequestration environments vary considerably—not only because of temperature and pressure differences, but also because of impurities. Conditions involving the source of CO_2, for example flue gas, can introduce more complicated mechanisms such as the presence of contaminants from products of combustion. An aqueous phase—containing elemental sulfur, sulfuric acid, and nitric acid, in addition to O_2 and H_2S—can form when the CO_2 stream contains water, nitrogen dioxide (NO_2), sulfur dioxide (SO_2), H_2S, and O_2 in concentrations within many published limits for maximum impurity concentrations for CO_2.

These acids and other species can corrode carbon steel if water is present; therefore, the most efficient method of corrosion control is to eliminate the amount of water.[12] Field experience shows that dry, pure CO_2 and pure CO_2 that contains water below the saturation limit in the pure CO_2–H_2O system is noncorrosive to carbon steel. However, a strategy should be developed to handle accidental water ingress because the corrosion rate of carbon steel will be unmanageably high (3–40 mm/y [120–1600 mpy]) if allowed to persist over long-term operations.

1.8.1 Coal Seam Gas

A type of carbon capture project that has offered unique challenges is recovering coal seam gas. These relatively low-pressure systems often contain acid gases (CO_2 and H_2S), and many have reported microorganism (bacteria) problems, particularly sulfate reducing bacteria (SRB). If part of the gathering system is operated at a vacuum, oxygen can cause greatly accelerated corrosion in parts of the system downstream of the point of oxygen entry.

1.9 Offshore Operations

From the corrosion and corrosion control standpoints, offshore oil and gas production has many similarities to onshore operations. Internal corrosion environments are essentially the same as onshore operations—the oil, gas, and water do not know whether they are onshore or offshore. A major difference between onshore and offshore is the use of seawater injection for enhanced oil recovery offshore, and principally produced water injection onshore. However, one example of onshore seawater injection is the Greater Prudhoe Bay Field in Alaska. Seawater injection involves pumping raw, aerated seawater through a mechanical (usually a vacuum or inert gas) deaerator and filters; and injecting one or more biocides, such as sodium hypochlorite and/or nonoxidizing biocides to control microbiologically influenced corrosion (MIC), along with a chemical oxygen scavenger to eliminate dissolved O_2 left by the deaerator.

The other major difference between onshore and offshore operations is the marine atmosphere affecting offshore and coastal locations. Coatings are the major barrier to atmospheric corrosion. Marine and coastal atmospheres require more coating maintenance than most onshore locations.

1.9.1 Reservoir Souring

Seawater injection is often associated with subsequent "reservoir souring" i.e., the production of H_2S in a reservoir that was initially "sweet" (CO_2-only) prior to seawater injection. This may take several years to occur after seawater injection commenced.

The production of H_2S due to reservoir souring results in safety and corrosion concerns, including exposing personnel to H_2S in surface equipment, and potential sulfide stress cracking (SSC) of susceptible carbon and low-alloy steel components that were not constructed to sour service requirements. Tensile and armor wires for subsea flexible pipe that are made from high strength steel are a concern for SSC failure because of their high strength. H_2S may introduce the potential for localized corrosion of carbon steel from the formation and subsequent breakdown of weak iron sulfide (FeS) films as well as from MIC-generated pitting. MIC and SSC are discussed further in Chapter 2, Section 2.5, and Chapter 3, Sections 3.13 and 3.5.4.

1.9.2 Corrosion Control

When it comes to corrosion control, the basic differences between offshore and onshore are associated with space for people, space for equipment, logistics, weight considerations, and economics—not necessarily in that order:

- Space for equipment, such as tanks and pumps for chemical injection, or space for a deaeration tower to remove dissolved oxygen from seawater to be used as water flood source water.
- Logistics of getting supplies and equipment to the offshore structure when it is needed.
- Equipment weight is an important consideration for offshore structures.
- Economics (including risks) of the increased costs of operation at a remote location and of repairs or remedial work if corrosion failures occur.

The technical aspects of internal corrosion control are not much different offshore or onshore.

1.9.2.1 Control of External Corrosion

Control of external corrosion can be quite a challenge offshore where the facilities will see a variety of external environments. External corrosion on offshore structures may be divided into three zones of attack. These zones overlap somewhat, and some differences in corrosion rate may be expected within the same zone as illustrated in Figure 1.9.

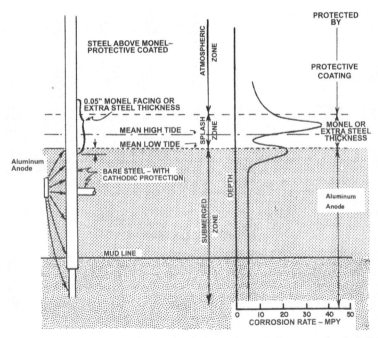

Figure 1.9 Corrosion zones, relative corrosion rates, and typical corrosion control measures for steel offshore structures.

The three zones, as defined by NACE SP0176 are as follows: [13]

- Submerged Zone—The zone that extends downward from the splash zone and includes that portion of the platform below the mud line.
- Splash Zone (or tidal zone)—The portion of the structure that is alternately in and out of water because of tides, winds, and seas. Excluded from this zone are surfaces that are wetted only during major storms.
- Atmospheric Zone—The portion of the platform that extends upward from the splash zone and is exposed to sun, wind, spray, rain, and seawater during firewater system testing, and moisture condensation due to night-day temperature cycling. It is where the metal appears to be dry most of the time, but experiences thin-film wetness and chloride deposits.

Submerged Zone

In the submerged zone, the general corrosion tends to be rather uniform. Shallow localized corrosion is also common. Corrosion rates will vary around the world in different seawater environments. This is discussed further in the section on offshore structures in Chapter 7, Section 7.6.6, and in the appendices of NACE SP0176.[12] CP is the usual method of corrosion control in the submerged zone throughout the world's offshore industry. As mentioned in the section on bolting, high strength subsea steel bolts have failed because of hydrogen embrittlement. CP is believed to play a role in charging hydrogen into the steel, although certain coatings operations may also be culpable of charging hydrogen into the metal. Operators are specifying controlled hardness bolts and hydrogen bake-out procedures after electrolytic application of coatings.

Splash Zone

Corrosion damage to offshore structures is most severe in the splash zone. Here the washing action of well-aerated saltwater removes corrosion products before a protective coating can be formed, resulting in deep pits. The vertical extent of the splash zone depends on the tidal range and the normal height of waves. A greater metal area will be exposed in this zone in the Pacific Coast (USA) area and in the North Sea than in the Gulf Coast (USA) where the waves are not as high and the tidal range is not as great. A large tidal zone increases the size of the splash zone. High tidal currents and suspended solids increase splash zone corrosion. The splash zone problem is aggravated by rubbing and scouring of the legs of offshore structures by large ice floes in the winter. Additional steel thickness (also referred to as "wear plates") can be a method to minimize corrosion failures in the splash zone. A popular alternative is to use a highly corrosion-resistant metal wrap made of nickel-copper alloy Monel®[(4)] 400 (UNS N04400) over the steel in the splash zone, as shown in Figure 1.9. Other alternatives listed in NACE SP0176 include: vulcanized chloroprene, high-build organic coating, high-performance platform coatings 250–500 μm (10–20 mils), heat-shrink sleeves, thermal-sprayed aluminum, and petroleum and/or wax-based tape systems.[12]

The design of the structure, and the method used to join the members, can reduce corrosion damage in this zone. Because this zone is the most difficult to protect, it is desirable to limit the number of cross members to a minimum. Ideally, the cross members are located in the submerged zone where they can be protected by CP. In winter ice areas, cross members need to be well below the depth of submerged ice.

---

4. Monel is a registered trademark of the Special Metals Corporation, Huntington, WV.

It cannot be over-emphasized that the corrosion protection in the splash zone must be well thought out. For example, the protective devices must be brought high enough to the atmospheric zone to ensure that splashes of seawater do not enter the interface between the termination of the splash zone protection and the less durable protection used in the atmospheric zone. Riser clamps also need to be suitably protected in the splash zone, which requires special consideration for designing Monel sheathing.

Atmospheric Zone

The atmospheric zone usually appears dry, but it has a thin layer of salt and water on its surface. As the structure cools at night and the humidity increases, the salt absorbs water from the air forming a salty film of moisture on the surface. The sun dries the film of moisture and reduces the rate of corrosion. The corrosion products tend to flake off and this irregular spalling of rust promotes pitting. High-performance coatings systems are the usual corrosion control technique for the atmospheric zone. Conventional paints, surface preparation, and application methods should not be used in extremely corrosive offshore environments. Some of the details are discussed further in Chapter 6, Section 6.7.1.1. NACE SP0108 provides detailed coverage of offshore coatings systems, surface preparation, coating application, and inspection.[14]

1.10 Corrosion Under Insulation

CUI was introduced earlier in Section 1.3.1.2. It is an environment often overlooked until a problem occurs because the environment is under the thermal insulation. Because of this, in many cases it is not visible during external visual inspections with the insulation in place.

External thermal insulation, also referred to as "lagging," can create a unique environment that approaches full immersion in water. The environment under insulation theoretically should be a dry area without water. Most thermal insulation on pipes and vessels is "jacketed" to prevent water entry; wet insulation loses its insulating properties. Jacketing consists of galvanized steel, stainless steel sheet metal, or polymer wrapped around the insulated equipment to keep the insulation intact and hopefully free from mechanical damage. However, experience has shown that with time, jackets leak because of punctures, caulking failure at seams, mechanical damage, and insulation exposed to atmospheric conditions and then it gets wet. Breaks in the insulation weather jacketing allow water from the atmosphere, or extraneous sources such as fire water testing, to enter the inside steel surface of the pipe, vessel, or tank, where they are held against the metal surface and cause oxygen concentration cell corrosion wherever prolonged moisture contact occurs. As thermally insulated equipment ages, water entry is inevitable because no jacketing systems are impervious to damage and breakdown. Depending on the corrosivity of the external atmosphere, CUI may take 10–20 years before becoming a serious problem (considerably shorter in offshore marine or coastal atmospheres). Oxygen that enters through the same openings in the jacket as the water will cause severe oxygen corrosion (Figures 1.10–1.12).

In some cases, soluble materials contained within the insulation, such as chlorides, can be leached out, which can lead to chloride pitting and/or chloride stress corrosion cracking of stainless steel equipment. The insulation can trap and hold water against the metal structure.

Coating the pipe prior to insulation is a preferred method of controlling CUI. Special temperature-resistant, immersion-service external organic or thermal spray coatings are required. These systems are discussed in NACE SP0198.[2]

Further information on CUI is found in Chapter 3, Section 3.3, and Chapter 9, Section 9.3.6.

Figure 1.10 CUI—Moisture and oxygen, entering through an opened jacket seam, led to this corrosion failure of a pipe in a coastal facility. Corrosion was not visible until insulation was removed after the leak occurred.

Figure 1.11 A leak in a 100 mm (4 in) diameter pipe located in an offshore facility resulting from CUI. Corrosion was not visible until insulation was removed.

Figure 1.12 CUI of top strake (course) of a refrigerated storage tank located in a coastal environment. CUI occurred where water entered below the roof and the body of the tank in seams in the insulation. Note the corrosion near the top and bottom of the photograph.

1.11 Summary

Corrosion can present problems in all parts of oil and gas production operations, wherever water is produced, condensed, or collected; and inside and outside wells, piping, and equipment. Corrosion control is a continuous job in production operations. The selection of control approaches is a technical/economic decision dependent on many considerations. The remainder of this book is devoted to a more detailed examination of these considerations along with the methods used to detect, monitor, and control corrosion in oil and gas production.

References

1. ISO 9223 (latest edition), "Classification of the Corrosivity of Atmospheres" (Geneva, Switzerland: International Standards Organization, 2012).
2. NACE SP0198 (latest edition), "Control of Corrosion Under Thermal Insulation and Fireproofing Materials—A Systems Approach" (Houston, TX: NACE International, 2016).
3. NACE MR0176 (latest edition), "Metallic Materials for Sucker Rod Pumps for Corrosive Oilfield Environments" (Houston, TX: NACE International, 2012).
4. ANSI/NACE Standard MR0175 (latest edition), "Petroleum and Natural Gas Industries—Materials for Use in H_2S-Containing Environments in Oil and Gas Production (Houston, TX: NACE International, 2015).
5. API RP 11BR (latest edition), "Recommended Practice for Care and Handling of Sucker Rods" (Washington, DC: American Petroleum Institute, 2008).
6. API Spec 6A (latest edition), "Specification for Wellhead and Christmas Tree Equipment" (Washington, DC: American Petroleum Institute, 2011).
7. NACE SP0178 (latest edition), "Design, Fabrication, and Surface Finish Practices for Tanks and Vessels to be Lined for Immersion Service" (Houston, TX: NACE International, 2007).
8. API RP 14E (latest edition), "Recommended Practice for Design and Installation of Offshore Production Platform Piping Systems" (Washington, DC: American Petroleum Institute, reaffirmed 2013).

9. Sani, F.M, "The API RP 14E Erosional Velocity: Origin, Application, Misuse, Limitation, and Alternative," CORROSION 2019, paper no. 51319 (Houston, TX: NACE International, 2019).
10. ASTM A320 (latest edition), "Standard Specification for Alloy-Steel and Stainless Steel Bolting for Low-Temperature Service" (West Conshohocken, PA: ASTM International, 2018).
11. Byars, H.G., and B.R. Gallop, "Injection Water + Oxygen = Corrosion and/or Well Plugging Solids," *Materials Performance* V13, N12 (Houston, TX: NACE International, 1974): p. 34.
12. IEAGHG, "Corrosion and Selection of Materials for Carbon Capture and Storage," 2010/03, April 2010 (Cheltenham, United Kingdom: IEAGHG, 2010).
13. NACE SP0176 (latest edition), "Corrosion Control of Submerged Areas of Permanently Installed Steel Offshore Structures Associated with Petroleum Production" (Houston, TX: NACE International, 2007).
14. NACE SP01018 (latest edition), "Corrosion Control of Offshore Structures by Protective Coatings" (Houston, TX: NACE International, 2008).

Bibliography

American Petroleum Institute. *Introduction to Oil and Gas Production*, 5th ed.—Book One of the Vocational Training Series (Washington, DC: API Publishing Services, 1996).

Pollock, W.I., and C.N. Steely, eds. *Corrosion Under Thermal Insulation* (Houston, TX: NACE International, 1990).

APPENDIX 1.A: Internal Corrosion and Corrosion Control in Oil and Gas Production

System or Equipment (Environment)	Usual Problem Areas	Evaluation Methods (Detection and Monitoring)	Common Corrosion Control Methods
OIL WELLS			
Natural Flow	· Usually no problem downhole at low water cuts (percentages) · May encounter water dropout in horizontal or highly deviated wells · Tubing-casing annulus: due to packer fluid or high acid gas	· Failure history · Inspection · Calipers · Coupons	· Downhole Equip: inhibition, coatings, and/or materials selection · Packered annulus: fluid selection + inhibitor and/or biocide · Open annulus: inhibition and neutralization
Artificial Lift			
Sucker Rod Pump	· Rods, tubing, pumps, annulus · Rods usually fail by corrosion fatigue	· Failure history · Inspection · Sometimes: coupons; iron counts (sweet fluids only)	· Inhibition · Pump metallurgy · Closed annuli
Hydraulic Pump	· Tubing, downhole pump	· Failure history · Inspection	· Inhibition · Pump metallurgy · Clean power fluid
Submersible Pump	· Tubing, annulus, pump, cables	· Failure history · Inspection	· Inhibition · Coat pump case · Pump metallurgy
Gas Lift	· Tubing, annulus, gas lift valves	· Failure history · Calipers · Inspection · Occasionally: coupons; iron counts (sweet fluids only)	· Tubing: coating, CRA metallurgy · Inhibition (inhibitor must get to bottom of hole) · Annuli: gas lift dehydration · NOTE: Lift gas should be O_2-free
OIL FLOWLINES AND GATHERING SYSTEMS	· Along bottom of pipe where free water flows and in low places where water collects	· Failure history · Inspection · Coupons, probes · Iron counts (sweet systems only) · Inhibitor residuals in water phase	· Inhibition—downhole may be sufficient, or inject downstream of wellhead choke · Internal lining or coating · Nonmetallic pipe

System or Equipment (Environment)	Usual Problem Areas	Evaluation Methods (Detection and Monitoring)	Common Corrosion Control Methods
OIL WELL PRODUCED FLUID HANDLING			
Oil and Gas Separators	· Free water section (bottom and under deposits) · Wet gas area (where water condenses) · Water dump valves and downstream piping	· Inspection: internal visual, external ultrasonic thickness · Failure history	· Internal immersion-grade coating · Cathodic protection in free water zones · Upgrade metallurgy of internals and water dump valve · Periodic flushing/ cleanout of sand
Free Water Knockout (FWKO)	· Free water section, shell, baffles, piping · Along bottom under deposits · Water dump valves and piping	· Inspection: internal visual, external ultrasonic thickness · Failure history	· Internal immersion-grade coating · Cathodic protection in free water zones · Periodic flushing and cleanout of sand
Heater Treaters	· Free water and treating sections: shell, baffles, bottom under deposits · Fire tube—particularly under scale deposits · Water dump valves and water lines · Gas section where water condenses	· Inspection: internal visual, external ultrasonic · Failure history	· Free water and treating section: Internal immersion grade coating and cathodic protection · Fire tube: cathodic protection and/or scale control chemical · Water dump valves: metallurgy · Gas section: Internal coatings · Siphons and water lines: coatings and/or nonmetallic materials
Gun Barrels (Wash Tanks, Settling Tanks)	· Free water section · Gas boot and piping in high H$_2$S areas · Water dump valves · Under side of decks	· Inspection: internal and external visual, external ultrasonic thickness · Failure history	· O$_2$ exclusion · Internal immersion grade coating · Cathodic protection in free water zone · Nonmetallic materials for gas boots and piping · Water dump valve: metallurgy · Aluminum decks
Lease Tanks	· Under side of deck · Bottom and lower portion of bottom ring	· Inspection: internal and external visual, external ultrasonic · Failure history	· O$_2$ exclusion · Internal immersion grade coating · Cathodic protection in free water zone if water level maintained to cover anode · Aluminum deck · Routine flushing and cleanout

System or Equipment (Environment)	Usual Problem Areas	Evaluation Methods (Detection and Monitoring)	Common Corrosion Control Methods
Hydraulic Pumping Equipment			
Power Fluid Tanks	• Power oil tanks – same as lease tanks • Power water tanks – same as salt water disposal and water injection tanks	• Inspection • Failure history	• O_2 exclusion • Routine flushing and cleanout • coating • Cathodic protection where appropriate
Power Fluid Pumps			• O_2 exclusion • Pumps: metallurgy
Power Fluid Lines			• O_2 exclusion • Inhibition
Gas Lift Systems	• Along bottom of line where free water flows or collects • On vessel walls where water condenses	• Inspection • Failure history • Monitoring: coupons; iron counts (sweet gas systems only)	• Dehydration • O_2 exclusion • Inhibition
Gas Wells	• Tubing interior, downhole equipment (such as subsurface safety valves, gas lift gas valves and mandrels), fittings • Wellhead and Christmas tree • Annulus above packer, depending on packer fluid	• Tubing: calipers, coupons; iron counts (sweet systems only), inspection • Wellhead and Christmas tree: inspection (internal visual, external radiographic and ultrasonic) • Casing internal: calipers, inspection, failure history	• Tubing: metallurgy, internal coating, and/or inhibition • Downhole valves, etc.: metallurgy • Wellhead and Christmas tree: metallurgy, internal CRA cladding, internal coating • Annulus: packer fluid selection and/or inhibition
Gas Well—Flowlines	• Along bottom of line where free water flows or collects • Along top where water condenses • Bends, tees and elbows	• Inspection visual, radiographic, ultrasonic, intelligent pigs • Monitoring: Coupons, probes, iron content (sweet only) • Failure history	• Inhibition (downhole carryover, flowline injection) • Routine maintenance pigging • Well site dehydration
Gas Gathering Systems and Gas Pipelines	• Along bottom of line where free water flows or collects • Along top where water condenses	• Inspection visual, radiographic, ultrasonic, intelligent pigs • Monitoring: Coupons, probes, iron content (sweet) • Failure history	• Dehydration • Inhibition and pigging • O_2 exclusion (especially for vacuum systems)

System or Equipment (Environment)	Usual Problem Areas	Evaluation Methods (Detection and Monitoring)	Common Corrosion Control Methods
GAS COMPRESSION and HANDLING EQUIPMENT			
Coolers	• Where water condenses	• Inspection: visual, tube calipers, external ultrasonic • Failure history	• Tube metallurgy • pH neutralization or inhibitor injection
Accumulators (liquid knock-out drums)	• Where water collects	• Inspection: internal visual, ultrasonic • Failure history	• Internal immersion grade coating • Metallurgy: solid or clad CRA • pH neutralization • cathodic protection in free water phase
Pressure vessels	• Where water condenses and in free water phase	• Inspection: internal visual, external ultrasonic	• Internal immersion grade coating • Cathodic protection in free water phase • Metallurgy: solid or clad CRA • Inhibitor or pH neutralizer injection
Compressors	• Reciprocating compressors: valves, cylinders, rods, bottles – fatigue and corrosion • Corrosion product fouling of centrifugal compressor impeller	• Inspection: internal visual	• Improved bracing to mitigate fatigue • Improved design to minimize fatigue • Improved water elimination to minimize corrosion • Metallurgy upgrade
Glycol Dehydrator	• Absorber (contactor) tower: corrosion of shell and trays in rich glycol area • Hydrogen Induced cracking (HIC) of absorber tower in sour gas • Regenerator (stripper) tower: corrosion in hot rich glycol; reboiler return to tower, reboiler tubes	• pH of recirculating rich and lean glycol • Frequency of filter plugging • Analysis of glycol chemistry—hydrocarbon, salt, iron, suspended solids • Monitoring: probes and coupons • Inspection: internal visual or external ultrasonic	• O_2 elimination – gas blanket atmospheric pressure glycol vessels • Metallurgy upgrade (solid or clad) • HIC-tested steel • pH neutralizer injection • Filtration frequency
Dry Bed (Mol Sieve) Dehydrator	• Wet gas handling areas: inlet gas and regen tower overhead condenser • Thermal fatigue of regen gas inlet piping to dehydrator tower	• Inspection: internal visual, external ultrasonic; crack inspection • Failure history	• HIC-tested steels in sour service (wet areas) • Internal coating where temperature allows

System or Equipment (Environment)	Usual Problem Areas	Evaluation Methods (Detection and Monitoring)	Common Corrosion Control Methods
Gas Sweetening (MDEA and Similar Amine Systems)	• CO_2 and H_2S gas breakout from rich amine solution • Rich amine piping exiting absorber tower • Lean amine circuit associated with regen tower reboiler • Buildup of amine degradation products and/or chlorides (heat stable salts) increases corrosion of regenerator tower • Potential HIC of absorber vessel in sour gas • Cracking of as-welded steel in amine service	• Monitoring of amine chemistry for heat stable salt content • Coupons and/or corrosion probes • Inspection: internal visual, external ultrasonic; crack inspection	• Control acid gas loading (mol acid gas/mol amine) • Operate below maximum allowable reboiler temperature • Reduce heat stable salt content • Metallurgy upgrades (solid or clad CRA) in select areas such as gas breakout, regen tower reboiler return and above rich amine entry • Postweld heat treatment to prevent cracking of steel welds • HIC-tested steels for absorber tower in sour gas • O_2 elimination—gas blanket atmospheric pressure amine vessels • Corrosion inhibition of regen tower overhead system
WATER HANDLING and INJECTION (DISPOSAL and FLOOD)			
Vessels and Tanks	• Shell when water very corrosive • Bottom under deposits • Vapor space when sour	• Monitoring: O_2 detection (galvanic probe), bacteria activity, iron counts (sweet systems) • Coupons, probes • Inspection: internal visual, external ultrasonic • Failure history	• O_2-free operation (exclusion and/or removal) • Cathodic protection • Internal coatings • Biocides and/or inhibitor injection • Nonmetallic materials • Periodic cleanout
Filters	• Bimetallic connections • At filter media (sand) level)	• Same as Vessels	• Same as Vessels
Gathering and Injection Lines, and Plant Piping	• Along bottom under deposits	• Same as Vessels	• O_2-free operation • Coatings and linings (including cement) • Maintenance pigging of lines • Biocides and/or inhibitor injection
Injection and Transfer Pumps	• Wetted parts • Seals leaking air	• Monitoring: O_2 probe • Inspection: visual • Failure history	• O_2-free operation • Upgrade metallurgy
Injection Wells	• Tubing interior, downhole valves, fittings, etc. • Wellhead and Christmas tree		• O_2-free operation • Coatings and linings (including cement, fiberglass) • Biocides and/or inhibitor injection • Metallurgy upgrade (e.g., tubing mandrel) if converted producing to injection well

System or Equipment (Environment)	Usual Problem Areas	Evaluation Methods (Detection & Monitoring)	Common Corrosion Control Methods
UTILITIES **Instrument Air**	• Internal corrosion of air receiver vessel where water accumulates • External corrosion of tubing located in marine atmosphere	• Periodic ultrasonic inspection • Visual inspection of lines	• Internal coating • Upgrade metallurgy of tubing • External coating
Heating and Cooling Water: **Glycol-Water Heating and Cooling (closed, recirculating system)**	• Fouling associated with corrosion products • Corrosion anywhere where deposits accumulate	• Monitoring water chemistry: pH, inhibitor residual, suspended solids, bacteria counts • Inspection of heat exchanger surfaces and pressure vessels: internal visual or external ultrasonic • Corrosion coupons	• O_2-exclusion • Inhibition (normally inorganic nitrite, molybdate, etc.) • Periodic biocide injection for nitrite treated water • Flushing heavily fouled system
Open, Recirculating Cooling Water (Cooling Tower)	• Fouling associated with bacteria growth, corrosion products • Under-deposit corrosion • In-leakage of hydrocarbons from tubing leaks creates bacteria growth and fouling	• Monitoring water chemistry, pH, inhibitor residual, bacteria counts, • Corrosion coupons and probes	• Inhibition (normally inorganic and organic phosphates, copper inhibitor) • Scale and deposit dispersants • Biocide treatment: chlorination and non-oxidizing biocide • Repair leaks and clean system
Boilers and Steam Systems	• Boiler tubes—internal corrosion under deposits, and/or O_2 contamination. • Overheating of boiler tubes due to deposition, poor combustion control • Carryover of dissolved solids from water into steam, fouling downstream steam turbine, control systems • Internal corrosion of steam condensate piping by dissolved CO_2	• Frequent monitoring boiler water chemistry: pH, phosphate residual, O_2 content, alkalinity • Check flame pattern in combustion chamber • Check superheater outlet temperature	• Mechanical deaeration (deaerator tower) • Chemical deaeration (O_2 scavenger) • pH control of boiler water using phosphate and alkali • pH control of steam condensate by injecting volatile amine blends • Scale control chemical (normally phosphate-based)

Corrosion Basics

Robert J. Franco

2.1 Introduction

The majority of oil field equipment and facilities are constructed of a metal—usually steel. The most common steel used, both downhole and surface, is called "carbon steel." NACE defines carbon steel is defined as: "An alloy of carbon and iron containing up to 2 mass percent carbon and up to 1.65 mass percent manganese and residual quantities of other elements, except those intentionally added in specific quantities for deoxidation (usually silicon and/or aluminum)."[(1)] Carbon steels used in the petroleum industry usually contain less than 0.8% carbon and even lower for steels to be fabricated by welding, and/or subjected to toughness requirements and maximum hardness restrictions per NACE MR0175.[1] Thus, oil field steels are often referred to interchangeably as low carbon steels, mild carbon steels, carbon-manganese steels, and low-alloy steels. Metals other than low carbon steel are used for special applications in oil and gas facilities—usually for their resistance to specific corrosive environments. These metals are discussed in the sections on corrosion-resistant materials in Chapter 4 (Sections 4.5.3, 4.6.4, and 4.7). Unless otherwise noted, the following discussion of the fundamentals of corrosion and its reactions apply to oil field steels.

2.2 Why Iron and Steels Corrode

Corrosion is a natural act of metals trying to return to their lowest level of energy. In the case of iron, it is iron ore. Iron ore consists of iron oxides, iron sulfides, iron carbonate, and similar iron compounds. This iron ore is mined, and its energy level increases as it is processed and converted to steel and steel products (pipe, plate, etc.); and from that time on, it is trying to return to the lowest level of energy. Thus, the tendency is for iron to corrode—to become "corrosion products." Corrosion products are iron oxides, iron sulfides, iron carbonates, and similar iron compounds. The energy released when the iron converts to corrosion product is, in fact, the energy stored in the metal during

---

1. NACE Corrosion Glossary, https://www.nace.org/glossary/corrosion-terminology-c, Accessed 9 May, 2019. Accessible to NACE International members only.

the refining and steelmaking process. This energy supplies the driving force for corrosion. Because corrosion is a natural act, it cannot be prevented—only slowed down. Therefore, corrosion control programs are efforts to postpone the inevitable for the life of the project.

2.3 Corrosion Reactions

As stated in Section 1.1, corrosion is defined as an electrochemical reaction, and the chemical reaction that occurs produces an electric current. As noted in Figure 1.1, corrosion requires an anode, cathode, electrolyte, and metal path. In oil and gas production facilities, the electrolyte is water, (or water-based fluids or water-wet solids), and the metallic path is the steel pipe or equipment. Electrical current flows during the corrosion process. For current to flow, there must be a driving force, or a voltage source, and a complete electrical circuit.

The source of voltage in the corrosion process is the energy stored in the metal by the refining process. Different metals require varying amounts of energy for refining and, therefore, have different tendencies to corrode, see Table 2.1. Note that the standard hydrogen electrode (SHE) is an electrode whose potential is arbitrarily defined as zero. The difference in potential between the SHE and a metal's potential is the value listed in the table.

Table 2.1 Electromotive force series of metals.

	Metal	Volts(*)	
Most energy required for refining ↑ ↓ Least energy required for refining	Magnesium	−2.37	Greatest tendency to corrode ↑ ↓ Least tendency to corrode
	Aluminum	−1.66	
	Zinc	−0.76	
	Iron	−0.44	
	Tin	−0.14	
	Lead	−0.13	
	Hydrogen	0.00	
	Copper	+0.34 to +0.52	
	Silver	+0.80	
	Platinum	+1.20	
	Gold	+1.50 to +1.68	

(*) Versus standard hydrogen electrode (SHE).

The size of the driving voltage generated by a metal in a water solution is called the "potential of the metal." It is related to the energy released when the metal corrodes. The absolute value of the potential of a given metal is influenced by the water composition, temperature, velocity, and many other factors. However, their relative values remain about the same in most production waters.

Besides a source of voltage, there must be a complete electrical circuit. As noted earlier, in addition to the metallic path, the electrical circuit of the corrosion cell consists of three parts—an anode, a cathode, and an electrolyte.

The anode is the area of the metal surface that corrodes—where the metal dissolves (goes into solution). When a metal dissolves, a metal atom loses electrons, and it goes into solution as a metal ion (a positively charged particle). The chemical reaction for the corrosion of iron follows:

$$Fe \rightarrow Fe^{+2} + 2e^-$$

$$\text{Iron atom} \rightarrow \text{iron ion (ferrous) + electrons.}$$

The iron ion goes into solution, and two electrons are left behind in the metal. The iron in solution can react with other ions to form corrosion products (such as iron oxide, iron sulfide, iron carbonate, or iron chloride).

The cathode is the area of the metal surface that does not dissolve. The electrons from the anode travel through the metal to the cathodic area. At the cathodic area, the electrons, which carry a negative charge, react with ions in the water that carry a positive charge. A typical reaction at the cathode is:

$$2H^+ + 2e^- \rightarrow H_2 \uparrow$$

$$\text{Hydrogen ions + electrons} \rightarrow \text{Hydrogen gas}$$

Or, if oxygen is present, two other reactions may occur:

In acid solutions

$$O_2 + 4H^+ + 4e^- \rightarrow 2H_2O$$

In neutral and alkaline solutions

$$O_2 + 2H_2O + 4e^- \rightarrow 4OH^-$$

The role of dissolved oxygen is described in Section 2.4.3.1.

The reaction at the anode areas produces electrons, and the reaction(s) at the cathode areas consume the electrons. Electrical current is the passage of electrons from one point to another. Convention says the electrical current flows from + to −, or in the opposite direction of electron travel. (Historical note: electron theory was developed after the sign convention for "conventional" current was assigned.) Thus, as electrons flow from the anode area to the cathode area, electrical current by convention is said to flow in the opposite direction, from the cathode to the anode. This current flow is within the metal (current flow requires a metallic path). Therefore, the metallic path between the anode and the cathode is always a conductor of electricity. Ionic flow occurs in the opposite direction in the electrolyte.

To permit ionic flow, the metal surface (both the anode and the cathode) must be covered with electrically conductive solution. Such a solution is called an "electrolyte." Pure water is a poor electrolyte, but electrical conductivity increases rapidly with the addition of dissolved salts.

This combination of anode, cathode, metallic path, and electrolyte is referred to as a "corrosion cell." The schematic of the corrosion process in Figure 1.1 (Chapter 1), is shown merely as an illustration. Metal atoms do not necessarily dissolve at a single point on a metal surface, nor are cathode areas restricted to one area on the surface. In fact, local anodes and cathodes are distributed over the entire surface. As corrosion progresses, a local cathode can become a local anode (and vice versa) as local conditions in the metallurgical structure or solution composition change.

2.4 Corrosion Variables

The corrosion reactions and the corrosion reaction rate (corrosion rate) are affected by the many variables that make up the corrosion environment. For example, the electrolyte's conductivity and pH, dissolved gases (particularly oxygen, carbon dioxide [CO_2], and hydrogen sulfide [H_2S]), microbiological conditions, and physical variables (such as temperature, pressure, and velocity) all play a significant role in the corrosion of a specific situation.

2.4.1 Conductivity

As previously stated, the metal surface must be wetted by an electrically conductive solution (the electrolyte) to conduct electrical current from the anode to the cathode of the corrosion cell. Thus, the more conductive the electrolyte, the easier current can flow, resulting in a higher corrosion rate if nothing else slows down the corrosion reaction. The less conductive the electrolyte, the greater the resistance to current flow becomes and the slower is the corrosion reaction rate. It is important to realize that the amount of metal that dissolves is directly proportional to the amount of current that flows between the anode and the cathode. [Note: One ampere of current flowing for one year represents a loss of 20 pounds (9.1 kg) of iron.]

Distilled water is not very conductive and is not very corrosive. In contrast, produced brine and seawater are quite conductive and can be very corrosive depending on what else is dissolved in the water. Brine and seawater may be virtually noncorrosive if they contain no dissolved gases or other corrosive agents. The importance of conductivity is its effect on the ease of current flow from the anode to the cathode.

2.4.2 pH

The pH of any water is a measure of the acidity or alkalinity of an aqueous solution; pH is expressed as a number between 0 and 14. The midpoint of the pH scale is 7; a solution with this pH is neutral. Numbers below 7 denote acid solutions with pH 0 being the most acidic and pH 6 being the least acidic; those above 7 denote alkaline (often called "basic") solutions, with pH 14 being the most alkaline and pH 8 the least alkaline.

The pH is the negative logarithm of the hydrogen ion concentration as shown here in Equation 2.1:

$$pH = -\log[H^+] \tag{2.1}$$

The greater the concentration of hydrogen ions, the more acidic the solution and the lower the pH value. Because pH is a logarithmic function, solutions having a pH of 6.0, 5.0, and 4.0 are 10, 100, and 1,000 times more acidic, respectively, than one with a pH of 7.0. Hydrogen ions make a solution acidic and, therefore, force the pH toward zero. Hydroxyl ions (OH$^-$) make a solution basic or alkaline and push the pH upward. An aqueous solution with pH 14 is 10 times more alkaline than one with pH 13.

The corrosion rate of steel usually increases as the pH of the water decreases (becomes more acidic), although extremely high pH solutions can also be corrosive, particularly at elevated temperatures. The general variation of the corrosion rate of steel with pH is shown in Figure 2.1. The actual variation of corrosion rate with pH is dependent on the composition of the water or electrolyte. In many oil field waters, protective scales, such as iron hydroxides or carbonate scales, may form on the steel surface and slow down the corrosion rate.

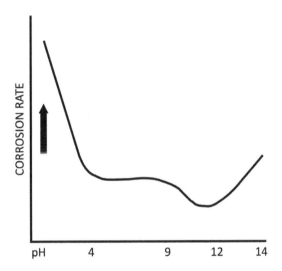

Figure 2.1 Steel corrosion rate curve typical of variations related to hydrogen ion concentration (pH) in the electrolyte.

2.4.3 Dissolved Gases

Oxygen, CO_2, or H_2S dissolved in water, drastically increases its corrosivity. In fact, dissolved gases are the primary cause of most oil and gas corrosion problems. If they could be excluded and the water maintained at a neutral pH or higher, most oil field waters would cause very few corrosion problems.

Dissolved oxygen (DO) serves as a strong oxidizer to promote the corrosion reaction. CO_2 and H_2S ionize in condensed water to form acids that lower the pH (Figure 2.2A). The presence of buffers, such as bicarbonates from formation water, moderates the pH and makes the system less corrosive (Figure 2.2B).

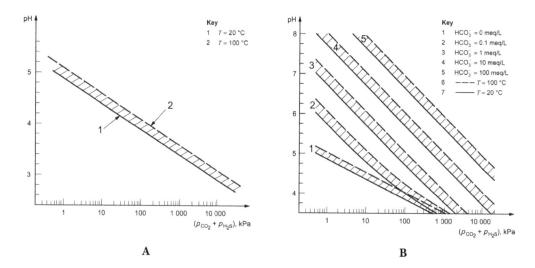

Figure 2.2 (A) The effects of CO_2 and H_2S partial pressures on the pH of condensed water (water free of dissolved salts), as well as produced water including bicarbonates (from Reference 2, Fig. D.1). (B) The pH of condensate water (wet gas) or formation water containing bicarbonate (undersaturated in $CaCO_3$) under CO_2 and H_2S pressure at two temperatures (20°C, 70°C), from Reference 2, Fig. D.2.
Figures provided courtesy of NACE International.

2.4.3.1 Dissolved Oxygen (DO)

Of the three dissolved gases mentioned (CO_2, H_2S, O_2), oxygen is by far the most corrosive of the group. It can cause severe corrosion at very low concentrations. Less than 0.05 parts per million (ppm) (50 parts per billion [ppb]) can cause accelerated corrosion. If either or both of two gases (CO_2 and H_2S) are present, it drastically increases their corrosivity. Oxygen usually causes pitting type attack. A concentration in excess of 50 ppb is usually an oxygen corrosion problem, and less than 10 ppb is usually not. A concentration between 10–50 ppb may be corrosive depending on the CO_2 and H_2S concentration.

Oxygen can change protective scales to nonprotective ones. As will be noted in Chapter 3, Section 3.5.3, oxygen concentration cells (also known as "differential aeration cells") can cause localized attack or pitting. Any time there is a difference in the oxygen content of water in the presence of an immersed metal, attack will take place on that metal preferentially in the area exposed to the lowest oxygen concentration.

Sources of Dissolved Oxygen

It should be noted that oxygen is not naturally present in subsurface formations. Most formations contain several compounds that react readily with oxygen. Presumably, in geologic time any oxygen originally present in the reservoir has been consumed by those compounds. Thus, oxygen should not contribute to corrosion downhole or in surface equipment. Oxygen should be rigorously excluded from producing oil and gas wells, surface facilities, and water supply and injection systems. When it is present, the fluids can contact air through an inadvertent leak or by exposure to the atmosphere. Water flood source waters from lakes, oceans, or streams will contain dissolved oxygen, and water from shallow wells may contain some dissolved oxygen.

Sometimes, however, oxygen corrosion is found in downhole equipment. When this occurs, it usually is caused by incorrect operating techniques or faulty equipment. A common cause of oxygen entry into pumping wells is operating with the annulus vented to the atmosphere, thus reducing backpressure on the producing formation. Air quality regulations have eliminated this practice in many areas by eliminating the vented annulus. The probable cause of oxygen entry today in rod pumping and ESP wells is leaking wellhead seals and packing glands. Oxygen can cause problems in gas lifted wells if the lift gas compressor suction drops below atmospheric pressure, or if the lift gas is return gas from a vacuum gathering system.

Often, aerated waters must be handled. Surface waters will contain some oxygen whether it is a water flood source water, fire water, cooling water, or water for the fresh water "water maker" on an offshore platform or floating production unit. Corrosion control (oxygen scavenging, inhibition, pH control) or materials selection (coatings, linings, nonmetallic materials, or corrosion resistant alloys) should be considered when handling aerated surface waters.

Tuberculation

The limited supply of oxygen may cause the formation of distinct lumps called "tubercles" (Figure 2.3). The corrosion product is a soft, jelly-like material when wet, and deep sharp-bottom pits may occur under the deposits. It is often red or red-brown in color. Tubercles become hard and crusty when dry. Rapid perforation of the metal, obstruction of flow, and equipment plugging by the corrosion products can be expected. The volume of corrosion products compared with the amount of oxygen can be surprisingly large and may cause more damage than the level of oxygen seemingly would indicate. Small quantities of corrosion products can cause plugging of equipment and plugging of the formation at injection wells.

The solubility of oxygen in water is a function of pressure, temperature, and dissolved solids content (primarily chlorides). Oxygen is less soluble in saline water than in fresh water. Appendix 2.A, Figure 2.A.1 presents oxygen solubility data for fresh water and for saline waters of various chloride contents at various temperatures, at atmospheric pressure.[3] An increase in pressure will increase oxygen solubility; however, most produced waters are contaminated with oxygen at atmospheric pressure. Similar data for very heavy brines (e.g., weighted packer brines) indicates ambient temperature oxygen saturation as low as 2–3 ppm.

Oxygen can change protective scales to nonprotective ones. Typical examples are corrosion at water–air interfaces, crevices, under deposits or scales, and "oxygen tubercles" in water systems (Figure 2.3).

Figure 2.3 Oxygen corrosion tubercles—a large tubercle that covered the localized corrosion was dislodged during pipe sectioning (see also Figure. 3.16, in Chapter 3, for another photo of tuberculation).[4]
Photograph provided courtesy of NACE International.

2.4.3.2 Dissolved Carbon Dioxide (CO_2)

When CO_2 dissolves in water, it forms carbonic acid, decreases the pH of the water, and increases its corrosiveness (Figure 2.2). CO_2 is not as corrosive as oxygen, but usually results in localized corrosion (Figures 2.4 and 2.5). Corrosion primarily caused by dissolved CO_2 is often referred to as "sweet" corrosion, as opposed to corrosion caused by H_2S, commonly called "sour" corrosion. CO_2 corrosion takes different forms depending on many variables such as flow rate, temperature, presence of welds, and CO_2 partial pressure. Figures 2.4 and 2.5 and figures in Chapter 3 show various forms of CO_2 corrosion.

Figure 2.4 (A) CO_2 corrosion of a gas condensate flow line well flow line. Note the sharp edges of the localized corrosion. (B) Closeup of the sharp-edged corrosion pattern in Figure 2.4A.

Figure 2.5 Two examples of localized CO_2 corrosion. (A) Note the sharp edges and flat bottom of the localized corrosion, adjacent to uncorroded metal. This is an example of localized corrosion referred to as "mesa corrosion" (described further in Chapter 3). (B) Localized corrosion in the heat affected zone of a weld.
Photographs provided courtesy of Nalco Champion, an Ecolab Company.

Because CO_2 plays such a prominent role in oil field corrosion, some of the factors governing its behavior should be considered. The factors governing the solubility of CO_2 are pressure, temperature, and composition of the water. Pressure increases the solubility, which lowers pH; increasing temperature decreases the solubility, which raises pH (Figures 2.2A and B). Many dissolved minerals prevent pH reduction (that is, they buffer the water). Most notable of these ions in oil field water is the bicarbonate ion (HCO_3^-).

In a gas condensate well with few dissolved minerals and at relatively high temperatures, pressure is the controlling factor influencing CO_2 solubility. In fact, the partial pressure of CO_2 can be used as a yardstick to predict corrosion problems in gas condensate wells. The partial pressure of CO_2 can be determined by the formula:

$$P_i = P_t * X_i \tag{2.2}$$

Partial pressure (P_i) = total pressure (P_t, absolute) * mole fraction of CO_2 in the gas (X_i)

For example, in a well with an absolute bottomhole pressure of 24.1 MPa (3,500 psi) and a gas containing 2% CO_2:

CO_2 partial pressure at the bottom of the well = 24.1 × 0.02 = 0.5 MPa (70 psi).

Using the partial pressure of CO_2 as a prediction for the likelihood of CO_2 corrosion, the following estimations have been used in industry:

- A partial pressure above 0.2 MPa (30 psi) indicates CO_2 corrosion will cause sufficient problems to justify corrosion control. (Some operators consider 0.1 MPa (15 psi) as the break-over partial pressure.).
- At a partial pressure between 0.02–0.2 MPa (3–30 psi), CO_2 may cause corrosion, but other factors such as temperature, pH, and turbulence may have more of a predominant effect than partial pressure (refer to Figure 3.4 and discussion).
- A partial pressure below 0.02 MPa (3 psi) indicates CO_2 corrosion should not be a problem, and the gas may be considered noncorrosive. Corrosion may cause problems, but CO_2 is not likely to be the primary cause.

The CO_2 partial pressure guideline does not predict corrosion rate, rather it indicates the need for corrosion control. Corrosion modeling has superseded the above guideline approaches for design and prediction.

Much work has been done, and work continues to develop computer models to predict corrosion rates, and the references are a small sample of the literature.[5-7] Several models are being used by various companies in the industry. Because this is a developing technology, if more detailed estimates than the guideline are required, the latest literature should be consulted.

Although originally developed for gas condensate wells, the CO_2 partial pressure guideline has proven to work reasonably well in gas gathering systems, gas handling facilities, and gas pipelines. Its accuracy depends on the concentration of buffering ions and organic acids present in the water phase, and the degree of turbulence. Organic acids reduce pH and can make even low CO_2 partial pressure streams corrosive. A high amount of turbulence can make lower CO_2 partial pressures more corrosive than expected. The corrosion rate in CO_2 service generally increases with temperature. An iron carbonate scale forms, but under predictable conditions of temperature and water composition, a protective iron carbonate scale can grow and reduce the corrosion rate. However, film damage from flow or erosion effects can prevent the protective film from forming or can remove the film after it has formed.

On the other hand, oil wells with CO_2 usually produce a brine that contains dissolved minerals that buffer pH and reduce corrosivity, and the foregoing relationship may not always apply. In addition, oil wetting and the potential for forming oil external-water internal emulsions may strongly reduce the corrosion rate in produced water from oil wells (refer to Chapter 1, Section 1.4). However, corrosion most often is encountered when the CO_2 content is high. As a first approximation, the partial pressure of CO_2 is useful in predicting corrosion of oil wells. The calculated solubility of CO_2 in a typical well is shown in Figure 2.6. This figure shows that at bottomhole conditions, CO_2 is more soluble in water than at the wellhead pressure. As oil and dissolved gases rise up the tubing string, pressure and temperature are lowered. When reservoir pressure is less than the bubble point pressure, fresh water may condense from the associated gas while being produced and cause corrosion if buffers are not present (i.e., if there is a low amount of produced formation water that contains bicarbonates). Condensed water has a more acidic pH than produced water at the same temperature and CO_2 partial pressure. CO_2 dissolved in water at temperature and pressure can result in CO_2 corrosion along the inside of the tubing string.

Figure 2.6 should not be interpreted as indicating that CO_2 corrosion is worse at bottomhole pressure because of the higher CO_2 solubility, for two reasons:

1. In a gas and gas condensate well that does not produce formation water, there may not be any liquid water phase at all at bottomhole conditions. However, a separate liquid water phase will exist in a crude oil well if formation water is present.
2. If liquid water is present and it contacts the steel surface, the corrosion reaction with steel forms iron carbonate, which is more stable and can provide a level of corrosion protection at elevated temperature (approximately 80°C/176°F and above), depending on the degree of turbulence.

These competing factors are accounted for in CO_2 corrosion models that are based on thermodynamic mechanisms, as opposed to empirical factors. The release of CO_2 from solution can cause local turbulence and increase corrosion rate. Until the produced fluids reach a separator, the released CO_2 gas forms a gas phase in equilibrium with the water phase; this will keep the water phase saturated with CO_2 under the temperature and pressure conditions present and allows corrosion to continue if the water phase contacts the steel surface.

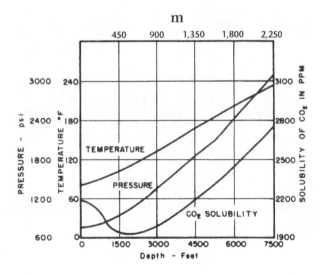

Figure 2.6 Solubility of CO_2 at various depths of a typical oil well.

2.4.3.3 Other Factors Affecting CO_2 Corrosion

Besides temperature, pressure, fluid velocity, and CO_2 concentration, there are diverse factors that can strongly influence the CO_2 corrosion rate, such as the concentration of volatile organic acids, steel composition, and steel heat treatment.

Volatile Organic Acids

In addition to acidity produced by CO_2, low molecular-weight organic acids (such as acetic acid) contribute to corrosion by reducing pH below the value caused by dissolved CO_2 and H_2S. Sometimes these acids play a key role in determining the corrosion rate in a specific system.

The in situ pH can also be influenced by the presence of dissolved volatile organic acids, such as acetic acid, propionic acid, etc. (and their salts), which are not considered in Figures 2.2A and B. These figures consider pH caused by dissolved CO_2 and H_2S. Organic acids may be present in produced fluids and can exert a strong acidifying effect on water pH, especially in waters that contain a low concentration of bicarbonate ions. This effect has been observed in produced gas wells, flowlines, and pipelines, and it must be accounted for to predict an in situ pH at temperature and pressure.

If the bicarbonate ion in produced water is measured by titration, the acetate ion will be included in the titration results, resulting in a false higher concentration of bicarbonate than is present. The calculated pH is more alkaline than the actual pH and could lull an operator into believing that corrosion is not a concern. Where organic acids are suspected, bicarbonate should be measured by ion chromatography, or total inorganic carbon by infrared spectroscopy. An accurate in situ pH should be calculated under pressure and temperature conditions using the true concentration of bicarbonate and organic acid ions.

2.4.3.4 Steel Chemistry and Microstructure

It is widely reported in the literature that small amounts of chromium (Cr) to the steel melt produce a product that is more resistant to localized CO_2 corrosion. It has been known for years that 9% Cr steel was similar in CO_2 corrosion resistance to 13Cr stainless steel. More recently, it has been shown by many researchers that 1–5% Cr steel is more resistant to localized CO_2 corrosion than ordinary carbon steel with no chromium addition. Besides the increased cost of the low Cr steel, consensus is that the increased corrosion resistance imparted by chromium is insufficient to eliminate the need for corrosion inhibition, and many of these same studies indicate the presence of chromium adversely affects inhibitor performance. Users should test the effect of chromium content on the inhibited rate of steel in the laboratory under field conditions before applying low chromium steel alloys.

Steel microstructure also plays a role in determining the corrosion rate in CO_2. Microstructure is a result of heat treatment, which affects the distribution of carbon in the steel microstructural phases (ferrite, cementite, martensite). Chapter 4, Section 4.4 discusses heat treatment and its effect on microstructure.

Many operators select standard API grades for steel downhole tubing and pipelines and ignore the effect of microstructure on corrosion rate. Only certain API grades require a quenched and tempered martensite microstructure. Other grades may have a pearlite-ferrite microstructure in a normalized or hot-worked condition. The ferrite-pearlite microstructure has been reported to be more resistant to CO_2 corrosion than the quenched and tempered microstructure. Therefore, steel coupons used in laboratory corrosion testing should have the same microstructure as the component to be constructed. For example, if a pipeline is to be constructed of a quenched and tempered martensite steel grade, then laboratory specimens should be quenched and tempered steel and not made from rolled strips of carbon steel with a ferrite-pearlite microstructure.

2.4.3.5 Dissolved Hydrogen Sulfide (H_2S)

A "sour system" is the most common term used to identify an environment when H_2S is present. Besides safety and health issues, the presence of H_2S presents three potential problems to materials of construction and equipment:

1. H_2S corrosion
2. Sulfide stress cracking (SSC) failure
3. Plugging with iron sulfide and possibly elemental sulfur

Three Forms of H_2S Corrosion Product

H_2S, when present in oil or gas, will dissolve in any accompanying water and lower the pH in much the same manner as CO_2; however, H_2S is not as potent as CO_2 in lowering the pH of the water. The overall corrosion reaction is H_2S dissolved in water reacts with iron to form iron sulfide (FeS) and atomic hydrogen.

There are three types of FeS commonly found in oil field corrosion: pyrite, pyrrhotite, and mackinawite. Pyrite, an ordered solid solution of FeS and elemental sulfur, is found only when elemental sulfur is present in the system. Pyrrhotite is a form of FeS that forms in most oil field sour environments. With pyrrhotite, iron is substoichiometric to sulfur ($Fe_{1-x}S$ where $0 < x \leq 0.2$); that is, iron is below a simple molar ratio of 1:1 with sulfur. Mackinawite is a metastable form of FeS that forms under the special conditions of low H_2S activity in the environment, which is referred to herein as "slightly sour."

Effect of CO_2

In almost all produced fields where H_2S is present, CO_2 is also present, so the corrosion reaction is driven by either CO_2 corrosion or H_2S corrosion, depending on the relative quantities of each of these gases. The presence of H_2S, even in very low concentrations, can control the corrosion and reduce the corrosion rate to be less than it would be when caused by CO_2 alone. This occurs when there is sufficient H_2S present to form an FeS corrosion film. In "slightly sour corrosion" it is possible that corrosion products from both CO_2 corrosion (iron carbonate) and H_2S corrosion (iron sulfide) coexist.

Need for Corrosion Inhibition

Most oil field steels require corrosion protection by inhibition even if they are resistant to SSC. Weight loss and localized corrosion because of H_2S is a very complicated situation. Under ideal conditions, H₂S will form an FeS pyrrhotite film on the steel surface and protect the steel from further corrosion. In general, corrosion inhibition may still be required even when a protective film forms; however, corrosion in a limited number of sour fields has been adequately controlled by the protective film alone. Localized corrosion can occur in (a) gas and gas condensate, (b) oil plus associated gas, and (c) produced water environments when the environmental conditions cause a disruption or change in the crystalline structure of the iron sulfide film. Localized corrosion on tubing exterior due to H₂S in the annulus of an open annulus well is shown in Figure 2.7.

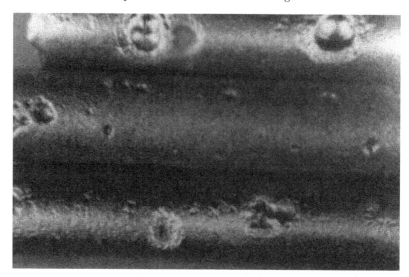

Figure 2.7 H₂S corrosion resulted in localized corrosion of external surface of tubing in sour service—annulus was open (i.e., no packer).
Photograph provided courtesy of Nalco Champion, an Ecolab company.

Conditions That Lead to Localized H₂S Corrosion[8-12]

- *Presence of oxygen*—Potential sources of oxygen contamination have been previously described in this chapter. In low-pressure oil wells being artificially lifted, the presence or absence of oxygen may be critical to minimize corrosion. In the presence of H₂S, oxygen may react with the H₂S to form elemental sulfur. Wet elemental sulfur is very corrosive to steel.
- *Presence of elemental sulfur*—Sour gas wells with high H₂S contents (5–35+ mol%, the so-called "super-sour" wells) may produce elemental sulfur along with the gas and greatly increase the corrosion potential. Deep, hot, sour gas wells that do not produce hydrocarbon liquids have been very corrosive to wells completed with carbon and low-alloy steel tubulars, especially near the bottom of the well. These wells require protective measures such as corrosion resistant alloy (CRA) construction or continuous corrosion inhibition. Produced hydrocarbon liquids dissolve elemental sulfur and reduce the corrosion rate by preventing the formation of sulfur deposits.
- *Slightly sour conditions*—The form of iron sulfide corrosion film formed on steel surfaces under slightly sour conditions (at low ppm levels of H₂S in the gas) is called mackinawite. However, this film may not be stable enough to reduce the corrosion rate and prevent localized corrosion

because local film breakdown occurs readily. Film stability increases with increasing H_2S (to form pyrrhotite), decreasing temperature (below 40°C/104°F) and higher pH levels (above 6). Industry guidelines have not been proven to be accurate as to what concentration of H_2S or what ratio of CO_2/H_2S is required to form a stable FeS film.

- The minimum partial pressure of H_2S required to form a stable FeS is estimated by the author to be between 7–35 kPa (1–5 psi).
- The demarcation between sweet (CO_2-only) and sour ($H_2S + CO_2$) corrosion has been at a reported CO_2/H_2S ratio = 500, but this has been shown to be inaccurate.[8, 10] Also, the implication is that at this point of demarcation, an iron sulfide film will form, but that it may not be stable. An equation that calculates the minimum partial pressure of H_2S required to form the mackinawite form of FeS is given in Reference 8 and is repeated in Appendix 2.B.[8]

- *Elevated temperature*–Above approximately 93–120°C (200–248°F), mackinawite becomes unstable and its crystal structure begins to transform into pyrrhotite. Scale stability is reduced until a continuous pyrrhotite film forms.
- *Under-dosing of corrosion inhibitor*–In corrosion-inhibited sour systems, an insufficient concentration of corrosion inhibitor can result in iron sulfide film breakdown. This ensuing localized corrosion occurs at a higher rate than in the absence of the inhibitor, thus making under-dosing an accelerator of corrosion.[11]
- *Excessive turbulence*–Iron sulfide resists breakdown from flow effects and high wall shear stress better than iron carbonate. Many operators have been able to successfully operate at, and somewhat above, the critical velocity determined in API RP14E (refer to Chapter 1, Section 1.5.5) when operating in a sour environment.
- *Sand-erosion*–The presence of sand production, can either prevent the formation of a protective FeS film, or locally destroy it and prevent it from re-forming if the abrasive continues to be produced at a sufficient quantity and velocity.
- *Low pH*–Lower pH is more corrosive and requires more H_2S to form a protective FeS film in a produced fluid. Lower pH can be a result of high partial pressure of ($H_2S + CO_2$), the presence of volatile organic acids, the absence of buffering bicarbonate ions (condensed water), or lower temperature.
- *Mechanical abrasion*–Rod wear or wireline work can damage the iron sulfide film. Wireline work is typically of short duration, and the iron sulfide film should rapidly re-form when the well returns to service. Corrosion inhibition will help minimize film damage. The use of centralizers helps minimize rod wear and film damage.

The temperatures, pressures, and H_2S concentrations cited here are approximations. Unlike CO_2 corrosion, corrosion modeling of H_2S corrosion is not as well developed. Therefore, conducting corrosion testing is the best means of determining corrosion rates, addressing if localized corrosion is a concern, and if corrosion inhibition is required. Test durations of two or more weeks may be necessary to allow areas of local attack to nucleate and grow to the point of being visible and measurable.

Sulfide Stress Cracking (SSC)

SSC causes a rapid metal failure. SCC failures are brittle in appearance and plastic deformation is absent. An example of a tubing failure due to SSC is shown in Figure 2.8.

Figure 2.8 SSC of tubing—high hardness tubing that failed while being run into a well (circa 1950s).

In many instances, SSC may occur with very little wall loss or localized corrosion. The SSC of metallic materials in the presence of H$_2$S (partial pressure of H$_2$S is greater than 0.3 kPa [0.05 psia] has been recognized by NACE MR0175 since 1975.[13] Carbon and low-alloy steels satisfactorily resist SSC if they meet the latest NACE MR0175 requirements. Following are the steel requirements described in MR0175:

- Heat treatment—Only certain heat treatments or metallurgical conditions are allowed: hot rolled; annealed; normalized; normalized and tempered; normalized, austenitized, quenched, and tempered; and austenitized, quenched, and tempered.
- Hardness—In general, the maximum permitted hardness is 22 Rockwell C (22 HRC). Higher hardness limits apply to certain Cr-Mo low-alloy steels (UNS G41XX0, where XX = 30 or 40) if restrictions on heat treatment and strength are met. Additional hardness restrictions exist for certain forging and wrought pipe grades.
- Nickel concentration in the steel is less than 1%—Higher concentrations are acceptable if the steel is qualified according to the standard.
- Free machining steels are not permitted.
- Cold working is less than 5% of outer fiber stress.

NACE MR0175 has made it possible for the industry to operate very successfully with a minimum of SSC problems. The failures that occurred have been a result of inadequate metal selection, inadequate inspection, or damage during operation (examples are cold working, slip marks from tubing tongs, etc.). Selection of corrosion-resistant alloys also can be affected by the presence and concentration of H$_2$S. This topic is discussed later under CRAs in Chapter 4, Section 4.6.

It is very important to remember that resistance to SSC does not mean resistance to other forms of corrosion. Most oil field steels that are resistant to SSC require corrosion protection, such as inhibition. The use of plastic coatings and linings, as well as electroplated or electroless plated metallic coatings should not be considered adequate for preventing SSC because their protective layer may be inadvertently damaged.

Other forms of damage to steels from wet H_2S exposure are described in MR0175. These are hydrogen induced cracking (HIC) and stepwise cracking (SWC). Annex B of MR0175 provides guidance on test methods and acceptance criteria to evaluate resistance to HIC and SWC.

Injection Well and Producing Well Plugging

The presence of H_2S in injection water or produced formation water may result in injection well plugging. Suspended iron sulfide corrosion products may cause a significant injection well plugging problem. Plugging by elemental sulfur deposits may be caused by oxygen contamination; however, elemental sulfur can also form in producing wells in the absence of oxygen in wells with high concentrations of H_2S in the gas phase, typically more than 5%. ("super-sour" wells). The absence of liquid hydrocarbon production increases the tendency for plugging because liquid hydrocarbons can dissolve elemental sulfur.

Mitigation of plugging has included a number of methods:

- Reducing the corrosion rate by injecting more effective corrosion inhibitors. This reduces the amount of iron sulfide available for plugging.
- Adding a biocide. If H_2S generation is caused by bacterial action, then biocides can control bacteria and reduce the amount of H_2S and iron sulfide formed.
- Adding an H_2S scavenger (see Chapter 9, Section 9.9.2).
- Maintaining an oxygen-free environment helps prevent elemental sulfur from forming in most sour injection and production wells. Super-sour gas wells may require batch treatment of a sulfur solvent.

2.5 Microbiological Influences

Certain types of bacteria contribute to a form of corrosion called microbiologically influenced corrosion (MIC) in produced fluids containing water.[14-16] There are three broad categories for these bacteria:

1. Sulfate reducing bacteria (SRBs)
2. Acid producing bacteria (APBs)
3. Metal (iron and manganese) oxidizing bacteria (MOBs)

2.5.1 SRBs

H_2S can be generated by microorganisms. In oil and gas production, one of the primary source of H_2S-generation problems is the family of bacteria commonly known as sulfate reducing bacteria

(SRBs). There are many types of SRBs; one well-known one is *Desulfovibrio desulfuricans*. It was formerly believed that SRBs caused most MIC problems. Now it is recognized that mixed populations of bacteria contained in a biofilm are the most damaging. During the formation of a biofilm, aerobic organisms (bacteria requiring free oxygen) in the outer layer create an anaerobic (oxygen-free) environment at the base of the film where their colony is attached to the metal surface.

SRBs are anaerobes; for the most part they require a complete absence of oxygen and a highly reduced environment to function efficiently. Investigators have noted that SRB circulate (probably in a resting state) in aerated waters until they find a site to their liking. Some SRB strains can tolerate low levels of oxygen.[14]

SRBs contribute to corrosion because of their ability to change sulfate ions in the water to sulfide ions and to generate H_2S. Then the H_2S reacts with the steel surface and any iron in solution to form FeS precipitate and scale. The FeS scale is cathodic to steel, resulting in corrosion of scale-free areas, which in turn, adds more iron to the solution. SRBs contribute to corrosion in oil and gas operations—usually in water flood or disposal systems and, at times, external corrosion of buried pipelines due to the bacteria being present in moist soil.

2.5.2 APBs

APBs generate acid by-products. These short-chain fatty organic acids (acetic, formic, and lactic) are themselves corrosive and they provide food for SRBs. APBs promote the removal of electrons from the cathode by hydrogen, which results in cathodic depolarization and the continuation of an unabated corrosion rate.

2.5.3 MOBs

MOBs are usually found in fresh water but can grow in produced water. They normally are aerobic, but only require a small concentration of oxygen. Typically, MOBs cause fouling by metal oxides.

2.5.4 Field Monitoring

Guidelines for field monitoring of bacteria growth in oil and gas systems are provided in NACE TM0194.[17] This standard describes sampling procedures, culturing techniques, results interpretation, and the evaluation of chemicals for control of bacteria that cause MIC. MIC monitoring in pipeline systems is described in two references.[18-19]

Horacek et al.,[15] stated in 1993 that

> "The reasons the petroleum industry suffers from MIC fall into three related categories: industry operational practices, industry failure to recognize the impact of MIC, and the lack of standardized methods to diagnose, control, and monitor MIC."

However, now that several industry standards *do* exist, probably the most critical of these are the operational practices. The industry has tended to ignore methods to avoid microbiological concerns until a system has become contaminated. Poor practices are responsible for many MIC problems. Examples of these include mixing bacteria-contaminated open water sources (e.g., rain water, surface water) with produced water; lack of maintenance pigging in pipelines; operating with dead legs where water, bacteria, and solids (such as sand, corrosion products) can settle in quiescent zones; allowing sludge buildup in water tanks; and many more. MIC control and monitoring are further discussed in Chapter 9, Sections 9.8 and 9.9, and Chapter 10, Section 10.3.3.

Scale analysis is not a sufficient determinant if a corrosion failure was caused by MIC. The presence of iron sulfide scale in the corrosion product of carbon steel in a sweet system is a strong indicator; however, many factors, including operating practices and locating sources of contamination, must also be considered.

2.5.5 Rapidly Evolving Understanding

This chapter provides a simplified introduction to the complex subject of MIC. Modern techniques that use molecular microbiological methods (MMMs) are advancing science's understanding of the role of microorganisms in corrosion.[20] Different MMMs provide various types of information about microbial diversity, abundance, activity, and function—all of which are quite different from the culture-based (e.g., serial dilution) results that are familiar to oil and gas industry corrosion professionals. Traditional techniques that count the number of bacteria (e.g., SRBs, APBs) are supplemented by MMMs that identify the species of bacteria. Yet, even with industry's rapidly evolving technology, a process for establishing clear links between microbiological data, corrosion mechanisms, and corrosion rates has not yet evolved. There are no industry-accepted corrosion models that can be used to quantify the corrosion threat of MIC. Industry's understanding of the mechanisms of MIC and the use of MMMs are rapidly evolving. Readers are therefore encouraged to review recent corrosion literature to keep abreast of that latest advances.

2.6 Physical Variables

Physical variables such as pressure, temperature, and fluid velocity play a strong role in determining the corrosivity of a system.

2.6.1 Temperature

Like most chemical reactions, corrosion rates generally increase with temperature. A rough guideline suggests that the reaction rate doubles for every 10°C (18°F) rise in temperature. Corrosion rates do increase with temperature in closed systems. However, in a system open to the atmosphere, corrosion rates may increase at first as the temperature is increased and then decrease with a further temperature increase. As the solution gets hotter, the dissolved gases come out of solution, decreasing the corrosivity of the water. CUI is an example where corrosion rate increases with temperature until a point is reached where it decreases. Another example occurs in open cooling water systems.

In a closed system, corrosion rates do not exhibit this behavior because system pressure prevents the gases from escaping. However, in a closed system containing CO_2, corrosion rate may be reduced with an increase in temperature if a stable iron carbonate film can form on the metal surface. In this situation, an increase in temperature may result in an insoluble corrosion product scale that can be protective.

Corrosion rate increases with increasing temperature in strong acids (e.g., hydrochloric), which are used to recomplete or clean up wells of scale. In hydrochloric acid, protective corrosion films do not form on carbon and low-alloy steel.

2.6.2 Pressure

Pressure affects the rates of chemical reactions, and corrosion reactions are no exception. In oil field systems, the primary importance of pressure is its effect on the solubility of dissolved gases. More gas goes into solution as the pressure is increased. This may, in turn, increase the corrosivity of the solution. This factor is included in the discussion of CO_2 and H_2S corrosion (Section 2.4).

2.6.3 Velocity

Velocity, or lack thereof, affects corrosion in several ways. Stagnant or low velocity fluids usually give low corrosion rates. However, dead (no flow) areas in piping and vessels can serve as incubation sites for bacteria and collection areas for suspended solids. In such instances, corrosion may be accelerated, and localized corrosion under deposits is more likely.

Corrosion rates usually increase with velocity. High velocities and/or the presence of suspended solids or gas bubbles can lead to flow-assisted corrosion, erosion-corrosion, impingement, or cavitation (refer to Chapter 3, Sections 3.2 and 3.7). Operators of production facilities often concentrate internal inspection of production equipment in high turbulence and dead flow areas.

2.7 Corrosion Product and Scale Analysis

Analysis of corrosion products is an important technique in determining the corrosion mechanism. Likewise, analysis of inorganic scales can lead to an understanding of their formation and potential prevention. The sections in this chapter describe the types of corrosion products expected to be found in O_2, CO_2, and H_2S corrosion mechanisms. It is not the purpose of this book to describe the myriad analytical tools that are available to corrosion engineers. Suffice it to say that bulk analysis of corrosion products that are scraped from equipment that has failed consists of elemental and/or compound analysis. In cases where microbiologically influenced corrosion (MIC) is suspected, microbiological analysis of corrosion products (types and number of bacteria) are also useful. Elemental analysis provides which elements are present, but not how they are combined into compounds. What worsens this situation is the tendency of labs to report elements as if they are compounds; for example, Fe reported as FeO (wüstite) or some other iron oxide. This can mislead an untrained corrosion engineer into believing that an elemental analysis is reporting actual compounds present. For H_2S corrosion, elemental analysis reveals S and Fe, but not FeS. The analysis may report Fe in the form of an oxide and sulfur in the form of a sulfate. Elemental analysis is useful but may leave the corrosion engineer pondering what corrosion mechanism has occurred.

2.7.1 Elemental Analysis

Elemental analysis can provide qualitative (major/minor) or quantitative results, depending on the technique. Examples of qualitative elemental analysis techniques include, but are not limited to, x-ray spectrometry (XRS), ion chromatography (IC), optical emission spectroscopy (OES), and inductively-coupled plasma atomic emission spectroscopy (ICP-AES). Depending on the x-ray detector used, light elements such as carbon and oxygen can be detected or entirely undetected. For example, an elemental detector that is incapable of detecting carbon will not show signs of carbon in iron carbonate, which is a CO_2 corrosion product. Only iron will be detected, leaving the corrosion engineer uncertain whether the corrosion product is iron carbonate or iron oxide.

2.7.2 X-Ray Diffraction

X-ray diffraction (XRD) reveals the crystal and phase structure of the corrosion product and can resolve many of these uncertainties if crystalline compounds are present in sufficient quantity, usually more than a few percent by weight. XRD will reveal the type of iron sulfide present, or if iron carbonate or iron oxide is present. Sample preservation is critical to ensure that the sample has not reacted with water or air prior to analysis. The following techniques are examples of quantitative analysis methods used on bulk samples: atomic absorption spectrometry (AAS), IC, ICP-AES, OES, and XRS.

2.7.3 Original Manufacturing Defects

The corrosion engineer should always be aware that failures may be due to original manufacturing defects that grew, or can grow, to failure in service. Such defects in steel often contain high-temperature oxides formed during steelmaking and welding, which are not found in products of aqueous corrosion. These include wüstite (FeO), magnetite (Fe_3O_4) and hematite (Fe_2O_3). Magnetite can also be naturally formed in steam boilers. Hydrated iron oxides such as $Fe_2O_3 \cdot nH_2O$, iron(III) oxide-hydroxide [FeO(OH) and $Fe(OH)_3$] are common aqueous corrosion products in "rust" and are distinctively different from high-temperature scales formed during steelmaking, welding, or during overheating in service (e.g., in boiler water walls and superheater tubes, flares, and furnace tubes, or during exposure to fire).

If a sample from a failed component is examined under an optical or scanning electron microscope (typically, a destructive analysis), elemental analysis using x-ray techniques is possible on a very small scale. Techniques such as Raman spectroscopy can play a role in determining the compound present in small quantities. The author has used these techniques to prove to a regulatory agency that crack-like flaws from the outside diameter surface detected during inspection of an old, buried carbon steel pipeline that transported sour gas had originated from manufacturing defects, and were not due to external SCC from the soil environment (external SCC is discussed in Chapter 3, Section 3.17.1.2). In Chapter 3, Figure 3.4 analysis revealed the presence of iron carbonate on the fracture surface. It was present in small quantities because the rupture and subsequent exposure to seawater (the pipeline rupture occurred subsea) destroyed or oxidized the original corrosion products.

2.7.4 Chemical Analysis Inside Cracks

Analysis of corrosion products inside cracks is difficult because the surrounding metal overwhelms the analysis. With microscopy, however, the corrosion engineer can determine if the scale or corrosion product inside the crack goes from the surface to the crack tip. An oxide-filled, blunted crack tip may indicate the crack is not active. However, corrosion fatigue cracks are often shaped like an inverted carrot and are oxide-filled, and they can be active because crack growth can continue under cyclic stress. A difference in appearance between the corrosion product at the mouth and tip of the crack indicates the mouth has potentially been exposed to aqueous corrosion that has not penetrated to the crack tip. Investigation at the crack tip is important, not just at the crack mouth.

2.7.5 Quick Field Analysis Methods

Quick analyses methods can be performed in the field if the corrosion engineer has access to dilute acid. A 5% dilute hydrochloric acid (HCl) works well. Dilute sulfuric acid (H_2SO_4) also works but it may form insoluble reaction products.

- A quick test to determine if H_2S corrosion is operative is to add a few drops of dilute hydrochloric acid onto the corrosion product. Release of a pungent, rotten-egg odor of H_2S indicates release of H_2S gas, meaning iron sulfide, a corrosion product of steel exposed to wet H_2S, is a component of the scale. X-ray diffraction and other laboratory techniques are useful for confirmation, and quantitative and compound analyses.
- A quick test for CO_2 corrosion is to add a few drops of dilute hydrochloric acid onto the scale. Effervescence with no odor release indicates release of CO_2 gas, indicating iron carbonate (a corrosion product of steel exposed to wet CO_2), is a component of the scale. As with the test for H_2S corrosion, X-ray diffraction and other laboratory techniques are useful for confirmation, and quantitative and compound analyses to verify CO_2 corrosion.
- The same field test for iron carbonate is used to determine if calcium carbonate scale is present, with the same results. The difference between iron carbonate and calcium carbonate is the latter is white in color, whereas the former is gray.

Color plays a role in other corrosion scales. A greenish color often indicates the presence of corrosion products from copper and its alloys.

Table 2.2 is a reference guide to determine the likely composition of scales and corrosion products using simple field tests.

Table 2.2 Field test reference guide to corrosion and scale analysis.

Reaction with Dilute HCl at Room Temperature	Deposit Color*	Possible Composition
Effervesces with no odor	White or light gray	Calcium carbonate; can also appear dark brown if sufficient iron is included
Effervesces with no odor	Gray or dark gray	Iron carbonate
Effervesces with odor of rotten eggs	Black, dark brown	Iron sulfide (instead of smelling for H_2S, exposing released vapors from test to wet lead acetate paper will turn paper black because of formation of lead sulfide)
Soluble in acid but no odor and no effervescence	Red or red-brown	Iron oxide (from oxygen corrosion) if metal of interest is carbon steel
	Black	Iron oxide (magnetite) if metal of interest is carbon steel and scale is magnetic

Insoluble (or minimally so) in acid, no odor, and no effervescence	Various: depends on other elements present, but usually white or gray	Water scales such as
		  Gypsum (calcium sulfate, $CaSO_4$)
		  Barite (barium sulfate, $BaSO_4$)
		  Strontium sulfate ($SrSO_4$)
		  Silica (SiO_2)
		require detailed analysis

Soluble in acid but no odor and no effervescence (except copper sulfide)		If metal of interest is copper-based or is a copper alloy
	Red	Copper oxide (CuO)
	Green	Copper sulfate ($CuSO_4$), typical corrosion product in water or the marine atmosphere
	Black	Copper sulfide (CuS), typical corrosion product of copper alloy in wet H_2S. Effervesces H_2S in acid

* Color can be strongly influenced by small quantities of iron in corrosion products, turning a white or gray scale to red or brown. Exposure to moisture and air after the sample is removed can also result in oxidation and a color change.

References

1. ANSI/NACE Standard MR0175 (latest edition), "Petroleum and natural gas industries—Materials for use in H_2S-containing environments in oil and gas production" (Houston, TX: NACE International).
2. ANSI/NACE Standard MR0175 (latest edition), "Petroleum and natural gas industries—Materials for use in H_2S-containing environments in oil and gas production. Part 2 Annex D (Houston, TX: NACE International).
3. *Standard Methods for the Examination of Water and Waste Water*, 18th ed. (Washington, DC: American Public Health Association, American Waterworks Association, Water Environment Federation, 1992): pp. 4–101.
4. Herro, H.M., "Deposit Related Corrosion in Industrial Cooling Water Systems," CORROSION'89, paper no. 197 (Houston, TX: NACE International, 1987).

5. Franco, R.J., J.L. Pacheco, W. Sun, D.V. Pugh, K. Geurts, and G.A. Robb, "Applications of Corrosion Models to Oil and Gas Production," CORROSION 2010, paper no. 10370 (Houston, TX: NACE International, 2010).
6. Nyborg, R., "Guidelines for Prediction of CO_2 Corrosion in Oil and Gas Production Systems," Report IFE/KR/E-2009/003 (Kjeller, Norway: IFE Institute for Energy Technology, 2009).
7. Wei, S., and S. Nesic, "A Mechanistic Model of H_2S Corrosion of Mild Steel," CORROSION 2007, paper no. 07655 (Houston, TX: NACE International, 2007).
8. Smith, S.N., and M.W. Joosten, "Corrosion of Carbon Steel by H_2S in CO_2 Containing Oil Field Environments," CORROSION 2006, paper no. 06115 (Houston, TX: NACE International, 2006).
9. Bonis, M.R., "Weight Loss Corrosion with H_2S: From Facts to Leading Parameters and Mechanisms," CORROSION 2009, paper no. 09564 (Houston, TX: NACE International, 2009).
10. Smith, S.N., "The Carbon Dioxide/Hydrogen Sulfide Ratio—Use and Relevance, *Materials Performance* 54, 5 (Houston, TX: NACE International, 2015): pp. 64–67.
11. Sun, W., D.V. Pugh, S.N. Smith, S. Ling, J.L. Pacheco, and R.J. Franco, "A Parametric Study of Sour Corrosion of Carbon Steel," CORROSION 2010, paper no. 14700 (Houston, TX: NACE International, 2010).
12. Smith, S.N., and J.L. Pacheco, "Prediction of Corrosion in Slightly Sour Environments," CORROSION 2002, paper no. 02241 (Houston, TX: NACE International, 2002).
13. ANSI/NACE Standard MR0175 (latest edition), "Petroleum and natural gas industries—Materials for use in H_2S-containing environments in oil and gas production." Part 2, "Cracking-resistant carbon and low alloy steels, and the use of cast irons." (Houston, TX: NACE International, 2018).
14. Tatnall, R.E., "Introduction," in *A Practical Manual on Microbiologically Influenced Corrosion*, G. Korbin, ed. (Houston, TX: NACE International, 1993): p. 6.
15. Horacek, G.L., E.S. Littmann, and R.E. Tatnall, "MIC in the Oil Field," in *A Practical Manual on Microbiologically Influenced Corrosion*, G. Korbin, ed. (Houston, TX: NACE International, 1993): p. 45.
16. NACE TPC 3, "Microbiologically Influenced Corrosion and Biofouling in Oil Field Equipment" (Houston, TX: NACE International, 1990).
17. NACE TM0194 (latest edition), "Field Monitoring of Bacterial Growth in Oil and Gas Systems" (Houston, TX: NACE International, 2014).
18. NACE TM0212 (latest edition), "Detection, Testing and Evaluation of Microbiologically Influenced Corrosion on Internal Surfaces of Pipelines" (Houston, TX: NACE International, 2012).
19. NACE TM0106 (latest edition), "Detection, Testing and Evaluation of Microbiologically Influenced Corrosion on External Surfaces of Pipelines" (Houston, TX: NACE International, 2016).
20. Eckert, R.B., and T.L. Skovhus, "Advances in the Application of Molecular Microbiological Methods in the Oil and Gas Industry and Links to Microbiologically Influenced Corrosion," *International Biodeterioration & Biodegradation*, 126 (New York: Elsevier Ltd, 2018): pp. 169–176.

Bibliography

Atkinson, J.T.N., and H. Van Droffelaar, *Corrosion and its Control–An Introduction to the Subject*, 2nd ed. (Houston, TX: NACE International, 1995).

Borenstein, S., *Microbiologically Influenced Corrosion Handbook* (Houston, TX: NACE International, 1994).

Eckert, R., *Introduction to Corrosion Management of Microbiologically Influenced Corrosion* (Houston, TX: NACE International, 2015).

Landrum, R.J., *Fundamentals of Designing for Corrosion–A Corrosion Aid for the Designer* (Houston, TX: NACE International, 1989).

Roberge, P.R., *Corrosion Basics: An Introduction*, 2nd ed. (Houston, TX: NACE International, 2006).

Skovhus, T.L, D. Enning, and J.S. Lee (editors), "Microbiologically Influenced Corrosion in the Upstream Oil and Gas Industry" (Boca Raton, Florida: CRC Press Taylor and Francis Group, 2017).

Appendix 2.A: Solubility of Oxygen in Water

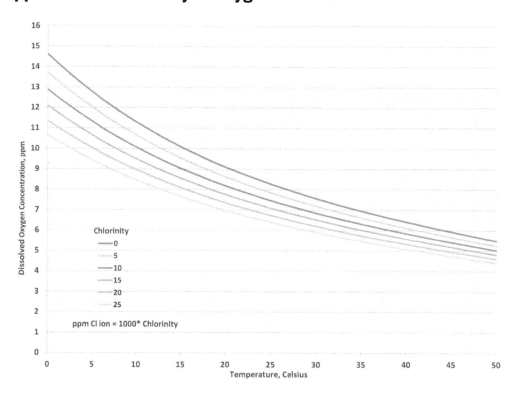

Figure 2.A.1 Solubility of oxygen in water exposed to water-saturated air at atmospheric pressure (101.3 kPa) Adapted from Table 4500-0:1, Reference 3.

Appendix 2.B: Minimum Partial Pressure of H₂S to Form Mackinawite[8]

$$ppH_2S = \frac{\left(10^{\frac{\Delta G_T^0}{4.575*T} - 2*pH}\right) * P}{\gamma} \tag{2.B.1}$$

where

$$\Delta G_T^0 = 961.1 + 93.473 * T - 11.297 * T * \ln(T) - .135 * T^2 - \frac{4.196 * 10^5}{T} \tag{2.B.2}$$
$$+ \frac{2.147 * 10^7}{T^2} + 0.0196 * T^2 * \ln(T) + 7.2 * 10^4 * \frac{\ln(T)}{T}$$

ppH_2S is the minimum partial pressure of H₂S required to form mackinawite,

γ is the fugacity coefficient for H₂S is the gas phase,

T is temperature in Kelvin, and

P is the system pressure.

Inputs are temperature, pressure, and pH of the water phase. Whatever unit is used for pressure will be the unit for partial pressure.

It is critical to point out that the pH value used in this calculation is the pH of the corrosive water *at the temperature and pressure of exposure*, not those measured under atmospheric conditions.

The equation is limited to approximately 6900 kPa (1000 psig) total pressure.

The fugacity coefficient for H₂S ranges between 0.4 and 1.0. An approximate value that can be used in the absence of data = 0.8. A more precise value can be determined if a detailed gas analysis is available.

Forms (or Types) of Corrosion

Robert J. Franco

3.1 Introduction

Failures due to corrosion take many forms, and the various causes and mechanisms of corrosion can result in several types of metal loss with different appearances. Table 3.1 lists the most common forms found in oil field operations and their respective mechanisms.

General corrosion (Figure 3.1) is the uniform loss of metal over a large area of the corroded vessel, pipe, or tank. The overall impression is "thinning of the metal." General corrosion is not the usual form occurring in oil field operations. It is more prevalent in the process industries and refining operations. In oil field operations, metal loss is usually localized in the form of discrete pits, grooves, gouges, crevices, or larger localized areas. In addition, metals may crack because of corrosion without perceptible loss of material.

TABLE 3.1 Types or forms of corrosion.

CORROSION THAT CAN RESULT IN METAL LOSS AND POTENTIAL COMPONENT FAILURE:
    General—overall metal loss, general thinning

    Localized Corrosion—pits, corrosion grooves, etc.
    Localized attack may be caused by or show up in these forms:

    — Concentration Cells
      under deposits, in crevices, in cracks

    — Erosion-Corrosion
      affected by flow with or without solid particles in the liquid

    — Galvanic Corrosion
      dissimilar metals are connected in an aqueous environment

    — Microbiologically Influenced Corrosion (MIC)
      affected by presence or activity of certain bacteria

    — Preferential Weld Corrosion
      related to galvanic corrosion

    — Wear-Corrosion
      affected by rubbing corrosion product off the metal surface

CORROSION THAT CAN RESULT IN COMPONENT FAILURE WITHOUT CAUSING SIGNIFICANT METAL LOSS:
    Environmental Cracking—includes stress corrosion cracking, sulfide stress cracking, and corrosion fatigue

    Intergranular Corrosion—related to preferential corrosion along grain boundaries

OTHER TYPES OF LOCALIZED ATTACK NOT NORMALLY FOUND IN OIL AND GAS PRODUCTION:
    Dealloying—removal of an alloying element from an alloy because of corrosion

    Thermogalvanic Cell Corrosion—related to galvanic effect caused by temperature differences

---

In some cases, metallurgical factors become predominant. Discussion of forms of corrosion that fall in this category necessarily involve the use of basic metallurgical terms. A discussion of metallurgy is presented in Chapter 4, and familiarity with that material will help understand of some of the following discussion. In addition, as noted in Table 3.1, certain forms of corrosion are not commonly encountered in oil field operations, however, they are included for the sake of thoroughness.

3.2 Definitions and Classifications

Proper classification of corrosion damage aids in understanding the corrosion mechanism and how to mitigate it. The terms general corrosion, uniform corrosion, localized corrosion, pitting, and cracking/crack-like flaws, and many other terms used in this chapter are generally defined in textbooks and are below. However, corrosion engineers have not numerically defined these terms, and so some engineers may call the same corrosion damage by different terms. Wherever possible, the reader should properly identify the form of corrosion damage; therefore, definitions and examples are provided in this section from the NACE International Corrosion Technology website.[1] (Accessible to NACE International members only.)

3.2.1 Mechanisms Involving Wall-Loss Damage

- *Concentration cell corrosion*—corrosion caused by an electrochemical cell, the electromotive force of which is caused by a difference in concentration of some component in the electrolyte. (This difference leads to the formation of discrete cathodic and anodic regions.)
- *Crater*—a metal surface anomaly consisting of a bowl-shaped cavity with the minimum dimension at the opening greater than the depth. (Contrast this with "Pit.")
- *Crevice corrosion*—localized corrosion of a metal or alloy surface at, or immediately adjacent to, an area that is shielded from full exposure to the environment because of the proximity of the metal or alloy to the surface of another material or to an adjacent surface of the same metal or alloy
- *Erosion*—the progressive loss of material from a solid surface resulting from mechanical interaction between that surface and a fluid, a multicomponent fluid, or solid particles carried with the fluid.
- *Erosion-corrosion*—a conjoint action involving erosion and corrosion in the presence of a moving corrosive fluid or a material moving through the fluid, leading to an accelerated loss of material.
- *Galvanic corrosion*—accelerated corrosion of a metal because of an electrical contact with a more-noble metal or nonmetallic conductor (e.g., graphite) in a corrosive electrolyte.
- *General corrosion*—corrosion that is distributed more-or-less uniformly over the surface of a material. General attack does not have to be uniform and can result in having areas of greater and lesser metal loss (see Figure 3.1).
- *Impingement corrosion*—a form of erosion-corrosion generally associated with the local impingement of a high-velocity, flowing fluid against a solid surface.
- *Localized corrosion*—corrosion at discrete sites (e.g., pitting or crevice corrosion).
- *Microbiologically influenced corrosion* (MIC)—corrosion affected by the presence or activity, or both, of microorganisms, such as sulfate reducers and acid producers.
- *Pit*—a surface cavity with depth equal to or greater than the minimum dimension at the opening of the cavity. (Contrast this with "Crater.")
- *Pitting*—localized corrosion of a metal surface that is confined to a small area and takes the form of cavities called pits.
- *Thermogalvanic corrosion*—corrosion resulting from an electrochemical cell caused by a thermal gradient.
- *Uniform corrosion*—corrosion that proceeds at the same rate over the surface of a material. (This is the assumption used when calculating corrosion rate or corrosion loss from mass loss or electrochemical measurements. This term should not be used instead of general corrosion to describe an observed surface distribution of corrosion).
- *Wear-corrosion*—damage caused by synergistic attack of wear and corrosion when wear occurs in a corrosive environment.

3.2.2 Mechanisms Not Necessarily Involving Significant Wall-Loss Damage

- *Chloride stress corrosion cracking* (CSCC)—cracking of a metal under the combined action of tensile stress and corrosion in the presence of an electrolyte containing dissolved chlorides. CSCC is a form of environmental cracking.
- *Corrosion fatigue*—the process wherein a metal fractures prematurely under conditions of simultaneous corrosion and repeated cyclic loading at lower stress levels or fewer cycles than would be required to cause fatigue of that metal in the absence of the corrosive environment. Corrosion fatigue is a form of environmental cracking.
- *Dealloying*—a corrosion process whereby one constituent of an alloy is preferentially removed, leaving an altered residual structure (also known as "parting," "selective dissolution," or "selective leaching." In the case of brass alloys, dealloying of zinc is called "dezincification."
- *Environmental cracking*—cracking of a material wherein an interaction with its environment is a causative factor in conjunction with tensile stress, often resulting in brittle fracture of an otherwise ductile material (also known as "environmentally assisted cracking"). Environmental cracking is a general term that includes the terms associated with specific environments, e.g., caustic cracking, chloride stress corrosion cracking, corrosion fatigue, hydrogen embrittlement, hydrogen induced cracking (stepwise cracking), hydrogen stress cracking, liquid metal cracking, stress corrosion cracking, sulfide stress cracking.
- *Hydrogen induced cracking* (HIC)—stepwise internal cracks that connect adjacent hydrogen blisters on different planes in the metal, or to the metal surface—also known as "stepwise cracking" (SWC).
- *Intergranular corrosion*—preferential corrosion at or adjacent to the grain boundaries of a metal or alloy.
- *Stress corrosion cracking* (SCC)—form of environmental cracking of a material produced by the combined action of corrosion and sustained tensile stress (residual or applied).
- *Sulfide stress cracking* (SSC)—cracking of a metal under the combined action of tensile stress and corrosion in the presence of water and hydrogen sulfide (a form of hydrogen stress cracking).

3.2.3 Alternative Definitions API 579-1/ASME FFS-1, "Fitness-For-Service"[2]

API 579-1/ASME FFS-1 "Fitness-For-Service" provides some alternative, definitions that contribute to quantifying some forms of corrosion damage.

- *Crack-like flaws*—planar flaws that are predominantly characterized by a length and depth, with a sharp root radius. Crack-like flaws may be surface-breaking, embedded, or through-wall. Examples of crack-like flaws include planar cracks, lack of fusion and lack of penetration in welds, sharp grove-like localized corrosion, and branch-type cracks associated with environmental cracking (Note: this definition includes corrosion fatigue, CSSC, and SSC).
- *General metal loss*—relatively uniform thinning over a significant area of the equipment (same as the NACE term "general corrosion").
- *Gouge*—elongated local mechanical removal and/or relocation of material from the surface of a component, causing a reduction in wall thickness at the defect; the length of a gouge is much greater than the width and the material may have been cold worked in the formation of the flaw. Gouges are typically caused by mechanical damage, for example, denting and gouging of a section of pipe by mechanical equipment during the excavation of a pipeline (no NACE analog).

- *Groove*—long, elongated, thin spots caused by directional erosion or corrosion; the length of the metal loss is significantly greater than the width (no NACE analog).
- *Hydrogen induced cracking* (HIC)—characterized by laminar (in-plane) cracking with some associated through-thickness crack linkage. This is sometimes referred to as stepwise cracking (SWC).
- *Local thin area*—local metal loss on the surface of the component; the length of a region of metal loss is the same order of magnitude as the width (no NACE analogue; different than crater).
  - Note that the rates of localized metal loss can vary significantly within a given area of equipment.
- *Pitting*—localized regions of metal loss that can be characterized by a pit diameter on the order of the plate thickness or less, and a pit depth that is less than the plate thickness (more quantitative than the NACE definition of pitting).

3.2.4 Applications of Definitions

When these definitions are applied to real world applications, it must be concluded that corrosion damage classifications are not well enough defined to ensure that everyone will agree with the same classification. For example, the author asked six experienced corrosion engineers to classify the corrosion damage in three real examples as being uniform or localized. There were no unanimous classifications!

1. **Case 1**: Horizontal piping is corroded uniformly in the water phase between the 5–7 o'clock position (Figure 3.2). The remainder of the pipe above the 5–7 o'clock position is unaffected.
   a. Five out of the six engineers classified this as uniform corrosion mainly because where the electrolyte (water) resided, corrosion was general, albeit confined to the water-wetted area only.
   b. The engineer who selected localized corrosion did so because for it to be uniform, that engineer required uniform corrosion damage to be reliably detected and provide the corrosion engineer with high confidence that the damage detected is representative everywhere.
   c. Some corrosion engineers refer to this type of corrosion as grooving corrosion (refer to Section 3.2.3 for the definition of a groove).
2. **Case 2**: A piping system is mildly corroded along the straight sections but has severe corrosion only at the outer sweep (extrados) of a pipe elbow, where the elbow is uniformly thinned. Photographic examples are found in Section 3.7 Figures 3.40 A and B.
   a. Four out of the six corrosion engineers classified this as general corrosion, and two as localized corrosion.
   b. The same engineer argued for localized corrosion as in Case 1b, above.
3. **Case 3**: An amine regenerator tower is severely corroded at the area where fluid exiting the reboiler enters the tower (Figure 3.3). Corrosion consists of large patchy areas of uniform corrosion (many square millimeters [mm^2] in area), with patches of unaffected metal between them (using ASME/API 579-1/ASME "Fitness for Service" criteria,[2] the length separating the adjacent corroded areas has been determined to require the corroded areas be treated as one large area of corrosion). The remainder of the tower does not exhibit any corrosion damage.
   a. Five out of six of the engineers classified this as uniform corrosion that is confined to a certain area of the tower. By NACE definition, however, this should be classified as general corrosion, but this was not offered as an option, and frankly, most corrosion engineers do not make a distinction between the two categories.

b. The one dissenting vote for localized corrosion was based on the corroded area being presumably less than 10% of the total tower area, but this was the engineer's personal definition.

The conclusion is that the definitions of forms of corrosion are useful, but there are grey areas in using them. The reader may be wondering why classification of corrosion damage is important at all. As stated earlier, the corrosion engineer should properly classify corrosion damage to aid in understanding the mechanism and how to mitigate it, and to guide investigators examining the root cause of any failure. Photographs and measurements made of the corrosion damage are more important for assessing corrosion damage than the definitions used to describe them. Supportive data such as fluid and corrosion deposit analyses provide confirmatory information regarding the mechanism. For these reasons, corrosion engineers should strive to use the same terminology to describe the same form of corrosion damage.

3.3 General Corrosion

While localized attack is the most usual form of corrosion in oil field environments, general corrosion occasionally may be noted (Figure 3.1), particularly on metals exposed to uninhibited acid. It also may be observed in some types of process equipment, such as carbon steel trays from contactor (absorber) towers in amine sweetening service.

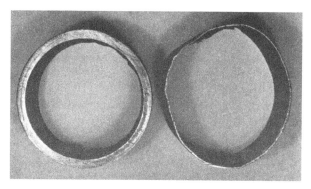

Figure 3.1 Carbon steel pipe before (left) and after (right) general corrosion.
Photograph provided courtesy of GPA Midstream Association.

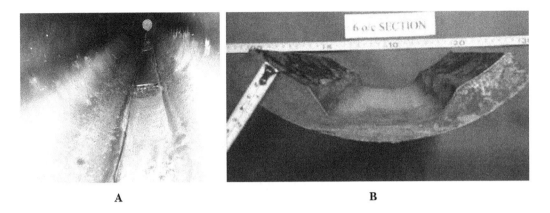

Figure 3.2 (A) Internal general corrosion at the 6 o'clock position of a produced water injection pipeline exhibiting grooving corrosion along the bottom of the line where water had flowed. This is similar to Case 1, Section 3.2.4, where the reason for describing it as general corrosion is stated. (B) Close-up of groove at 6 o'clock position.
Photographs provided courtesy of Chevron Corporation. ©2019 Chevron

General corrosion has also been observed on carbon steel in marine and other atmospheres, corrosion in mineral acids (e.g., hydrochloric), and glycol dehydration units. Corrosion under insulation (CUI) has many characteristics of general corrosion in areas that may be separated by unaffected metal. However, CUI can also display characteristics of highly localized corroded areas. General corrosion has also been found to occur in high temperature equipment and is not caused by water, but by an oxidizing gas at elevated temperature. Examples include high temperature oxidation of fired equipment (e.g., furnace tubes, boiler tubes), and high temperature sulfidation, the reaction of a metal or alloy with a sulfur-containing species to produce a sulfur compound that forms on or beneath the surface of the metal or alloy. Sulfidation of carbon steel has occurred during high temperature regeneration of mol sieve dehydration units if the inlet gas contains hydrogen sulfide (H_2S).

Although uniform and general corrosion may cause the greatest amount of mass loss of metal, they can pose less threat to equipment failure compared to localized corrosion or environmental cracking. If routine, thorough, internal visual and/or external nondestructive testing is practiced, general and uniform corrosion can be detected in time to prevent failure. However, the three cases described above show that even uniform and general corrosion might exist only in discreet parts of a component, leaving the remainder of the component unaffected. Knowledge of both the environment and corrosion mechanism are required to select areas for inspection. The rate of corrosion caused by general and uniform corrosion is more predictable than localized corrosion, using data from repeat inspections. Figure 3.3 shows broad areas of general corrosion in an amine regenerator (stripper) tower in sour gas service. This is the same as Case 3, described in Section 3.2.4.

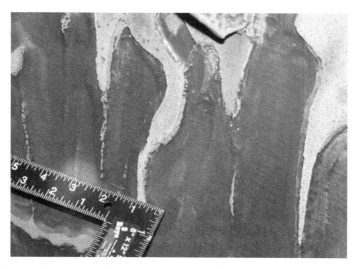

Figure 3.3 Broad areas of general corrosion in an amine regenerator tower in sour gas service, limited to the area where the reboiler fluid returned to the tower.

Uniform or general corrosion of pressurized equipment can result in rupture failure of that equipment if the corrosion damage is left undetected and unabated. Corrosion can progress to such an extent that a large patch of thinned metal blows outward, i.e., it is no longer able to contain the internal pressure so it ruptures. Note that it does not necessarily have to leak before the rupture happens. Figure 3.4 shows a ruptured 406 mm (16 in) diameter subsea, high-pressure gas condensate pipeline that contained carbon dioxide (CO_2). Although the carbon steel line was corrosion-inhibited, high velocity prevented the formation of a protective iron carbonate film and general corrosion occurred. The pipeline was constructed with API 5L grade X52.[3] Failure occurred after 14 years of service with an average corrosion rate of 0.67 mm/y (27 mpy). Operating pressure was 9.7 MPa (1400 psi) and temperature was 74 °C (165 °F) with 1–1.5% CO_2.

Figure 3.4 Rupture of a 406 mm (16 in) diameter high-pressure gas condensate pipeline caused by internal, general CO_2 corrosion. Metal loss progressed undetected until sudden failure occurred.

Sudden failure could also occur in structural equipment if the structural member is covered in concrete or fireproofing, and corrosion continues unabated as the load-bearing area is reduced. This type of corrosion cannot be seen by visual inspection. The likelihood of sudden failure occurring with uniform and general corrosion are real but are lower than having a leak from localized corrosion occurring at a point where the wall thickness is first perforated. The owner/operator company should develop an inspection strategy that addresses detection of the type of corrosion expected; the likelihood of that type of corrosion occurring; and the safety, health, environmental, and financial consequence if that type of corrosion were to lead to loss of pressure or structural containment. Acceptance criteria for corrosion damage found—be it either uniform, general, or localized—are described in API 579-1/ASME FFS-1.[2]

3.4 Localized Corrosion

Localized corrosion occurs when the corroding metal suffers metal loss at localized areas rather than over its entire surface. The entire driving force of the corrosion reaction is concentrated at these localized areas. The corrosion rate at the areas being attacked will be many times greater than the average corrosion rate over the entire surface. The resultant dimensions of localized corrosion damage may be large and shallow, or narrow and deep. The shape of the damage may be nearly perfectly round or elliptical or have an irregular shape. Failure in the form of fluid leakage is typical of localized corrosion.

3.4.1 Pitting Corrosion

"Pitting" is the term used by many field people to describe localized corrosion of carbon and low-alloy steel even if the depth is not equal to or greater than the minimum dimension at the opening. In recent years, however, corrosion engineers prefer to use the more precise term "localized corrosion" in lieu of "pitting corrosion" if the material affected is carbon and low-alloy steel and the environment is a typical produced brine with CO_2 with or without H_2S. The term "pitting corrosion" is then reserved for describing localized corrosion in stainless steels or other corrosion-resistant alloys that forms a passive, thin protective oxide film.[4] The stainless steels and nickel alloys with chromium are especially susceptible to pitting by breakdown of the film at local sites, but aluminum alloys, titanium alloys, and copper alloys are also susceptible because they rely on a passive oxide film for corrosion protection. Pitting in these corrosion-resistant metals is usually manifested as displaying the depth of corrosion greatly exceeding its diameter at the opening (mouth) of the pit. However, pits in passive alloys come in many shapes, including some that do not meet the diameter/depth definition. These are described in ASTM G46.[5]

It is difficult to design a new component, or assess corrosion damage in an existing component, knowing that localized corrosion or pitting is, or will be, the primary mode of failure. Therefore, the designer should avoid failure by using proper materials selection, ensuring correct fluid velocity, controlling operating parameters that are conducive to causing localized or pitting corrosion, and installing and maintaining corrosion barriers (e.g., mitigators such as chemical inhibition).

Some engineers quantify the rate of pitting corrosion. The "pitting rate" is derived by taking pit depth measurements, choosing the deepest pit, and converting the deepest pit depth to an annual penetration rate by dividing depth by the exposure time. This is a misleading number because the rate of pitting corrosion often decreases with time, like a power law with the time exponent less than unity. This means of determining the "pitting rate equivalent" typically results in an equivalent annual penetration rate that is higher than a rate made by repeating measurements taken over time. In fact, sometimes shallow pits do not grow much deeper and their rate of growth may level out, thereby becoming nonthreatening to equipment integrity. Some engineers use the term "pitting factor"—the ratio of the depth of the deepest pit resulting from corrosion divided by the average penetration as calculated from mass loss, but this is a term best suited to the laboratory and is not a useful term for determining continued fitness for service. Pitting can be much more important than general corrosion because the pitted area can penetrate a pipe or vessel wall in a time shorter than the design life of the equipment, and because the isolated nature of pitting, and its relatively small size, make it difficult to detect using nondestructive testing.

A review of Table 3.1 shows that localized pitting corrosion, grooving corrosion grooves, and crevice corrosion are the results of several different corrosion mechanisms or phenomena. Most of these occur frequently in oil field environments on carbon steel equipment and with other alloys depending on circumstances that will be discussed in this chapter. The most prevalent form of attack of carbon and low-alloy steel shows up as localized corrosion, sometimes as isolated areas of wall loss, and other times as groups, clusters, or in lines. The configuration depends on the specific details of the environment, and sometimes on the manufacturing process. For example, corrosion of carbon steel caused by either dissolved oxygen or MIC is often described as "isolated pitting." Corrosion of carbon steel caused by dissolved CO_2 is often described as localized corrosion because the wall loss occurs in connected clusters, or as in wide or narrow corrosion grooves adjacent to relatively uncorroded metal. The latter form is referred to as "mesa corrosion," which is a term unique to oil field CO_2 corrosion. The term is derived from the geographic term that describes an isolated flat-topped hill with steep sides; the top of the hill being the uncorroded area, and the steep sides and reduced wall thickness area immediately next to it being the corroded area (see Chapter 2, Figure 2.5).

Figures 3.5 and 3.6 show the sharp-edged localized corrosion caused by CO_2 corrosion of steel compared to the rounded localized corrosion of steel caused by H_2S corrosion, shown in Figures 3.7 and 3.8. Well conditions for Figure 3.8 were a maximum temperature of 43°C (109°F), a CO_2 partial pressure of 14 KPa (2 psia), and H_2S partial pressure of 28 KPa (4 psia). The through-wall pit was attributed to MIC. The other, circled areas of localized corrosion were attributed to H_2S corrosion. Analysis confirmed the presence of iron sulfide and elemental sulfur in the corrosion product.

Bottom of the line corrosion in a production header caused by CO_2 is shown in Figure 3.9. Operating pressure was 1 MPa (145 psi) and 7% CO_2 content. Inhibition was ineffective, possibly because of the presence of sludge deposits along the bottom of the line. Along the bottom, below the water line, corrosion could be categorized as general corrosion that is confined to the bottom of the header. A horizontal band of localized corrosion formed above the water line. Above that, the pipeline is unaffected.

Figures 3.10–3.13 show pitting attack of alloys that develop passive oxide films. Cavern-shape pitting is typical of 304, 316, and 410 stainless steel in stagnant, aerated, brackish water or seawater, possibly also associated with MIC.

Figure 3.5 Localized corrosion: Isolated metal wall loss inside carbon steel tubing. Note the sharp edges and flat bottom of the corroded area, indicative of CO_2 corrosion.
Photograph provided courtesy of Nalco Champion, an Ecolab Company.

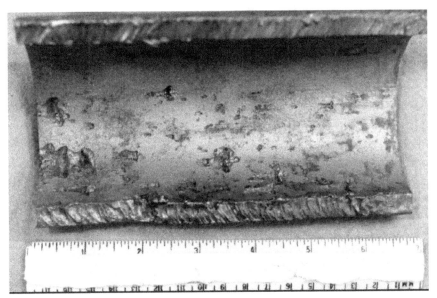

Figure 3.6 Localized corrosion: Internal localized corrosion of carbon steel gas condensate well tubing from CO_2 corrosion (photo taken after cleaning).
Photograph provided courtesy of Nalco Champion, an Ecolab Company.

Figure 3.7 Localized corrosion: Clusters of external pits on a carbon steel rod pump well tubing exposed to sour produced fluids (photo taken after cleaning).
Photograph provided courtesy of Nalco Champion, an Ecolab Company.

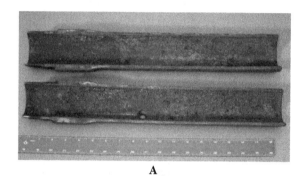

Figure 3.8 Internal localized corrosion of carbon steel tubing string in sour produced fluid. (A) Plan (surface) view of tubing showing one penetrating localized corrosion area and several areas of significant penetration of the tubing wall (circled). (B) Cross section of a partial penetrating localized corrosion area showing significant metal loss in the form of broad, rounded areas typical of H_2S corrosion.
Photographs provided courtesy of Chevron Corporation. ©2019 Chevron

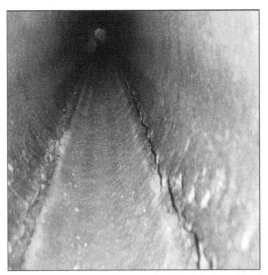

Figure 3.9 Bottom-of-line corrosion confined to the 5–7 o'clock position in a production header due to increasing water cut, water pooling, and CO_2 corrosion.

Pitting attack can occur in a variety of metals used in oil and gas facilities. Figures 3.10–3.12 show pits in martensitic-, austenitic-, and superduplex stainless steels, and Monel. Pitting has also been observed in aluminum alloys and copper alloys in brine service and offshore.

Figure 3.10 Pitting: Cross section of a pit in 410 stainless steel initiated on the inside diameter of a 3 mm-thick heat exchanger tube. Pitting failure occurred before the tubing went into service because it had been hydrotested with brackish water that was left undrained for a few months.

CHAPTER 3: Forms (or Types) of Corrosion

Figure 3.11 Pitting corrosion: Plan view of pitting corrosion of a 316 stainless steel in petroleum production equipment. *Photograph provided courtesy of Nalco Champion, an Ecolab Company.*

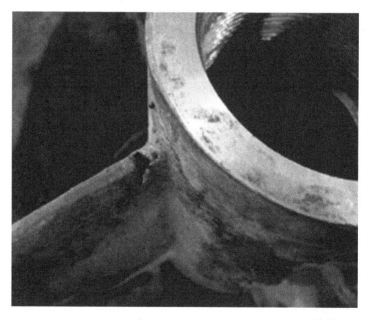

Figure 3.12 Pitting corrosion: Pitting of 2507 SDSS (superduplex stainless steel) with a defective microstructure (sigma phase) in a submersible seawater lift pump.

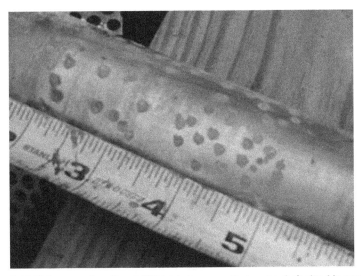

Figure 3.13 Pitting corrosion: Monel 400 (nickel-copper alloy) pump shaft pitted in seawater.

3.5 Concentration Cells

Theoretically, there are many causes of localized corrosion in metals and alloys, and all of them center around localized differences in the metal and/or in the electrolyte in contact with the metal. Localized differences in electrolyte composition generally are referred to as "concentration cells." Depending on the particulars of the metal and corrosive fluid(s) involved, concentration cells are referred to in several ways: differential aeration cells, metal-ion cells, under-deposit attack, or crevice corrosion, to name a few. Figure 3.14 is a schematic cross section of a differential aeration cell, in which the dissolved oxygen concentration in the electrolyte contacting the metal surface varies. The same type of electrochemical cell can occur with species other than oxygen in the electrolyte.

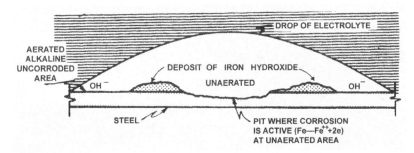

Figure 3.14 Concentration cell corrosion: Schematic of reactions in a differential aeration cell. Edges of the drop are high in oxygen, while the center is low in oxygen.

Figure 3.15 shows concentration cell corrosion on the external surface of a buried carbon steel trunkline. The pipe carried warm produced crude. External corrosion was mitigated by an outer tape coating and impressed current cathodic protection, but the pipe failed after nine years of service because of deficiencies in these corrosion barriers.

Figure 3.15 Concentration cell corrosion: External localized corrosion of a buried carbon steel trunkline because of a differential aeration cell.
Photograph provided courtesy of Chevron Corporation. ©2019 Chevron

3.5.1 Crevice Corrosion

Corrosion in crevices is a serious form of deterioration that can be minimized by careful design and fabrication. This type of corrosion is a unique form of concentration cell corrosion. Because it is very common, crevice corrosion generally is considered a class of corrosion by itself. A crevice can be formed in many ways, such as by a metal contacting another metal or a nonmetal (such as a gasket). Most metals and alloys, except a few highly corrosion-resistant alloys, are susceptible to this type of attack.[6]

Figure 3.16A shows crevice attack of a superduplex stainless steel seawater lift pump flange. This material had a defective microstructure containing sigma phase, a precipitate that is rich in iron and chromium. A crevice is created in a low-alloy steel raised face flange at the bore of the pipe between the raised face and gasket (Figure 3.16B). An improved gasket design can eliminate flange face corrosion. This requires having the gasket inner diameter match the flange bore diameter. This design eliminates crevice corrosion caused by a dead-flow area, and flange face erosion-corrosion caused by turbulent flow. Clamps also create crevices (Figure 3.17). Except for a few special configurations, tubular good joints have crevice areas at the threaded joints. Crevice corrosion occurs at the pin end of the joint. A crevice can cause localized corrosion by providing a situation where the concentration of the electrolyte inside the crevice is different than the concentration outside the crevice.

Crevice corrosion can be especially serious in oxygenated systems (e.g., raw seawater, raw freshwater systems, or raw-water contaminated produced liquids) where the oxygen in the crevice may be consumed more rapidly than fresh oxygen can diffuse into the crevice. Oxygen-depletion results in a decrease in pH in the fluid within the crevice, resulting in a more acidic environment that accelerates corrosion.

Figure 3.16 Crevice corrosion of flanges: (A) 2507 superduplex stainless steel flange with defective microstructure (sigma phase) in submersible raw seawater lift pump. Mating flanges created a tight crevice. Corrosion is associated with a differential aeration cell with high oxygen concentration outside the flange, and low oxygen inside the crevice (also see Figure 3.20). (B) Low-alloy steel raised face flange from a shutdown valve showing crevice corrosion on the flange face at the pipe bore. The pipe was in gas condensate service with wet CO_2. Corrosion is associated with formation of a concentration cell inside the crevice.

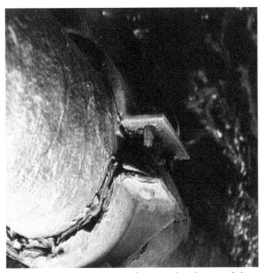

Figure 3.17 Crevice corrosion: Example of crevice attack of a coated carbon steel riser pipe on an offshore platform underneath the riser clamp. Volumetric expansion of corrosion product, combined with atmospheric corrosion of the clamp bolts above the splash zone, have broken the clamp.

3.5.2 Oxygen Tubercles

This is a form of localized corrosion that results from the same type of mechanism as crevice corrosion. It is encouraged by the formation of a porous layer of iron oxide or hydroxide, which partially shields the steel surface.[7] Figure 3.18 illustrates one form of this attack. Figure 3.14 shows it schematically.

Figure 3.18 Tuberculation corrosion of carbon steel ship mooring structure exposed to a marine atmosphere in the Caribbean.

3.5.3 Differential Aeration Cells

An air-water interface is one example of a differential aeration cell. The water at the surface contains more oxygen than the water slightly below the surface. The difference in concentrations can cause preferential attack just below the water line. A comparable situation can occur in process vessels and tanks at an interface between an electrolyte (water in most oil field situations) and a gas, or between two liquids when there is a difference in concentration of the corroding species (e.g. iron ions) at the interface. Another example occurs where a pipeline is buried in a soil containing different amounts of oxygen at certain areas along the line. For example, freshly excavated soil is more aerated than compacted soil. Under these conditions, the pipe in the well-aerated soil will be cathodic, and the pipe in the poorly aerated soil will be anodic and corroded.

In Figure 3.19, pipe is shown passing under a paved road where it is in contact with soil having a restricted oxygen supply as compared to pipe on either side of the road in soil not sealed by surface paving. Corrosion of the pipe under the paving can be severe. Another pipeline situation can occur when the soil's oxygen content on the top of the line is quite different from that at the bottom. Cathodic protection plus coating can mitigate this problem (refer to Chapter 7, Section 7.6.3).

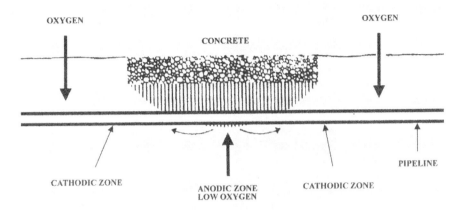

Figure 3.19 Concentration cell corrosion: Drawing showing how different oxygen concentrations in the soil can affect a pipeline.

In all differential aeration cells, it is the oxygen-starved area that is anodic and corrodes preferentially to the area that has an adequate supply of oxygen (Figure 3.20). Chapter 6 Figure 6.10 depicts one means of minimizing differential aeration cell corrosion at the contact point of offshore piping by inserting a nonmetallic half-round rod between the pipe and the pipe support beam. The half-round bar sheds water and prevents moisture being held against the pipe, and it is reinforced to resist sagging under the pipe weight.

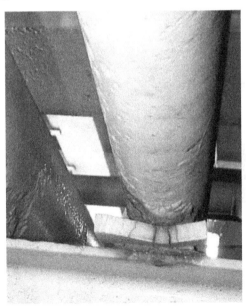

Figure 3.20 Differential aeration cell created by a pipe resting on a rubber pad on a pipe support. The pad also served to hold aerated water against the pipe.

CHAPTER 3: Forms (or Types) of Corrosion

3.5.4 Scale and Deposits

Scale, corrosion products, and deposited solids can have the following physical characteristics:

- poorly adhered to the metal surface; or
- porous and partially, or completely, covers the surface; or are
- nonporous and incompletely cover the surface.

The outcome is accelerated corrosion due to the formation of a differential concentration or a galvanic cell, and corrosion typically occurs underneath the deposit. Even tight, adherent scales can create problems if they form only in spots rather than uniformly over the metal surface. Figures 3.21–3.23 are examples of corrosion under deposits. In Figure 3.21, sand deposits prevent access of a corrosion inhibitor to the metal surface. In Figure 3.22, iron sulfide was transported into a produced water softener, and deposited in the softener resin bed and the 2 in (50 mm) diameter outlet piping. The yellow vertical bar crosses the pipe's electric resistance long seam weld and highlights preferential weld corrosion (PWC). Figure 3.23 shows corrosion at the extrados of a pipe elbow under sand deposits.

Figure 3.21 Under-deposit corrosion (UDC) caused by formation of concentration cells under sand deposits at the bottom of a horizontal primary separator. A pit gauge measures the depth of localized corrosion after sand has been removed.

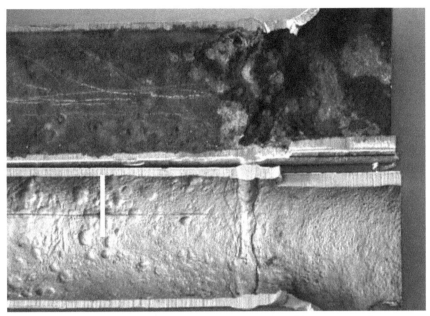

Figure 3.22 UDC: The 2 in (50 mm) diameter Schedule 80 carbon steel pipe before and after cleaning. The deposit is iron sulfide.
Photograph provided courtesy of Chevron Corporation. ©2019 Chevron

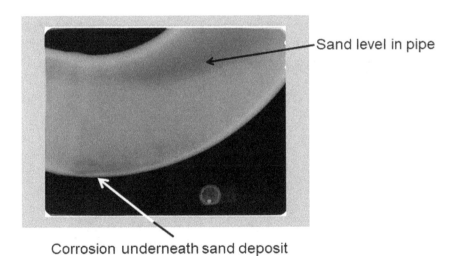

Figure 3.23 Radiographic image of a pipe elbow showing sand deposition and corrosion under sand deposits at the bottom of the elbow. The darker area at the extrados of the elbow indicates metal loss at the inside surface.

3.4.5.1 MIC

MIC is not typically classified as a form of concentration cell corrosion. However, MIC is included in Section 3.5.4 because bacteria (sulfate reducing bacteria [SRB] and others) thrive under scales, sludge, formation solids, and other debris, thus creating H_2S and organic acids that cause localized corrosion. Figures 3.24–3.26 show the results of such bacterial activities. The pit in the center and bottom left of Figure 3.25 is a "classic" MIC pit; that is, it has the "stair step" or terraced configuration often referred to resembling an open pit mine. Pits within pits are a common feature of MIC. However, other mechanisms such as CO_2 corrosion can also cause pitting within a MIC pit, as is believed to have occurred in Figure 3.26.

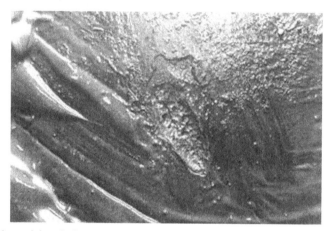

Figure 3.24 UDC under mud deposits in a seawater ballast tank on a floating production storage and offloading (FPSO) vessel. MIC likely contributed to the observed corrosion; coating and cathodic protection can provide corrosion barriers.

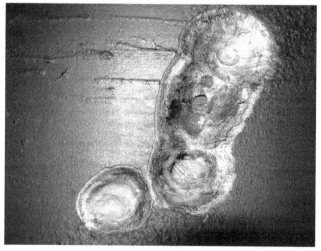

Figure 3.25 MIC of carbon steel (after cleaning). It caused a leak in the sales quality crude oil pipeline that contained sand and other deposits which settled in the bottom of the low-velocity line.

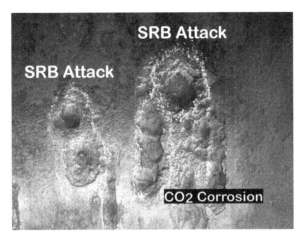

Figure 3.26 MIC of carbon steel (after cleaning). Like Figure 3.25, this photograph shows terraced localized corrosion that occurred under colonies of SRB. Some evidence of CO_2 corrosion also appears.

Alloys that depend on a passive oxide film for corrosion protection can also fail by concentration cell corrosion under deposits. Figure 3.27 shows localized pitting at the bottom half of a 9.5 mm diameter 316 stainless steel instrument line in wet stabilized crude after five years of service. Low fluid velocity contributed to deposition. Pits are located under deposits.

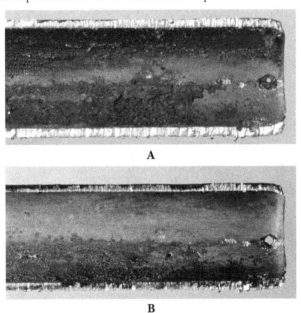

Figure 3.27 UDC pitting of 316 stainless steel: (A) deposits located in bottom half of tubing, before cleaning; (B) additional pits revealed after cleaning.

3.5.5 Localized Corrosion and Pitting of Metals

Localized corrosion occurs at discrete sites (e.g., pitting or crevice corrosion). Unlike general or uniform corrosion, corrosion damage may occur discontinuously on the metal surface separated by areas where no corrosion has occurred. Because of its discontinuous nature, localized corrosion is difficult to detect.

3.5.5.1 Carbon and Low-Alloy Steel

Localized corrosion of carbon and low-alloy steel is of importance to oil production because the majority of oil production equipment is in fact made of carbon and low-alloy steels. Localized corrosion is a problem on sucker rods, production tubing and casing, and drill pipes, as well as on other downhole and surface equipment. It is also the predominant corrosion mechanism in pipelines, well lines, gathering lines, and surface facilities including piping, pressure vessels, and tanks.

A common cause of localized corrosion in carbon steel is the formation of local cells due to the partial breakdown or destruction of protective scales. When a corroding metal becomes covered with a corrosion product that is dense and adherent, the product protects the metal from its environment, and corrosion slows down and may ultimately cease. If the protective scale is removed from some areas, those areas become anodic to the areas beneath the scale that are still protective. The anodic areas corrode preferentially and result in pitting. H_2S and CO_2 are two corrosives that frequently cause localized corrosion of oil field equipment. Oxygen is also a major cause of pitting even when present in extremely small quantities. Figure 3.28 shows localized corrosion of a carbon steel downhole tubular that was generally protected by an iron carbonate film—except at an area of local film breakdown. Better inhibition practice (concentration, frequency) could have prevented the failure. In pipelines, improving pipeline cleanliness by maintenance pigging also helps to prevent failure.

Figure 3.28 Localized corrosion of carbon steel tubular in CO_2 gas condensate service at breakdown of the protective iron carbonate film. Rust occurred after pipe sample removal and exposure to air.

3.5.5.2 Corrosion-Resistant Alloys (CRAs)

The oil and gas industry currently uses a wide selection of CRAs to solve the difficult corrosion problems associated with deep hot wells and CO_2-enhanced recovery projects. These materials range from martensitic stainless steel to a variety of expensive nickel-chromium alloys. The presence of H_2S and chlorides complicate the selection of alloys for specific service. Part 3 of NACE MR0175[8] provides guidance on material selection for resistance to SCC and SSC in the presence of H_2S. The primary concerns for corrosion failure of the CRAs are SCC, SSC, localized corrosion, and often crevice corrosion or pitting attack. General or uniform corrosion is rarely encountered.

Most Common CRA Grades

The most common CRAs are martensitic stainless steel (UNS S41000 and S42000 [grades 410 and 420]), austenitic stainless steels (UNS S30400, S31600 [304 and 316]), duplex stainless steels (UNS S31803), and superduplex stainless steels (UNS S32750). More resistant, high-cost alloys also are available and are used in some severe environments, which are usually highly sour. These CRAs include both the high-nickel-chromium alloys containing molybdenum and tungsten, and cobalt alloys. Selection of alloy grades is described further in Chapter 4 and service envelopes are provided in Reference 8 (this chapter).

Austenitic Stainless Steels

The corrosion resistance of austenitic stainless steels, such as 304 and 316, are widely used in surface facilities, but rarely downhole. Applications include instrumentation tubing, chemical injection lines, piping, valves, machinery, and pressure vessels. These stainless steels offer limited strength and are affected by the presence of chlorides in the fluid environment and temperature. Type 304 is less resistant than 316, because it contains no molybdenum (Mo). Crevice corrosion resistance of 304 is also less than 316; however, both grades are susceptible to crevice corrosion in marine atmospheres. Both grades are increasingly susceptible to SCC when chlorides are present at a temperature above approximately 60°C (140°F). Increasing fluid acidity (lower pH), increasing dissolved oxygen concentration, increasing temperature, and increasing H_2S concentration increase the susceptibility to cracking. This is addressed further in this chapter under Stress Corrosion Cracking (Section 3.17).

Austenitic stainless steels probably are the most susceptible of all ferrous alloys to pitting corrosion chiefly because of the property that makes them "stainless." This property is the ability to form a thin oxide film that protects the metal from further corrosion in many environments. When this film is locally destroyed, that area becomes an anode in a corrosion cell, and pitting is initiated. The pitting process tends to be autocatalytic in that the acidity in the pit does not permit the protective oxide film to reform, so the pits grow deeper. Pitting generally initiates in areas of stagnant flow, such as in crevices or under deposits, and is promoted by the presence of chloride ions. Pitting is also observed in a marine atmosphere (Figure 3.29). Resistance to pitting increases with increased molybdenum content. For example, 316 is more resistant than 304; however, with only 2% Mo, its resistance is marginal. Some of the highly alloyed austenitic stainless steels are very resistant to pitting in chloride systems. Pitting and cracking can be prevented by cathodic protection, but this is not a commonly used method due to the limitations of applying CP to the inside of piping and tubular goods, and CP is ineffective in preventing atmospheric corrosion. Instead, the engineer must select alloys that are inherently resistant to the environment and to well work fluids.

Welded austenitic stainless steel equipment can experience localized corrosion at welds that did not employ adequate shielding gas to prevent oxidation of the weld area, for example, heat tinted welds. Iron contamination of the surface of stainless steel can lead to localized corrosion in a marine atmosphere. Contamination derives from using grinding tools that have been previously used on carbon steel, welding splatter, and grinding debris from nearby carbon steel construction.

Figure 3.29 Pitting corrosion of 316 austenitic stainless steel 50 mm (2 in) diameter hydraulic lines in a marine atmosphere. Pits are visible by the rust stains.

Duplex Stainless Steel

Duplex stainless steels are more corrosion resistant than the austenitic or martensitic stainless steels. These alloys offer higher strength than austenitic stainless steels and improved resistance to pitting and to SCC. Duplex is susceptible to low pH corrosion in mineral acids, such as hydrochloric acid (HCl), and to SCC/SSC in the presence of H_2S.[8] Localized corrosion of superduplex stainless steel is shown in Figures 3.12 and 3.16A. Corrosion was unexpected in this alloy and was attributed to defective processing that resulted in an undesirable sigma phase microstructure. Guaranteeing quality control (QC) is important in manufacturing and processing duplex stainless steel and more highly alloyed CRAs. QC may result in higher cost, but good QC produces CRAs that ensure they can deliver their intended corrosion resistance.

3.5.5.3 Other Alloys

Many other CRAs are available and have been used for severely corrosive conditions. These specialty alloys can be made to very high strengths and are used for special items, such as springs.[8] Tungsten alloys, such as tungsten carbide, are used for wear resistant metallic coatings or in solid form.

All CRAs commonly used in oil field applications that involve corrosion are subject to pitting, localized corrosion, or stress corrosion cracking under some conditions. As with the ferrous alloys, the most common causes of pitting are concentration cells beneath deposits and at crevices, local breakdown of protective scales, and the presence of aggressive ions that cause loss of passivity, notably the chloride ion.

3.6 Galvanic Corrosion

When two different metals are placed in contact in an electrolyte containing an oxidizing agent, the more reactive metal will corrode and the other will not. This coupling of dissimilar metals is referred to as a "bimetallic couple," and the resultant corrosion referred to as "galvanic corrosion." It can be extremely destructive, drastically accelerating the corrosion rate of the more reactive of the two metals (Figure 3.30). This principle is utilized in a beneficial way in cathodic protection (Chapter 7). When steel is connected to a more reactive metal in the same electrolyte, such as magnesium, the steel is "protected" (does not corrode). The steel becomes a cathode, and the more reactive metal an anode. This is how galvanic corrosion can be put to practical use, and this is referred to as "cathodic protection."

Note that galvanic corrosion requires an electrolyte that must be corrosive to at least one member of the dissimilar metal couple. Galvanic corrosion does not occur in noncorrosive fluids such as sales quality crude and other hydrocarbon liquids, diesel, other nonpolar fluids, and dehydrated gas if they are not contaminated with a separate water phase.

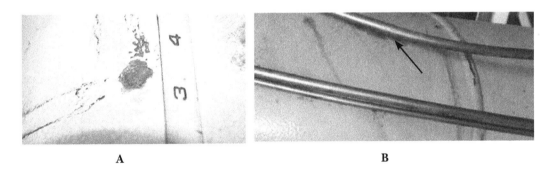

A   B

Figure 3.30 Galvanic corrosion between coated carbon steel and austenitic stainless steel in (A) seawater immersion, and (B) marine atmosphere, offshore. In (A), the coated carbon steel seawater ballast pump discharge column was galvanically corroded by a stainless steel clip. In (B), 316 austenitic stainless steel instrumentation tubing is directly touching a coated carbon steel pressure vessel in a marine atmosphere offshore. Corrosion occurred at the point of contact for the same reason as in (A), and streaks of rust bleeding are visible.

Galvanic corrosion is most severe in aerated seawater environments because of the high conductivity of the electrolyte and the presence of oxygen, which depolarizes the cathodic reaction. In downhole equipment, galvanic corrosion has occurred in rod pump ball and seat components and at breaks (porosity) in noble metal coatings on carbon steel subsurface valves such as electroless nickel, hard chromium, flame sprayed stainless steel, or Monel in high water cut (brine) oil production. Noble metal coatings are cathodic to the underlying substrate metal (steel). Acid jobs accelerate galvanic corrosion of the anodic metal in mixed-metal well workover components. A common occurrence of galvanic corrosion occurs in water service between brass and steel (e.g., steel nipple screwed into a brass water meter in a water flood, or steel in brass in offshore crew quarters fresh-water systems). However, galvanic corrosion can be found elsewhere, particularly if oxygen is present in the water.

3.6.1 Area Principle

One general rule indicating the acceleration of damage in a bimetallic couple is the "area principle." This rule states that the total corrosion is proportional to the ratio of the total cathodic area divided by the total anodic area exposed to the corrosive electrolyte. If there is a marked tendency for one metal to corrode in preference to another, such as iron over copper when both are exposed in aerated salt water, the less resistant metal suffers the entire corrosion. Thus, steel bolts (anodic) in copper plate (cathodic) corrode very rapidly in seawater; copper alloy bolts in steel plate cause insignificant damage in the same environment. The total corrosion in terms of metal weight loss is proportional to the area relationship exposed. An estimate of the corrosion rate that can be expected for the anodic metal is

$$C_a = C_n [1+(A_c/A_a)] \tag{3.1}$$

where

C_a = anode corrosion rate in presence of the galvanic couple

C_n = anode corrosion rate in absence of the galvanic couple

A_c = cathode surface area

A_a = anode surface area

Thus, if a metal has a corrosion rate of 0.1 mm/y (4 mpy) in the environment and is coupled to a cathode 10 times its size, the estimated corrosion rate will be 1.1 mm/y (44 mpy). Measurements in laboratory tests in simulated produced environments demonstrate that galvanic corrosion is not as severe as theory and the above equation indicate, particularly between low-alloy steel and 13Cr martensitic stainless steel. It is, however, very sensitive to temperature because of corrosion product formation. A protective corrosion product on the anode can reduce or eliminate galvanic corrosion.

In more ordinary terms, the guideline is

Large cathode and small anode = severe corrosion

Small cathode and large anode = minor corrosion

In many cases, the area principle explains why there are no problems with some bimetallic couples in field operations. For example, it has been common to use 6 or 12.5 mm (¼ or ½ in) diameter stainless steel needle valves screwed into carbon steel nipples, even in brine service, with very few corrosion failures. In most cases, examination shows the area of stainless is small relative to the area of carbon steel. The area principle, plus the lack of available oxygen, may also explain why galvanic corrosion has not been a major problem when martensitic stainless steel (13Cr) specialty items are used in carbon steel tubing strings.

It is not always obvious what area relationship to consider. For example, in a shell and tube heat exchanger, where long tubes are rolled into a tube sheet, the area of the tubes should not include the entire tube length. Usually the tube length affecting galvanic corrosion extends to 3–10 tube diameters from the tube sheet. This is because current flow inside a small diameter tube is not efficient down the length of the tubing because resistance increases with length. In general, galvanic corrosion effects are normally within 0.5 m (20 in) from the junction. This enables the use of heavy wall thickness spool pieces in situations where galvanic corrosion cannot be eliminated in the design. Other means of mitigating galvanic corrosion include installing electrical isolation (using nonconductive, nonmetallic material) at the junction point; coating the cathode in the bimetallic couple with a nonconductive coating to reduce the effective cathode area; and bringing the bimetallic junction outside of the electrolyte. The area relationship between anode and cathode plays an important role if one of the electrodes is coated with a nonconductive coating. Figure 3.30 shows the hazard of coating the anode in a galvanic couple. In Figure 3.30A, the outline of a stainless steel clip is visible by the rust stains on the coated carbon steel. Local coating breakdown resulted in a small anode (steel) area and a larger cathode (stainless steel) area, even though the area of stainless steel was small. Coatings often contain small "holidays," which are discontinuities in the coating that exposes a small area of the coated metal to the electrolyte (corrosive fluid). If the anode is coated, the holidays result in a miniscule exposed anode area. When electrically coupled to a more noble metal (the cathode) in the electrolyte, the small anode area/large cathode area results in pitting corrosion of the anode. Practitioners should coat the cathode—but never just the anode. Some owners coat both the anode and cathode for convenience.

3.6.2 Galvanic Series

Corrosion engineers refer to the galvanic series to determine which metal is anodic to another metal in a bimetallic couple. The galvanic series lists the half-cell potential of many alloys in an electrolyte. They are freely available in the literature, but they are almost exclusively developed for room temperature seawater as the electrolyte. Refer to Chapter 7, Table 7.2 for an example. The presence of CO_2 and H_2S in brine, a typical produced fluid, changes the galvanic series developed for seawater. Galvanic corrosion information in produced fluids and packer fluids can be found in this chapter's references.[9-12] A guideline is a potential *difference* of at least 200 mV between the two metals is required for galvanic corrosion. Values for potential can be found in galvanic series charts. However, galvanic corrosion rate is also not directly proportional to the potential difference between the anode and cathode. The galvanic series informs the corrosion engineer *if* the anode can corrode (i.e., 200 mV difference); however, it does not predict the rate at which the anode *will* corrode. Weight loss corrosion rate from galvanic corrosion is not directly proportional to the position of the metals in the galvanic series. It depends on the type and composition of the electrolyte.

3.6.3 Oxygen and Depolarizers

When corrosion-resistant alloy steels were first used in gas condensate wells, it was feared that coupling chromium-nickel or chromium stainless steels with carbon and low-alloy steel (such as API 5CT Grades N80 or L80[13]) casing would result in severe corrosion of the casing. However, experience in the field demonstrated little corrosion damage to the casing. The point to be learned from this example is that coupling dissimilar metals together in a neutral, nonaerated electrolyte will not necessarily cause galvanic corrosion. An oxidant, such as oxygen, is required to continue the reaction. Oxygen depolarizes the cathodic reaction and promotes higher corrosion rates because it reacts with hydrogen ions that accumulate at cathodic sites to form water instead of allowing hydrogen ions to combine as hydrogen gas molecules and polarize the cathodic reaction. The higher the conductivity of the electrolyte, the longer the distance from the junction of the two metals the anode will be adversely corroded. Aerated seawater is one of the most aggressive electrolytes in promoting galvanic corrosion. However, in an oil field water environments, several factors can reduce galvanic corrosion including oil-wetting, condensed water consisting of low dissolved salts content, or lack of dissolved oxygen. These factors will reduce, or even eliminate, the concern.

3.6.4 Polarity Reversal

Although not common in petroleum production, there can be cases where the anode–cathode relationship changes compared to the galvanic series developed for seawater. As an example, a copper alloy placed in an H_2S-containing (sour) brine will initially be cathodic to carbon steel but will become the anode in this bimetallic couple because of H_2S attacking copper. In clean seawater, copper would be the cathode in this same couple. Another example is zinc reversing its polarity and becoming cathodic relative to the potential of carbon steel as water temperature rises to approximately 60–70°C (140–160°F). At and above this temperature, zinc (used as either a sacrificial anode or as a metallic coating in galvanized steel) can accelerate corrosion and not cathodically protect steel.

3.6.5 Mill Scale

The area principle can be applied to localized corrosion of steel pipe. As it comes from the steel mill, pipe may be covered with mill scale (high-temperature formed iron oxides such as wüstite, FeO; magnetite, Fe_3O_4; and hematite, Fe_2O_3) unless purchased otherwise. Mill scale is a fair conductor of electricity and, at the same time, it also is insoluble in water and weak acid. This means that the areas covered with mill scale are protected and act to concentrate corrosion at breaks (cracks) in the scale, wherever there is an absence of mill scale. After a while, especially when exposed to a moist atmosphere, the mill scale oxidizes, loosens, falls off, and corrosion continues at a slower rate as the area relationship changes because the area of mill scale is small compared to the area of steel that used to be underneath it. Using Equation 3.1, the corrosion rate due to galvanic corrosion is not greater than the corrosion rate without mill scale. Consequently, the acceleration of attack is important in the early life of the pipe, tank, or pressure vessel when the area of broken mill scale (the anode) is small compared to the area of plate covered in mill scale (the cathode).

Figure 3.31 shows the galvanic corrosion effects at breaks in the mill scale.

Figure 3.31 Galvanic corrosion: Bimetallic effect of mill scale versus bare steel.
Breaks in the mill scale led to localized corrosion.
Photograph provided courtesy of Nalco Champion, an Ecolab Company.

New sections of pipe in an old line are sometimes attacked because the old pipe is covered with heavy layers of oxide and rust that are cathodic to the new pipe. This effect is depicted in Figure 3.32.

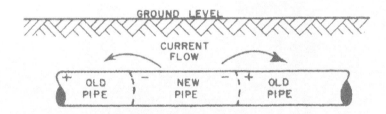

Figure 3.32 Electrochemical Corrosion: New pipe connected to old pipe.
New clean pipe will often be anodic to the old corroded pipe covered with corrosion product.

3.6.6 Ringworm Corrosion and Preferential Weld Corrosion (PWC)

Ringworm corrosion and PWC are both forms of galvanic corrosion caused by either differences in the steel microstructure as a result of heating, or differences in composition between base metal and weld metal.

3.6.6.1 Ringworm Corrosion

Another type of galvanic corrosion sometimes found in carbon steel and low-alloy tubing installed in oil or gas wells (particularly those of high CO_2 content in the gas phase) is known as "ringworm corrosion." As the term implies, the corrosion occurs in a ring a few inches from the forged upset (Figure 3.33).

This corrosion may take the form of severe localized or grooving corrosion near the upset. The cause of ringworm corrosion has been traced to the hot upset forging process that is used to make the threaded end connection thicker (Figure 3.33A). The heat required in upsetting the end of the tubing causes the heated end to have a different microstructure from the rest of the pipe. A transition zone in microstructure near the upset runout usually is susceptible to corrosion. This condition can be overcome by fully normalizing the tubing after upsetting. Normalizing is a heat treatment that gives uniformity to microstructure and grain size (refer to Chapter 4, Section 4.4.2).

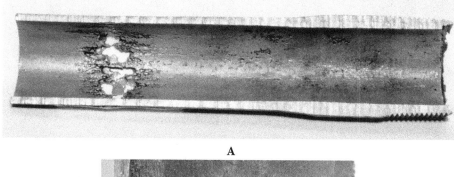

A

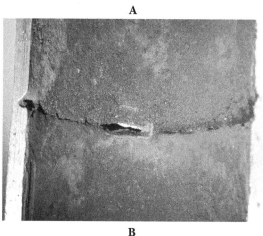

B

Figure 3.33 Ringworm corrosion: Two examples (A and B) of galvanic corrosion known as "ringworm corrosion" because of the microstructural difference between a steel downhole tubing pipe body and the hot-upset, forged end used to make the end connection.
Photograph provided courtesy of Nalco Champion, an Ecolab Company.

3.6.6.2 PWC of Steel

When a metal is welded, the welding process can create microstructures at the weld, which differ from the microstructure of the parent steel. The result can be three different microstructures: parent metal, weld metal, and the heat-affected zone (HAZ) between. The areas can act as different metals (bimetallic couples), and PWC can occur (Figures 3.34–3.37). The corrosion may show up in the weld metal, at the HAZ, or in the parent metal.

PWC of carbon and low-alloy steels has been observed in produced fluids containing a high partial pressure CO_2 in flow lines just downstream of the choke, especially in gas condensates. Conditions in Figure 3.34 were CO_2 partial pressure 126.2–168.2 kPa (18.3–24.4 psi), operating temperature 50–75 °C, flow velocity up to 12.2 m/s (40 ft/s). In addition, PWC of carbon steel has been observed in high-pressure, hot CO_2-containing gas downstream of the compressor interstage cooler, and in aerated seawater. In steel, other than differences in microstructures, the cause of PWC has also been attributed to weld metal chemistry (associated with nickel composition in the weld, but not the parent steel), and an increase in fluid turbulence at the weld versus the straight pipe. The parent pipe is cathodic to the weldment, and the resultant groove is quite substantial (Figure 3.34).

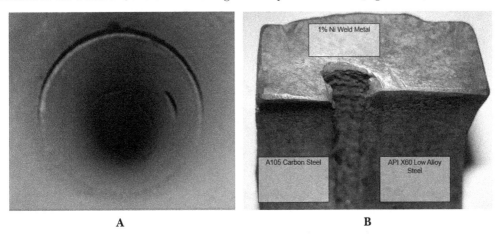

Figure 3.34 Preferential weld corrosion (PWC): Two views of PWC attack of a low-alloy steel, high-pressure gas condensate flowline caused by microstructure and chemistry differences between the weld metal, the base metal, and the HAZ: (A) pipe bore view; (B) cross-sectional view of the weld area. Note the weld filler metal contained 1% Ni.

PWC of carbon steel is avoided by good corrosion inhibition practices and proper welding practices. Reducing the height of the weld root at the inside surface helps to minimize flow turbulence. The concentration of corrosion inhibitor required to control preferential weld corrosion might be higher than that required to control corrosion of the uncoupled base metal and should be determined by laboratory or field testing. Some oil companies specify the use of low (<1% nickel) electrodes for welding consumables, at least in the root deposit, for high-pressure CO_2 service. However, in seawater service, nickel addition to the welding electrode reduces the tendency of preferential weld corrosion.

3.6.6.3 PWC of Austenitic Stainless Steel

PWC of austenitic stainless steel is well documented as occurring in aerated, chloride-containing waters (Figure 3.35). It has the appearance of a continuous array of corrosion along the HAZ and is known as "knife-line attack." Weld-line corrosion occurs because of precipitation of chromium carbide along grain boundaries (refer to "sensitization" in this chapter in the section on intergranular corrosion [Section 3.9]). This resultant localized, depleted concentration of chromium content of the alloy forms a galvanic cell. "Weld-line" is intergranular corrosion, which can be prevented by specifying low carbon or stabilized grades of austenitic stainless steel base and weld metals. Figure 3.36 shows the appearance of both weld line corrosion and erosion-corrosion of the adjacent base metal due to turbulence associated with a high-low misalignment at the weld.

PWC of austenitic stainless steel also can occur in marine atmospheres and in water service and is associated with oxidation of the weld (heat tinting) during welding due to insufficient shielding gas.[14] A dark heat tint color (Figure 3.37) indicates oxidation and resulting loss of alloying elements such as chromium.

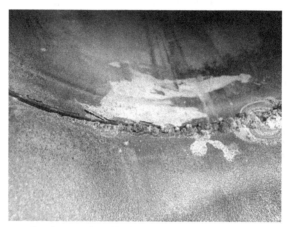

Figure 3.35 Weld line corrosion: Improperly welded austenitic stainless steel is subject to attack at the weld.
Photograph provided courtesy of Nalco Champion, an Ecolab Company.

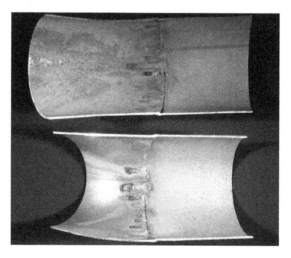

Figure 3.36 Weld line corrosion of an austenitic stainless steel elbow. An alignment mismatch created a step that caused a flow disturbance. The effect of fluid flow is evident in the corrosion pattern in the base metal.
Photograph provided courtesy of Nalco Champion, an Ecolab Company.

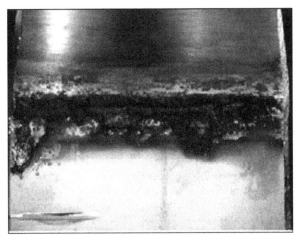

Figure 3.37 Preferential weld corrosion at the weld root of 316 stainless steel heat-tinted weld.

3.6.6.4 PWC Caused by MIC

In austenitic stainless steel welds, MIC can preferentially occur at welds (Figure 3.38). Sensitization of the weld heat affected zone is not required, but its presence increases the likelihood of MIC attack (refer to Section 3.9 for a discussion of sensitization). In austenitic stainless steel, pitting from MIC attack occurs in the weld metal or at the interface between the weld and base metal (originating at the HAZ). MIC appears to preferentially attack the ferrite phase of the weld metal initially. Austenitic stainless steel welds intentionally contain about 3% by volume percentage ferrite phase to avoid solidification cracking. The appearance of MIC in stainless steel is different from knife-line attack. MIC occurs in unconnected areas of the weldment, whereas knife-line attack occurs along the length of the weld seam. Many MIC pits are cavernous and have small openings at the initiating surface, and wide, subsurface caverns that are not visible (Figure 3.38) unless radiographed or cross-sectioned.

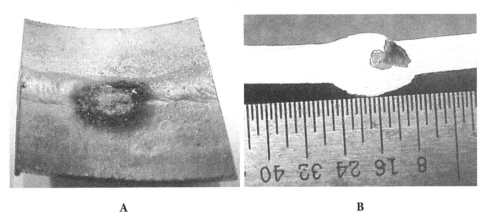

A   B

Figure 3.38 (A) Preferential weld corrosion of austenitic stainless steel plate because of MIC.
(B) Cross-section of butt weld showing undercutting nature of pit.

3.7 Erosion-Corrosion, Impingement, and Wear

Increased flow rate often increases the corrosion rate of a metal in a corrosive environment. This is often attributed to bringing fresh corrosive species to the metal surface and reducing or removing concentration gradients that can polarize the metal surface. Depolarization allows the initial, high corrosion rate to remain; however, flow can accelerate corrosion by other means.

3.7.1 Erosion-Corrosion

Because most metals owe their corrosion resistance to the formation and maintenance of a protective scale, removal of this scale at local areas can lead to accelerated attack. High-velocity flow or turbulence frequently will erode the protective scale to expose fresh metal, which can be corroded. "Erosion-corrosion" is this combination of erosion of the scale and corrosion of the underlying metal.

Some engineers distinguish between erosion-corrosion and flow-assisted corrosion by stating that erosion-corrosion requires the presence of a solid phase (e.g., sand) and a corrosive fluid; whereas in flow-assisted corrosion, the metal corrosion film is removed by the wall shear stress from flow. The NACE definition provided earlier in this chapter combines these into one definition. Both are a common cause of failure of oil field equipment. Carbon and low-alloy steels are particularly susceptible in environments that form scales that are easily removed, such as iron carbonate, which was formed by corrosion occurring below approximately 70°C. Erosion-corrosion normally occurs at certain areas, such as changes of sections, at threaded and welded connections where there is turbulence from flow, or at bends, elbows, and tees. Figure 3.39 shows a ball and seat cage from a sucker rod pump where the flow lines are clearly visible. Figure 3.40 shows two examples of erosion-corrosion in pipe elbows.

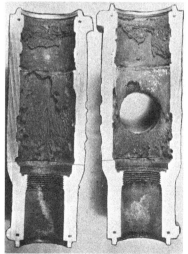

Figure 3.39 Erosion corrosion: Ball and seat cage from a sucker rod pump.
Photograph provided courtesy of GPA Midstream Association.

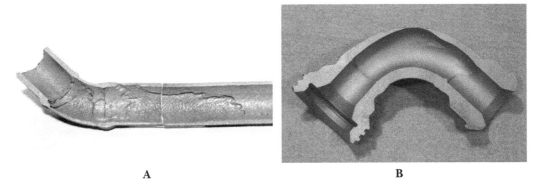

Figure 3.40 (A and B) Erosion-corrosion at elbows caused by highly turbulent flow (flow direction is left to right in both photographs). Note the absence of metal loss immediately upstream of the elbow and a short distance downstream after the flow became less turbulent (photos taken after cleaning).
Photograph A provided courtesy of Nalco Champion, an Ecolab Company

When looking to detect erosion-corrosion and impingement corrosion, many operators focus their inspections at changes in flow direction, such as elbows and tees, and after large pressure drops. Figure 3.41 shows erosion-corrosion of an API 5L Grade X65[3] low-alloy steel flowline downstream of a wellhead choke. The high-pressure drop and high CO_2 partial pressure (approximately 413 kPa [60 psi]) resulted in erosion-corrosion. In addition, corrosion inhibitor injection was poorly maintained.

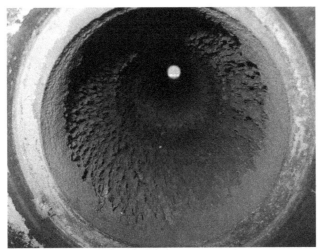

Figure 3.41 Erosion-corrosion of a X65 low-alloy steel flowline downstream of a choke. Note the pattern of flow.

3.7.2 Wear Corrosion

A form of corrosion closely related to erosion-corrosion is "wear-corrosion." Rather than high-velocity or turbulence-removing corrosion products and protective scales to expose the base metal, mechanical wear (rubbing of one metal over another) does the damage. Figures 3.42–3.44 present examples of wear-corrosion. Figure 3.42 presents a section of failed tubing from a sucker rod pumping well showing what is usually referred to in the field as "rod wear." Figure 3.43 shows a severely damaged tubing collar from a rod pumping well. The well was not equipped with a tubing hold down; thus, the tubing with the collar moved up and down with each pump stroke. The collar rubbed on the casing, the annulus contained corrosive well fluids, and wear-corrosion took place. Gas wells can also experience wear-corrosion. Figure 3.44 shows wireline wear that was initiated during a workover operation and the resulting corrosion along it during production. The clue that these damages are a combination of wear and corrosion is the appearance of the thinned and damaged surfaces. Note that the surfaces in these figures are not smooth "wear" surfaces; they are basically rough, pitted, corroded-looking surfaces with a few smooth spots. Pure wear surfaces are smooth.

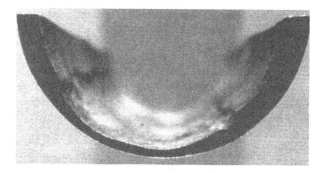

Figure 3.42 Wear-corrosion: End view of a sucker rod coupling. Rubbing corrosion product from this pumping well tubing resulted in this so-called "rod wear" form of wear-corrosion. End view showing thinning.

Figure 3.43 Wear corrosion: Pumping well tubing collar. Note that the well did not have a tubing anchor, thus the collar could rub on the casing with each pump stroke.

Figure 3.44 Wear corrosion: Wirelines rubbing corrosion product from this gas well tubing resulted in the so-called "wireline cut" form of wear corrosion.

3.7.3 Impingement Corrosion

Another manifestation of erosion-corrosion, and often a more localized attack, is known as "impingement corrosion." This localized erosion-corrosion is caused by turbulence or impinging flow. It occurs when a fluid stream impinges upon a metal surface and breaks down protective films at very small areas. The resulting attack is in the form of pits that are characteristically elongated and undercut on the downstream end of the flow direction. It is a problem particularly in copper and copper-based alloys but has been observed in carbon steel and even in 410 martensitic stainless steel. Figure 3.45 shows 90-10 copper nickel alloy condenser tubes with this type of attack. Figure 3.46 illustrates erosion corrosion in the cross section of a 304 stainless steel flange, weld and pipe in an alkaline, liquid Stretford®[1] solution, which selectively removes H_2S from gas streams containing CO_2. Solids were not reported to be in the liquid. The sharp contour of the corroded surface is a result of flow effects and is often found in erosion-corrosion. Figure 3.47 shows martensitic stainless steel, ASTM A182 Grade F6a[15] in high-pressure, slightly sour gas condensate with 20% CO_2. Service temperature was 57°C. Solids found in the production separator suggest particulates were in the flow stream, making this an example of erosion-corrosion. Typically, austenitic and martensitic stainless steels have high resistance to both erosion-corrosion and impingement type attack.

---

1. The Stretford process and solution are registered trademarks of British Gas Corporation, United Kingdom.

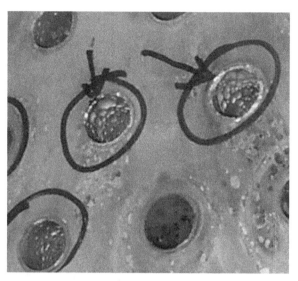

Figure 3.45 Erosion-corrosion (impingement attack) in 90-10 copper-nickel heat exchanger tube ends in seawater due to excessive velocity and turbulence. The tube ends exhibit a flow-shaped metal loss pattern where the tube was rolled into the tube sheet.

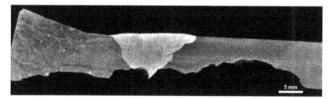

Figure 3.46 Erosion-corrosion of 304 stainless steel in an alkaline "Stretford®" solution, shown in cross section. From left to right: flange, weld (light area), and pipe. This gas plant experienced widespread erosion-corrosion near welds. Flow direction is left to right.

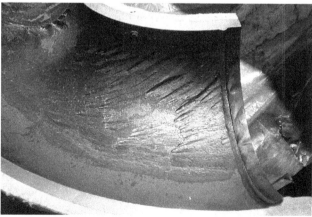

Figure 3.47 Erosion-corrosion of 410 martensitic stainless steel at the intrados of an elbow. Metal loss was 5 mm deep. Flow direction is left to right.

3.7.4 Erosion

Like pure wear, pure erosion is also smooth, and is typically observed in sand production at the wellhead choke and downstream elbows (Figures 3.48 and 3.49). However, high velocity gas flow carrying liquid droplets can also result in erosion in the absence of a corrosive fluid (Figure 3.50). Erosion has also occurred in nominally dry export and sales gas pipelines at valves due to impingement of "black powder," erosive particles consisting of corrosion products iron oxide, iron sulfide, and iron carbonate.[16-18] Mitigation of sand erosion is complex and may involve downhole completion and sand consolidation strategies (gravel packing and sand screens), controlling well production rates, installing cushion tees, and applying hard facing coatings to susceptible materials. Erosion from liquid droplets may require controlling fluid velocity and installing hard facings. Pressure and/or temperature conditions may require adjustment to operate in the single-phase gas region. Corrosion inhibition is used to mitigate black powder erosion.

Figure 3.48 Sand erosion of a retaining sleeve (right) in a wellhead choke component. Note the smooth, polished appearance of the metal loss.

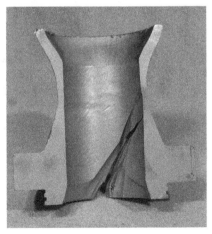

Figure 3.49 Sand erosion of a wellhead choke component.

Figure 3.50 Erosion of a valve plug from a pressure control valve at the inlet to a cryogenic gas plant. The high fluid velocity caused erosion in the absence of corrosion. The gas occasionally carried liquid hydrocarbon droplets.
Photograph provided courtesy of Nalco Champion, an Ecolab Company.

3.8 Cavitation

Cavitation is the formation and collapse of vapor bubbles in fluids (e.g., water) because of rapid changes in pressure. It can occur whenever the absolute pressure at a point in the liquid stream is reduced to the vapor pressure of the fluid so that bubbles form, and this is followed by a rapid rise in pressure, resulting in bubble collapse. Cavitation damage is the wearing of a metal from repeated impact from collapse of bubbles within a fluid. Corrosion usually plays a minor role in the rate of cavitation damage. Pump impellers (Figure 3.51) often show cavitation damage effects. Maintaining the correct net positive suction head prevents cavitation of centrifugal pumps.

Figure 3.51 Cavitation attack: A nickel-aluminum-bronze pump impeller with severe cavitation attack.

3.9 Intergranular Corrosion

As the name implies, intergranular corrosion is the preferential attack of a metal's grain boundaries. (Intergranular corrosion often is confused with SCC. However, intergranular corrosion can occur in the absence of stress; stress corrosion cracking occurs only while the metal is under tensile stress.) Intergranular corrosion reduces the load-bearing capability of the metal even though to the naked eye there may not appear to be metal loss. However, when the cross section is viewed under a microscope, a loose assemblage of grains is obvious. Intergranular corrosion has occurred in many alloys including austenitic stainless steels, copper alloys, aluminum alloys, and nickel alloys. Figure 3.52 is a photograph of intergranular attack. In most cases, intergranular corrosion results from a metallurgical structure that causes the grain boundaries to be more susceptible to attack than the grains themselves. Proper heat treatment or control of the chemistry of the alloy generally can eliminate the susceptible grain boundary constituent and render the alloy resistant to intergranular attack.

3.9.1 Sensitization of Austenitic Stainless Steel

The most serious occurrence of intergranular corrosion involves austenitic stainless steel. When austenitic stainless steels are held in or cooled slowly through a temperature range of 427–871 °C (800–1600 °F), chromium carbides precipitate in the grain boundary, and the stainless steel is said to be "sensitized." Sensitization becomes a problem whenever stainless steels are welded or when they are exposed to the above temperature range. Subsequent exposure of the sensitized stainless steel to nonaggressive solutions, such as weak acids, results in intergranular attack. The following list shows corrosives that cause intergranular corrosion of sensitized austenitic stainless steels. Note that produced fluids are not included in this list.[19]

Caustic soda
Cerium oxide
Cu sulfate-Al sulfate
Fe sulfate-Al sulfate
High temperature, high purity water
Hydrochloric acid
Hydrofluoric acid
Hydrogen sulfide

Lithium
Nitric acid
Oxalic acid
Polythionic acid
Sea water
Sodium
Sulfuric acid + oxidizer

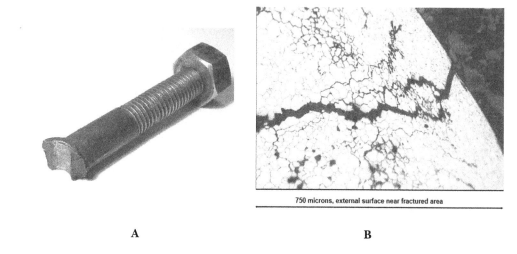

Figure 3.52 Intergranular attack of a sensitized austenitic stainless steel bolt in a marine atmosphere: (A) bolt failure; (B) photomicrograph of the fracture surface showing the intergranular nature.

There are several theories to account for sensitization of austenitic stainless steels. The most popular is that precipitation of chromium carbides depletes areas immediately adjacent to grain boundaries of chromium. The depleted zones become anodic to the grains and corrode preferentially.

3.9.2 Preventing Intergranular Corrosion

To prevent intergranular corrosion, two methods are used: heat treatment, and control of chemical composition. Sensitized stainless steel can be solution annealed by heating to 1066–1,121°C (1950–2050°F) and cooling rapidly, which does not allow time for precipitation. Another treatment is to hold them at 871–899°C (1600–1650°F) for several hours, thus allowing chromium to diffuse from the grain boundaries back to the depleted zones. Heat treatment in the field is often deemed to be impractical, and control of chemical composition is most often practiced.

Precipitation of carbides can be controlled in two ways. First: use a stainless steel with very low carbon (below 0.03%) so that not enough carbon is available to tie up the chromium. The "L" grades—where L represents "low carbon"—of austenitic stainless steels restrict carbon content to a maximum of 0.03%. These are the grades specified if the stainless steels will be welded, for example, 304L (UNS S30403) and 316L (UNS S31603). The second approach is to use a stabilized grade (e.g., 321 and 347 stainless steel). Stabilized grades contain either titanium or niobium, both of which have a greater affinity for carbon than chromium. This results in precipitation of niobium or titanium carbides, rather than chromium carbides. The most commonly used stabilized grades do not contain molybdenum, and therefore, their resistance to pitting corrosion is inferior to 316L stainless steel. The best method depends on the application. Even though most produced fluids do not result in intergranular corrosion of austenitic stainless steel, regardless, it is standard practice in oil field operations to specify low carbon grades for welded components.

There are several tests to detect the susceptibility of austenitic stainless steel to intergranular corrosion.[20] Tests are specified by the user to ensure that the welded stainless steel components to be purchased are not sensitized, and will be delivered in the proper heat treatment and chemical composition.

3.10 Hydrogen-Induced Failures

Hydrogen atoms may be produced on a metal surface in an aqueous environment by a corrosion reaction, cathodic protection, electroplating, or acid pickling. Some of the hydrogen atoms combine to form gaseous molecular hydrogen (H_2) on the submerged metal surface. These gas bubbles escape into the electrolyte as it flows past the metal. A portion of the hydrogen atoms remain atomic (elemental); they will enter the metal, and may cause problems, such as blistering, cracking, and hydrogen embrittlement. They enter the metal lattice because they are very small and fit in the interstices of the iron lattice. Certain substances in the electrolyte, such as sulfide ions, phosphorus, and arsenic compounds, retard the formation of molecular hydrogen and thereby increase the entry of atomic hydrogen into metal. In upstream operations, the presence of H_2S provides sulfide ions, thus creating conditions favoring hydrogen-induced failures.

3.11 Hydrogen Blistering and Hydrogen-Induced Cracking

Hydrogen entry into ordinary-strength steels can result in hydrogen blisters if there is a macroscopic defect in the steel, such as a lamination or significant content of nonmetallic inclusions. Any void or discontinuity in the steel provides a place for hydrogen atoms to combine, form hydrogen gas, and build sufficient pressure internally to cause blistering. Blistering results in a surface bulge on the surface that is closest to the source of hydrogen (the side contacted by the sour [H_2S] fluid), as shown in Figure 3.53. Besides blistering, this figure also shows internal fissuring known as "hydrogen induced cracking" (HIC). HIC, or stepwise cracking, has also occurred along the plane of nonmetallic inclusions, but the cracking jumped to another parallel plane in a stair-step fashion in the through-thickness direction. The wet sour-fuel gas scrubber tower was in service for six years at an operating temperature of 38°C (100°F).

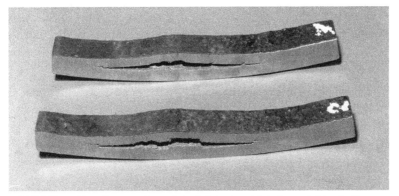

Figure 3.53 HIC cracking: Cross sections of steel plate show blistering and delamination occurred by accumulated pressure of hydrogen gas forming at nonmetallic inclusions in the metal wall. In addition to surface blistering, HIC (or stepwise cracking), occurred mid wall.

In oil field operations, hydrogen blistering and HIC are problems in sour environments for tanks, pressure vessels, and pipe. They mostly occur in rolled and seam-welded plate or pipe made of lower strength, ductile steels such as ASTM A516 Grade 70.[21] As manufactured, there is a sufficient concentration of manganese sulfide inclusions in commonly used grades of steel. Unless steels are specified to contain very low sulfur and phosphorous and incorporate inclusion shape control to avoid elongated inclusions after rolling, they are susceptible to hydrogen blistering and cracking. Although blistering rarely results in a failure, it is a sign of adverse hydrogen activity. Blisters and HIC can cause leakage, but rarely, rupture. However, other forms of hydrogen damage, such as SSC combined with stress-oriented HIC, have caused rupture. For newly constructed equipment, the best defense against blistering and HIC is to require HIC testing (per NACE TM0284[22]) and specifying the use of clean steel (low sulfur and low phosphorus content) plus control of the shape of these nonmetallic inclusions. For existing equipment, the operator should control the corrosion process, which decreases the rate of hydrogen generated and should periodically inspect the equipment for damage. If hydrogen blisters or HIC have been detected in the operating equipment, they can be assessed for continued service or repair/retirement using the criteria in API 579-1/ASME FFS-1.[2] Internal coatings, metallic overlays, or cladding that have been qualified for sour service help prolong the life of equipment.

Hydrogen probes detect hydrogen that is internal to the steel structure. They are shown in Figure 3.54 and used to monitor hydrogen gas buildup. Data from the probes is supplemented with data from process variables that are more conducive to promoting hydrogen generation, such as H_2S concentration, temperature, pressure, and flow rate. There are essentially two types of hydrogen probes shown in the figure: intrusive and patch (nonintrusive). There is also variation in probe design with access fittings to allow the probe to be withdrawn under pressure onstream; a thermometer can be added so temperature, as well as pressure, may be recorded to allow gas pressure to be temperature compensated.

3.11.1 Intrusive Hydrogen Probes

Atomic hydrogen is generated at the cathode of a corrosion cell in neutral or acid pH fluids. An intrusive hydrogen probe can detect the buildup of hydrogen gas inside the probe as corrosion proceeds. The probe penetrates the pipe wall and into the flowing production fluid. It consists of a hollow tube. When the probe is placed in a sour system and allowed to corrode, hydrogen atoms form at the cathode. Some of these atoms diffuse through the wall of the sealed hollow tube where they combine into hydrogen gas molecules. The molecules are too large to diffuse back through the tube and pressure increases as the number of hydrogen gas molecules increases. This increase in pressure is recorded. Pressure buildup is directly proportional to the flux of atomic hydrogen, which can be determined by implementing the ideal gas law. If the volume of the hydrogen probe cavity and the cross-sectional area across in which diffusion is occurring is known, then the hydrogen flux can be calculated. This style of hydrogen probe is constructed of a sealed, hollow steel tube with a pressure gauge attached to the top (shown in Figure 3.54).

3.11.2 Patch (Nonintrusive) Hydrogen Probes

Patch probes are placed on the external surface of the pipeline of a sour service pressure vessel and do not intrude through the pressure boundary wall. The internal corrosion reaction generates hydrogen on the cathodic sites, and atomic hydrogen permeates through the wall of the vessel or pipeline. The patch probe is an electrochemical device, which is tightly attached to the outside surface of the pressure vessel or pipeline that electrochemically reacts with the hydrogen and generates a measurable current. This type of device works by polarizing a palladium foil held to the wall of the metal by a transfer medium such as wax; as the palladium foil is polarized, it acts as a working electrode quantitatively oxidizing the hydrogen as it emerges from the wax. The current induced by the instrument is directly equivalent to the hydrogen penetration rate.

Hydrogen probes have seen less usage in petroleum production environments compared to corrosion monitoring coupons and probes. Regardless of whether the probe is intrusive or nonintrusive, operators have reported successes and failures using them. Success occurs when probe readings can be correlated to process changes, including corrosion inhibitor dosage. Failures occur when there is no correlation to these changes or with other types of corrosion monitoring. Hydrogen probes can be useful tools when their data is combined with process data (temperature, pressure, flow rates), sour gas compositional changes, and the probe data correlates with other forms of corrosion monitoring.

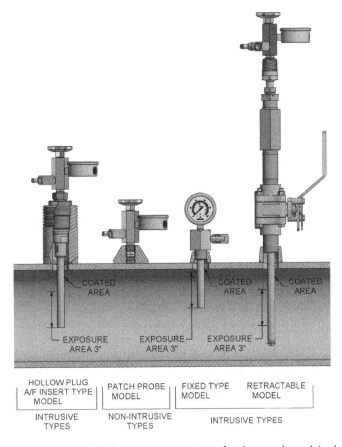

Figure 3.54 Hydrogen probes: The most common types of probes are shown: intrusive, nonintrusive, insert, and retractable under pressure.
Drawing provided courtesy of Cosasco.

3.12 Hydrogen Embrittlement

Hydrogen entry into high-strength steels can result in hydrogen embrittlement; in such a state, the steel can fail in a brittle manner at stresses considerably below its yield strength. Limiting this phenomenon to high-strength materials may be because only such materials can have tensile stresses high enough to initiate the mechanism, or to the greater chance these steels contain undesirable microstructures that are susceptible to cracking. Hydrogen leading to embrittlement may be entrapped during pouring of the molten metal, absorbed during electroplating or pickling, or generated by corrosive action (hydrogen ions are present at the cathodic sites of the corrosion cell). Cathodic protection can also charge hydrogen into high strength steel. Welding practices that minimize hydrogen generation should be used. Welding practices should also avoid forming hard zones of untempered martensite in both the weld and weld HAZ, either of which can act as a crack-initiation point.

If the metal is under a high tensile stress, brittle failure can occur. The path of failure may be either intergranular or transgranular, and it often is extremely difficult to distinguish failures caused by hydrogen embrittlement from SCC and SSC. Hydrogen embrittlement failures usually occur only with high-strength steels—generally those having yield strengths of 620 MPa (90,000 psi) or higher. Some practitioners use a break point of 552 MPa (80,000 psi), and some even lower. Fasteners (bolts) may have strength of this magnitude unless specified otherwise. High-strength drill pipe, tubing, and casing that is not designed for sour service (e.g., API 5CT grades P110 and Q125[13]) also exceed these strength levels. The susceptibility to hydrogen embrittlement failure increases with increasing strength and hardness.

As mentioned in Chapter 1, Section 1.5.5.4 evidence suggests that recent failures of critical high-strength steel bolts and fasteners in subsea service are associated with environmental cracking in high-strength (hardness) materials by hydrogen embrittlement because of either (a) internal hydrogen from manufacturing or electrolytic application of protective metallic coatings, and/or (b) external hydrogen from in-service cathodic protection (CP) from galvanic and/or impressed current systems. There is also an additional, yet only partially assessed, concern that other environmental cracking mechanisms (e.g., chloride stress corrosion cracking, sulfide stress cracking, liquid metal embrittlement) can also induce embrittlement in certain high-strength bolting materials under subsea conditions. Operators are specifying hardness control and hydrogen bake-out procedures after the application of an electrolytic coating to remediate these failures, and, as of 2020, NACE International is considering developing standards to control cracking of these critical components.

Failures due to hydrogen embrittlement do not always occur immediately after application of the load or exposure to the hydrogen-producing environment. Usually, there is a period during which no damage is observed, followed by sudden, catastrophic failure. This phenomenon is referred to as "delayed failure." The time prior to failure is the "incubation" period during which hydrogen is diffusing to points of high triaxial stress (such as at crack tips). The time-to-failure decreases as the amount of hydrogen absorbed, applied stress, and strength level of the steel increases.

Until a steel containing hydrogen forms cracks or blisters, there is no permanent damage. In many cases, dissolved hydrogen can be baked out by suitable heat treatments, and the original properties of the steel can be restored. This is frequently done after electroplating high-strength steel parts. It is also done when preheating a hydrogen-charged steel exposed to sour oil field fluids, prior to welding. In early oil field operations, sucker rod strings that had been installed in sour wells, some of which had been experiencing brittle failures, were not immediately installed downhole, but left outdoors to "weather," or were heated to bake out hydrogen before being run in another well. Now, operators specify standards that require thermal bake-out, which drives out dissolved hydrogen quicker than weathering.

3.13 SSC

Spontaneous brittle failures that occur in steels and other high-strength alloys when exposed to moist H_2S and other sulfide environments have been called by many names. Among these are SCC, H_2S cracking, wet H_2S cracking, sulfide cracking, SSC, and sulfide stress corrosion cracking. The possible explanation for the number of names is the number of theories about the mechanism of failure. These theories range from SSC is a form of hydrogen stress corrosion cracking, to that it is a form of hydrogen embrittlement. Regardless of the mechanism, the fact remains that SSC can cause failures in sour oil field environments. Figures 3.55–3.57 show some examples of SSC failures from the field. As the industry handles more sour gas around the globe, more materials are developed to be resistant to SSC. For this reason, ANSI/NACE Standard MR0175, "Petroleum and Natural Gas Industries—Materials for Use in H_2S-Containing Environments in Oil and Gas Production," Parts 1–3, are updated periodically.[8, 23-24]

3.13.1 SSC Susceptibility and Steel Strength

SSC failures have been noted in highly susceptible (high strength) steels that are "not in sour service." Figure 3.57 shows SSC of low-alloy steel ASTM A193 Grade B7 stud bolts.[25] These bolts attached counterweights to the balance beam of a rod-pumped oil well and were not directly exposed to sour produced fluids; the bolts were in an onshore, atmospheric environment. However, their high hardness (HRC 32), the stress concentration of the thread, and the presence of small quantities of H_2S, and moisture in the surrounding atmosphere made them susceptible to cracking. Stress was primarily the result of torque-imparted bolt loading. Here, cracks were initiated by SSC and propagated by fatigue. The presence of sulfur was found by elemental analysis, and iron sulfide corrosion product was confirmed using the spot acid test described in Chapter 2, Table 2.2. Bolts were replaced with ASTM A193 B7M studs with controlled hardness for sour service.

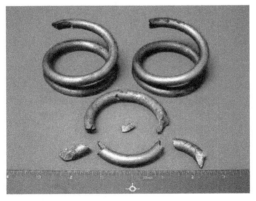

Figure 3.55 SSC of 17-4 PH stainless steel springs in a subsea valve. Spring hardness was verified as HRC 49-50. Note the fractures lack plastic deformation. Photograph provided courtesy of Chevron Corporation. ©2019 Chevron

Figure 3.56 SSC of a high-strength, low-alloy steel kill string.
Photograph provided courtesy of Nalco Champion, an Ecolab Company.

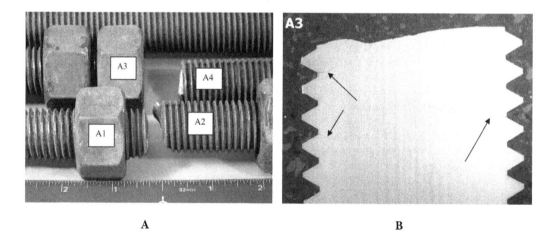

A   B

Figure 3.57 SSC of low-alloy steel stud bolts: (A) fractured stud bolts labeled A1–A4. The helical fracture surface shows effect of torsional stress associated with bolt makeup. (B) Cross section of stud bolt A3 showing the fracture surface at the top and three cracks initiated at the root of threads, shown by arrows.
Photographs provided courtesy of Chevron Corporation. ©2019 Chevron

3.13.2 SSC Is Typically Intergranular

The mechanism of SSC is not completely understood. Cracking is typically intergranular cleavage with no plasticity. Figure 3.58 shows SSC cracks in a cross section of a full-thickness, welded test specimen of API 5L Grade X60[3] microalloyed steel line pipe exposed to a laboratory four-point bent beam SSC test. SSC was initiated at isolated local hard spots with hardness values above Vickers hardness 250 HV_{10}, even though the average hardness of the weldment was below this value. SSC cracks are jagged because of their intergranular path but are not branched.

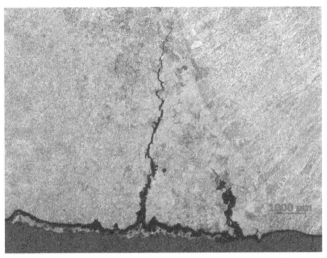

Figure 3.58. Cross section of a weldment that failed laboratory SSC testing.
Photographs provided courtesy of Chevron Corporation. ©2019 Chevron

3.13.3 Conditions Required for SSC

It is generally accepted that the following conditions must be present before cracking can occur.

- H_2S must be present.
- Water must be present—even a trace amount of moisture is sufficient.
- A "high-strength" steel must be involved; the exact strength level varies with the composition and microstructure of the steel. (Although other materials are susceptible to SSC, this discussion is confined to carbon and low-alloy steel.)
- The steel must be under tensile stress or loading (stress may be residual or applied).

If all these conditions are present, SSC may occur after some time. It is important to realize that SSC usually does not occur immediately after exposure to the sour environment—it may take place after hours, days, or years of service. In severe cases where local hard microstructure (untempered martensite) is exposed to a high concentration of H_2S, the author has observed SSC failure of major equipment (pressure vessels, valves) at welds in one to two days. In other circumstances, equipment that has been in service for years can suddenly experience SSC if conditions have changed and the H_2S concentration has significantly increased or the fluid pH became more acidic.

3.13.4 Factors Affecting SSC Susceptibility

The susceptibility of a material to failure by this mechanism is primarily determined by the following variables.

1. Strength or Hardness: The hardness of carbon steels is proportional to strength, and strength is related to the microstructure. Thus, hardness has become a very important parameter for the susceptibility of steel to SSC because it is relatively easy to measure. Carbon steels with hardness values of 22 HRC (hardness, Rockwell C scale) or less are generally considered to be immune to SSC. Steels with hardness above this level are susceptible to cracking. For weld procedure qualification, and in failure analysis of welds, Vickers hardness with a 10 kg load (HV_{10}) is typically used because of its smaller indenter size compared to Rockwell C scale. The 22 HRC corresponds to approximately 250 HV_{10}. The higher the hardness (strength), the shorter will be the time to failure. If steel is alloyed with other materials, such as nickel, failure can occur at hardness levels less than 22 HRC.[23] Conversely, certain heat treatments and steel compositions can raise the maximum permissible hardness level above this value.
2. Stress Level (either residual or applied): The time-to-failure decreases as the stress level increases. In most cases, the stress results from a tensile load or from the application of pressure, or both. However, residual stresses and hard spots can be created by welding or by cold working the material (i.e., cold bending, wrench marks, etc.).
3. Hydrogen Sulfide Concentration: The time-to-failure decreases as the H_2S concentration increases. Delayed failures can occur at very low concentrations of H_2S in water (0.1 ppm), and at partial pressures as low as 0.1 kPa (0.015 psia)—although the time-to-failure may become very long depending on the level of the steel strength. At high concentrations (5–10% by volume) of H_2S in the gas phase, the time-to-failure can be minutes in susceptible steels. In early high H_2S exploration in areas of Wyoming and Canada, steel tubing strings were reported to crack and fail while being run.
4. pH of Solution: Cracking tendency increases as the pH decreases (becomes more acidic). The tendency to failure can be drastically reduced if the pH of the solution is maintained above 9.0. The pH of most oil field produced waters resides in the 4–7 range. The effect of pH is graphically shown in ANSI/NACE MR0175, Part 2, Figure 1.[23]
5. Temperature: There is a considerable amount of data indicating that cracking susceptibility decreases above approximately (60°C [140°F]). Guidance on this is given in NACE Standard MR0175.[23]

3.14 High Temperature Attack

At temperatures above 232°C (450°F), low-strength steels can be embrittled by exposure to hydrogen gas. A water phase is not required. At these temperatures, hydrogen dissolved in the steel reacts with iron carbides in the steel microstructure to form methane. Methane occupies a larger volume than the iron carbide and, therefore, will cause small cracks or voids. As a result, the steel can withstand very little deformation without cracking, that is, it becomes embrittled. The diffusion rate increases with an increase in hydrogen gas partial pressure and with temperature. In refinery operations, high-temperature hydrogen attack is a concern. In oil and gas production, however, temperatures above 232°C (450°F) are seldom seen, if ever; likewise, large quantities of hydrogen gas are seldom seen in produced streams. Because it is a form of hydrogen-induced failure, it is mentioned for thoroughness.

> **JUST BECAUSE IT RESISTS SULFIDE CRACKING, DOESN'T MEAN IT WON'T CORRODE.**
>
> It is not unusual to hear someone say: "Oh, it won't corrode; it's a NACE valve," or "It's a sour service valve," or, "It was ordered to MR0175."
>
> Remember that NACE MR0175 addresses the selection of metallic materials that are resistant to sulfide stress cracking (SSC). It does NOT necessarily address materials that are corrosion resistant. Yes, many corrosion-resistant metals can and do meet the requirements of MR0175, but many crack-resistant carbons and low-alloy steels, and some corrosion resistant alloys (CRAs) are not resistant to general or localized metal loss corrosion. If a CRA is resistant to cracking, it may still fail by pitting corrosion; therefore, operating within the acceptable service envelope for that alloy is important. For example, carbon steels such as API L-80 are designed for sour service and will resist SSC, but they will corrode in a sour well unless protected by corrosion inhibition.
>
> When a well or facility is handling sour fluids, use materials that meet MR0175 (latest edition) to avoid cracking. Plus, select the most appropriate method for controlling other forms of corrosion.

3.15 Fatigue of Metals in Air

Metal fatigue is the cause of many costly failures in the petroleum production industry. "Fatigue failure" is the name given to failures that occur when metals are repeatedly stressed far below the yield strength in a cyclic manner in air, and they fail in a brittle-appearing mode with no visible signs of deformation. By this definition, fatigue is not a form of corrosion. However, as will be discussed in the next section, a corrosive environment can greatly accelerate fatigue failure.

"High cycle" fatigue (>10^5 cycles) is the most common form of failure in rotating equipment, e.g., reciprocating compressor rods, and centrifugal pump shafts. Common factors that have been attributed to "low-cycle" fatigue are high stress levels at <10^4 number of cycles to failure. There is, however, with some metals (notably steels), a stress below which the material may be cyclically stressed indefinitely without failure. This stress level is called the "endurance limit" and is always lower than the yield and tensile strengths. The performance of materials subjected to high cyclic stressing is normally described by plotting the stress at failure against the logarithm of the number of cycles to failure for a series of stress levels. This type of plot is known as an "S-N curve" (Figure 3.63); S-N curves are not useful in low-cycle fatigue where the component experiences plastic strain. Some components, such as coiled tubing, are subjected to cyclic plastic strain when reeled and unreeled, low-cycle fatigue is plotted as plastic strain amplitude versus number of cycles.

The endurance limit for nonwelded ferrous metals usually is 40–60% of their tensile strengths, depending on the microstructure and heat treatment. Because the tensile strength of ferrous metals is roughly proportional to hardness over a wide range, the endurance limit is also proportional to hardness over this range. However, this relationship only holds true for nonwelded components. Components fabricated by welding contain small weld discontinuities and stress concentrations (notches) that act as stress risers and fatigue crack initiation sites. Because of these factors, high strength welded components do not have proportionally higher fatigue strength and endurance limit. Initiation sites include crack-like flaws such as weld undercut, lack of root fusion, lack of side-wall fusion, and other linear discontinuities. For welded components, joint design plays a significant role in determining the fatigue strength because joint design influences the stress concentration factor.

Metallurgical microstructure plays an important role in fatigue strength. Rapid quenched and tempered steels normally have better fatigue properties than slow-cooled ferrite-pearlite carbon steels. However, when considering welded joints, the effect of weld quality and joint design often is greater than the base metal microstructure. For example, fillet welds are the most detrimental joint design to fatigue strength, irrespective of base metal strength, hardness, and microstructure. By their design, fillet welds create notches.

Fatigue cracks usually start at the metal surface, and the fatigue performance of any item is drastically affected by surface conditions. In many cases, fatigue initiates at a geometric discontinuity that causes a stress concentration effect, including at threads, sharp corners, incorrect weld joint designs, and excessive weld cap height or root penetration. Notches, pits, or metal inhomogeneities, such as a lack of weld fusion anomalies, inclusions, and porosity act as stress raisers, and the actual stress at the root of a notch may be many times the nominal applied stress. Heat-treating processes can decarburize the surface of a metal part. The decarburized surface is lower in carbon concentration than the base steel and, therefore, it has a lower hardness and strength. The fatigue strength is dependent on the strength of this lower carbon content surface layer.

Fatigue in air affects components such as mining trucks and shovels in oil sands[2] operations, and reciprocating compressor pistons and connecting rods in gas production operations. Small diameter connections, particularly threaded or socket-welded, to large diameter piping to attach instrumentation are a chronic source of fatigue failures in plants (Figure 3.59). The figure shows the failure of a ½ in (12.5 mm) diameter 316L threaded pipe nipple that attached a pressure gauge and a bleed valve to a progressive cavity pump. In cases where there is more than one initiation site, multiple cracks can initiate (in this case from the OD) and grow progressively through the cross section. Figure 3.59 shows multiple points of crack growth along parallel planes. If the planes are close, one crack can join with another crack on a nearby plane. The resultant radial step is referred to as a "rachet line."

Vibration leading to cyclic stress and fatigue failure occurs because of internal fluid flow conditions, pressure pulsations, and in some cases, rotating machinery or wind-induced vibration provides the source. Remediation requires proper connection design (welded not threaded, avoiding socket welds), proper weld design, good fit-up for welding, and pipe bracing.

---

2. Oil sands are a natural mixture of sand, water, and bitumen (oil that is too heavy or thick to flow on its own). They exist in Western Canada and are mined, not drilled.

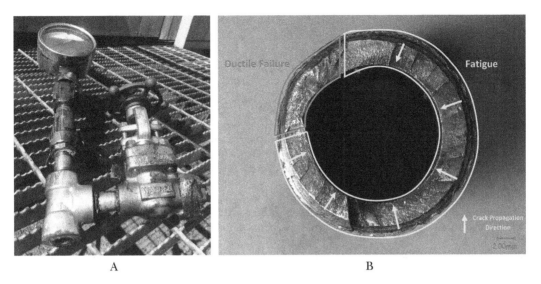

Figure 3.59 Mechanical fatigue failure of a threaded branch connection. (A) Flat fracture surface at a nipple threaded into tee at lower left. (B) Fatigue cracks initiated at the thread root of the last engaged thread of the nipple, and the progressive crack growth is shown in the close-up photograph of the fracture surface.
Photographs provided courtesy of Chevron Corporation. ©2019 Chevron

3.16 Corrosion Fatigue

Fatigue that occurs in the presence of a corrosive fluid, even if only mildly corrosive, is described as corrosion fatigue. Most fatigue failures in the oil and gas producing industry are corrosion fatigue. In corrosion fatigue, the cumulative adverse effects of environment and fatigue are more than what might be expected from each effect occurring separately. It is the most common cause of failure for sucker rods and drill pipes. Saltwater and produced brine are the most common corrosive environments. Brine may also contain corrosive gases such as O_2, CO_2, and H_2S. Sucker rods experience cyclic loading during operation of rod pumped wells. Sucker rods are manufactured from hot rolled and forged steel, are threaded, and do not contain welds. Corrosion fatigue causes most sucker rod failures because the fatigue life of a metal is substantially reduced when the metal is cyclically stressed in a corrosive environment. Sucker rods reside inside the tubing string and are exposed to corrosive produced fluids. The presence of the corrosive medium adds to the fatigue mechanism and shortens the time to failure. Hydrogen embrittlement can also contribute to fatigue failure in high-strength steels—some sucker rod steels are more susceptible to corrosion fatigue than others, depending on the strength level and manufacturing processes used.

Remediation of rod failures by corrosion fatigue involves proper torqueing of the rods to connections to minimize over- or under-stressing the rod, and by utilizing good corrosion inhibition practices. It may also require changes in the materials specifications regarding strength level to prevent hydrogen embrittlement. A materials upgrade to a corrosion resistant alloy with a high pitting resistance index has been used to mitigate pitting and subsequent corrosion fatigue failure of polished rods—the topmost portion of a string of sucker rods that goes through the stuffing box and is alternately exposed to air and the produced fluid.

The appearance of the rod break shown in Figure 3.60 is very typical of corrosion-fatigue failures. Four distinct areas characterize the appearance of the fracture surface. Visual inspection of the failure surface can often identify the initiation site for fatigue cracks. A small pit or notch on the surface serves as a stress riser to initiate the fatigue crack (Label A). The crack opens and closes as the rod cycles from compression on the down stroke, to tension on the upstroke (crack growth occurs only in tension). As the crack grows through many cycles, the hammering together of the side of the crack produces a smooth area—this moon-shaped surface appears to have been peened (Label B). At some point in its life, as stress intensity increases, the growth rate of the crack becomes too rapid for the peening to smooth the surface, and the fracture surface will appear rougher (Label C). Stress intensity at Label C at the crack tip is much higher than at the slower, progressive crack growth stage (Label B). Failure is imminent, and it occurs when there is too little sucker rod cross-section area remaining to support the load, and the rod pulls apart, leaving a ductile fracture typical of a tension failure with a shear lip, 45° angle (Label D). This area of final ductile fracture is the only portion of the break that shows any sign of plastic deformation. Otherwise, the fatigue failure portion of the break is flat. It may show "beach marks" or striations that indicate progressive crack growth over many stress cycles. Figure 3.61 shows a similar rod break from corrosion fatigue, but at least two additional corrosion fatigue cracks initiated but did not grow to failure. Corrosion fatigue failures often contain multiple cracks running parallel to the fracture surface. Cracks tend to be filled with corrosion product.

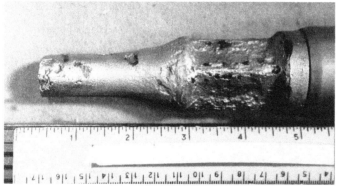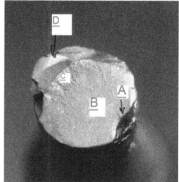

Figure 3.60 Corrosion fatigue failure of a corroded sucker rod.
The fracture surface shows four distinct areas, labeled A–D, as described in the text.
Photographs provided courtesy of Nalco Champion, an Ecolab Company.

Figure 3.61 Corrosion fatigue failure of a sucker rod.
Note, in addition to the rod break, two additional cracks initiated in the area denoted by arrows.
Photograph provided courtesy of Nalco Champion, an Ecolab Company.

Figure 3.62 shows a corrosion fatigue failure of a drill-pipe connector. Two independent circumferential cracks initiated at the inner diameter threads and separately grew to failure. The cracks progressed towards the OD (flat areas) until they penetrated the pipe wall, resulting in a pressure leak. The presence of two shear lips (raised, 45° angle areas of final overload fracture) indicates at least two cracks had formed and grown. Multiple cracks can grow independently but link into one wider crack front on the same fracture plane. Widening the crack increases stress intensity at the crack tip and shortens the time to failure. Multiple cracks may also grow independently on parallel planes, join when the ligament between cracks is insufficient to sustain the load, and result in the fracture surface having small jogs, or deviations, from planar flatness called rachets.

Figure 3.62 Corrosion fatigue failure of a heavyweight drill pipe connector.
Two independent circumferential cracks initiated at the inner diameter threads.

CHAPTER 3: Forms (or Types) of Corrosion

Endurance limits do not occur in corrosion fatigue, as shown in the schematic curves in Figure 3.63. Here, S-N curves are graphical presentations of cyclic stress data. Stress-to-failure is plotted against the logarithm of the number of cycles to failure. These curves are used for high cycle fatigue where stress is in the elastic range. The figure shows schematic S-N curves for steel in air and in saltwater. In aerated seawater, no endurance limit is observed. Therefore, the component must be designed to either withstand the requisite number of cyclic stresses over its life or be protected from the corrosive environment. For example, fixed offshore platforms are designed to withstand normal and storm-related wave loads and cycles, and cathodic protection is applied below the water line to enhance corrosion fatigue resistance of structural members.

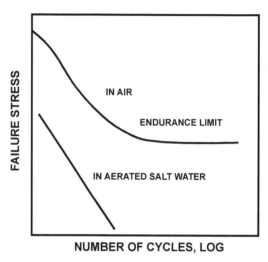

Figure 3.63 Schematic S-N curves for steel in air and in saltwater. S-N curves are graphical presentations of cyclic stress data. Stress-to-failure is plotted against the logarithm of the number of cycles to failure.

In the absence of an endurance limit, corrosion fatigue performance is normally characterized by the "fatigue limit"—an arbitrarily defined quantity of the number of cycles. The corrosion fatigue limit is commonly defined as the maximum value of stress at which no failure occurs after 10^7 cycles. (At a pumping speed of 15 strokes/minute, 10^7 cycles are reached in approximately 463 days of pumping.)

In corrosion fatigue, the corrosiveness of the environment is extremely important. The presence of dissolved gases, such as H_2S, CO_2, or oxygen, causes a pronounced increase in corrosiveness and results in decreased corrosion fatigue life (Table 3.2).

Table 3.2 Corrosion fatigue of steel in brine with dissolved gas.

Dissolved Gas	% Decrease from Air Endurance Limit
Hydrogen sulfide (H_2S)	20
Carbon dioxide (CO_2)	41
Carbon dioxide + air	41
Hydrogen sulfide + air	48
Hydrogen sulfide + carbon dioxide	62

NOTE: Table data is for information only and should not be used for design. The user should perform lab testing representative of the environment to establish fatigue limits.

Pitting or localized attack is most damaging from the standpoint of corrosion fatigue because of its stress concentration effect, but even slight general corrosion will substantially reduce fatigue life when compared to a smooth surface.

Variables that affect corrosion rates are important. The fatigue life of carbon steel in saltwater or drilling mud with a pH above 11 has been shown to markedly increase corrosion fatigue life by reducing corrosivity. Cathodic protection increases the corrosion fatigue life of structural members below water. Remember: CP raises the pH of the electrolyte at cathodic sites by forming hydroxyl ions (from Chapter 2, Section 2.3). However, CP can result in hydrogen embrittlement of susceptible materials.

Unlike fatigue in air, the corrosion fatigue performance of carbon and low-alloy steels is independent of strength. This has been shown both for aerated fresh water and for brine containing H_2S, CO_2, air, or some combination. Hence, heat treatment and alloying are more important to improving corrosion resistance than to mechanical strength.

3.17 Stress Corrosion Cracking (SCC)

SCC is caused by the "synergistic" action of corrosion and applied tensile stress; that is, the combined effect of the two is greater than the sum of the single effects. In the absence of stress, the metal would not corrode (in such a cracking manner), and in the absence of corrosion, the metal easily supports the stress. The result of the combined effect is a brittle-appearing failure of a normally ductile metal, with no signs of plastic deformation.

The stress leading to SCC is always a tensile stress. It can be either applied or residual. Frequently, residual stresses are more dangerous because they may not be considered in evaluating the overall stresses. When a metal suffers SCC, metal loss from corrosion is generally very low, although pitting is frequently observed. In many cases, pitting precedes cracking, with stress corrosion cracks developing from the bases of the pits. Cracking may be either intergranular or transgranular but is always in a direction perpendicular to the highest tensile stresses. Sometimes crack growth will relieve stresses and change the direction of the highest stress. The cracks will not be straight—they will continue at right angles to the highest stresses.

All alloys are subject to SCC, and only pure metals appear to be immune. In every case, however, stress corrosion results from exposing an alloy to a particular corrosive. No single corrosive environment causes stress corrosion in all alloys, and most alloys are subject to attack in only a few specific environments.

In general, the time-to-failure and extent of cracking will vary with corrosive concentration, temperature, and stress intensity ("stress intensity" is a term that accounts for stress and crack size). Cracking tendency always increases with increasing stress level, while time-to-cracking decreases. One approach to preventing failures is to minimize tensile stresses. Areas that are particularly important are areas of stress concentration, differential expansion, geometric mismatches, and other spots where local stresses would be high. Stress relief after welding is an important method of mitigating SCC in some alloy–environment combinations. One example of this is welded carbon steel exposed to rich alkaline amines in gas sweetening operations. In some cases, peening, burnishing, or rolling surfaces to induce compressive stresses will reduce cracking tendency.

Increasing solution concentration usually increases SSC tendency and decreases the time to cracking. This is most important in normally weak solutions that can become concentrated at local spots, such as liquid levels, hot spots, and dead ends. For instance, 304 stainless steel will crack in solutions containing only a few parts per million chloride ion concentration in the bulk liquid; however, the chloride concentration usually is found to be much higher at the point where cracking occurs because of concentration mechanisms such as an alternating wet-dry environment, local boiling, or evaporation. Oxygen can increase the chances of SCC even at low temperatures.

At higher temperatures, cracking tendency in many systems is greater, and time-to-failure shorter, if there is liquid water present. Failures will develop at local areas where the temperatures are higher than those of the system in general. SCC does not occur in dry gases because of the absence of an electrolyte.

3.17.1 Carbon and Low-Alloy Steels

These are the most common corrosives causing SCC of mild steels:

- Certain alkaline environments such as sodium hydroxide and alkaline amines containing dissolved acid gas (e.g., CO_2 from a gas plant)
- High pH carbonate + bicarbonate solutions found in soils in contact with underground pipelines
- Near-neutral pH bicarbonate + CO_2 environment found in soils in contact with underground pipelines
- CO–CO_2 combinations found in inert gas flooding
- Nitrate solutions

3.17.1.1 Alkaline SCC of Carbon and Low-Alloy Steels

SCC of carbon and low-alloy steel in production operations can occur in either sodium hydroxide (NaOH) or in caustic and amine liquids. Although now known as "alkaline SCC," it was formerly known as "caustic embrittlement" and "amine SCC," respectively. Examples of the caustic and amine liquids include monoethanolamine and methyl diethanolamine-containing dissolved acid gas (CO_2

with or without H_2S). Alkaline SCC was once a very serious problem. However, enough is now known about the phenomenon that proper stress relief after welding (postweld heat treatment) is widely used to avoid this type of SCC.

Although written for refinery service, NACE SP0403.[26] addresses this mechanism in production services as well. API RP 945 [27] addresses prevention of alkaline SCC in amine plants.

3.17.1.2 External SCC of Carbon and Low-Alloy Steel Underground Pipelines

There are two forms of external SCC on carbon and low-alloy steel underground pipelines: high pH SCC and near-neutral pH SCC, both of which are associated with different environments that develop at the external pipeline surface in the areas of disbonded coatings.[28] Both environments are associated with the presence of CO_2 in the soil, probably a result of decaying organic matter.

High pH SCC is most often associated with disbonded coal tar coatings, and the environment that develops at the pipe surface has a pH of 9–13. High pH SCC occurs at temperatures >40 °C (104 °F). Elevated temperatures can promote fracture by increasing the crack growth rate, and, because of this, most of the SCC failures have occurred near the discharge (hot) side of compressor stations. Applied CP causes the pH of the electrolyte beneath the disbonded coating to increase, and CO_2 within the soil dissolves in the elevated pH electrolyte (wet soil), resulting in a potent high pH-cracking environment containing carbonate and bicarbonate.

Near-neutral pH SCC is most commonly associated with tape and asphalt coatings. The environment that develops in this case has a pH of 5–7 because of shielding of the CP current or inadequate CP. CO_2 dissolves in the near-neutral pH electrolyte resulting in a second type of potent cracking environment containing bicarbonate and carbonic acid.

In addition to disbonded coatings, both SCC types require a cyclic stress, particularly near-neutral pH SCC. Gas and liquid pipelines are exposed to service stresses of varying amplitudes. In pipelines, cyclic loading arises from two principal sources: (1) approximate daily pressure fluctuations during operation, which are on the order of ±10% of the nominal operating pressure; and (2) shutdowns and startups for regular service or because of an upset.

For new pipelines, specifying coatings that do not disbond and shield CP, such as fusion bond epoxy, is necessary but not sufficient. As of the time of writing, no SCC-resistant steels have been developed. Therefore, pipeline operators practice direct assessment[29] to assess the presence and extent of SCC in existing pipelines, including crack inspection using inline inspection tools and hydrotesting.

Additional distinguishing features between the two types of external SCC of buried carbon and low-alloy steel pipelines are in Table 3.3. Notice the electrochemical potential ranges where SCC occurs. Effective CP is more negative than the values shown and can bring the pipeline out of the cracking potential range if the applied coating does not disbond and shield the pipeline from receiving protective current.

Table 3.3 Factors Involved in External SCC of Carbon Steel Pipelines[a]

Factor	High-pH SCC	Near-Neutral SCC
pH range	9–13	5–7
Location	Usually gas transmission piping within 20 km of a compressor station with operating stress >60% SMYS[c]	Locations of cyclic stress
Temperature	40°C	No apparent correlation
Soil chemistry	Concentrated carbonate/bicarbonate solution	Dilute bicarbonates intensified by sulfate reducing bacteria
Potential range[b]	–600 to –750 mV with generally effective CP	–760 to –790 mV with locally ineffective CP
Crack morphology	Intergranular—narrow	Transgranular—wide

a. Adopted from Reference 28.
b. With respect to Cu–CuSO$_4$ reference electrode
c. Specified minimum yield strength

3.17.1.3 Other SCC Environments Affecting Carbon and Low-Alloy Steels

Certain CO–CO$_2$ combinations that can be encountered in inert gas flooding can cause transgranular cracking of carbon and low-alloy steels.[30] Note that SCC does not occur in water containing only CO$_2$; therefore, CO appears to be important in inhibiting general corrosion but results in SCC. No SCC was found in gas compositions with a CO partial pressure below 9.8 kPa (1.4 psia). SCC can be prevented by maintaining the operating conditions above the water dew point.

Cracking by ammonium, calcium, and sodium nitrates are minimized by preventive measures, such as avoiding conditions that can cause localized concentration, controlling temperature, and reducing stresses. Because the use of these materials that can cause SCC is minimal in oil field operations, this type of SCC is not typically encountered.

3.17.2 High Strength Steels

For oil field applications, high-strength steel is loosely defined as having a yield strength of 550 MPa (80 ksi), and ultrahigh-strength steel as a yield strength of 965 MPa (140 ksi) or above. Such steels may be found in connectors, fasteners, drill pipe, casing and tubing, the reinforcement wires in subsea flexible pipe, and wirelines. Cracking of high-strength steels is more widespread than that of mild steels, i.e., carbon steel with a more restrictive carbon content, usually less than 0.25% (also known as "plain carbon steel"). It occurs in salt solutions, moist atmospheres, and, with the ultrahigh-strength steels, even in tap water. While many such failures are due to SCC, there is evidence that much of the cracking of high-strength steels is caused by hydrogen embrittlement. Cracking tendency increases with the strength of steels; this accounts for the extreme susceptibility of the ultrahigh-strength steels. Cathodic protection, especially impressed current CP, can charge atomic hydrogen into cathodic sites of high-strength steel and embrittle it. This is a concern in subsea CP with equipment that contains high-strength steel fasteners. Electroplating introduces atomic hydrogen that should be baked out by thermal treatment before the component is put into service. Experience or actual tests are necessary to determine the corrosives and conditions that will cause cracking of these steels.

3.17.3 Austenitic Alloys

The most commonly used austenitic alloys in production operations are the austenitic stainless steels—the 300 series (304 and 316, UNS S30400 and S31600, respectively). They are used in water handling, instrumentation tubes, valves, pumps, chemical handling equipment, and many more services. Of special concern is chloride SCC (CSCC) of pressure equipment (vessels and piping), fasteners, and shackles, because failures have resulted in hydrocarbon releases and/or equipment failure.

The service conditions that favor CSCC are complex. They include temperature, chloride concentration, electrolyte pH, and dissolved oxygen. The metallurgical condition, especially strain hardening, also plays a significant role. A sensitized microstructure increases susceptibility to pitting and CSCC. Acidic pH electrolytes below pH 4.5 and those containing chlorides are extremely potent SCC-promoting fluids.

3.17.3.1 Temperature

The threshold temperature for CSCC of solution-annealed austenitic stainless steel is approximately 60°C (140°F). Below the threshold temperature, CSCC is unlikely. However, strain hardening can promote CSCC at ambient temperature. Cracking is not a concern when the electrolyte temperature is high enough to ensure there is no water phase. However, alternating wet-dry conditions are the worst conditions for promoting CSCC because any chloride on the metal surface during the wet phase evaporates and forms a high-chloride concentration before total evaporation occurs. Alternating wet-dry conditions causes external CSCC in thermally insulated austenitic stainless steel where water entry may occur. Chloride ions come from a marine atmosphere, seawater deluge testing, and even from the insulation itself.

3.17.3.2 pH

Electrolyte pH plays a key role. CSCC can occur at room temperature in fluids with a pH below 1. At alkaline pH above 10, which is not a typical production environment, CSCC is unlikely to occur. Cathodic protection can reach this high pH as can alkaline amines. This explains why amine plants constructed with the 300 series austenitic stainless steels have not experienced many CSCC problems internally in the high pH rich or lean amine solution, at least up to 1000 ppm chloride content.

CSCC of austenitic stainless steel has a unique appearance under the microscope. One or more cracks start from the surface then branches into many different cracks (typically but not exclusively transgranular), and these branches have their own branches (Figure 3.64). In Figure 3.64A, the 304 austenitic stainless steel nut was specified to ASTM A194 Grade 8.[31] It was installed onto strain-hardened stud bolts and exposed to a marine atmosphere in the North Sea. The service temperature was ≥30°C, and application was in an instrumentation system that included small diameter bolts (≤22.2 mm or 7/8 in).

Figure 3.64B shows a cross section of a wellhead bonnet valve stud bolt. Material was specified to ASTM A453 Grade 660,[32] a precipitation hardened austenitic stainless steel. The bolt temperature was approximately 60°C.

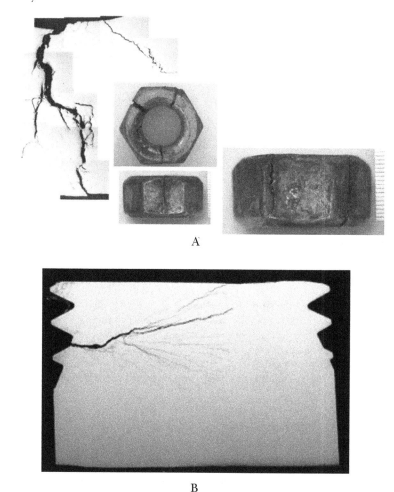

Figure 3.64 Chloride stress corrosion cracking (CSCC) of austenitic stainless steel fasteners in marine atmospheres: (A) montage showing entire crack length in 304 stainless steel nut; (B) cross section of a wellhead bonnet valve stud bolt showing CSCC of a precipitation hardened austenitic stainless steel (ASTM A453 Grade 660 [32]).

3.17.3.3 Oxygen

Below pH 4.5, dissolved oxygen does not play a significant role in promoting CSCC, because these environments already are potent CSCC promotors if chlorides are present. At a pH range of 4.5–8.5, which includes many production environments, dissolved oxygen promotes CSCC; however, if the dissolved oxygen concentration is below 10 ppb, the chance of CSCC is very limited. This is one reason that CSCC is not commonly encountered in produced fluids. However, when hydrotesting stainless steel equipment with aerated, chloride-containing water, the equipment must be fully drained and dried before the equipment is returned to service, to avoid pitting and CSCC. Even oxygen-scavenged water using chemicals may not reduce dissolved oxygen below 10 ppb, depending on the rate of reaction.

3.17.3.4 Strain Hardening

Otherwise known as "cold working," this increases the susceptibility of austenitic stainless steels to CSCC. Fasteners and marine shackles made of these materials are strain hardened to increase their strength. Bolts manufactured as strain hardened are used at higher pressure classes (> ANSI 150 class flange rating) than annealed bolts. Application of 300 series austenitic stainless steel and precipitation hardened austenitic stainless steel bolts (ASTM A453 Grade 660[32]) in marine atmospheres has resulted in bolt failure because of CSCC issues caused by the naturally occurring presence of chlorides in the marine atmosphere environment and the use of seawater deluge testing in offshore and/or coastal facilities. In some instances, the service temperature was hot (for example, wellhead bolts). In many cases, however, the temperature was below the commonly used threshold 60°C for solution-annealed 316 stainless steel. Failures have occurred suddenly, and without warning with strain-hardened instrumentation (level bridles, transmitters, etc.), and 3-piece ball valves with small bolt diameters (≤22 mm (0.875 in), which are prone to be overtorqued. Solution-annealed stainless steel should be resistant to CSCC up to 60°C, but it can be strain hardened and made susceptible to CSCC at ambient temperature, particularly if overstressed because of overtorqueing. The internal production environment has not generally played a role in initiating the external failures, except where the service temperature is hot. However, the operator should consider the consequence of the escaping production fluids in the risk assessment.

3.17.3.5 Thermal Insulation

Thermally insulated stainless steel exposed to a coastal or marine atmosphere, or any environment that contains airborne chlorides, is another application that promotes external CSCC. Even if low chloride-content insulation is installed, open seams and cutouts in the weather jacketing allow chloride-laden air and moisture to enter and contaminate the insulation, increasing its chloride content, and leading to external CSCC. Mitigating external CSCC is described in NACE SP0198.[33]

3.17.3.6 Sensitized Microstructure

The precipitation of chromium carbides along grain boundaries, and subsequent depletion of chromium in the bulk alloy along grain boundaries can result in intergranular SCC, whereas in most cases, transgranular cracking is found.

Where conditions promoting CSCC of 300 series stainless steels are present, higher nickel-containing austenitic alloys may be considered. These alloys include Alloy 28 (UNS N08904), Alloy 825 (UNS N06975) and duplex stainless steels (UNS S31803 or S32750). ANSI/NACE MR0175 or ISO 14156 Part 3 provides service envelopes for higher-alloyed austenitic alloys in sour service.[8] Some operators install CSCC-susceptible stainless steel and use protective coatings to prevent CSCC. Coating small diameter components offshore, such as instrumentation lines, is difficult, and coatings must be maintained to be effective. In some cases, thermal spray aluminum coatings have been applied to prevent CSCC of austenitic stainless steel because aluminum acts as a source for cathodic protection at breaks in the coating.

For completeness purposes, there is a real possibility that the 300 series grades may lose their passivity and undergo rapid general corrosion in hot 40–50% caustic solution, and temperatures exceeding 100°C (212°F) in these alkaline environments may result in transgranular SCC, irrespective of chloride concentration. Such conditions could occur if a concentration mechanism exists, but these conditions are not commonly encountered in production operations.

3.17.3.7 Machining Grades

Free-machining grades of austenitic stainless steel are very susceptible to SCC and pitting corrosion failure in upstream oil and gas production facilities. Numerous failures have occurred onshore and offshore, even far away from coastal locations. The grades are made to be easily machinable by additions of sulfur, selenium, and phosphorous, and include grades 303 and 303Se. Many operators write specifications that prohibit the use of these grades; however, they continue to enter service in small, assembled parts, notably valve components.

3.17.4 Copper-Based Alloys

The most common use of copper alloys in production operations is in water handling, including piping and heat exchanger tubes; 90Cu-10Ni containing 1–2% iron is a common alloy for raw seawater service. Nickel-aluminum bronze is used in valve and pump bodies in produced water. A nickel-copper alloy, Monel K-500 (N05500) is a heat treatable alloy capable of achieving high strength for valve stems and other components, such as Bourdon tubes in pressure gauges.

Pure copper, like other pure metals, apparently is immune to SCC. Alloying copper with even small amounts of phosphorous, zinc, antimony, aluminum, silicon, or nickel makes it susceptible to serious intergranular SCC by ammonia and ammonia solutions. In some cases, cracking is both intergranular and transgranular. The presence of oxygen, in addition to ammonia, is necessary to cause cracking of the copper alloys, but even very small amounts of ammonia can be damaging. In oil and gas systems, ammonia is not usually present, so SCC has not been a problem. Even 90Cu-10Ni pipe in polluted seawater has no reported instances of SCC, although laboratory testing shows this alloy is susceptible to SCC in solutions of sodium sulfide, which is supposed to represent polluted seawater.[34] Theoretically, some of the amines used in corrosion inhibitors could cause cracking, but a variety of stainless steels, not copper alloys, are typically used for chemical handling. Experience and testing are necessary to verify the suitability of an alloy for a particular exposure.

Monel K-500 (a nickel-copper precipitation hardenable alloy) has been shown to be susceptible to hydrogen embrittlement in wet H_2S. This alloy has been used successfully in mildly sour service, but when conditions of temperature and H_2S partial pressure are severe, operators specify precipitation-hardenable nickel alloys such as Alloy 718 (UNS N07718). Refer to Chapter 4, Section 4.7.2 for more information.

3.17.5 Aluminum Alloys

Aluminum alloys are used in oil and gas production facilities because of their cold-temperature resistance or high strength-to-weight ratio in the following applications:

- reciprocating compressor piston heads;
- cryogenic service in liquefied natural gas processing facilities as piping and heat exchangers in cold boxes and in brazed aluminum heat exchangers;
- offshore platform structural components for helidecks (with proper fire-fighting equipment added); and in
- offshore topsides buildings

Aluminum fins are used to improve heat transfer of carbon steel tubes used in air-cooled heat exchangers. Because of its low melting point and low strength, aluminum alloys are rarely used in handling produced fluids, in separation facilities, and in processing facilities except as noted. Aluminum drill pipe has had limited application and is considered for directional drilling. In that aluminum depends on an oxide film for its corrosion protection, and without oxygen it is susceptible to chloride pitting, its use in the production industry has been limited to where its protective oxide film is not in jeopardy.

Because of its limited use, SCC of aluminum in production facilities is uncommon. In compressor and in cryogenic services (e.g., LNG liquefaction facilities), there is no electrolyte to cause SCC. The application of aluminum is generally confined to alloys such as 3000 series (Al-Mn); 5000 series (Al-Mg-Mn-Cr) and 6000 series (Al-Mg-Si) alloys. The high-strength alloys found in the aerospace industry, such as the 20XX series (aluminum-copper) and the 70XX series (aluminum-zinc-magnesium-copper) are susceptible to SCC in seawater and are rarely used in production facilities. Where susceptible alloys are used, coatings or some secondary protection is necessary.

3.17.6 Nickel and Nickel-Based Alloys

Nickel and nickel-based alloys are susceptible to stress corrosion cracking in a few specific corrosives, but generally have excellent resistance to all types of production environments. The most commonly used nickel-based alloys include Alloy 2550 (UNS N06975), Alloy 925 (UNS N09925), Alloy 718 (UNS N07718), and Alloy 625 (UNS N06625). MR0175 Part 3[8] describes service envelopes for high-nickel alloys to prevent SCC and hydrogen embrittlement in wet H_2S service. Refer to Chapter 4, Sections 4.6 and 4.7 for more information.

3.17.7 Titanium and Titanium Alloys

Because of their excellent corrosion resistance to a wide variety of environments, use of titanium alloys is increasing. Applications include raw seawater piping for cooling and firewater offshore (a competitor to 90Cu-10Ni pipe), seawater strainers, heat exchanger tubing in water and seawater service, and plate frame heat exchanger plates in many types of service fluids. Mostly low-strength and low-alloyed ASTM grades 1 (UNS R50250) or 2 (UNS R50400) are specified. Titanium has been found to be susceptible to intergranular SCC in anhydrous methanol. Although methanol is used in upstream operations to control hydrate plugging in gas production, 5% water or more added to the methanol mitigates SCC.

Titanium and its alloys are rapidly attacked by environments that contain hydrofluoric acid (HF). Exposure to HF can occur if mud acid is injected downhole for scale control.

3.18 Liquid Metal Embrittlement

Another form of brittle failure that can result from corrosion is liquid metal embrittlement (LME). The metal attacked reacts with another metal or a corrosion product, which is under stress, and penetrates the metal along grain boundaries. This intergranular phase is either much weaker or more brittle than the base alloy and, therefore, cannot support the load, or else melts at the temperature of service. Liquid metal embrittlement is another form of attack not usually found in oil field operation, but it can occur.

Ambient temperature attack is limited to that caused by mercury on brass and other copper alloys, and on high-strength aluminum alloys. The copper alloys are the more susceptible of the two, and whenever stressed brass or other copper alloys contact mercury or mercury salts, serious cracking can result. Fortunately, neither liquid mercury nor the susceptible copper alloys are commonly found together in production facilities.

Mercury has been found to exist in some produced natural gas as mercury vapor carried with the gas stream. Once the gas is cooled in a cryogenic facility, liquid mercury can condense and accumulate in contact with equipment made from aluminum or Monel (Ni-Cu). Broken instruments, such as manometers and thermometers containing mercury, are another potential source of liquid mercury. The commonly used aluminum 3000 series (Al-Mn), 5000 series (Al-Mg-Mn) and 6000 series (Al-Mg-Si) alloys,[35] and Monel (Ni-Cu) that might be found in cryogenic facilities are susceptible to LME if they contact liquid mercury. One way to mitigate LME is to substitute more LME-resistant alloys or remove mercury in the natural gas in filters added upstream of the cold boxes.

According to The Welding Institute (TWI) the relative rank order of alloys from most-to-least susceptible to mercury or zinc LME is Monel 400, Monel K-500, Alloy 625, Monel R405, Alloy X750, Alloy 718, Alloy 600, pure nickel, Alloy 825, and Alloy 800.[36]

Zinc is most commonly found as a thin protective layer in galvanized steel or in zinc-coated steel. Zinc melts at 420°C (787°F). Two situations are known to melt zinc: welding and fire.[36-37] When zinc melts, it could drip onto adjacent stainless steel equipment. At elevated temperatures above the zinc melting point, molten zinc penetrates stainless steel and embrittles it, causing it to fail under stress. Some operators specify that no zinc-coated equipment be allowed to be in contact or near stainless steel equipment.

Steels are embrittled by molten zinc above 420°C. Prior to welding galvanized steel, the welder must remove the galvanized layer near the weld so that molten zinc will not embrittle the carbon steel weld.[36]

Austenitic stainless steel is embrittled by molten copper adjacent to welds.[36] Copper will melt during welding, so copper should be removed from the work area prior to welding.

References

1. NACE International Corrosion Terminology website (accessed December 5, 2019), https://www.nace.org/resources/general-resources/corrosion-glossary. Accessible to members of NACE International only.
2. API 579-1/ASME FFS-1 (latest edition), "Fitness-for-Service," (Washington, DC: American Petroleum Institute, 2016).
3. API Spec 5L, "Line Pipe" (latest edition) (Washington, DC: American Petroleum Institute, 2018).
4. Jones, D.A., *Principles and Prevention of Corrosion*, 2nd ed. (Saddle River, NJ: Prentis Hall, 1996): p. 15.
5. ASTM G46 (latest edition), "Standard Guide for the Examination and Evaluation of Pitting Corrosion" (West Conshohocken, PA: ASTM International).
6. Landrum, R.J., *Fundamentals of Designing for Corrosion–A Corrosion Aid for the Designer* (Houston, TX: NACE International, 1989): p. 49.
7. Dillon, C.P., editor, *Forms of Corrosion–Recognition and Prevention*, NACE Handbook 1, Case History 2.2.2.6. (Houston, TX: NACE International, 1982): p. 27.
8. ANSI/NACE Standard MR0175 (latest edition), "Petroleum and Natural Gas Industries—Materials for Use in H_2S-Containing Environments in Oil and Gas Production," Part 3: Cracking-Resistant CRAs (Corrosion-Resistant Alloys) and Other Alloys (Houston, TX: NACE International, 2015).
9. Hara, T., H. Asahi, and H. Keneta, "Galvanic Corrosion in Oil and Gas Environments," CORROSION'96, paper no. 63 (Houston, TX: NACE International, 1996).
10. Mendez, C., and C. Scott, "Laboratory Evaluation and Modeling of API-L80/13Cr Galvanic Corrosion in CO_2 Environment," CORROSION 2008, paper no. 08325 (Houston, TX: NACE International, 2008).
11. Wilhelm, S.M., "Galvanic Corrosion in Oil and Gas Production: Part 1— Laboratory Studies," CORROSION'92, paper no. 480 (Houston, TX: NACE International, 1992).
12. Efird, K.D., "Galvanic Corrosion in Oil and Gas Production," in *Galvanic Corrosion*, H. Hack, Ed., ASTM STP 978 (West Conshohocken, PA: ASTM International, 1988): pp. 260–282.
13. API Spec 5CT (latest edition), "Specification for Casing and Tubing" (Washington, DC: American Petroleum Institute, 2018).
14. Velu, B., "Failure Analysis of Stainless Steel Piping at an Offshore Platform," *Materials Performance*, 47, 5 (Houston, TX: NACE International, 2005): pp. 62–65.
15. ASTM A182/A182M, "Standard Specification for Forged or Rolled Alloy and Stainless Steel Pipe Flanges, Forged Fittings, and Valves and Parts for High-Temperature Service" (West Conshohocken, PA: ASTM International, 2018).
16. Sherik, A.M., and A.L. Lewis, "Corrosion Inhibition of Sales Gas Transmission Pipelines," *Materials Performance*, 52, 9 (Houston, TX: NACE International, 2013): pp. 52–56.
17. Sherik, A., and E. El-Saadawy, "Erosion of Control Valves in Gas Transmission Lines Containing Black Powder," *Materials Performance*, 52, 5 (Houston, TX: NACE International, 2013): pp. 70–73.
18. Olabisi, O., S. al-Sulaiman, A. Jarragh, Y. Khuraibut, and A. Mathew, "Black Powder in Export Gas Lines," *Materials Performance*, 56, 4 (Houston, TX: NACE International, 2017): pp. 50–54.
19. Staehle, R.W., B.F. Brown, J. Kruger, and A. Agrawal, Eds., *Localized Corrosion, NACE Reference 3* (Houston, TX: NACE International, 1974): p. 262.
20. ASTM A262 (latest edition), "Standard Practices for Detecting Susceptibility to Intergranular Attack in Austenitic Stainless Steels" (West Conshohocken, PA: ASTM International, 2015).

21. ASTM A516/A516M (latest edition) "Standard Specification for Pressure Vessel Plates, Carbon Steel, for Moderate- and Lower-Temperature Service" (West Conshohocken, PA: ASTM International, 2017).
22. NACE Standard TM0284 (latest edition), "Standard Test Method—Evaluation of Pipeline and Pressure Vessel Steels for Resistance to Hydrogen-Induced Cracking" (Houston, TX: NACE International, 2016).
23. NACE Standard MR0175 (latest edition), "Petroleum and Natural Gas Industries—Materials for Use in H_2S-Containing Environments in Oil and Gas Production," Part 2 "Cracking-Resistant Carbon and Low-Alloy Steels, and the Use of Cast Irons," Figure 1: "Regions of environmental severity with respect to the SSC of carbon and low-alloy steels" (Houston, TX: NACE International, 2015).
24. NACE Standard MR0175 (latest edition) "Petroleum and Natural Gas Industries—Materials for Use in H_2S-Containing Environments in Oil and Gas Production," Part 1 "General Principles for Selection of Cracking-Resistant Materials (Houston, TX: NACE International, 2015).
25. ASTM A193A193M (latest edition), "Standard Specification for Alloy-Steel and Stainless Steel Bolting for High Temperature or High Pressure Service and Other Special Purpose Applications" (West Conshohocken, PA: ASTM International, 2017).
26. NACE SP0403 (latest edition), "Avoiding Caustic Stress Corrosion Cracking of Refinery Equipment and Piping" (Houston, TX: NACE International, 2015).
27. API RP 945 (latest edition), "Avoiding Environmental Cracking in Amine Units" (Washington, DC: American Petroleum Institute, 2013).
28. Jacobson, G.A., "Pipeline Stress Corrosion Cracking: Detection and Control," *Materials Performance*, 54, 8 (Houston, TX: NACE International, 2015): pp. 30–37.
29. NACE SP0204 (latest edition), "Stress Corrosion Cracking (SCC) Direct Assessment Methodology" (Houston, TX: NACE International, 2015).
30. Kowaka, M., and M. Nagata, "Stress Corrosion Cracking of Mild and Low-Alloy Steels in CO-CO_2-H_2O Environments," *CORROSION*, 32, 10 (Houston, TX: NACE International, 1976): pp. 226–231.
31. ASTM A194 (latest edition), "Standard Specification for Carbon Steel, Alloy Steel, and Stainless Steel Nuts for Bolts for High Pressure or High Temperature Service, or Both" (West Conshohocken, PA: ASTM International, 2017).
32. ASTM A453 (latest edition), "Standard Specification for High-Temperature Bolting, with Expansion Coefficients Comparable to Austenitic Stainless Steels" (West Conshohocken, PA: ASTM International, 2017).
33. NACE SP0198, "Control of Corrosion Under Thermal Insulation and Fireproofing Materials—A Systems Approach" (Houston, TX: NACE International, 2016).
34. Islam, M., W.T. Riad, S. al-Kharraz, and S. Abo-Namous, "Stress Corrosion Cracking Behavior of 90/10 Cu-Ni Alloy in Sodium Sulfide Solutions" *CORROSION*, 47, 4 (Houston, TX: NACE International, 1991): pp. 260–268.
35. Coade, R., and D. Coldham, "The Interaction of Mercury and Aluminum in Natural Gas Plants," *International Journal of Pressure Vessels and Piping*, 83 (Atlanta, GA: Elsevier Ltd., 2006): pp. 336–342.
36. TWI, "What is Liquid Metal Embrittlement and What are Common Embrittling Metal Couples?" https://www.twi-global.com/technical-knowledge/faqs/faq-what-is-liquid-metal-embrittlement-and-what-are-common-embrittling-metal-couples (Cambridge, United Kingdom: The Welding Institute, May 2017). (Jan. 2019).
37. ICE, "Guide Notes on the Safe Use of Stainless Steel in Chemical Process Plant" (Manchester, Great Britain: Institute of Chemical Engineers, 1978).

Bibliography

Atkinson, J.T.N., and H. Van Droffelaar, *Corrosion and Its Control–An Introduction to the Subject*, 2nd ed. (Houston, TX: NACE International, 1995).

Hack, H.P., Ed., *Galvanic Corrosion in Oil and Gas Production*. ASTM STP 978 West Conshohocken, PA: ASTM International, 1988.

Heidersbach, R., *Metallurgy and Corrosion Control in Oil and Gas Production*, (Somerset, NJ: Wiley Publishing, 2011).

Kaesche, H., *Metallic Corrosion–Principles of Physical Chemistry and Current Problems*, R.A. Rapp, Translator (Houston, TX: NACE International, 1985).

Palacios Tenreiro, C.A., *Corrosion and Asset Integrity Management for Upstream Installations in the Oil/Gas Industry* (self-pub., Basenji Studio, LLC, 2016).

Roberge, P.R., *Corrosion Basics: An Introduction*, 2nd ed. (Houston, TX: NACE International, 2006): pp. 394–395.

Szklarska-Smialowska, Z., *Pitting Corrosion of Metals* (Houston, TX: NACE, 1986).

Metallic Materials Selection

Robert J. Franco

4.1 Introduction

Materials selection is a strategy to understand the key issues in selecting materials associated with CO_2, H_2S, produced water, marine atmosphere, and submerged and cold temperature environments; and to use guidelines to select a cost-appropriate material for a given production environment. Materials selection requires an understanding of the requirements demanded of the component in service. These requirements include its desired mechanical properties (e.g., strength, toughness, hardness, elongation); necessary fabrication requirements (e.g., weldability, forgeability, and formability); corrosion and cracking resistance to the internal fluid environment, and to the external atmospheric or immersion environment; and wear and/or erosion resistance. To completely understand the materials selection criteria, several additional factors must be known or established:

- Design life of the component
- Hydrocarbon production profile (field life)
- Material supplier qualification
- Procurement issues associated with the location of the production facility (including regulated
- local content)
- Any safety, health and/or environmental requirements
- Planned corrosion mitigation

Included in a materials-selection strategy is the minimization of the number of materials used onsite, ease of fabrication in shop and field, and the ability to procure materials from qualified suppliers. In summary, materials selection must balance cost and longevity of the component.

There are two types of costs associated with materials selection:

1. "Capital costs" (CAPEX)—These are fixed, one-time expenses incurred on the purchase of the material and its construction. It is the total cost needed to bring a project to a commercially

operable status but not to operate it day-to-day. Examples for materials selection are the cost of a component that is new or an upgrade from a corroded carbon steel component to a corrosion resistant alloy (CRA), and the cost for a new installation of cathodic protection equipment.

2. "Operating costs" (OPEX)—An operating cost is an expense required for the day-to-day functioning of a business. Examples of materials selection-type operating costs are the cost of corrosion inhibition, routine coating maintenance, equipment inspection, the costs to maintain and provide cathodic protection, and the cost associated with replacing corroded or worn components with the same material. Replacement of corroded carbon steel with new carbon steel is OPEX. Replacement with CRA is partly CAPEX (the cost differential between CRA and carbon steel) and partly OPEX (the cost of carbon steel).

These two expense categories have different requirements for project funding, taxes, and depreciation. There are many textbooks that describe how to calculate recurring costs over time to a present value and provide further information on capital and operating expenses. The goal is to minimize the *lifecycle cost* of the component, balancing the high initial cost of a new component with the recurring cost of maintaining that component in a functioning condition over its life. In practice, however, some projects are built on the premise of minimizing capital costs, instead of a full lifecycle assessment, thereby transferring a higher operating cost to the operator. A reduction in capital cost may allow a project to reach profitability hurdles for funding, but may increase OPEX over the life of the facility.

There are standards and papers published on materials selection. Examples are listed below:

- Sucker rods[1]
- Sour service[2]
- Oil and gas production systems[3-4]
- Offshore topsides process and utility piping and systems[5]
- CRA selection guidelines[6]
- Offshore pipelines[7]
- Materials application limits[8]

In addition, the American Petroleum Institute (API) publishes standards for materials for metallic components such as drill strings, production tubing and casing, wellheads, sucker rods, storage tanks, line pipe, machinery, and structural members. ASTM International publishes specifications for metallic materials, their chemical composition, and required mechanical properties. Most metallic components in a production facility are constructed of materials that conform to ASTM standards. Design codes for constructing pipe, pipelines, and pressure vessels, are available through ASME. DNV-GL has standards for metallic components in subsea and maritime equipment. The International Organization for Standards (ISO) has many standards like ASTM and API that are used outside of and within the United States.

The purpose of this book is focused on corrosion control; therefore, in this chapter, we will discuss metallic alloys from this perspective. Mechanical properties and toughness are not the focus of the discussion but are mentioned for completeness. A well-rounded materials engineer must consider these and all factors in selecting metals. Chapter 5 addresses selection of nonmetallic components.

Because oil and gas production piping and facilities are constructed primarily of metals, it is important to understand some of the principles of metallurgy. The terminology, various microstructures, compositions, mechanical properties, and characteristics of metals used in the oil field operations are reviewed in this chapter. Note: When specific metals are mentioned, they will be identified by the appropriate Unified Numbering System (UNS) designator. UNS only specifies composition. Mechanical properties are specified in ASTM, API, or other standards. Common (commercial trade) names are excluded unless necessary. Exceptions are listed with the registered trademark symbol, ®. A detailed listing of UNS numbers and common names can be found in most engineering and metallurgy handbooks, including the *NACE Corrosion Engineers Reference Guide*.[9] An internet search provides much information on the UNS-designated metals and alloys.

4.2 Basic Metallurgy

The goal of learning basic metallurgy, particularly steel and ferrous alloys, is to recognize how properties such as strength and toughness relate to microstructure, or the appearance of the metal under microscopic examination. Metals are unique because they have a crystalline structure with a regular, ordered array of atoms.

4.2.1 Crystalline Structure

Nearly all metals and alloys exhibit a crystalline structure. The atoms, which make up a crystal, exist in an orderly three-dimensional arrangement. Figure 4.1 is a schematic representation of the unit cells of the most common crystal structures found in metals and alloys. The unit cell is the smallest portion of the crystal structure, which contains all the geometric characteristics of the crystal; it can be considered the smallest building block of the crystal. The crystals, or grains, of a metal are made up of these unit cells repeated in three-dimensions; the atoms are represented by balls. The crystalline nature of metals is not readily obvious: it is not visible and must be studied using x-ray diffraction or electron diffraction techniques.

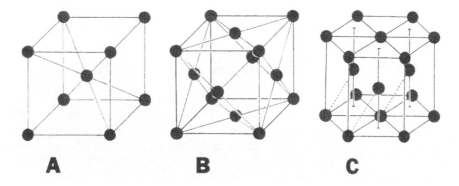

Figure 4.1 Schematic representation of unit cells of the most common crystal structures found in metals: (A) body-centered cubic (BCC), e.g., carbon steel; (B) face-centered cubic, FCC, e.g., austenitic stainless steel and aluminum; and (C) hexagonal close packed HCP, e.g., titanium.

4.2.1.1 Grain Boundaries

When a metal solidifies, the atoms, which are randomly distributed in the liquid state, arrange themselves into crystalline blocks. As the blocks of crystals or grains meet, there is a mismatch at their boundaries. When the metal has solidified and cooled, there will be numerous regions of mismatch between each grain. These regions are called "grain boundaries." Because the most stable configuration of a metal is its crystal lattice, grain boundaries are higher-energy areas with more open space to trap impurities (undesirable elements in the alloy) and are more chemically active. Hence, grain boundaries are usually attacked slightly more rapidly than grain faces when exposed to a corrosive environment. In fact, metallographic etching depends on just this effect to develop contrast between grains and grain boundaries, grains and precipitates, and grains and second phases.[1]

Controlled etching with selected electrolytes will normally show the granular characteristics (grains and grain boundaries) of metals and alloys. Normally, metal grains are so small they can only be observed with a microscope. These features are typically invisible to the naked eye. The general range of grain size (the mean diameter of the aggregate of crystals) usually runs from 0.01–2.5 mm (0.0004–0.1 in) in diameter. To determine the grain size or microstructure of a metal or alloy, it is usually necessary to prepare a sample for microscopic study by special grinding and polishing techniques. The polished surface is then reacted with a suitable etching reagent. The etchant attacks the grain boundaries and grains differently. This reveals the grain boundaries and the distinguishing features of the microstructure. The resulting photograph is called a "micrograph." Even though grains are very small, each is still made up of thousands on thousands of unit cells. Examples of etched samples showing various grain structures are shown in later photomicrographs such as Figures 4.2–4.5.

4.2.1.2 Isotropic and Anisotropic Polycrystalline Metals

Metallic components consist of many crystals, often randomly oriented. This is known as a "polycrystalline metal." The random crystal orientation results in the metal having mechanical properties that are more-or-less equal in all directions, a property known as "isotropic." In petroleum production, almost all metallic components are polycrystalline.

In unique circumstances, a metallic component may consist of only one crystal. Individual crystals have anisotropic properties, i.e., their properties are different in different directions. A unique example of a single crystal application is a high-temperature gas turbine blade (although most manufactured turbine blades, however, are polycrystalline, so single crystal blades are rarely used in petroleum production facilities).

In a polycrystalline metal, if the crystals are not randomly oriented but aligned along the direction of certain planes in the crystal structure, this is known as "preferred orientation" and the product exhibits anisotropic properties. Anisotropy results in different properties along different directions. This occurs when a metal has been highly cold or hot worked. An example of an anisotropic metal is a cold-worked corrosion resistant alloy manufactured as a downhole tubular product. Another example is found in wrought[2] products manufactured by hot rolling or forging. Pressure vessel plate man-

---

1. Section 4.2.1.1 describes grain boundaries. Sections 4.4.4 and 4.5.4 describe precipitates; and Section 4.5.1 describes second phases.
2. Wrought products are forms of a metal that are worked into sheet, foil, plate, extrusions, tube, forgings, rod, bar, and wire. Wrought products exclude castings.

ufactured by hot rolling has the mechanical properties of strength, ductility, and toughness, which is highest in the longitudinal direction—i.e., the direction of working. In some circumstances, this direction may show evidence of grain flow, depending on the working temperature. The other two directions are transverse to the roll working direction, and they have properties that are somewhat inferior. Anisotropy does not generally apply to castings, which are likely to have random crystalline structure (exceptions exist).

4.2.1.3 Effect of Grain Size on Mechanical Properties

Grain size affects the strength and toughness of steel: steels with fine (very small) grain size are tougher and stronger. For example, steels that will be used at cold temperature require fine grain for improved toughness. Fine grain size can be achieved by adding a minute quantity of aluminum to the melt because aluminum oxide particles act as nucleation sites for solid crystals to form, thereby creating more, smaller, grains. Because aluminum oxide inclusions adversely affect fatigue resistance, fine grains can also be achieved through rapid solidification, and/or by the addition of niobium (Nb) and vanadium (V). Rapid solidification is a process in which the molten metal is solidified at a cooling rate $>10^6 \,°C/s$. During this process, many nuclei simultaneously form with limited time to grow before they collide each other. Nb and V additions increase the recrystallization temperature during hot working in the austenite zone and prevent abnormal grain growth. Austenitizing steel at a temperature slightly above Ac_3[3] by ≈ 20–$30\,°C$ produces a fine grain structure. This controlled austenitizing is commonly used in double quenched and tempered heat treatment for oil country tubular goods (OCTG). The yield strength of fine grain steels is also higher than coarse grain steels; this effect is proportional to the inverse square root of the grain diameter.

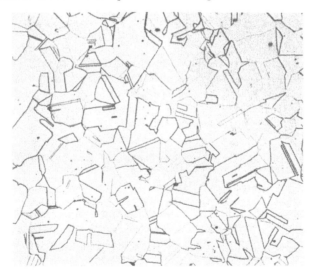

Figure 4.2 Photomicrograph of austenitic stainless steel (Type 304). Note the characteristic uniform and equiaxed (equal size in all directions) grains typical of solution-annealed austenite. Original magnification 250X.

---

3. Ac_3 is the upper critical temperature on heating at which ferrite + austenite transforms to 100% austenite.

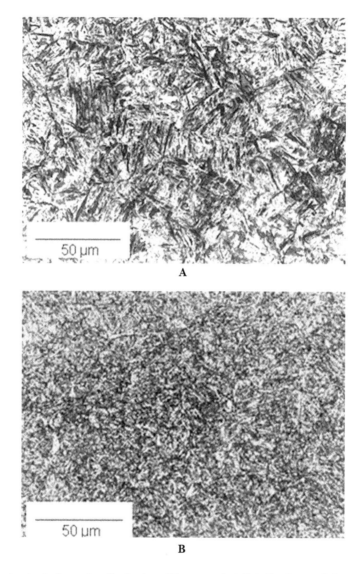

Figure 4.3 Photomicrograph of a martensitic structure: (A) as quenched. Note the characteristic needle-like (acicular) and lath (narrow platelet) structure of as-quenched (untempered) martensite in carbon steel. (B) The structure, after tempering, is too fine to resolve at this magnification; it consists of iron carbide precipitates in equiaxed ferrite grains. Depending on tempering temperature, residual austenite and/or untempered martensite may also be present.

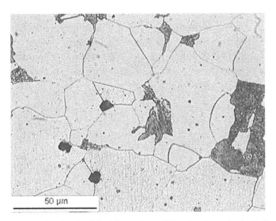

Figure 4.4 Photomicrograph of a dual ferritic and pearlitic structure, typical of mild carbon steel (carbon content less than approximately 0.25%). The ferrite grains are the light-colored ones. The pearlite phase shows up as the dark areas. Within the dark areas are alternating plates of ferrite and iron carbide, which are too fine to be resolved at this magnification.

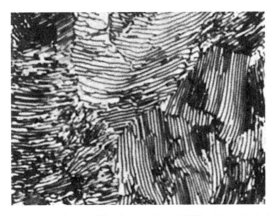

Figure 4.5 Photomicrograph of a pearlitic structure in an 0.8% carbon steel showing the typical alternating ferrite iron—iron carbide plates that make up the dark (pearlite) phase in Figure 4.4. The white phase is ferrite and the dark phase is iron carbide.

4.2.2 Alloys

Metals listed as "pure" or "commercially pure" contain a variety of impurities and imperfections. These impurities and imperfections are inherent causes of corrosion in an aggressive environment. However, high purity metals generally have low mechanical strength, and are rarely used in engineering applications. It is necessary, therefore, to work with metallic materials that are stronger and are usually formed from a combination of several major elemental metals. One metal element usually serves as a "major" or "base metal" to which other metallic elements or nonmetallic constituents are added. These metallic mixtures are called "alloys."

While alloys can exist in an almost unlimited number of combinations, only a portion of these combinations are useful. The common alloys are those that have a good combination of mechanical, physical, and fabrication qualities, which tend to make them structurally, as well as economically, useful. Of course, an alloy's behavior in corrosion environments is also very important. This chapter discusses the major alloys used in petroleum production.

4.3 Steels

As mentioned in the introduction (Section 4.1), most oil field equipment is constructed of steel, and steel is primarily an alloy of iron and carbon. Pure iron is a relatively weak, ductile material. When it is alloyed with small amounts of carbon (usually <1.0%, and often a lot less), a much stronger material is created. However, because of reacting part of the iron with carbon, now the metal has two components: pure iron (called "ferrite") and iron carbide, which is the product of the iron–carbon reaction. Ferrite is iron that contains a small amount of carbon in solid solution. In a solid solution, one metal melts into the other and remains dissolved upon solidification. The atoms of the secondary metal either replace some of the atoms of the primary metal in the crystal structure, or they fit into the spaces between the atoms of the primary metal (i.e., in the interstices).

In steel, carbon is located between iron atoms; therefore, carbon is called an "interstitial alloying element." In steel, there is too much carbon present to all fit into the interstices, so a separate phase, iron carbide, appears and is distributed alongside the ferrite as microscopic islands, layers (platelets), or other shapes. The higher the carbon content, the more quantity of a separate phase appears alongside or within the ferrite phase. The microstructure—or morphology—of the steel is dictated by the quantity of carbon and the way the steel was heated and cooled during its manufacture. This heating and cooling processes are referred to as the heat treatment of the steel. Its primary purpose is to alter the mechanical properties of the metal by controlling the microstructure, although it may also influence its corrosion resistance.

4.3.1 Structure of Steel

The microstructure (crystalline structure) of steel takes many forms. The most common—austenite, ferrite, martensite, and pearlite—are defined in Table 4.1. Photomicrographs of these structures are shown in Figures 4.2-4.5.

Table 4.1 Types of microstructure of steels.

Steel Type	Description
Austenite	The face-centered crystalline phase of iron-based alloys.
Ferrite	A body-centered cubic crystalline phase of iron-based alloys. In steel, ferrite is essentially pure iron with a very low quantity of dissolved carbon.
Martensite	A supersaturated solid solution of carbon in iron characterized by an acicular (needle-like) microstructure.
Pearlite	A mixture of ferrite and cementite (an intermetallic compound of iron and carbon, Fe_3C), characterized by a microstructure of alternate plates.

4.3.2 Names for Steels

Steels are often referred to by common names or descriptions. The common ones used in today's oil field literature are listed in Table 4.2, along with the chemical compositions and microstructures used to describe various steels. Additional terms and definitions used in discussing metals and metallurgy are given in Table 4.3.

Table 4.2 Definitions of the steels used in the oil and gas industry.[10]

Steel Type	Description
Austenitic	A steel whose microstructure at room temperature consists predominately of austenite.
Carbon	An alloy of carbon and iron containing up to 2% mass fraction carbon, up to 1.65% mass fraction manganese, and residual quantities of other elements, except those intentionally added in specific quantities for deoxidization (usually silicon and/or aluminum). Carbon steels used in the petroleum industry usually contain less than 0.8% carbon. [Author's note: Downhole tubing and casing is limited to 0.35% maximum carbon for the high strength grade Q125 per API 5CT.[11] Steels used for welded API 5L PSL 2 pipelines typically should have restricted steel chemistry (CE ≤ 0.43 and Pcm ≤ 0.25) to improve weldability and prevent weld cracking.[12]] CE is "carbon equivalent," and Pcm is the "critical metal parameter." Equations for them can be found in welding standards.
Duplex (austenitic or ferritic) stainless steel	A stainless steel whose microstructure at room temperature consists primarily of austenite and ferrite.
Ferritic	A steel whose microstructure at room temperature consists predominately of ferrite.
Low-alloy	A steel with a total alloying element content of less than about 5% mass fraction, but more than specified for carbon steel.
Mild carbon	A subset of carbon steel with a more restrictive carbon content, usually less than 0.25%. Also known as "plain carbon steel."
Martensitic	A steel in which a microstructure of martensite can be attained by quenching at a cooling rate fast enough to avoid the formation of other microstructures.
Stainless	A steel containing 10.5% or more chromium. Other elements may be added to secure special properties.

Table 4.3 Definitions used when discussing metals and metallurgy for oil and gas production tubing and equipment

Term	Description
Age Hardening	Hardening by aging, usually after cooling or cold working.
Aging	A change in metallurgical properties due to precipitation of a metastable phase that generally occurs slowly at room temperature (natural aging) and more rapidly at higher temperature (artificial aging).
Anisotropic	A metal having mechanical properties that are unequal in three directions.
Annealing:	Heating to, and holding at, a temperature appropriate for the specific material for a controlled period and then cooling at a suitable rate, to reduce hardness, improve machinability, or obtain desired properties. (Also see Solution Heat Treatment.)
Case Hardening	Hardening a ferrous alloy so that the outer portion, or case, is made substantially harder than the inner portion, or core. Typical processes are carburizing, cyaniding, carbonitriding, nitriding, induction hardening, and flame hardening.
Hardenability	The depth of hardening a steel may be hardened upon quenching from high temperature.
Heat Treatment	Heating and cooling a solid metal or alloy in such a way as to obtain desired properties. Heating for the sole purpose of hot working is not considered heat treatment. (See also Solution Heat Treatment.)

Table 4.3 (cont.) Definitions used when discussing metals and metallurgy for oil and gas production tubing and equipment

Term	Description
Isotropic	A metal having mechanical properties that are equal in all three directions.
Lower Critical Temperatures	In ferrous metals, the temperatures at which austenite begins to form during heating or at which the transformation of austenite is completed during cooling.
Mechanical Properties	Properties that are measured in mechanical tests such as tensile strength, yield strength, elongation (ductility), toughness, and hardness.
Normalizing	Heating a ferrous metal to a suitable temperature above the transformation range (austenitizing), holding at temperature for a suitable time, and then cooling in still air to a temperature below the transformation range.
Polycrystalline	A metal that consists of many crystals, often randomly oriented, with each crystal separated from another by a grain boundary.
Precipitation Hardening	Hardening caused by the precipitation of a constituent from a supersaturated solid solution.
Quench and Temper	Heating a steel above the Upper Critical Temperature, hardening it by cooling it rapidly (quenching in water or oil), followed by reheating it below the Lower Critical. Temperature (tempering).
Solution Heat Treatment (Solution Anneal)	Heating a metal to a suitable temperature and holding at that temperature long enough for one or more constituents to enter into solid solution, then cooling rapidly enough to retain the constituents in solution.
Tempering	Heat treatment by heating to a temperature below the lower critical temperature to decrease hardness and increase toughness of hardened steel, hardened cast iron, and sometimes normalized steel.
Tensile Strength (also called "ultimate strength")	In tensile testing, the ratio of maximum load to the original cross-sectional area.[13]
Toughness	The ability to resist fracture, measured as the energy absorbed by a material as it fractures.
Transformation Ranges	Those ranges of temperature for steels within which austenite forms during heating and transforms during cooling. The two ranges are distinct, sometimes overlapping, but never coinciding.
Upper Critical Temperature	In ferrous metals, the temperatures at which austenite is completely formed during heating or at which the transformation of austenite to ferrite begins during cooling.
Yield Strength	The stress at which a material exhibits a specified deviation from the proportionality of stress to strain. The deviation is expressed in terms of strain by either the offset method (usually at a strain of 0.2%) or the total-extension-under-load method (usually at a strain of 0.5%).[13]

4.4 Heat Treatment of Steels

Plain or mild carbon steels typically contain 0.04–0.25% carbon content and are normally found in one of four heat treatment conditions: annealed, normalized, spheroidized, or quenched and tempered. Each of these heat treatments results in different microstructures and different mechanical properties. If steel were not a common metal, its ability to transform to many different forms with different properties would be called miraculous! Heat treatment does this because the crystal structure of the main component of steel—iron—changes from body-centered cubic at room temperature to face-centered cubic at elevated temperatures. In pure iron, this crystal structure change occurs at a single, defined temperature (912°C/1674°F); however, in steel, it occurs over a range of lower temperatures, starting with the lower critical temperature (727°C/1340°F) and ending with the upper critical temperature, which varies with the carbon and alloy content. Between the two critical temperatures, both crystal structures coexist.

4.4.1 Annealing

Annealing involves heating steel to a temperature above its upper critical temperature of approximately 871°C (1600°F) and slow cooling in the furnace; cooling in this manner may take 8–10 h. The resulting microstructure is relatively coarse pearlite. It may be noted from Figure 4.5 that the iron carbide platelets in an annealed microstructure have significant thickness and are rather widely spaced. Annealed steel is not typically used in petroleum production wells and facilities.

4.4.2 Normalizing

A steel is normalized by heating it to a temperature above its upper critical temperature, then removing it from the furnace, and allowing it to cool in still, ambient air. Under such conditions, a piece of steel will likely cool to room temperature within 10–15 min, or longer depending on mass. In comparison to annealed carbon steel with its longer cooling time, this somewhat more rapid cooling will produce iron carbide platelets that are much thinner and are dispersed on a much-finer spacing. A normalized microstructure is usually described as ferrite containing "fine pearlite." Fine pearlite cannot be resolved microscopically at normal magnification (e.g., 100X) to its two components of ferrite and iron carbide.

Normalizing is used to refine the grain structure (provide finer grain size) and to reduce segregation of chemical elements in steel by allowing these elements to diffuse throughout the steel at the elevated temperature. Normalizing can harden certain low-alloy steels, and therefore, could be followed by tempering. Downhole tubing is normalized to unify the microstructure of the steel tubular to prevent ringworm corrosion in upset-end tubular goods (refer to Chapter 3, Section 3.6). Pressure vessel plate is often made of normalized steel, especially when the thickness exceeds 25 mm (1 in). Normalizing provides a uniform grain size throughout the thickness and improves toughness.

4.4.3 Spheroidizing

Spheroidizing of carbon steel involves soaking the steel at a relatively high temperature below the lower critical temperature [approximately 649°C (1200°F)] for 24 h or more. Under these conditions, the iron carbide platelet phase has an opportunity to coalesce into its most stable globular form. This is not a very common heat treatment and is not apt to be encountered in most major steel products used in petroleum production. However, it can be found in steels exposed to high temperature for many hours and is a good indicator of that steel being overheated. Examples include boiler tubes and flare equipment. It can only be detected by microscopic examination. Microscopic analysis is generally a destructive laboratory assessment, but it can be performed in the field nondestructively using a method called "surface replication."

4.4.4 Quenching and Tempering

Quenching and tempering is a two-step treatment in which the steel is austenitized by heating to a temperature above its upper critical temperature, rapidly cooled (quenched) by immersing in water or oil, and then reheated (tempered) and held at the elevated tempering temperature, which is below the lower critical temperature. The time needed for tempering depends on the thickness of the product. Quenching and tempering obtains the highest strength and toughness of carbon and low-alloy steel. The rapid cooling rate keeps carbon dissolved in solution and prevents iron carbide platelets from precipitating. The microstructure that results after quenching is called "martensite," which is hard and brittle, making the steel component useless in this state (Figure 4.3A). Not only is the steel brittle, it has a lot of locked-in residual stress that promotes cracking. The microstructure also contains retained austenite—austenite that did not transform to martensite or any lower temperature phase. Therefore, steel is tempered (reheated) to transform martensite to tempered martensite, and transform retained austenite to ferrite and pearlite (ferrite plus a lamellar structure of ferrite + iron carbide). Tempered martensite produces steel that is useful for components, unlike as-quenched martensite.

Tempered martensite has fine precipitates of iron, chromium, and molybdenum carbides within ferrite grains, and it has a unique appearance under the microscope (Figure 4.3B). Needles or "laths" (narrow plates of martensite that form in low- and medium-carbon steels) may be observed as a remnant of the acicular and/or lath, as-quenched martensite, if the tempering temperature is low (below ≈300°C/572°F). Certain tempering-temperature ranges are avoided by manufacturers because they result in embrittlement.

Quenched and tempered steel has the optimum combination of strength, ductility, and toughness of any steel microstructure. It is used extensively in downhole tubing and casing (e.g., API 5CT Grade L80[11]), machinery parts (e.g., shafts), fasteners, valve stems, and in some high strength pipeline steels built to API Spec 5L.[12] It is not common to use quenched and tempered steel in large, complex components such as pressure vessels. These components typically do not require high strength steel and they would be difficult to heat treat because of their size. Plates for tanks and pressure vessels, if heat treated, are typically either normalized, or normalized and tempered.

4.4.4.1 Carbide Distribution and Corrosion Resistance

Heat treatment has a rather striking effect on the distribution of iron carbide, Fe_3C. As might be expected, this variation in distribution affects the corrosion behavior of steel. While the effect is relatively small, it has been demonstrated that a platelet-type distribution of the carbide phase, such as characterizes normalizing, results in greater corrosion resistance than the globular or particulate carbide distribution, characteristic of the spheroidized or quenched and tempered steel. This somewhat superior corrosion resistance of carbon steels having platelet-type carbide distribution seems to be due to either

- the development of a nonreactive iron carbide platelet barrier when the corrosion reactions extend to a significant depth within the steel; or
- to the development of a more protective iron carbonate ($FeCO_3$) corrosion product that forms because of the higher carbon content of the iron carbide platelet.

An example of this is the reported improvement in CO_2 corrosion resistance of API 5CT Grade J55 (ferrite and pearlite) versus API L80 (quenched and tempered) in modest CO_2 partial pressure oil production environments.[14] Standard practice, however, is to inject a corrosion inhibitor regardless of the steel microstructure rather than rely on the improved corrosion resistance of a particular microstructure.

4.4.5 Localized Heat Treatments

A piece of steel in the annealed condition and a similar piece of steel in the quenched and tempered condition, exposed to a similar corrosive environment, exhibit a relatively small difference in corrosion rates in most production environments. Thus, from this standpoint, the effect of heat treatment is generally not of major consequence. There are circumstances, however, in which this difference becomes exaggerated and the net result is severe corrosion. Serious effects from heat treatment variations can be expected when marked differences of metallic structures produced by heat treatment exist within short distances of one another. Two results of this variation are ringworm and weld line corrosion. Examples of these types of corrosion can be found in Chapter 3, Section 3.6.5.

4.4.5.1 Ringworm Corrosion

The term "ringworm corrosion" is used to describe the severe attack of tubular goods along a narrow band of demarcation between two heat treatments. A common example of this is found in forged-upset end tubing that has not been full-body heat treated (normalized) after upset forging. In this procedure, the tube is raised to an elevated temperature on both ends for hot forging the upset end. Where these heated regions run into the body of the tube, marked variations in microstructure can develop. These bands of severe microstructural gradients, when exposed to corrosive environments, corrode at much more rapid rates than steel on either side of the bands (Chapter 3, Figure 3.33). The affected zones normally have more globular iron carbide distributions than the normalized ferrite + pearlite structures on either side. As noted in Chapter 3, Section 3.6.5, full length normalizing after upsetting will provide a uniform microstructure and eliminate ringworm corrosion.

4.4.5.2 Weld Line Corrosion

A similar condition to ringworm corrosion's microstructural gradient can be observed at welds when the proper welding procedure was not used. Both the longitudinal seam of welded pipe and field joint circumferential welds can exhibit accelerated corrosion attack in the microstructural gradient regions (Figure 3.34 for steel and Figures 3.35–3.38 for stainless steel). As discussed in Chapter 3, Section 3.6.6, corrosion may take place in the weld metal, the heat-affected zone (HAZ), or the base metal depending on the specifics of the situation. Developing welding specifications that address the chemical composition (e.g., limits on nickel concentration) of the welding consumable, and adherence to the proper welding specifications by welders qualified to those specifications will provide the uniform microstructure necessary to avoid weld line corrosion. Laboratory testing has shown that corrosion inhibition can reduce or eliminate weld line corrosion; however, the inhibitor and the correct concentration should be qualified by laboratory testing in the simulated field environment.

In summary, the effect of heat treatment on corrosion resistance is generally of a low order of magnitude, and normally it can be ignored when selecting fully heat-treated steels for corrosive environments. It is worth expending some effort to obtain uniform heat treatment of parts to be exposed in corrosive environments, thus avoiding ringworm and weld line-type corrosion attack. It is also worthwhile, when dealing with materials for service in sour systems, to ensure that the heat treatment results in the proper hardness as specified in NACE MR0175/ISO 15156.[2]

4.5 Alloys

Alloys are mixtures of two or more metals or elements. Thus, carbon steel is an alloy because it contains carbon and minor amounts of other materials in addition to the major constituent, iron. However, the word "alloy" is normally reserved for materials other than carbon steels.

4.5.1 Homogeneous and Heterogeneous Alloys

There are two general kinds of alloys—homogeneous and heterogeneous. Homogeneous alloys are solid solutions, that is, the elements are completely soluble in one another, and the material has only one phase. An example of a homogeneous solid-solution alloy is 18-8 (300 series austenitic 304 or 316) stainless steel, which consists of approximately 18% chromium, 8% nickel, a fraction of a percent carbon, and the balance iron. The iron, chromium, nickel, and carbon are dissolved completely (like milk in coffee, or sugar in water), and the alloy has a uniform composition. Other alloy examples are austenitic [Alloy 825 (UNS N08825)] and super austenitic [2550 (UNS N06975)]. Additional examples of homogeneous alloys are 90Cu-10Ni pipe and titanium, both of which are used offshore to transport seawater cooling medium or firewater. Microscopically, solid solution alloys appear as polycrystalline metals where the grains are all similar in appearance.

Heterogeneous alloys are mixtures of two or more separate phases. Some of the elements of such alloys are preferentially soluble in one phase, and others are preferentially soluble in other phases. The composition and structure of these alloys are not uniform. Perhaps the most familiar examples of heterogeneous alloys are the carbon steels. Examination of the equilibrium phase diagram for iron–carbon alloys indicates that alloys with even a fraction of a percent carbon will be heterogeneous at ambient temperatures and consist of a mixture of ferrite (a solid solution of carbon in BCC iron) and cementite (iron carbide). In annealed steels, the carbide appears in lamellar form called pearlite (Figure 4.4 shows dual pearlite–ferrite microstructure steel). Another example of a heterogeneous alloy is duplex stainless steel, which consists of roughly equal amounts of two phases: (1) ferrite and (2) austenite (Figure 4.6).

Therefore, in a heterogeneous alloy such as plain carbon steel, there are two distinct phases; thus, a galvanic couple is built right into the alloy. Fortunately, ferrite and iron carbide have electrode potential values sufficiently close together that the corrosion rate of plain carbon steel is ordinarily lower than that for a steel-brass galvanic couple, for example.

4.5.2 Ferrous and Nonferrous Alloys

Another way to classify alloys is to divide them into two major categories because of commercial importance—ferrous and nonferrous alloys. The major element of ferrous alloys is iron, with significant quantities of other elements such as manganese, nickel, chromium, and molybdenum. "Nonferrous" alloys may still contain iron, but it is no longer the major constituent. Examples of nonferrous systems are titanium, aluminum alloys, brasses (copper alloys), and super austenitic nickel or cobalt alloys.

4.5.3 Corrosion Resistant Alloys (CRAs)

The stainless steels and nickel alloys fall into a category of materials referred to as: "corrosion-resistant alloys" in oil field terminology, and "high-performance alloys" elsewhere. The CRAs have a wide range of corrosion resistance, depending on chemical composition, manufacturing method, and heat treatment. The CRAs gain their strength by either heat treatment, cold working, or a combination of cold working and heat treatment.

The corrosion control effectiveness of the CRAs depends on the severity of the environment. The H_2S partial pressure, temperature, chloride content, and the presence or absence of free (i.e., elemental) sulfur are critical factors in the final decision. These factors may affect both the SCC resistance and the crevice or pitting potential. Careful consideration of the anticipated environmental conditions should be made before the final decision is made. The cost factor compared to carbon steel may vary from 1.5–20 depending on the alloy chosen. NACE MR0175/ISO 15156[2] describes the sour service envelopes for many corrosion resistant alloys in production environments.

4.5.4 Heat-Treated Alloys

Many homogeneous alloys get their strength from solid solution hardening; that is, the mismatch of atomic sizes between the matrix element and the minor alloying element strains the lattice and increases resistance to yielding under load. These alloys can only gain further strength by cold working (mechanically working the alloy below its recrystallization temperature[4]). Microscopic examination of cold worked alloys shows the single-phase grains are elongated in the direction of mechanical working that is performed at relatively modest temperatures at which built-in strain hardening is not relieved. Alloys that can only be strengthened by cold-working cannot be used for components that are too thick or complex-shaped. These components must be made from alloys that can gain strength from heat treatment.

In most alloys, heat treatment is a two-step process, the first step being to get all alloy constituents into solid solution by heating and holding the alloy at high temperature. After cooling, the second step is called aging, during which the supersaturated solid solution decomposes with time or temperature as the alloying elements form small precipitate clusters. The formation of these clusters acts to significantly strengthen the material if the precipitates remain small and dispersed. In oil field alloy systems, these precipitates require heat treatment to form and harden the material, in a process referred to as "artificial aging." The precipitate is a fine dispersion of a second-phase particle. Examples are precipitation hardening (PH) stainless steels, e.g., 17-4 PH and ASTM A453 Grade 660,[15] and nickel alloys, e.g., alloys 718, 925, and 725. In 17-4 PH, the matrix is tempered martensite and the fine precipitate is rich in copper and niobium. For sour service, 17-4 PH is double-aged. In nickel-based alloys, artificial aging precipitates a nickel-aluminum-titanium particle. Heat-treatable alloys include 17-4 PH (UNS S17400) and A 286 (ASTM A453 Gr 660, UNS S66286) stainless steels, and nickel based alloys 718 (UNS N07718), 725 (UNS N07725), and 925 (UNS N09925). These alloys are often used for valve components, wellhead equipment, tubing hangers, bolting, and other items requiring both corrosion resistance and strength, and are discussed further in this chapter.

---

4. Regarding mechanically working the alloy below its recrystallization temperature, think of a jeweler hammering gold or silver. After enough hammering, the jeweler must heat the metal above its recrystallization temperature to eliminate the cold work and allow further hammering to continue without cracking the jewelry.

4.6 Ferrous Alloys

Ferrous alloys contain iron in a concentration exceeding 50%.

4.6.1 Carbon and Alloy Steels

By far the largest single classification of metals used in oil field equipment are carbon steels and alloy steels. They offer a wide variety of properties, are easily formed and fabricated, and in general, are the least expensive metals commonly used. As defined in Table 4.2, carbon steels are those that contain up to about 2% carbon and only residual quantities of other elements (except those added for deoxidization as part of the melting process). Thus, carbon steels may contain as much as 1.65% manganese and 0.60% silicon without being considered alloy steels. Alloy steels, on the other hand, are those containing significant quantities of alloying elements (other than carbon, manganese, and silicon added for deoxidization, and sulfur and phosphorus present as residual elements). In common practice, steel is considered to be an alloy steel when the maximum content of the alloying elements exceeds one or more of the following limits: manganese 1.65%, silicon 0.6%, copper 0.6%—or when a definite range of any of the following elements is required: aluminum, boron, chromium, cobalt, niobium (columbium), molybdenum, nickel, titanium, tungsten, vanadium, zirconium, or other alloying elements added to obtain a desired effect. Frequently, steels will be referred to as low-alloy or high-alloy steels. These designations are somewhat arbitrary, but low-alloy steels are generally regarded as those having 5% or less metallic alloying elements.

Above 5%, steels are considered high-alloy steels, or are given other specific designations such as "chrome-moly" or "stainless" steels. The mechanical properties of carbon and alloy steels cover the full range of properties used for oil field equipment. The properties depend not only on composition, but also on other factors such as forming, fabrication, and particularly heat treatment. Alloying elements influence the corrosion resistance of steels as well as the response to heat treatment, formability, machinability, weldability, and other engineering properties. In addition, alloying increases the cost of the steel.

4.6.2 Chromium-Containing Steels

Chromium-containing low-alloy steels are used for several purposes in petroleum production facilities. High strength quenched and tempered steel grades used for heavy wall couplings, tubing, and casing (e.g., API 5CT Grade T-95[11]) intended for sour service contain 0.4–1.5% Cr and less than 1% Ni. Other grades that are used in fasteners, valve components, wellheads, and machinery include Grade 4130 (UNS G41300) and Grade 8630 (UNS G86300). Grade 4130 contains about 1% Cr; Grade 8630 contains about 0.5% Cr, 0.45% Ni, and 0.2% Mo. The "hardenability" of steel, that is, the size or depth of the steel that can be hardened uniformly, is increased with an increase in Cr, Ni, and Mo concentration. This allows wellheads, which are a substantial mass of casting or forging, to be heat treated to uniform strength throughout their wall thickness. Steels of low hardenability, such as ordinary carbon-manganese steels, can only be uniformly hardened in small diameter or size parts. Thick components made from steel with an inadequate hardenability will at best have the requisite strength only in a thin layer at the outer surface, leaving the bulk of the mass lower strength.

Carbon steel has not traditionally performed well in produced water or treated seawater injection service. Despite the low concentration of dissolved CO_2 and H_2S, downhole injection tubing often fails by localized corrosion caused by inadvertent oxygen contamination of the produced water. Sources of oxygen ingress are described in Chapter 3. Industry has found that adding about 1% Cr minimizes pitting and spreads the damage to more uniform corrosion.

4.6.2.1 Chromium-Molybdenum Steels

Chromium-molybdenum steels are used at high-temperature oxidizing environments because of their improved oxidation resistance and high temperature mechanical properties compared to carbon steel. Grades of progressively increasing resistance to oxidation at temperatures above 500°C (932°F) are

- 1.25% Cr, 0.5% Mo
- 2.25% Cr, 1% Mo
- 5% Cr, 0.5% Mo (note that ASTM refers to this as "4 to 6% Cr" and omits the Mo, although the spec requires 0.44–0.65% Mo)
- 6–8% Cr
- 9% Cr, 1% Mo

These grades are used as tubes and pipes for heat recovery steam generators and superheater tubes in boilers because of their better resistance to creep and high-temperature oxidation. They are also used in high-temperature (nonaqueous) H_2S services because of their better sulfidation resistance in high temperature refinery services. They are not typically required in aqueous petroleum production services. However, "1.25% Cr, 0.5% Mo" and "2.25% Cr, 1% Mo" superheater tubes may be found in petroleum production service, mainly in high-pressure steam boilers onshore and offshore. Note that "9% Cr, 1% Mo" steel has improved resistance to CO_2 corrosion than ordinary steel and has been installed as a lower cost alternative to 13Cr martensitic stainless steel downhole tubing and accessories.

4.6.3 Nickel-Containing Steels

Ordinary carbon and low-alloy steels lose toughness at cold temperatures. Beginning at about −29°C (−20°F), steel may have insufficient toughness to prevent brittle fracture. However, carbon steel can exhibit brittle behavior at warmer temperatures when wall thickness exceeds above 25 mm (1 in) and service temperature is 0°C (32°F) or colder. Brittle behavior is especially a concern for steels that were not originally manufactured to meet toughness requirements and were then subjected to tensile stress in cold service. The toughness transition temperature depends on heat treatment, steel composition, melting practice, and component thickness. Quenched and tempered microstructures provide the toughest steels when all other variables are held constant. Melting practice can improve toughness. Adding a small quantity of aluminum, niobium, or vanadium to the melt makes the grain size of the cooled product finer, which improves toughness.

Many users specify that steels to be installed in cold temperature service contain 3.5% or 9% nickel. Nickel imparts cold temperature toughness and allows these steels to be used in liquefied petroleum gas (LPG) and liquefied natural gas (LNG) service. The 3.5% nickel steels are tough to −101°C (−150°F); 9% nickel steels are applicable to −196°C (−320°F). Cold temperature steels are used for piping, tanks, pressure vessels, pumps, and valves. Specifications for these steels require confirmation of their impact toughness through testing. These steels are not normally used in corrosive services. In fact, resistance to sulfide stress cracking (SSC) is deteriorated by adding 1% or more of nickel. Chapter 3, Section 3.13 discusses SSC in detail.

4.6.4 Stainless Steels

The stainless steels are not one set of alloys, but they are a group of alloy systems with widely differing microstructure and mechanical properties. The one thing stainless steels have in common is corrosion resistance due to chromium. To be classed as a stainless steel, an alloy usually contains at least 10.5% chromium. With the various chemistries, microstructures, and mechanical properties of the stainless steel systems come variations in corrosion resistance. All stainless steels are not suitable for all services in all environments. It is extremely important that the specific stainless steel be specified for a specific service. Many types of stainless that are excellent in other industries should not be used in oil and gas production applications without testing.

The corrosion resistance of stainless steels is associated with the ability to build a passive layer on the surface of the alloy. Chromium is a reactive element, but it and its alloys passivate in oxidizing environments and have excellent corrosion resistance. The passivation process is believed to involve the formation of an electrically insulating monolayer chromium-oxide film on the surface of the alloy.

The stainless steels may be divided into distinct classes based on their chemical content, metallurgical structure, and mechanical properties: martensitic, ferritic, austenitic, precipitation hardening, and duplex.

4.6.4.1 Martensitic Stainless Steels

The martensitic stainless steels contain chromium as their principal alloying element. The most common types contain approximately 12% chromium; although for the different grades, the chromium may be as high as 18%. The carbon content ranges from 0.03–1.10%, and other elements such as nickel, columbium, molybdenum, silicon, selenium, and sulfur are added in small quantities for special purposes in certain grades. Oil field applications should avoid using free-machining grades of martensitic stainless steel that have intentional additions (minimum required concentration) of sulfur, phosphorous, or selenium because these grades—such as 403, 416, 416Se—are prone to pitting and cracking.

The most important characteristic that distinguishes martensitic stainless steels from other grades is their response to heat treatment. The martensitic stainless steels can be hardened by the same heat treatments used to harden carbon and low-alloy steels. As their name implies, the microstructure of these stainless steels consists of tempered martensite and resembles that of quenched and tempered steel (Figure 4.3B). All of the martensitic stainless steels are strongly magnetic under all conditions of heat treatment. If a user finds that a grade of martensitic stainless steel is not magnetic, then the supplier has provided a different alloy, likely austenitic stainless steel.

The martensitic stainless steels comprise part of the 400 series of stainless steels. The most common of the martensitic stainless steels are 410 (UNS S41000) and 420 (UNS S42000). The Cr content of 410 is 11.5–13.5; whereas for 420, Cr lies between 12.0–14.0%. Because both 410 and 420 nominally contain around 13% Cr, these grades are referred to as "13Cr" in upstream applications, even though 410 stainless steel does not meet this composition as an average. Type 410 fulfills the requirements of the standard for a material for tubes and wellheads (cast and forged), and type 420 meets the requirements for a forging material. Type 420 has a higher carbon content and is used for downhole tubing and couplings because it can meet the strength and hardness limitations in MR0175, while offering improved resistance to SSC.[16]

Service Envelopes

Martensitic stainless steels have had the widest range of use of any of the available CRAs, and account for about two-thirds of CRA usage in petroleum production, mainly in sweet crude and gas services. Through heat treatment, martensitic stainless steel may be manufactured into tubular products with acceptable yield strengths for downhole tubing. Many millions of feet of tubing (API 5CT grade L-80 type 13Cr[11], maximum hardness of HRC 23 [Hardness Rockwell C scale]) are in corrosive well service. L-80 type 13Cr is considered the material of choice for many of the deep, sweet-gas wells, where the service temperatures are <149°C (300°F) and the produced water pH ≥3. NACE MR0175/ISO 15156 Part 3 provides service envelopes, hardness, and heat treatment restrictions for several grades of martensitic stainless steel and for different components such as tubulars, wellheads, and compressor components.[17] API 5CT grade L-80 type 13Cr tubing has been used in sour environments up to about 10 kPa (1.5 psia) H_2S partial pressure. Strength level plays a key role in the material's SSC susceptibility. Therefore, the hardness limitations stated in MR0175 should be followed. Many operators impose a strength and hardness limitation even in sweet service.

The passivity of 13Cr can be destroyed by chlorides, and some operators further restrict martensitic stainless steel's operating envelope to a maximum chloride concentration, as compared to "any combination of chloride and temperature occurring in the production environment" that is permissible in MR0175.[17] For these operators, their allowable chloride concentration depends on water pH, H_2S partial pressure in the gas phase and maximum temperature. An example of one operator's more restrictive guidelines for casing and tubing is

- H_2S = 0, Cl^- ≤160,000 ppm, T ≤149°C (300°F), pH ≥3.5
- H_2S ≤0.7 kPa (0.1 psia), Cl^- ≤60,000 ppm, T ≤149°C (300°F), pH ≥3.5
- H_2S ≤10 kPa (1.5 psia), Cl^- ≤160,000 ppm, T ≤121°C (250°F), pH ≥5.0

In addition to downhole tubing and casing, Type 410 and similar martensitic cast products (designated ASTM A487 Grade CA15[18]) and forged products (ASTM A182 F6A and F6NM[19]) have also been used extensively for valves and wellhead equipment, including valve bodies, tubing hangers, and chokes. Wellheads are specified to API 6A,[20] forgings to ASTM A182 F6A,[19] and castings to ASTM A487 CA15.[18] Wellheads are manufactured to lower strength than downhole casing and tubing; therefore, some operators impose fewer restrictions on H_2S partial pressure, temperature, chloride concentration, and pH for wellhead equipment in sour service. However, crevice corrosion in the gasket and gate valve seat pockets of wellhead equipment made from martensitic stainless steel has been observed. Therefore, when conditions exceed the recommended limits given in MR0175 for martensitic stainless steel, many operators specify internally CRA-clad wellhead equipment.

Cold Temperature Service

Cold temperature service is defined as exposure to temperature ≤ −29 °C (−20 °F). Examples include wellheads in cold climates, and running tubing and casing in cold weather. Martensitic stainless steel may not possess sufficient toughness for some cold temperature services. ASTM A182 Grade F6NM[19] (forged) or ASTM A487 Grade CA6NM[18] (cast) contains ≈4% nickel. This improves toughness, but at the expense of resistance to hydrogen stress cracking and SSC mechanisms. The threshold temperature for SSC varies with chloride and H_2S concentration.[21] SSC is reported to be caused by a hydrogen embrittlement mechanism, and often initiates at the root of corrosion pits.

Effect of Oxygen

Martensitic stainless steels are susceptible to oxygen corrosion in aerated water and have been known to pit externally while the tubular product is resting on supports in outdoor storage in a humid, coastal, or marine environment. Care is required during shut-ins and in well workovers to prevent oxygen ingress. Rod-pumped wells can aspirate oxygen into the annulus. Oxygen pitting is one reason why martensitic stainless steels are not usually specified for water injection well tubing. Oxygen can also enter with hydrotest water and cause pitting (Chapter 3, Figure 3.10). During well work, acidizing inhibitors must be selected to be compatible with martensitic stainless steels, because inhibitors that are used to protect carbon steel may not be effective. Some heavy weight packer brines are not compatible with martensitic stainless steel. Martensitic stainless steels are attacked by wet elemental sulfur, resulting in pitting and SSC.

Weldability

In general, 13Cr is not easily weldable because it forms hard, untempered martensite during welding fabrication; welds can crack either immediately, or in service because of their low toughness and susceptibility to hydrogen-related failure mechanisms. For these reasons, in petroleum production, it is often used in components that do not require welding. In operations where 13Cr downhole tubulars are used, such as in CO_2 service, operators were forced to use carbon steel with corrosion inhibition or to upgrade flowlines, gathering lines, etc., to weldable grades of CRAs such as duplex stainless steel. These weldable grades are more highly alloyed and more costly than what is required for the service conditions. A group of alloys referred to as "modified 13Cr" or "super 13Cr" (or "S13Cr") was developed to address this gap. These alloys contain 3.5–7% nickel. Some alloys also contain some molybdenum. NACE MR0175/ISO 15156 Part 3, Table D.6 provides the chemical analysis of several grades of super martensitic alloys, and Tables A.18–A.23 inclusive, describe environmental service envelopes.[17] Like 13Cr, super 13Cr alloys are martensitic. One typical grade of super martensitic stainless steel is UNS 41427. Applications for super 13Cr include downhole tubing, flowlines, gathering lines, and pipelines.

4.6.4.2 Ferritic Stainless Steels

The second class of stainless steels, the ferritic stainless steels, are like the martensitic stainless steels in that they have only chromium as the principal alloying element. The chromium contents of the ferritic stainless steels are, however, normally higher than those of the martensitic stainless steels, and the carbon contents are generally lower. The chromium content of the ferritic stainless steels ranges from about 13% to about 27%. They are not hardenable by heat treatment and are used principally for their good corrosion resistance and high-temperature properties. Ferritic stainless steels have limited use, however, in oil and gas service.

The ferritic stainless steels are also part of the "400" series—the principal types being types 405 (UNS S40500), 430 (UNS S43000), and 436 (UNS S43600). The microstructure of the ferritic stainless steels consists of ferrite, and they are also strongly magnetic. Ferrite is simply body-centered cubic iron, or an alloy based on this structure (Figure 4.1).

4.6.4.3 Austenitic Stainless Steels

The austenitic stainless steels have two principal alloying elements—chromium and nickel—and their microstructure consists essentially of austenite. Austenite is face-centered cubic iron, or an iron alloy based on this structure (Figure 4.1). They contain a minimum of 16–18% chromium (depending on grade) and 8% nickel, with other elements added for special purposes, and range up to as high as 25% chromium and 20% nickel. The austenitic stainless steels have the highest general corrosion resistance of any of the stainless steels, but their strength is lower than that of the martensitic and ferritic stainless steels. They are not hardenable by heat treatment (although they are hardenable by cold work) and are generally nonmagnetic unless cold-worked. The austenitic stainless steels constitute the "300" series—the most common ones used in oil field equipment are 304 (UNS S30400) and 316 (UNS S31600, high Cr and Ni, and also includes Mo). Other austenitic stainless steels used may include 347 (UNS S34700) and 321 (UNS S32100) (stabilized for welding and corrosion resistance, but these do not include molybdenum, which is required for improved resistance to pitting and crevice corrosion).

> **WARNING:** AVOID FREE-MACHINING STAINLESS STEELS. Never specify type 303 (UNS S30300) (free machining) or Type 303Se (UNS S30323) in any production environment. Often, these materials are supplied by manufacturers for items such as small diameter valve bodies, compression fittings, valve components, etc.—*often without the explicit knowledge or consent of the buyer.* "Free machining" means the metal contains a high sulfur or selenium concentration to aid in machinability. This results in pitting and cracking even under ambient atmospheric conditions. The author analyzed an externally cracked machined valve body made from free machining stainless steel that was exposed to a normally dry west Texas atmosphere! Free machining grades of martensitic and ferritic stainless steel are also available and should not be used; these grades include 403 (UNS S40300), 416 UNS S41600), and 416Se (UNS 41623).

The austenitic stainless steels are used for items not requiring high strength. The highly alloyed austenitic stainless steels can be made to higher strengths and are more corrosion resistant. Types 304 and 316 (and their low-carbon grades 304L and 316L, respectively) are used throughout petroleum production facilities in instrumentation tubing, valves, pipe, pressure vessels, heat exchanger tubes, subsea control lines, compression fittings, filters and screens, seal rings and gaskets in ring joint flanges, and many other applications. Used as an internal cladding on carbon steel, or by itself, austenitic stainless steel is used in amine sweetening units in the corrosive sections of the absorber (contactor) and regenerator (stripper) towers, piping, reboiler tubes, and heat exchangers.

Because of its exceptional toughness at low temperature, austenitic stainless steel is used in producing liquefied natural gas (-160°C/-260°F) and in helium production (-196°C/-320°F) at a temperature where most iron-based metals are brittle.

Chemical injection equipment is made of austenitic stainless steel. Downhole applications include gravel pack screens, which are made from cold worked wire resistance welded to a cage. Both cold working and welding reduce the material's SCC resistance, in which case higher grades of austenitic alloys (such as alloy 825) may be required.

The main differentiation between type 316 and 304 is that 316 contains molybdenum, which makes it more resistant to pitting corrosion, and as a result, type 316 (316L and cast CF-8M) is the workhorse austenitic stainless steel in petroleum production. The "L" or low-carbon grade is used when the material will be welded to avoid sensitization during welding (refer to Chapter 3, Section 3.9). Most often, the material is dual stamped, e.g., 316/316L, by manufacturers, and it meets either specification.

Corrosion in Marine Atmospheres

The 304 and 316 stainless steels are marginal performers in marine atmospheres, with 316 being better because of its Mo content. Despite this performance problem, operators continue to specify 316 stainless steel for applications such as instrumentation tubing. Argon Oxygen Decarburization has become the standard melting practice for making austenitic stainless steel. Because of higher accuracy in producing alloy concentration, the Mo content of 316 produced today tends to fall to the minimum 2.0% Mo required concentration, whereas formerly it would have been produced to the midpoint 2.5%. This has resulted in decreased service life of type 316 when it is installed in an environment where it is widely used but marginally resistant, namely the marine environment. Some operators specify 316 with 2.5% minimum Mo content or 317 stainless steel, which has 3% Mo minimum, to ensure obtaining an adequate Mo content; however, 317 is not as widely available.

A smooth, polished surface finish can improve the longevity of austenitic stainless steel in a marine environment because it reduces the amount of atmospheric salt deposited on the metal. When the surface finish becomes finer, the passive layer is thin, compact, and enriched with chromium oxide (Cr_2O_3), and molybdenum oxide (MoO_2). Some operators specify a bright finish (numbers 2B, BA, or 4) for instrumentation tubing. The problem is not in obtaining a bright finish, the problem is in keeping it bright. During heavy construction, welding and grinding work occurring above the stainless steel installation drops abrasive grit, steel particles, weld splatter, and dirt onto the bright finish and ruins it. This is called "iron contamination." The adhered and deposited iron particles form crevices on the stainless steel surface. When exposed to a corrosive environment, the iron particles will quickly be oxidized and accumulate concentrated corrosive species around it, creating a severe localized corrosive environment, breaking down surface passive film, and resulting in localized corrosion. Some operators apply a protective coating to extend the life of austenitic stainless steel in a marine atmosphere.

Heat-Tinted Welds

A common corrosion problem that occurs with 304 and 316 is oxidation of the weld root (so-called "heat tint") resulting in decreased pitting corrosion resistance to atmospheric, as well as the internal, fluid environment because the heat-tinted layer is depleted of chromium concentration. Piping welded in air may show a blue or straw-yellow heat tint from inadequate gas purge during welding and appears rusty after exposure to the environment. Excessive grinding can also produce this heat tint by overheating. If installed offshore or in a coastal facility, heat-tinted stainless steel will prematurely corrode. Proper welding guidelines are available.[22] Once heat tint has occurred, it is difficult to restore the passive film of stainless steel. Grinding or grit blasting the heat tint layer does not adequately restore the passive film. ASTM A380 describes several acid compositions that can remove heat tint and embedded iron.[23] These solutions are not readily applicable to a field environment because they pose safety and health issues and require proper disposal. Some operators resort to abrasive blasting and coating stainless steel equipment in marine environments to avoid problems with heat tinting.

SCC

Stress corrosion cracking is an important failure mechanism to avoid when using austenitic stainless steels. The most common constituent in the environments that cause SCC is the chloride ion, so much so that a "C" is often added before "SCC" to indicate chloride stress corrosion cracking (i.e., CSCC). Chapter 3, Section 3.17.3 provides information on the CSCC limits of austenitic stainless steel.

Crevice Corrosion

In addition to CSCC, the austenitic stainless steels are susceptible to crevice corrosion in seawater or a marine environment, particularly under clamps and in flange faces. Flush inside diameter piping at flanges help minimize crevices. Deposits also promote crevice corrosion. Clamps should be avoided, but if they must be used, they should be made of a material that is anodic to stainless steel, such as aluminum. Even electrically nonconductive materials (tapes, etc.) can result in crevice corrosion.

Other Corrosion Failure Mechanisms

- Galvanic Corrosion: Graphite and carbon-based gaskets in contact with stainless steel should be avoided because these are highly cathodic materials relative to austenitic stainless steel in seawater and other water services, and marine and coastal atmospheres.
- Microbiologically Influenced Corrosion (MIC): Stagnant water is particularly damaging to austenitic stainless steel because it promotes deposit formation and bacteria growth resulting in under-deposit corrosion and pitting from MIC.

Refer to Chapter 3 for more information on these corrosion mechanisms.

4.6.4.4 Precipitation Hardening (PH) Stainless Steels

Another type of stainless steel is a general class known as "precipitation hardening stainless steels," which contain varying amounts of chromium and nickel. They combine the high strength of the martensitic stainless steels with the good corrosion resistance of the austenitic stainless steels. Most of these stainless steels were developed as proprietary alloys, but now there is a wide variety of generic standards and commercially available compositions.

Martensitic PH Stainless Steels

A common name that may show up in field equipment is "17-4 PH stainless steel" (UNS S17400)—an approximately 17% chromium, 4% nickel alloy. Alloy 15-5 PH stainless steel (UNS 15500) is also found in production applications. Both are martensitic precipitation hardened alloys, and they have a tempered martensitic microstructure. The distinguishing characteristic of the precipitation hardening stainless steels is that by certain specified heat treatments, at relatively low temperatures, the steels can be hardened to various strength levels. Most can be formed and machined before final heat treatment, with the finished product being hardened. The most common application of these stainless steels is in corrosion-resistant and wear-resistant parts for equipment, such as valve stems, fasteners, compressor impellers, reservoir sampling bottles, downhole testing tools, subsurface safety valves, and gas lift gas valves. Most 17-4 PH fasteners manufactured for U.S. consumers are manufactured to ASTM F593 Grade 630.[24] Service envelopes and heat treatment requirements for martensitic PH stainless steels are included in MR0175 Part 3, Tables A.27–A.30 inclusive.[17] The document limits 17-4 PH to a maximum hardness HRC 33, an H_2S partial pressure of 3.4 kPa (0.5 psia) and a pH ≥4.5. Heat treatment is specified to be solution annealing and double precipitation hardened (artificially aged).

Notable failures of 17-4 PH that was heat treated in accordance to NACE MR0175/ISO 15156 have occurred in high H_2S gas fields. From the literature, successful equipment and applications in H_2S-containing producing environments have been either short duration exposure (e.g., sampling bottles) or low applied stress levels (≤50% of specified minimum yield strength).[25] However, examples exist of 17-4 PH stainless steel valve stems failing even if stressed below 50% SMYS in a high H_2S environment.[26] Bolts made of 17-4 PH have pitted and cracked in a marine environment in offshore Malaysia. Pitting was attributed to galvanic corrosion in contact with an antiseize thread lubricant that contained metallic copper. If properly heat treated (e.g., condition H1000 or higher), components made of 17-4 PH are reported to have adequate resistance to CSCC in a marine atmosphere.[27-28] Centrifugal compressor components made of 17-4 PH have been successful in sour gas service, most likely due to the low chloride concentration and limited exposure to a liquid water phase. Users of 17-4 PH stainless steel must consider means to prevent CSCC failure by controlling heat treatment, applied stress, cathodic protection (CP) potential (subsea immersion), or upgrade to a more CSSC-resistant alloy such as Alloy 718 (UNS N07718).

The SSC resistance of 15-5 PH stainless steel is reported to be similar, but somewhat inferior, to 17-4 PH.[29] NACE MR0175/ISO 15156 Part 3 contains service envelopes for this alloy.[17]

Austenitic PH Stainless Steels

Austenitic PH stainless steels have an austenite structure that contains alloying elements, which allow hardening by precipitation hardening and aging heat treatments. An austenitic PH stainless steel alloy used in bolts and fasteners is ASTM A453 Grade 660 (also known as Alloy A286, UNS S66286).[15] It is an iron-based PH alloy that contains about 25% Ni and 15% Cr. Its coefficient of expansion is like austenitic stainless steel, but its strength is significantly higher. As mentioned in Chapter 3, Section 3.17.3, this alloy severely cracked (CSCC) and pitted in a Gulf of Mexico marine environment on wellhead flange bolts, as well as on a wellhead located in offshore Nigeria (82°C/180°F flowing wellhead temperature). In H_2S environments, this alloy is restricted to 100 kPa (15 psia) H_2S partial pressure up to 65°C (150°F) if the maximum hardness is 35 HRC and the alloy is heat treated to solution annealed and aged, or solution annealed and double aged. These conditions are addressed in

NACE MR0175/ISO 15156 Part 3, Table A.26.[17] Where ASTM A453 Grade 660 has failed by CSCC, it has been replaced by high-strength PH alloys such as Alloy 718 (UNS N07718).

High Strength, SSC-Resistant PH Alloys

In services with more severe H_2S partial pressure, users specify more SSC-resistant precipitation hardened nickel-based alloys instead of 17-4 PH or Grade 660 for the applications stated above, especially in environments where highly alloyed CRAs are used for downhole tubulars. These alloys include precipitation hardened Alloy 625 Plus (UNS N07716), Alloy 718 (UNS N07718), Alloy 725 (UNS N07725), and Alloy 925 (UNS N09925). These are discussed in more detail in this chapter in Section 4.7.2 "Precipitation-Hardened Nickel-Based Alloys." Despite their improved corrosion and cracking resistance, these precipitation hardened nickel alloys can become embrittled by hydrogen charging under cathodic protection (CP) under subsea conditions.[30] Because it is the cathode in the circuit, hydrogen is generated at the alloy surface where it diffuses into the material and can result in failure, known as "hydrogen induced stress cracking" (HISC). Prevention requires limiting the strength of the alloy (same as following SSC prevention guidelines) and controlling CP potential. Controlling the applied stress is also desirable–keeping in mind that these alloys are used in highly stressed components like such as fasteners.

4.6.4.5 Duplex Stainless Steels

The class of alloys known as "duplex stainless steel" has a microstructure that is a mixture of austenite and ferrite (Figure 4.6). The volume fraction of the ferrite phase should be 35–65% per NACE MR0175/ISO 15156 Part 3.[17] Excessive ferrite content results in hydrogen cracking.[31] This alloy class was developed to provide a material with the corrosion resistance of the austenitic stainless steels and with the higher strengths of the ferritic stainless steels. Although more expensive than 13Cr or many other corrosion-resistant alloy materials used in oil and gas production, duplex stainless has found a place in severely corrosive service. Duplex is being used in downhole tubing, flowlines, facilities piping, and pressure vessels. Most duplexes have compositions in the range of 20–29% Cr and 1.5–7% Ni. A commonly used grade—22Cr—contains 22% Cr and 5% Ni (UNS S31803). Alloys referred to as "super duplex stainless" have a composition of approximately 25% Cr and 7% Ni—25Cr. Super duplex stainless 25Cr has improved resistance compared with 22Cr duplex stainless to CSCC in high-temperature marine environments such as might be experienced in risers, and a more generous operating envelope in sour service. This improvement is due to the higher pitting resistance equivalent number (PREN) discussed later in this section. The alloys contain nitrogen for strength and improved resistance to localized corrosion.

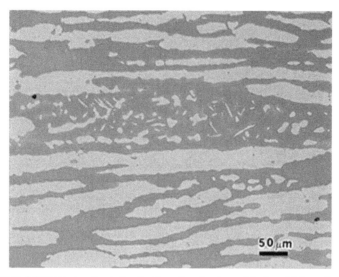

Figure 4.6 Photomicrograph showing the microstructure of duplex stainless steel. The lighter areas are austenite, and the darker areas are ferrite.

The duplex stainless steels and many of the CRAs gain higher strength through cold working. The yield strengths can be as high as 760–860 MPa (110,000–125,000 psi), which makes these products useful for deep-well tubular goods. Higher strength grades above 760 MPa are more susceptible to environmental cracking like SSC and CSCC. Cold-worked duplex stainless steel has been used as downhole tubing in sweet corrosive wells. Duplex tubing has also been used in wells containing as much as 10 kPa (1.5 psi) H_2S partial pressure. The maximum service temperature in NACE MR0175/ISO 15156 Part 3 depends on the H_2S partial pressure.[17] The temperature and H_2S partial pressure limitations do not necessarily mean that the alloy will fail above these conditions, only that the alloys successfully passed the test requirements at these conditions.

Annealed duplex (450 MPa/65 ksi specified minimum yield strength [SMYS] for 22Cr; 550 MPa/80 ksi SMYS for 25Cr) has been used as welded and seamless line pipe and in subsea and surface facilities. Forty miles (64 km) of welded annealed duplex line pipe have been effective in wet CO_2 service for several years on the North Slope of Alaska. Annealed duplex stainless steels are more resistant to SSC than their cold-worked versions.

The class of duplex stainless steels requires care in welding to ensure obtaining the proper amount of ferrite. Because welding removes cold working, welded components are provided in the solution-annealed condition. High ferrite concentration has been shown to promote hydrogen cracking from fabrication and CSCC in an offshore pressure vessel under insulation.[32] Welding must be done properly to avoid forming brittle intermetallic Fe-Cr-Mo phases such as sigma and chi, which cause low toughness and brittle fracture. These brittle phases form in the temperature range of 700–950 °C (1290–1740 °F) during improper fabrication (e.g., hot working, welding) or solution-annealing heat treatment. Some operators specify Charpy impact testing of duplex stainless steel at −46 °C (−50 °F) as qualification even if the service temperature is warmer, to ensure that brittle phases are absent in the base metal and weldment. NACE MR0175/ISO 15156 Part 3, Section A.7.3 requires that a microstructural analysis be performed to confirm there is the correct ferrite content, and that there are no continuous precipitates along grain boundaries.[17] It also imposes a maximum limit to the volume percent of undesirable intermetallic phases and sigma phase.

Duplex alloys have experienced failure because of hydrogen charging in subsea components exposed to CP (referred to as "hydrogen induced stress cracking"). Duplex stainless components are exposed to hydrogen charging if they are electrically connected to steel equipment under CP, or if CP is applied directly even in the absence of carbon steel.[32-33] These failures were not all due to excess ferrite content. DNV Recommended Practice DNV-RP-F112 provides guidelines to design duplex stainless steel subsea equipment that will be exposed to CP.[33]

A Word About PREN

PREN stands for "Pitting Resistance Equivalent Number." It is a way to rank Fe-Ni-Cr-Mo alloys' pitting resistance using their actual, not nominal, chemical composition. Alloys with a higher PREN factor (F_{PREN}) value are more resistant to pitting and crevice corrosion, and SSC in the presence of H_2S.
Per NACE MR0175/ISO 15156 Part 3, F_{PREN} is defined as

$$F_{PREN} = W_{Cr} + 3.3(W_{Mo} + 0.5W_W) + 16W_N$$

where
- W_{Cr} is the mass fraction of chromium (Cr) in the alloy as a percentage of mass fraction of the total composition;
- W_{Mo} is the mass fraction of molybdenum (Mo) in the alloy as a percentage of mass fraction of the total composition;
- W_W is the mass fraction of tungsten (W) in the alloy as a percentage of mass fraction of the total composition; and
- W_N is the mass fraction of nitrogen (N) in the alloy as a percentage of mass fraction of the total composition.

For duplex stainless steel alloys, MR0175/ISO 15156 Part 3 divides the class of alloys into two categories:
- $30 \leq F_{PREN} \leq 40.0$ with Mo $\geq 1.5\%$. Example: 22Cr, 2205
- $40.0 \leq F_{PREN} \leq 45$. Example 25Cr, 2507.

Higher values of F_{PREN} are more resistant to pitting and SSC. However, as F_{PREN} continues to increase above 40, the alloy becomes increasingly susceptible to forming brittle phases.

> **"We Don't Want It to Corrode—So Just Order 'Stainless'"—WRONG!**
>
> There are well over 100 metal compositions that meet the basic definition of stainless steel; that is, they contain at least 10.5% chromium. There are numerous other metals that are commonly referred to as "stainless steels." Each one of these materials is corrosion-resistant (i.e., they are "stainless") in certain environments. Only a very limited number of the "stainless" materials resist the corrosive conditions in the petroleum production applications—the various produced oils, gases, waters, and chemicals. Furthermore, no single stainless type will fit all the requirements in every service.
>
> It is very important to always specify the type of stainless steel required for an application. Therefore, DO NOT just order, "a Stainless Steel Valve" or "Stainless Steel Heat Exchanger Tubes." Be specific! If you do not know which stainless steel is needed, give the vendor the details of the environment including the fluids, corrosives, temperatures, pressures, flow rates, etc. The more information the vendors have, the better they can serve you. But, remember, vendors will try to give you their standard material grades. Sometimes these "off-the-shelf" grades of stainless steel are free-machining grades that are known to fail rapidly in marine environments (even in the West Texas environment described earlier in this chapter under Austenitic Stainless Steels). Free-machining grades exist for most of the stainless steel grades, and they are to be avoided in petroleum production applications. Vendors will also try to supply the lowest cost material to them, but you will likely pay the same for a more appropriate grade of stainless steel.
>
> So, remember, stainless steels are NOT alike!—stainless steels can (and do) corrode!

4.6.4.6 Highly Alloyed Austenitic Stainless Steels

The highly alloyed austenitic stainless steels are described in NACE MR0175 Part 3, Table D.2.[17] Their compositions vary significantly from alloy to alloy, but overall their chromium content lies between 19–25%, nickel from 17–38%, and molybdenum from 2.5–7%. Most of the alloys also contain nitrogen for strengthening and for enhancing localized corrosion resistance. They are used in the annealed condition for piping in petroleum production facilities. If the alloy is cold worked, its maximum hardness should not exceed 35 HRC. As a class, these alloys cannot be strengthened by heat treatment. They are used where higher resistance to CSCC and SSC are required. Like other highly alloyed (and expensive CRAs), this class of alloy is used where its corrosion and cracking resistance are required.

As an example, alloys in this class include the "6Mo" family (UNS S31254, N08925, N08367, and N08926), Alloy 20 (UNS N08007, N08020) and other alloys such as 654SMO and 254SMO (UNS S32654 and J93254, respectively). The value of F_{PREN} varies from 28–62, demonstrating the wide range in composition of this class of alloys. Because of this wide range, NACE MR0175/ISO 15156 Part 3 further divides the class into two additional types in the notes to Table A.9:[17]

- Type 3a—$(w_{Ni} + 2w_{Mo}) > 30$, where w_{Mo} has a minimum value of 2%.
- Type 3b—$F_{PREN} > 40$

NACE MR0175/ISO 15156 Part 3, Table A.9 provides service envelopes for highly alloyed austenitic stainless steel Types 3a and 3b as a function of maximum H_2S partial pressure, temperature, and chloride concentration.[17] The table is too large to summarize here, and the reader is referred to it for further information. The temperature and H_2S partial pressure limitations do not necessarily mean that the alloy will fail *above* these conditions, only that the alloys successfully passed the test requirements *at* these conditions.

Potential users of these alloys should obtain the assistance of a competent metallurgist because knowledge of melt practice, composition, fabrication and forming, and heat treatment are required to ensure high-quality products are obtained.

4.6.5 Cast Irons

For completeness, the last class of ferrous alloys discussed is cast iron. Cast iron is a family of alloys in which most of the carbon content is not in solution in the iron, or as pearlite, as it is in steel. Therefore, there is free carbon in the microstructure in the form of carbon flakes or nodules. As the name implies, equipment made of cast iron is cast to shape rather than being hot or cold formed. The cast irons as a family, have very low ductility as compared to steels, and the strength is generally lower than that of the higher strength carbon and alloy steels. The big advantage of cast iron is the variety of shapes that can be made from them and a relatively low cost. There are several types of cast iron such as gray cast iron, white cast iron, malleable cast iron, and nodular cast iron, all of which differ in composition, heat treatment, microstructure, and properties.

Some operators do not allow the use of the less ductile cast irons in hydrocarbon service because of the chance of a brittle failure in case of an accident. However, austenitic gray cast irons produced to ASTM A436 (called "Ni-Resist") have performed well in produced water service in pump casings and valves.[34] Ni-Resist Types 1 and 2 in the series of these cast irons contain about 15% Ni, which greatly improves their corrosion resistance. These alloys are lower cost alternatives to nickel-aluminum bronze and cast duplex steel and have been installed in noncritical applications.

4.7 Nonferrous Alloys

Although most materials used in the oil field are steels and ferrous alloys, nonferrous alloys are used for special purposes. Nonferrous alloys contain less than 50% iron.

4.7.1 Solid-Solution Nickel-Based Alloys

Some individual alloys in the class of highly alloyed austenitic stainless steel contain less than 50% iron and are considered nonferrous alloys. As the service environment becomes more severe, with increasing H_2S concentration and/or temperature, NACE MR0175/ISO 15156 Part 3 describes another classification of solid-solution nickel-based alloys, which are more highly alloyed than the highly alloyed austenitic stainless steels. Many of these alloys are proprietary and do not have UNS identifications. Potential users of these alloys should obtain the assistance of a competent metallurgist because knowledge of melt practice, composition, fabrication and forming, and heat treatment are required to ensure high quality products are obtained.

Table A.12 in NACE MR0175/ISO 15156 Part 3 establishes six types of alloys in this classification.[17] They differ in the concentration of chromium, nickel + cobalt, molybdenum, and molybdenum + tungsten. The highest alloy, Type 4f (UNS N07022), contains 20%Cr, 58% (Ni + Co) and 15.5% Mo (e.g., Alloy C-276, UNS N08276). Like the highly alloyed austenitic alloys, alloys falling within the classification of solid solution nickel-based alloys are expensive and justified when the service conditions warrant their use, mostly in deep, hot sour gas wells.

The alloys are solid solutions and their microstructure is austenitic. They are available in the solution-annealed condition, or cold worked where further strength is required, as in downhole tubulars. The environmental limits for these alloys in the solution annealed or cold worked conditions are described in Tables A.13 and A.14 in NACE MR0175/ISO 15156 Part 3, including hardness and yield strength restrictions.[17] Because of the complexity of the tables, the reader is referred to them for further information.

Alloy 825 (UNS N08825) is a SCC-resistant alloy that is often specified to replace 316 stainless steel where the latter alloy has failed from CSCC, or where CSCC of the 300 series austenitic alloys is likely to occur. Its composition is approximately 42% Ni-22% Cr-3% Mo. Alloy 825 generally falls into the Type 4c category in NACE MR0175/ISO 15156 Part 3 Table A.12, depending on the actual alloy composition.[17] Table A.14 contains five H_2S-temperature test conditions where the alloy type has successfully passed. One of these conditions is 177°C (350°F) at 1,400 kPa (200 psia) H_2S partial pressure. The temperature and H_2S partial pressure limitations do not necessarily mean that the alloy will fail above these conditions, only that the alloys successfully passed the test requirements at these conditions. Alloy 825 is used in welded piping and threaded and coupled downhole tubing.

Alloy 625 (UNS N06625) generally falls into the Type 4e category in Table A.12, depending on the actual alloy composition. The composition of Alloy 625 is approximately >58% Ni-21% Cr-9% Mo. Alloy C-22 (UNS N06022) contains approximately 56% Ni-13% Cr-13% Mo. Alloy C-276 also generally falls into the Type 4e category: its composition is approximately 58% Ni-16% Cr-16% Mo-2% Co. However, the molybdenum content of Alloy C-276 puts it closer to the Type 4f composition with respect to that alloying element.

Alloys 625, C-276 and C-22 are some of the most corrosion and crack-resistant solid solution nickel-based alloys used in petroleum production. Because of their alloy content they cost about 20X the cost of carbon steel, so they are used in severe environments such as hot sour gas. Alloy 625 is used in solid form for pressure vessels and piping, and as internal cladding or weld overlay for pipelines, risers and pressure vessels. Using a clad layer or weld overlay can reduce the overall cost of the component because the pressure-bearing member is made from lower cost carbon or low-alloy steel. In the form of downhole tubing, the solid-solution nickel-based alloys are solid (not clad), and cold worked for increased strength. In NACE MR0175/ISO 15156 Part 3 Table A.14, cold-worked Type 4e alloys have successfully passed a maximum temperature of 232°C (450°F) at a maximum H_2S partial pressure of 7,000 kPa (1,000 psia).[17] The temperature and H_2S partial pressure limitations do not necessarily mean that the alloy will fail above these conditions, only that the alloys successfully passed the test requirements at these conditions.

Included in the classification of solid-solution nickel-based alloys is a nickel-copper alloy commercially known as Monel 400 (UNS N04400). The composition of Monel 400 is approximately 65% Ni and 32% Cu, the remainder being Mn, Si, and Fe with lesser amounts of S and C impurities restricted.

It consists of a face-centered cubic atomic structure, and the microstructure of the 400 alloy shows a homogeneous solid solution. A major application is as an external cladding of hot offshore steel risers. The alloy is used in water of all types, and in petroleum production for pump, surface, and subsurface valve parts and specialty items. Other than restricting maximum hardness to 35 HRC, NACE MR0175/ISO 15156 does not impose limitation on H_2S and chloride concentrations and temperature. Operators have installed Monel 400 as downhole accessories in sweet service to a maximum temperature of ≈149°C (300°F). Alloys with copper can experience corrosion in high-temperature, wet H_2S. In addition, strength decreases above this temperature. This alloy is not typically specified for severe sour environments. A precipitation-hardenable version known as Monel K-500 (UNS N05500), is described below.

4.7.2 Precipitation-Hardened Nickel-Based Alloys

Monel K-500 (UNS N05500) is a precipitation hardened version of Monel 400. NACE MR0175/ISO15156 Part 3, Tables A.34–A.35 provide information on service limitations and hardness restrictions. MR0175 limits the hardness of Monel K-500 to 35 HRC, and H_2S partial pressure to a maximum of 3.4 kPa (0.5 psia) for Christmas tree components (excluding bodies and bonnets)[(5)] and valve and choke components (excluding bodies and bonnets).[17] Monel K-500 is susceptible to hydrogen embrittlement, with hydrogen charging caused by H_2S or in subsea applications from CP. For severe sour service, other, more-resistant nickel-based PH alloys are selected.

There are many precipitation-hardened nickel-based alloys that are used in highly corrosive service at high H_2S partial pressure and high temperature. These are used in downhole accessories, valve components, and fasteners. Like other PH alloys, the PH nickel-based alloys are first solution annealed then artificially aged to obtain the desired strength. Fine precipitates known as "gamma prime" and "gamma double prime," (separate phases rich in nickel, titanium, and aluminum) form during artificial aging to strengthen the alloy. They can also be cold worked and aged or hot finished and aged.

The microstructure of these alloys is a homogeneous, austenitic solid solution. Microstructural examination is used to ensure that components are free of a network of continuous second phase precipitates along grain boundaries, which embrittle the alloys. These networks can form through many stages of manufacturing: raw material, melting and remelting, homogenization, hot working, solution annealing, and age hardening. Therefore, the nickel-based austenitic PH alloys require quality control to avoid expensive failures. API 6A718 specifies the quality control for Alloy 718.[35] Limitations and restrictions are addressed in NACE MR0175/ISO 15156 Part 3, Tables A.31–A.37 for a variety of components.[17] Nickel-based austenitic PH alloys are susceptible to hydrogen embrittlement under CP, as might be encountered in subsea applications.[30] Embrittlement can occur if the alloy is electrically connected to steel equipment that is under CP, or if CP is applied directly in the absence of carbon steel. Increasing yield strength raises the susceptibility to embrittlement.

An early-adopted alloy (Alloy X-750, UNS N07750) has been supplanted by alloys that contain Ni-Cr-Mo and are more resistant to SSC and hydrogen embrittlement, such as Alloy 625 Plus (UNS N07716), Alloy 718 (UNS N07718), Alloy 725 (UNS N07725), and Alloy 925 (UNS N09925). NACE MR0175/ISO 15156 Part 3, Table A.32, for example, gives the maximum H_2S partial pressure for alloys 718 and 925 at six different maximum temperatures.[17] At 232°C (450°F), the maximum H_2S

---

5. The body is the main component of a valve in the wellhead. The bonnet is the top part of a valve, attached to the body that guides the stem and adapts to extensions or operators.

partial pressure is 200 kPa (30 psia), whereas at 149°C (300°F), the maximum H_2S partial pressure is 2800 kPa (400 psia). The temperature and H_2S partial pressure limitations do not necessarily mean that the alloy will fail above these conditions, only that the alloys successfully passed the test requirements at these conditions. Hardness restrictions vary by heat treatment method between 35–40 HRC.

As with other CRAs, these are highly specialized alloys that are used in demanding services, and the assistance of a qualified metallurgist is suggested when an alloy is being considered, specified, and manufactured.

4.7.3 Cobalt-Based Alloys

NACE MR0175/ISO 15156 Part 3 describes several cobalt (Co)-based alloys that can be used in sour service.[22] For example, one with the tradename Elgiloy®[6] (UNS R30003) is used in springs, pressure seals, and pressure gauge diaphragms. The composition of this Co-Cr-Ni-Mo alloy is about 30% Co, 20% Cr, 15% Ni, and 7% Mo. For springs, seals, and diaphragms, a maximum hardness of 60 HRC is imposed; for other components, it is 35 HRC. Another cobalt-based alloy that is used in springs and other applications in petroleum production is MP35N®[7] (UNS R30035). Its composition is approximately 29% Co, 20% Cr, 35% Ni, and 10% Mo. A maximum hardness of 55 HRC is imposed for springs made of this alloy.

These and other alloys described in NACE MR0175/ISO 15156 Part 3, Tables A.38–A.40 are resistant to corrosion and cracking in sour environments and possess good resistance to crevice corrosion and SCC in seawater.[17] Cobalt Alloy 25 (UNS R30605) contains tungsten and is wear resistant. Other alloys in this group resist galling, and some are used for bolts.

4.7.4 Titanium Alloys

The advantage of titanium alloys is their corrosion resistance and high strength-to-weight ratio. Most titanium applications in petroleum production—heat exchanger tubes, plates in plate frame heat exchangers, and piping in seawater service—are described in Chapter 3 Section 3.17.7. Generally, unalloyed titanium ASTM Grade 1 (UNS R50250) and Grade 2 (UNS R50400) are specified for these applications. In addition, titanium alloys are used subsea in flowline jumpers, manifold piping, stress joint for risers, subsea valves, umbilicals, and coiled tubing. In the Norwegian North Sea, a modified high-strength titanium alloy, Ti-6Al-4V extra-low interstitial, is used for drilling riser joints in a tension leg platform and for fasteners.[36] There has been only limited service experience with titanium alloys under harsh conditions of H_2S, CO_2, brine, and elemental sulfur. High-strength titanium alloys can be heat treated. They are considered for downhole tubing and drilling risers. NACE MR0175/ISO 15156 Part 3, Table A.41 provides a list of titanium alloys by their UNS designations, heat treatment, and hardness restrictions.[17]

Titanium and its alloys are adversely affected by fluorine and hydrofluoric acid. Hydrofluoric acid is used in some downhole acidizing treatments (mud acids). In addition, many plate frame heat ex-

---

6. Elgiloy® is a registered trademark of Elgiloy Specialty Metals, Elgin, IL.
7. MP35N® is a registered trademark of SPS Technologies, Jenkintown, PA

changers use fluoropolymer elastomeric gaskets. It has been reported that HF was released by chemical attack of the fluoropolymer elastomer that was in intimate contact with the titanium plates.[37] The elastomer was attacked by the presence of an amine corrosion inhibitor at the high operating temperature (82 °C [180 °F]). A more resistant gasket would have prevented this failure.

In this case history, the titanium plate also showed signs of iron contamination, probably occurring during manufacturing. Embedded iron reduces the corrosion resistance of titanium as well as promotes hydrogen uptake. Titanium requires special handling precautions to avoid contamination by embedded iron. Embedded iron can damage titanium and all highly alloyed CRAs. Handling tools should be made of stainless steel. Iron contamination can result in the formation of a brittle phase of titanium hydride, which can cause the titanium component to fail in a brittle manner. When present on the surface of titanium, iron tends to corrode and generate hydrogen, and titanium is permeable to the hydrogen generated by corrosion. Surface iron acts as a path through which hydrogen can diffuse into titanium. Iron can originate from using steel tongs, steel bundling wire, steel tube rolling tools to mechanically join titanium tools into tubesheets, steel grinding tools, etc. Titanium tubes in a shell and tube heat exchanger that is fitted with carbon steel baffles or tubesheets can cause hydriding when the titanium metal temperature exceeds 71 °C (160 °F). Hydriding is the formation of brittle titanium hydride needles in the titanium microstructure.

Chapter 3, Section 3.17.7 describes the means to inhibit SCC of titanium in pure methanol.

4.7.5 Copper-Based Alloys

Unalloyed copper is used because of its good corrosion resistance, mechanical workability, and thermal conductivity. Its most typical application in corrosive services is in handling waters of various compositions.

Copper may be alloyed with any of several other metals to improve certain properties. The most common copper-based alloys are the brasses, which contain zinc as the principal alloying element. The big advantages of brasses compared to copper alloys are their improved mechanical properties and their greater resistance to impingement. However, the brasses are sensitive to certain types of corrosion, such as dealloying (dezincification) and SCC. As with unalloyed copper, the brasses find their widest use in applications handling water. Zinc-less bronzes, such as aluminum bronze (UNS C61300 and C61400) and nickel-aluminum bronze, also are widely used in injection water service valves and pumps.

Another class of copper alloys is the copper-nickel alloys. These are alloys containing 10–30% nickel and the balance copper. The more common alloys have designations such as: 90/10 copper-nickel (UNS C70600) and 70/30 copper-nickel (UNS C71500). Both contain a small quantity of iron to improve corrosion resistance. These alloys in general, have better corrosion resistance in certain environments than other copper alloys, and resist SCC better than copper-zinc alloys. The copper-nickels are used for raw seawater piping in seawater floods and similar applications.

4.7.6 Aluminum and Aluminum Alloys

Aluminum alloys are used in oil and gas production facilities because of their cold-temperature resistance or high strength-to-weight ratio in the following applications:

- Reciprocating compressor piston heads
- Cryogenic service in liquefied natural gas processing facilities as piping and heat exchangers in cold boxes and in brazed aluminum heat exchangers
- Offshore platform structural components for helidecks (with proper fire-fighting equipment added)
- Offshore topsides buildings

Aluminum fins are used to improve heat transfer of carbon steel tubes used in air-cooled heat exchanger.

Wrought aluminum alloys are classified by a four-number designation, the first two numbers signify the alloy series (main alloying elements) and the second two numbers describe individual grades of the alloy in this series. Applications of aluminum are generally confined to alloys such as 3000 series (Al-Mn); 5000 series (Al-Mg-Mn-Cr), and 6000 series (Al-Mg-Si) wrought alloys and their casting equivalents. The high-strength alloys found in the aerospace industry, such as the 2000 series (aluminum-copper) and the 7000 series (aluminum-zinc-magnesium-copper) are susceptible to SCC in seawater and are rarely used in production facilities. Aluminum alloy 2014 has been used as drill string tubing for geotechnical surveys, and there is increasing interest in using it for drilling onshore horizontal wells.

Aluminum is a very reactive metal, and like chromium, it gets its corrosion resistance from a tenacious oxide layer. Therefore, it is most useful in aerated and atmospheric situations. Unlike chromium, aluminum finds most application in corrosive environments at a pH level between 5 and 7 and undergoes rapid corrosion under either highly acidic or highly alkaline conditions. Aluminum is alloyed mainly to improve its mechanical properties.

4.7.7 Tungsten Alloys

Tungsten is used as an alloying element in many CRAs and steels, and the major application of it is in nickel-based alloys for corrosion resistance. When it is in the form of tungsten carbide (WC), it imparts a hard, wear and abrasion-resistant surface in gate valves used in wellheads, pump components, rotating equipment, and in drill bit cutting tools. Tungsten carbide coatings can be applied by flame spraying (high velocity oxyfuel) or solid shapes can be created as a sintered component that uses a cobalt or nickel-containing alloy as a binder for the tungsten carbide particles. Sintered materials are referred to as cemented tungsten carbide. Cobalt is the superior binder in high temperature, high H_2S-concentration environments which can corrode a nickel alloy binder.

4.8 Specifications, Standards, and Codes

In general, oil field equipment is ordered to specifications that include all properties of the materials of manufacture, as well as dimensions, product composition, and test conditions. Specifications are a detailed, precise, written presentation of requirements for a component or a plan. Specifications are often written by the company that is purchasing the equipment and often include company-specific requirements, which may be revisions to an industry standard as additions, deletions, and exceptions. User specifications are often considered to be purchase documents. Many user companies write specifications that maximize the use of industry specifications and industry standards to control costs and improve availability.

Standards are established by the authority responsible for the writing and maintenance of the standard, as a model or example; they provide a rule for the measure of quantity, weight, extent, value, or quality. Standards are not the same as specifications but are often confused as being the same. Some technical organizations, such as API, NACE, and ASTM issue both specifications and standards covering various facets of manufacture and service conditions. Some technical organizations also write standards, standard practices, recommended practices, guidance documents, design practices, and procedures.

Codes are standards that are written to be included in regulations because they affect public safety. For example, pressure vessels are designed and constructed to meet the ASME Boiler and Pressure Vessel Code. In some instances, an industry standard is required by a local or federal law without being a code by itself but included in a regulation such as the U.S. Code of Federal Regulations or state regulations. NACE MR0175 is such an example.

Often, a specification references all relevant industry standards for a purchase order. Using the example of a CRA downhole tubular purchase order, a specification addresses specific requirements for this tubing order:

- List of industry standards and specifications that are part of this specification
- Dimensions (diameter, wall thickness, length) and tolerances
- Process of manufacture, including heat treatment and straightening
- Materials requirements, including chemical composition, mechanical properties, hardness, and testing parameters and tolerances for these requirements
- Special test requirements for SSC resistance, SCC resistance
- Inspection and testing requirements, including hydrostatic pressure testing and nondestructive testing, with accept/reject criteria
- Marking
- Shipping
- Documentation and traceability

Standards for this CRA tubular order may include the various ASTM, ISO, API, NACE, and other industry standards relevant to this and similar CRAs and to CRA tubing, but they would be in general terms, not specifically for this tubing order (this dimension, quantity, etc.).

There are many standards-writing organizations around the world. Most countries have their own national standards and code bodies. The big push in standardization is to develop international standards. This work is being done through the International Organization for Standardization (ISO). The American National Standards Institute (ANSI) represents the United States. The US effort for the oil and gas industry involves many people from American Petroleum Institute (API) and NACE International (from the corrosion control standards standpoint).

References

1. NACE Standard MR0176 (latest edition), "Metallic Materials for Sucker-Rod Pumps for Corrosive Oil Field Environments" (Houston, TX: NACE International, 2012).
2. ANSI/NACE Standard MR0175/ISO 15156 (latest edition), "Petroleum and Natural Gas Industries—Materials for Use in H_2S-Containing Environments in Oil and Gas Production" (Houston, TX: NACE International, 2015).
3. ISO 21457 (latest edition), "Petroleum, Petrochemical and Natural Gas Industries—Materials Selection and Corrosion Control for Oil and Gas Production Systems" (Geneva, Switzerland: International Standards Organization, 2010).
4. NORSOK Standard M-001 (latest edition), "Material Selection" (Oslo, Norway: Standards Norway, 2014).
5. Song, S., and A. Nogueira, "Materials Selection and Corrosion Mitigation for Topsides Process and Utility Piping and Equipment," CORROSION 2012, paper no. C2012-0001632 (Houston, TX: NACE International, 2012).
6. Craig, B.D., and L. Smith, "Corrosion-Resistant Alloys in the Oil and Gas Industry—Selection Guidelines Update," Nickel Institute Technical Series 10073 (Toronto, Ontario, Canada: Nickel Institute, 2011).
7. Marsh, J., "Materials Selection for Offshore Pipelines," CORROSION 2012, paper no. C2012-0001649 (Houston, TX: NACE International, 2012).
8. Skar, J.S., and S. Olsen, "Development of the NORSOK M-001 and ISO 21457 Standards—Basis for Defining Materials Application Limits," CORROSION 2016, paper no. 7433 (Houston, TX: NACE International, 2016).
9. Baboian, R., *NACE Corrosion Engineer's Reference Guide*, Fourth Edition (Houston, TX: NACE International, 2016).
10. NACE International, "Corrosion Glossary" https://www.nace.org/resources/general-resources/corrosion-glossary (accessed December 2019). Note: accessible to NACE members only.
11. API Spec 5CT (latest edition), "Specification for Casing and Tubing" (Washington, DC: American Petroleum Institute, 2018).
12. API Spec 5L (latest edition), "Line Pipe" (Washington, DC: American Petroleum Institute, 2018).
13. ASTM A370 (latest edition), "Standard Test Methods and Definitions for Mechanical Testing of Steel Products" (West Conshohocken, PA: ASTM International, 2017).
14. Russ, P.R., "Oilwell Batch Inhibition and Material Optimisation," SPE Asia Pacific Oil and Gas Conference, paper no. SPE Paper SPE-28810-MS (Richardson, TX: Society of Petroleum Engineers, 1994).
15. ASTM A453/A453M (latest edition), "Standard Specification for High-Temperature Bolting, with Expansion Coefficients Comparable to Austenitic Stainless Steels" (West Conshohocken, PA: ASTM International, 2017).
16. Klein, L.J., "H_2S Cracking Resistance of Type 420 Stainless Steel Tubulars," CORROSION'84, paper no. 211 (Houston, TX: NACE International, 1984).
17. ANSI/NACE Standard MR0175 (latest edition), "Petroleum and Natural Gas Industries—Materials for Use in H_2S-Containing Environments in Oil and Gas Production," Part 3, Cracking-Resistant CRAs (Corrosion-Resistant Alloys) and Other Alloys (Houston, TX: NACE International, 2015).
18. ASTM A487/A487M (latest edition), "Standard Specification for Steel Castings Suitable for Pressure Service" (West Conshohocken, PA: ASTM International, 2014).
19. ASTM A182/A182 M (latest edition), "Standard Specification for Forged or Rolled Alloy and Stainless Steel Pipe Flanges, Forged Fittings, and Valves and Parts for High-Temperature Service" (West Conshohocken, PA: ASTM International, 2018).

20. API Spec 6A (latest edition), "Specification for Wellhead and Christmas Tree Equipment" (Washington, DC: American Petroleum Institute, 2018).
21. Agrawal, A.K., W.N. Steigelmeyer, and J.H. Payer, "Corrosion and Cracking Behavior of a Martensitic 12Cr-3.5NiFe Alloy in Simulated Sour Gas Environments," *Materials Performance*, 26, 3 (Houston, TX: NACE International, March 1987): pp. 24–29.
22. Nickel Development Institute Reference Book, Series No. 11 007, "Guidelines for the Welded Fabrication of Nickel-Containing Stainless Steels for Corrosion Resistant Services" (Toronto, Ontario, Canada: Nickel Development Institute, 1992).
23. ASTM A380/A380M-13, "Standard Practice for Cleaning, Descaling, and Passivation of Stainless Steel Parts, Equipment, and Systems" (West Conshohocken, PA: ASTM International, 2013).
24. ASTM F593 (latest edition), "Standard Specification for Stainless Steel Bolts, Hex Cap Screws, and Studs" (West Conshohocken, PA: ASTM International, 2017).
25. Badrak, R.P., "Investigation of Limits for UNS S17400 in H_2S Containing Environment," CORROSION 2014, paper no. 3816 (Houston, TX: NACE International, 2014).
26. Gareau, F.S., N. Chambers, and A.M. Martinson, "Effect of Stress and Environment on Failures of 17-4 PH Stainless Steel Valve Stems," CORROSION'93, paper no. 146 (Houston, TX: NACE International, 1993).
27. Fink, F.W., and W.K. Boyd, "The Corrosion of Metals in Marine Environments," Defense Metals Information Center DMIC Report 245 (Columbus, OH: Battelle Memorial Institute, May 1970): p. 29.
28. Hasson, D.F., and C.R. Crowe, Eds., *Materials for Marine Systems and Structures*, (San Diego, CA: Academic Press, Inc., 1988): p. 156.
29. Gaugh, R.R., "Sulfide Stress Cracking of Precipitation Hardening Stainless Steels," in *H_2S Corrosion in Oil and Gas Production–A Compilation of Classic Papers*, (Houston, TX: NACE International, 1981): pp. 333–338.
30. Huang, W., W. Sun, A. Samson, and D. Muise, "Investigation of Hydrogen Embrittlement of Precipitation Hardened Nickel Alloys Under Cathodic Protection Condition," CORROSION 2014, paper no. 4248 (Houston, TX: NACE International, 2014).
31. Gunn, R.N., Ed., *Duplex Stainless Steels Microstructures, Properties and Applications* (Abington, Cambridge, England: Woodhead Publishing Ltd, 1997): p. 189.
32. Gunn, R.N., Ed., *Duplex Stainless Steels Microstructures, Properties and Applications* (Abington, Cambridge, England: Woodhead Publishing Ltd, 1997): p. 190,
33. DNV Recommended Practice DNV-RP-F112 (latest revision), "Design of Duplex Stainless Steel Subsea Equipment Exposed to Cathodic Protection" (Oslo, Norway: DNV-GL, 2008).
34. ASTM A436 (latest edition), "Standard Specification for Austenitic Gray Iron Castings" (West Conshohocken, PA: ASTM International, 2015).
35. API Spec 6A718 (latest edition), "Specification of Nickel Base Alloy 718 (UNS N07718) for Oil and Gas Drilling and Production Equipment" (Washington, DC: American Petroleum Institute, 2006).
36. Simonson, T., J. Murali, and J.O. Foss, "The Heidrun TLP Titanium Drilling Riser" *Stainless Steel World*, 11, 5 (June 1999): pp. 41-49.
37. O'Rourke, D.F., and R.J. Franco, "Environmental Cracking of a Titanium Heat Exchanger," CORROSION'83, paper no. 100 (Houston, TX: NACE International, 1983).

BIBLIOGRAPHY

Craig, B.D., *Practical Oil Field Metallurgy and Corrosion*, 3rd ed. (Denver, CO: MetCorr, 2004).

NACE Publication 1 FI92 (latest edition), "Use of Corrosion-Resistant Alloys in Oil Field Environments" (Houston, TX: NACE International, 2013).

Verhoeven, J.D., *Steel Metallurgy for the Non-Metallurgist* (Materials Park, OH: ASM International, 2007).

Nonmetallic Materials Selection

Robert J. Franco

5.1 Introduction

The term "nonmetallic materials," when mentioned in petroleum production, usually refers to the terms "plastic" or "fiber-reinforced plastic (FRP)"; FRP is also referred to as "fiberglass" and is used interchangeably with FRP in the petroleum production industry and in this book. Generally, that is a valid reference; however, the subject is much broader. To be complete, thermoplastics, elastomers, thermoplastic seals, internal thermoplastic liners, polymers in subsea flexible pipe, concrete (cement), and mortars must also be mentioned. For completeness, this chapter begins with the oldest known manufactured nonmetallic material—cement and concrete—but concentrates on modern thermoplastics, polymers, and elastomers, which are the backbone of today's nonmetallic materials.

5.2 Cement and Concrete

Cement materials show up in several forms, from reinforced concrete structures offshore and onshore, and foundations for pumping units, injection pumps, compressors, and other heavy equipment, to cement linings for tank bottoms, internal cement-lined pipe coatings, external concrete weight coatings on subsea pipelines, and concrete secondary containment basins. Sulfur pits are also made of cement, typically using sulfate-resistant reinforced concrete.

Concrete is basically resistant to seawater and brines. The main problem with reinforced concrete is corrosion of the reinforcing steel bars (rebar) or mesh. Normally, concrete provides excellent protection to steel—the highly alkaline Portland cement in concrete allows a stable, corrosion-mitigating, passive oxide film to form on the steel. If the film does not form, or is destroyed, corrosion can occur. Chlorides and oxygen are the major causes of such corrosion. Corrosion of rebar leads to cracking and/or spalling of the concrete (because the corrosion product requires more volume than the steel) and can eventually lead to structural failure (Figure 5.1).

Corrosion of reinforcing steel (called "rebar corrosion") can be minimized by selecting a concrete formulation that is the least permeable and is low in contaminants (e.g., chlorides), designing the structure to have several inches of concrete encasing the rebar, coating the rebar with protective coatings (e.g., fusion bonded epoxy [FBE]), and installing cathodic protection. NACE SP0187 addresses these issues.[1] NACE SP0395 covers procedures for coating rebar with FBE applied by the electrostatic spray method.[2] NACE SP0290 presents guidelines for the use of impressed current cathodic protection on new or existing installations.[3] On the other hand, NACE SP0390 has guidelines for the repair of deteriorating or deteriorated structures.[4] Because of the tremendous amount of reinforced concrete in the world's infrastructure, corrosion control of rebar is an area receiving continuing attention.

Sulfur pits are found in sour gas treating facilities. Sulfur is the final product of sulfur removal prior to making sales-quality natural gas. Sulfur is sold in solid or liquid form. Pits are typically made of sulfate-resistant reinforced concrete, per ASTM C150 Type V.[5] The pit is designed to prevent water ingress which will form a corrosive acid when mixed with molten sulfur. The vapor space is often lined with an acid-resistant nonmetallic to prevent acid attack due to moisture ingress.

Concrete floors and foundations exposed to corrosives may deteriorate with time unless they are coated with protective coatings specifically formulated for that purpose. Secondary containment structures also may be coated with similar materials to seal them to prevent seeping.

Figure 5.1 Concrete deterioration and spalling due to corrosion of the steel reinforcement imbedded in a reinforced concrete bridge beam on a quay. As the steel rebar corrodes, the corrosion product (rust) has more volume than the steel, which causes cracking and spalling of the concrete.
Photograph provided courtesy of John Broomfield.

5.3 Polymers

The term "polymers" is preferred to "plastics" to discuss engineered, nonmetallic materials referenced in this chapter. Polymers also refers to the material's molecular structure, which consists of a large chain of consistent, identical molecules known as "homopolymers." Although polymers are the most widely used nonmetallic materials in today's oil field operations, they are an underused corrosion control option. That is, polymers are often overlooked when seeking corrosion control for a specific field application.

Unfortunately, polymers have often been misapplied or mishandled as if they were a direct substitute for steel. Properties—such as thermal expansion, long-term creep, temperature limits, and resistance to specific chemicals—have been overlooked when selecting nonmetallic piping for an application. However, with today's design information and cumulative experience, polymers are quite satisfactorily used for downhole tubing, flowlines (as standalone or as a liner), facilities piping, tanks, and the internals for water handling vessels such as tanks, deaeration towers, or filters.

In many circumstances, polymers are the most cost-effective solution to corrosion problems. The key to successful engineering with polymers is to select the proper material for the application. There are many polymers available with different physical, chemical, and mechanical properties.

A polymer is a nonmetallic compound usually synthetically produced from organic compounds by a process known as polymerization whereby a repeated structure is created by bonding two or more unit molecules called "monomers." Unlike metals, polymers are amorphous (noncrystalline). Polymers fall into the following two (broad) categories:

- **Thermoplastic materials** (polymers and elastomers) are long chain polymers that are recyclable. Thermoplastics can be melted, re-formed, molded, or extruded with little or no change in their properties. Thermoplastics often have lower temperature limits than thermosetting materials, but they can provide impact and chemical resistance advantages.
- **Thermosetting materials** (polymers and elastomers) are a long chain of cross-linked polymers. Thermosetting polymers are cured to develop cross-linking and cannot be re-melted. The cross-links impart an elastic nature and provide resilient recovery characteristics; they are stable under heat or pressure. Because they do not melt, thermosetting materials have higher service temperature limits than the thermoplastics. Cross linking may provide more severe service ability.

5.4 Thermoplastics

A thermoplastic is a polymer that can be melted and resolidified an indefinite number of times. The most common thermoplastics used in production operations are polyethylene (PE), low density PE (LDPE), high density PE (HDPE), polyamide (nylon), polypropylene (PP). Polyvinyl chloride (PVC); chlorinated polyvinyl chloride (CPVC), which is used in water service.

Other thermoplastics found in petroleum production surface, subsurface, and subsea facilities—mostly as seals—include polytetrafluoroethylene (PTFE), polyetheretherketone (PEEK), polyphenylene sulfide (PPS), and polyvinylidene fluoride (PVDF). PVDF is found in oil field components such as liners and seals. Polymers, such as HDPE, are used as standalone products for low-pressure line pipe (Figure 5.2), pipe liners (new and reclamation), and as chemical storage drums for some types of chemicals. Note in Figure 5.2 that the pipe is resting on rocky ground and is not protected from traffic, which demonstrates the ruggedness of the pipe, although this is not a best practice.

CPVC (Figures 5.3 and 5.4) and PVDF are being used in United States Coast Guard (USCG) approved water piping systems that need to comply with fire requirements of International Maritime Organization (IMO) A.753(18) for nonmetallic installations in U.S. waters.[6] Examples of what may fall under this regulation are offshore floating production and tanker requirements. In Canadian waters, Transport Canada TP14612 approval is required.[7]

It is essential to understand that not all thermoplastics are created equal. In fact, many plastics of the same name can differ in their respective types. There are several grades of PE, PP, and CPVC, for example.

To successfully apply a plastic material to a given application, users must first understand the use-cases of each type. Fundamental differences can be explained by understanding what holds these polymer chains together.

5.4.1 Amorphous Thermoplastics

Amorphous polymers, such as CPVC, rely on molecular entanglement of the polymer to become rigid. This type of polymer has no melting point and is used below its glass transition temperature T_g (where T_g corresponds to the gradual transformation from a liquid [flexible, rubbery] to a rigid [glassy, rigid] solid). Below T_g, the amorphous polymer is in a glassy state, whereas above T_g it is flexible and rubbery. In the glassy state, amorphous polymers are relatively immobile, relatively impermeable, and relatively inflexible. The T_g is a property of each individual polymer type, and whether the polymer has glassy or rubbery properties depends on whether its application temperature is above or below T_g.

5.4.2 Semicrystalline Thermoplastics

Semicrystalline polymers, such as HDPE, PP, and PVDF, rely on partial crystallization of the polymer. These types of plastics have a defined melting temperature and are used above their glass transition temperature (but below their melting temperature). For example, HDPE has a T_g of −90°C (130°F), a melting temperature +137°C (279°F). The service temperature must lie between these two boundaries, where these molecules are relatively mobile, moderately permeable, and moderately flexible.

5.4.3 Service Considerations for Thermoplastic Materials

Table 5.1 presents service considerations for the most common thermoplastics. Requirements for design, materials, manufacture, fabrication, installation, inspection, examination, and testing of thermoplastic pressure piping systems can be found in ASME NM.1, "Thermoplastic Piping Systems."[8] This ASME standard has been issued to address nonmetallic pressure piping systems and addresses both pipe and piping components produced as standard products, and custom products designed for a specific application. Dimensions, materials, physical properties, and service factors for PE line pipe are covered in the API 15LE "Specification for Polyethylene Line Pipe (PE)."[9]

Table 5.1 Service considerations for thermoplastic materials.

Thermoplastic	Satisfactory Service	Not Satisfactory
Chlorinated Polyvinyl Chloride (CPVC)	−18°C (0°F) to 93°C (200°F) Most acids Many alkalis/bases Water and brines	Solvents Hydrocarbons Chlorinated solvents
Polyethylene (PE, HDPE, PE-RT, PEX) a	−34°C (−30°F) to 82°C (180°F) note a Hydrocarbons <38°C (<100°F) Acids Alkalis Water and brines	Oxidizing acids Hydrocarbons >38°C (>100°F) Solvents >120°F (>49°C) Aromatic hydrocarbons Sunlight without stabilizers
Polypropylene (PP)	−1°C (30°F) to 99°C (210°F) Hydrocarbons <38°C (<100°F) Alkalis Acids Water and brines	Same as PE
Polyamide (PA) (Nylon 6) b	−40 to 93°C (−40 to 200°F); applies to Nylon 6. Hydrocarbons <38°C (<100°F) Alkalis Organic acids Water and brines	Hydrochloric, sulfuric, nitric acids Solvents Alcohols Aromatic hydrocarbons

a. Newer grades of HDPE, known as PE-RT (polyethylene of raised temperature), have pressure ratings at 80°C. PEX (cross linked PE) has pressure ratings at 93°C (200°F). Both are gaining wider acceptance in upstream oil and gas applications. PE-RT is standardized in ISO 24033[10] and is being considered for inclusion in API 15LE as of 2019.[9]
b. Because of the wide range of available polyamides, consult with a specialist if the temperature exceeds 93°C (200°F).

Appendix 5.A lists standards concerning nonmetallic materials used in oil and gas production operations.

Figure 5.2 HDPE pipe used in the oil field.
Photograph provided courtesy of Performance Pipe, a division of Chevron Phillips Chemical Co, LP.

Figure 5.3 CPVC piping in a membrane process potable water maker system in an offshore floating production facility.
Photograph provided courtesy of Corzan Industrial Systems, a division of The Lubrizol Corporation.

Figure 5.4 CPVC piping with flanged connections and fittings in water service on an offshore floating production facility. *Photograph provided courtesy of SeaCor, a division of GF Piping Systems.*

5.5 Thermosetting

Thermoset polymers, such as FRP, rely on crosslinks. This typically requires some type of cure process to allow crosslinks to form. Crosslinking is accomplished by adding atoms or molecules that covalently bond to the long polymer chains and join individual polymer chains together into one single large molecule. Thermosetting polymers are usually found in production operations most often as FRP. This type of material, that is, one made by intimately melding two entirely dissimilar materials into a single material, falls into a broad category known as "composites." Like the FRP used to make boats, automobile bodies, and other consumer products, the FRP used for production is basically high-strength glass fibers embedded in a thermosetting resin. The most widespread use of FRP is for tubular goods, tanks, and internal fittings in water handling service (Figures 5.5–5.8).

Figure 5.5 FRP tubing is not a new product, as can be noted in this 1963 photograph showing FRP tubing being run into a water injection well.

Figure 5.6 FRP piping in a water flood plant. Note the flanged connection between the FRP pipe and the metallic valve.

Figure 5.7 FRP Pipe: high-pressure 8 in (203 mm), 2500 psi (17.2 MPa) rated pipe. Note the threaded, integral joint connection with an O-ring seal.
Photograph provided courtesy of NOV Fiberglass Systems.

Figure 5.8 FRP pipe with adhesive and mechanical joints used in an oil tank battery.
Photograph provided courtesy of NOV Fiberglass Systems.

5.6 Nonmetallic Pipe

Both thermoplastic and thermosetting polymers are used to make nonmetallic pipe. Thermoplastic pipe is made by heating raw powdered or pelletized materials for fusion and extruding through dies for shaping into pipe. The soft, hot pipe is immediately cooled and thus rehardened. Typical polymers are PE, PP, and CPVC. Thermosetting polymers such as epoxy are made into FRP, which consists of glass fibers wetted with a liquid epoxy resin. More details on FRP pipe manufacturing are provided in this chapter.

5.6.1 Advantages and Disadvantages of Nonmetallic Pipe

Nonmetallic piping systems are economically attractive when they can meet the technical requirements at a cost equal to, or less than, a comparable steel system with corrosion protection. However, it is essential that the user be aware of the major advantages and disadvantages inherent in nonmetallic piping.

5.6.1.1 Advantages

- Nonmetallic materials are immune to scaling, pitting, and corrosion by water.
- Some polymers resist corrosion from a broader range of acids, bases, and salts.
- Plastic pipe weighs approximately two to five times less than metal alternatives, resulting in lower freight costs and easier handling.
- Nonmetallic pipe is quickly and easily joined and installed. However, for pressure piping, joiners are typically prequalified for the job.
- No external corrosion protection, such as coating, wrapping, or cathodic protection, is required.
- The high Hazen-Williams C Factor (i.e., smooth internal surface) results in lower fluid friction loss.
- Nonmetallic joining typically does not require hot work.
- The lower thermal conductivity of nonmetallic piping (e.g., CPVC's heat transfer coefficient is approximately 1/300th that of steel) can save insulation and energy costs and create a safer work environment with cooler surface temperatures, for example, in hot water systems.
- ASTM standards exist for nonmetallic materials in pressure pipe applications that define a minimum long-term hydrostatic strength requirement for design basis.

5.6.1.2 Disadvantages

- Nonmetallic pipe has a lower working temperature and pressure limit than carbon steel and alloy pipe.
- The pressure and temperature limitations are interrelated (i.e., the higher the temperature, the lower the permissible working pressure). Furthermore, mechanical properties change with time under normal conditions, however, methods for accounting for these relationships are in design standards.
- Careful handling is required in transportation (loading, unloading), and installation. Nonmetallic pipe is not as forgiving as steel pipe when handled roughly. (This is also true, however, to

get the maximum from any material, even steel. Internally plastic-coated steel and CRAs require great care in handling.)
- Nonmetallic pipe can be adversely affected by fires because it loses strength and may spread the flame. Section 5.11 describes application of FRP pipe in offshore firewater systems.
- Nonmetallic pipe has low resistance to impact damage. However, HDPE and PA can be squeezed off, then rerounded and placed back into service at the original design pressure. Trucks drive over PE pipe laid on the ground without damage.
- Plastics can be notch-sensitive, so sharp notches can be failure initiation points and should be avoided in the field joints and pipe body.
- Plastic pipe can be UV sensitive.

5.6.1.3 Disadvantages Unique to FRP

- Nonmetallic pipe (particularly FRP) is prone to having manufacturing defects such as porosity and voids. Manufacturers must have quality assurance plans, and users should expect to specify and implement quality control, including inspection requirements. This concern can be addressed through proper application of standards and controls.
- Weeping can occur in low-quality FRP pipe because of porosity and voids.
- FRP has low resistance to vibration and pressure surging. FRP (reinforced epoxy) is commonly downrated 30–50% from the manufacturer's recommended maximum working pressure to allow for pressure surging. Water hammering, due to transient pressure surges, more easily destroys an FRP fiberglass piping system than it does to a steel or alloy system.

5.6.2 Factors to Consider Before Selecting Nonmetallic Pipe

In the application of nonmetallic piping, four primary factors must be considered:

- Chemical resistance
- Mechanical properties
- Fire performance
- Economics

5.6.2.1 Chemical Resistance

In oil field operations, chemical resistance to oil field fluids or gases, and commonly used treating chemicals, are of primary concern. All polymeric pipe materials discussed are sufficiently resistant (in their usage range) to attack from most oil field fluids and have been used successively over the years in these environments. However, that does not mean they are totally unaffected. Some of these materials will be weakened by oil field fluids because of absorption over time. Therefore, an environmental or service factor should be considered for long-term installations. Permeation of the pipe is of varying levels of concern for different plastic piping materials. Service factors are included where appropriate in some of the API documents listed in Appendix 5.A.

Table 5.1 provides some general guidelines and Appendix 5.A provides test standards. Clarity between grade of material and chemical resistance performance at specified pressures and temperatures is a must.

5.6.2.2 Mechanical Properties

The two major physical properties that dictate selection of nonmetallic piping materials for a specific application are the pressure and temperature ratings. The pressure rating is usually dependent on the service temperature and the amount of time the material is exposed to service conditions. Because of the pipe's derating factor, the higher the temperature, the lower the permissible working pressure.

Pressure ratings are based on a standard such as ASTM D1598 for CPVC piping.[11] Long-term theoretical performance is typically designated in the manufacturing standard, and usually incorporates a safety factor of 2 or greater regarding pressure-bearing capability.

It is not uncommon to encounter surge pressures many times the normal operating pressure in long lines, especially when control valves are used. Water hammer has been known to destroy piping systems, including nonmetallic pipe. An example of dynamic pressure changes is firewater piping, which occurs when firewater monitors are suddenly turned wide open. Dynamic operations are normally well tolerated by nonmetallic pipe systems if they are designed for it. Pressure ratings must consider pressure surging. Excursions above the rated design pressure can reduce the product's lifetime, and linear cumulative damage models and other cumulative damage models have been successfully applied to calculate remaining life at the service stresses.

5.6.2.3 Fire Performance

Upstream production operations located offshore tend to be more heavily regulated for fire performance and safety than onshore facilities. Fire performance and chemical resistance are both critical in material selection, and IMO A.753(18) provides fire performance guidelines.[6] Regulation of acceptable guidelines can differ between countries or "Flags" in a mobile drilling and offshore production facility under regulation. U.S. regulations require compliance with IMO A.753(18) for nonmetallic installations in U.S. waters. Compliance is typically confirmed through classification societies such as the American Bureau of Shipping (ABS), Lloyds, and DNV. IMO standards referenced in Appendix 5.A provide fire test procedure (FTP) protocols. For mobile drilling and production facilities operating in Canadian waters, Transport Canada approvals are required just like U.S. Coast Guard (USCG) approval is required for utilization in U.S. waters.

5.6.2.4 Economics

The corrosion resistance of polymers increases the lifespan of the system in environments where steel and other metals are susceptible to corrosion and scaling. Polymer pipe costs less than metal to transport to the jobsite and handle onsite. Polymers, such as CPVC, provide economic advantages on cost of base material, weight considerations, time to install, and life expectancy. Because of its lower weight, PE pipe also costs less to ship to the jobsite than steel. With PE pipe plants located around the world, the distance from plant to jobsite is often less than a day's drive.

In terms of installations, butt fusion and solvent welding do not require highly skilled welders, which also decreases costs. The procedures are standardized and so is the equipment. Additionally, joint inspection is visual; no radiography is needed. Natural gas utilities discovered almost 50 years ago that PE pipe is cheaper and quicker to install, and because it needs no cathodic protection, it is cheaper to maintain. PE pipe has an enormous market share of natural gas utility piping in its pressure range. At many onshore well sites, PE pipe is laid extensively above ground because it is the most cost-efficient product for the application.

For applications at pressures higher than can be tolerated by thermoplastic pipe, reinforced plastic pipe may be required. The economics of reinforced pipe must consider the additional care in handling and installation including training, qualifying fabrication, and installation crews. Quality nonmetallic pipe (e.g., API monogrammed[1]) may have to be shipped into the country from a longer distance than metallic pipe made from commodity carbon steel or 300 series stainless steel pipe. Buried FRP pipe requires better quality backfill than steel pipe, although good backfill quality is also required for externally coated steel pipe. Aboveground FRP pipe, when on supports, requires additional pipe supports than steel because of its lower modulus of elasticity. Conversely, FRP does not require external coatings or cathodic protection, and—because of its low weight, and threaded and coupled joining—installation costs may be lower than welded steel.

5.6.3 Joining Methods for Pipe

As previously discussed, both thermoplastic and thermosetting are used to make oil field pipe or tubulars. Not only do they have different properties, but they may require different joining methods to go with the different applications. The methods used to join several types of nonmetallic pipe include heat welding, solvent welding, and threads. Using spoolable pipe, a joint may be eliminated.

5.6.3.1 Butt Fusion (Heat Welding)

PE, PP, CPVC, and most other common thermoplastic pipes, can be joined by a form of heat welding called "butt fusion" because of their molecular structure. This joining method uses an electrically powered heating element to soften the ends of the joints, which are then butted together and held in position until the joint cools and rehardens. A slight amount of upsetting occurs but is usually not objectionable. To ensure the new line is leak free, heating and joining procedures must be carefully followed according to the type, size, and thickness of the thermoplastic pipe. Dissimilar wall thickness pipe should not be welded by butt fusion. ISO 12176-1 addresses equipment for butt fusion-jointed polyethylene pipe systems.[12]

---

1. An "API Monogram" refers to a label that certifies the pipe meets the requirements of an API standard that governs that pipe. The label is covered over with a transparent resin, so it will not fall off.

5.6.3.2 Solvent Welding

The terms "solvent welding" and "gluing" are often used interchangeably, but that is not accurate. Glue functions with adhesion and cohesion forces. The advantage of solvent welding, however, is that the two pieces of material (i.e., the pipe and fitting) chemically fuse together at the molecular level. In the end, using solvent cement does not just adhere two pieces together, but creates one continuous piece of thermoplastic.

For instance, CPVC solvent cement consists of CPVC resin, stabilizers, and fillers dissolved in solvents. The solvents soften and dissolve the top layer of the pipe and fitting material. A taper in the fitting socket creates an interference fit that ensures contact between the pipe and fitting. This allows the material to fuse to itself when the two pieces are connected; refer to Figure 5.9.

Once assembly is complete, the newly fused CPVC molecules polymerize and harden as the solvent flashes off or evaporates. When all the solvent is gone, the joint is considered fully cured, resulting in one uniform piece of CPVC.

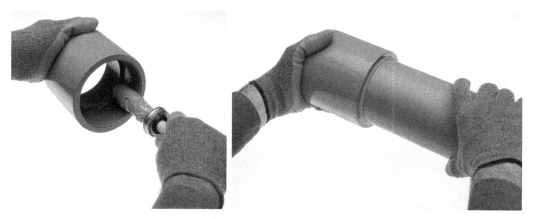

Figure 5.9 CPVC pipe measuring 50 mm (2 in) in diameter joined with solvent cement welding.
Photograph provided courtesy of SeaCor, a division of GF Piping System.

The piping installer must be familiar with the solvent welding procedures for the polymer being joined. As usual on any construction job, quality assurance is required to ensure a satisfactory installation. ASME B31.3 Chapter VII is available and recommended for all installers.[13] ASME B31.3 bonder qualified welders are required to comply with USCG type approval listings.

5.6.3.3 Threaded Connections

Threads can be used on most thermosetting nonmetallic pipe and provide good joints if the wall thickness is great enough. Schedule 80 is usually specified for threaded thermoplastic pipe. Operators have used threaded and coupled, high-pressure downhole FRP (thermosetting) tubing to make threaded flowlines or injection lines. For FRP pipe, the threads are either cut in an integral joint or molded onto standard pipe so there is no reduction in wall thickness (Figure 5.7). In threaded and coupled thermoplastic pipe, the working pressure rating may be reduced 50%.

5.6.3.4 Flange Connections

To connect plastic piping components, or to transition between different piping materials such as a thermoplastic and steel or copper-nickel pipe, flange connections are often recommended (Figures 5.4 and 5.10). Other uses include connecting pumps, valves, instrumentation, and more. Flanged joints incorporate an elastomeric gasket between the mating faces to provide the seal. Flat face flanges with elastomer gaskets and/or a steel-backing ring to stiffen the nonmetallic flange are used to prevent cracking of the plastic flange when bolted to a stiffer metallic flange.

Figure 5.10 CPVC pipe joined to a copper-nickel seawater line using flange connections on an offshore oil production facility.
Photograph provided courtesy of SeaCor, a division of GF Piping Systems.

5.6.3.5 Eliminate the Connection

"Coiled pipe" is HDPE pipe in diameters 8 in and smaller is often supplied as coiled pipe with hundreds of feet on a single coil (noting that larger diameter pipe has less on a coil than smaller diameter pipe).

"Spoolable reinforced pipe" (also called "Reinforced Thermoplastic Pipe," or RTP) can be reeled and brought to the jobsite, where it is unreeled and laid in place (Figure 5.11). The resulting line has no connections. For small diameter pipe, 59.7 mm (2.35 in outside diameter, a length of 1554 m (5100 ft) is available, with shorter distances for larger diameters. API Spec 15S addresses this type of reinforced plastic pipe.[14] The topic is addressed further in this chapter in Section 5.6.4.

Spoolable pipe using HDPE and glass fiber reinforcement can be cost competitive relative to pipe with connections especially because installation costs are lower. Upgraded reinforcement materials such as aramid fibers (Kevlar®[2]) and polyamides make spoolable pipe more expensive but more versatile. Pipe diameter is the limiting factor for RTP.

A

B

Figure 5.11 Spoolable reinforced pipe: pipe has been laid off the spool without connections.
(A) Pipe laid in a trench directly off the spool, before burial. (B) Spoolable 4 in (100 mm) diameter flowlines in a trench (before burial) connected to a carbon steel manifold.
Photograph A provided courtesy of NOV Fiberglass Systems. Photograph B provided courtesy of Polyflow LLC.

---

2. Kevlar® is a registered trademark of E.I. duPont de Nemours and Company, Wilmington, Delaware, USA.

5.6.4 Typical Applications of Pipe in Oil and Gas Production

Some typical applications of nonmetallic pipe in oil and gas operations are discussed in this section.

5.6.4.1 Thermoplastics

- Flow Lines are used principally for carrying produced fluid of oil or oil and water from the well to separating equipment and tank batteries
- Gathering Lines for oil and water or low-pressure gas
- Saltwater Lines for disposal systems and drains
- Liners for steel pipe in high-pressure operations such as flowlines, water lines, and water injection lines.
- Fuel Lines for gasoline engines
- Velocity strings
- Chemical injection strings
- Umbilicals
- Wastewater collection and treatment for fracking water
- Cooling lines (seawater and fresh) water
- Black and gray water lines on marine facilities (black is wastewater from toilets and gray is wastewater from sinks, kitchens, showers).
- Seawater lines for deck wash; feed pipe for water makers
- Potable water systems

In the Permian basin of West Texas, HDPE pipe (i.e., actual standalone pipe, not as a liner for steel pipe) is installed directly on the ground and used for low-pressure flowlines. Freestanding HDPE pipe is also used for drain lines and for providing firewater. Pipe diameters from 25–914 mm (1–36 in) are available, and joints are typically joined through butt fusion. A pressure rating up to 2558 kPa (400 psi) is available; however, if wet hydrocarbon liquid or gas is transported, the pressure rating may require a service factor of 0.5. The smooth surface and absence of any corrosion product buildup on HDPE pipe allows for the use of a lower friction factor than steel, and possibly a smaller diameter pipe can be used.

CPVC pipe is used to transport a variety of water qualities (potable, black, gray, seawater) onboard floating production systems. The typical range of diameters is 0.50 in (listed as "½ in" in the pipe charts) up to 8 in (12.5–203 mm) but can reach 24 in (610 mm).

Using nonmetallic pipe in petroleum production continues to rise in both onshore and offshore production applications. Plastics increasingly meet the stringent fire performance requirements of the Fire Test Procedures (FTP) Code that have been adopted by member countries of the IMO Safety of Life at Sea (SOLAS) Convention. Most requirements for fire performance of plastics on ships currently fall in the IMO Resolution A.753(18).[6] Figure 5.12 shows CPVC piping meeting this fire performance requirement. As more plastics are developed to meet stricter fire, smoke, and toxicity requirements, owners can increasingly use lighter building materials for constructing offshore platforms and ships in corrosive environments.

Figure 5.12 CPVC to copper-nickel transitions used for sodium hypochlorite injection into a marine sanitation device. The CPVC installed here was required to meet IMO A.753(18) guidelines for fire performance.
Photograph provided courtesy of Corzan Industrial Systems, a division of The Lubrizol Corporation.

5.6.4.2 FRP (Thermosetting)

- Uses: for all the thermoplastic applications given in Section 5.6.4.1.
- Tubing: for injection and disposal wells, and for producing high water-cut oil wells including rod-pumped wells.

FRP is addressed in several sections of this chapter; it is used both for downhole tubing and for surface flowlines in solid form, and as a liner grouted into steel pipe. FRP is the terminology used in the United States, and glass-reinforced plastic (GRP) and glass-reinforced epoxy (GRE) is used in Europe. Although ASME International uses "RTP" for reinforced thermoset plastic," in this chapter, the term RTP is reserved for spoolable pipe.

Because of the corrosion resistant nature of FRP, an FRP component can be made entirely from one glass-resin composite, or a second liner can be used. The inner, bonded liner is closest to the corrosive fluid and is made using different material properties than the structural portion. The liner, if made, is usually resin rich (e.g., 90% resin and 10% glass) and utilizes a different type of glass, called "C-Glass" (C for "chemical"), while the structural portion uses "E-Glass" (E for "electrical").

Well-production fluids typically do not attack FRP; however, certain chemicals that might be encountered in well workovers may have detrimental effects. Chemicals that are particularly damaging to FRP are hydrofluoric acid (including mud acid completion fluid containing hydrochloric and hydrofluoric acids), hot sodium hydroxide above 32 °C (90 °F), organic solvents, and chlorinated organic solvents. Data on the resistance of nonmetallic materials to various chemicals, solutions, and solvents can be found in the NACE book, *Corrosion Data Survey–Nonmetals Section*.[15] This also covers thermoplastics. A companion book on metals, *Corrosion Data Survey–Metals Section*,[16] covers alloy performance with many chemicals.

5.6.4.3 Spoolable Pipe

Spoolable pipe is used in flowing hydrocarbon and other produced fluids (flowlines, gathering lines, etc.). It may be used freestanding (Figure 5.11) or as a pull-through liner to increase the lifespan of corroded carbon steel piping.

Spoolable pipe consists of an inner thermoplastic liner surrounded by a helically wrapped steel or composite (carbon, polyaramid, polyester, or glass fibers) reinforcing elements and an outer cover. The reinforcing fibers may or may not be bonded in a thermosetting or thermoplastic matrix. Spoolable pipe is called "thermoplastic composite pipe" and is standardized in API 15S, API 17J, and DNVGL-RP-F119 (Refs. 15, 17–18, respectively). Pipe is continuously unspooled from a circular reel that can hold 50–150 mm (2–6 in) diameter pipe. Joints are eliminated.

Spoolable pipe can be cost competitive compared to welded or jointed pipe because of its lower installed cost (35–65% of welded steel pipe). No coating, cathodic protection, or radiographic inspection is required. When spoolable pipe is considered as a replacement for corroded steel pipe, the heavier wall thickness must be considered in hydraulics calculations. As a pull-through liner, application of spoolable pipe is restricted by the diameter of the outer steel pipe. For example, a 203 mm (8 in) diameter steel pipe has been successfully lined with 89 mm (3.5 in) OD spoolable liner.

Typical diameters for spoolable pipe are 50.8–203 mm (2–8 in) diameter, and they have pressure ratings of from 2.1–20.7 MPa (300–3000 psig). The liner can be matched to suit temperature and chemical environment, especially the presence of H_2S.

5.6.4.4 Flexible Pipe

Unbonded flexible pipe is used for static and dynamic risers, flowlines, jumpers, gas lift pipe, and other subsea applications. Unbonded flexible pipe is a multilayer construction whose design varies among several manufacturers. Unbonded flexible pipe is included in this chapter because it uses polymers. Specifications concerning unbonded flexible pipe include API Spec 17J.[17]

A typical construction (going from the inner bore to the outermost layer) consists of the following components:

- A metallic carcass (interlocked, spiral-wound, metallic segments). The metallic carcass is typically made of 316 stainless steel (UNS S31600); 22Cr duplex stainless steel (UNS S31803); or alloy 31 (UNS N08031), a super-austenitic, 27% chromium 6% molybdenum alloy. Alloy selection depends on the composition of the production fluid and the manufacturer.
- A nonmetallic polymer pressure sheath that covers the inner carcass. The pressure sheath is a corrosion-resistant polymer such as HDPE (thermoplastic or crosslinked), polyamide (PA-11 and PA-12), and PVDF, which is a fluorinated polymer.

- A layer of flexible tape outside the pressure sheath that is made from various polymers for wear resistance, including nylon (PA-11), PP, PE, or PA.
- Two armor layers reside outside of the tape, consisting of helically wound, high strength steel wires that provide strength (the strength level of these wires is reduced in sour service to minimize the concern for SSC).
- A polymer outer sheath covers the armor layers and keeps seawater out of the flexible pipe. The outer sheath is made from PA-11, medium density polyethylene (MDPE), or HDPE.

The role that nonmetallic materials play in unbonded flexible pipe is critical to its success. It is anticipated that the use of unbonded flexible pipe will increase with the extension of offshore projects to deeper water depths.

In addition to unbonded flexible pipe, bonded flexible pipe is available. It consists of an elastomeric matrix (such as nitrile rubber) reinforced with armoring layers or steel cables to give the pipe its required strength. The armoring layers are bonded to the elastomer material and the armoring is fully encapsulated by the elastomer. Bonded flexible pipe is covered by API Specification 17K.[19]

Applications of bonded flexible pipe include static and dynamic flexible pipes used as flowlines, risers, jumpers, and offshore loading and discharge hoses on docks and loading arms. The in-service experience with bonded flexible pipes used for production of hydrocarbons is mainly related to topside jumpers, drag chain hoses for floating production storage and offloading (FPSO) turrets, and short length riser systems. High-pressure bonded flexible hoses for exploration have been used in kill and choke jumpers, rotary hoses used in the derrick of the drilling rig, and cementing hoses. Many applications are usually in the form of hoses, often reeled. Offshore, bonded flexible pipe is used to offload crude oil from an FPSO vessel to a tanker. Bonded flexible pipe has had limited application as flowlines and well lines in petroleum production.

Compared to unbonded flexible pipe, bonded flexible pipe is manufactured in limited lengths. The large bore pipes, 406–610 mm (16–24 in) diameter, are only available in standard lengths of typically 12 m (40 ft). For moderate diameter, 100–254 mm (4–10 in) the length limitation is typically less than 100 m (328 ft). This is short compared to the length limitations for nonbonded flexible pipes, which are typically several kilometers.

5.6.5 Internal Polymer Liners for Pipes and Tanks

Internal polymer liners can be a cost-effective way to extend the life of corroded or eroded carbon steel pipe and tanks.

5.6.5.1 Polyethylene Liner

HDPE is a thermoplastic that is used as an internal liner to carbon steel pipe. Where pressure exceeds the load-bearing capability of solid HDPE, internal HDPE liners have been used successfully to transport high water-cut oil flowlines for new and for internally corroded steel lines. The use of HDPE liners exceeds that of PVC liners because HDPE is more flexible and can be pulled into a steel pipeline. Many high water-cut flowlines made of carbon steel that have corroded internally have been refurbished with a tight-fit HDPE liner. In these oil fields, corrosion inhibition may no longer be economical, and pipelines have become so corroded that frequent repairs or replacement are necessary.

NACE SP0304 addresses the design, installation, and operation of thermoplastic liners for oil field pipelines.[20] Essentially, the installation of a polyethylene liner consists of pulling an oversized, thin-wall polyethylene pipe into the steel pipe by a wireline unit and a diameter reduction system, and then expanding the polyethylene liner onto the ID surface of the steel pipe (Figure 5.13). Under normal circumstances, the liner OD slightly exceeds the inside diameter of the steel pipe and is sized to give a positive interference of up to 5% to ensure a tight fit. Scraper pigs are first run into the steel pipeline to thoroughly remove corrosion products and debris. To pull the liner into position, the wireline cable is first threaded through the pipeline by a launch pig and air compressor. The roller reduction system makes the oversized pipe liner fit into the steel pipe.

The liner is tight against the steel pipe wall to allow internal pressure to be transmitted to the steel. Loose liners are available, but are used less because they are susceptible to longitudinal buckling due to differences in thermal expansion between the liner and the steel pipe, and they have a considerably lower collapse pressure.

The use of a liner is cost-effective because it eliminates the need to excavate and replace corroded steel pipelines in marginal fields. Excavation is limited to where the flowline must be cut and flanged for the liner to be inserted. The distance between flanges is determined by the pipe diameter, the number of fittings, bends, tie-ins, and the liner thickness. A maximum distance of 760 m (2500 ft) is typical for 75–100 mm (3–4 in) diameter pipe; however, longer pulls of 2280 m (7480 ft) with 75–100 mm (3–4 in) diameter pipe have been reported.[21]

The service temperature is typically below 60 °C (140 °F) in hydrocarbon service, 80 °C (176 °F) in water-only service.[20] At higher temperatures, hydrocarbon contamination, and gas permeation may result in circumferential collapse of the liner. In gas service, a vent hole is required to bleed gas in service that has permeated the liner and has entered the annulus space. Although most connections of individually lined pipe sections are flanged, welded connections have been successful in subsea applications.[22]

A design life of >20 years is expected for HDPE-lined steel flowlines and produced water injection lines in conventional petroleum production.[23] NACE International documents provide an overview of thermoplastic liners for oil field applications.[20, 24] In service, liner integrity can be assessed by these methods:[25]

- Annulus pressure monitoring
- Annulus gas and fluid analysis
- Radiographic examination
- Visual inspection, aided by TV camera, borescope, etc.

A

B

C

Figure 5.13 Polyethylene (HDPE) installation shown in three sequences (note: photo sequence taken from three different pipelines). (A) HDPE liner being inserted into a diameter reduction system for insertion into a 20 in (508 mm) carbon steel outer pipe in water service. (B) Compressed HDPE liner being pulled into a 12 in (305 mm) carbon steel outer pipe for an oil-gas gathering system. (C) Liner has been brought over the steel flange face, and the lined section is ready to be bolted to another lined section; 8 in (203 mm) carbon steel pipe in produced water service.
Photographs provided courtesy of United Pipeline Systems, an Aegion Company.

5.6.5.2 Other Polymer Liner Materials for Pipelines

In addition to HDPE, polyamide-11 (nylon, PA-11) liners have been installed for more severe temperature (up to 80°C/176°F) and high gas-oil ratio (GOR) conditions than would be considered acceptable for HDPE.[26-27] PA-12 has also been used in sour multiphase gas production lines. Both polyamides (PA-11 and PA-12) extend liner life by >10X compared to HDPE in the same environment.

5.6.5.3 Tank and Slurry Pipeline Liners

Heavy oil production from Western Canadian fields has encountered abrasion and erosion problems associated with transporting slurries of sand, clay, and water. A polyurethane-based polymer developed for mining, quarrying, and other severely abrasive services was installed in a heavy oil production tailings pipeline to transport a slurry of solids (sand and clay) and water.[28]

The lining of tanks with sheet elastomers is more common in the chemical process industries than in oil and gas production. However, this approach has also been used in oil field operations. One application includes using rubber-lined steel pipe to galvanically isolate carbon steel pipe connected to titanium plate frame heat exchangers. The following are some common liner materials:

- Neoprene (polychloroprene)
- Butyl
- Ethylene propylene diene terpolymer (EPDM)
- Nitrile butadiene rubber (NBR) nitrile

The same service limits and remarks made above on seals also apply to liners. Successful tank lining requires an experienced applicator who will use procedures that ensure lap joint integrity, proper cure, and adhesion, while avoiding or at least minimizing bubbles, softening, and swelling.

As with all internal linings, preparation of the pressure vessel, tank or piping to be lined is critical to the success of the finished product. Two industry standards for rubber lining that address these issues are NACE SP0298[29] and British Standard BS 6374.[30]

5.7 FRP Pipe for Oil and Gas Production

Most FRP applications in oil and gas production are used in pipe form.

5.7.1 FRP Pipe Manufacturing

FRP consists of glass fibers wetted with a liquid resin, typically epoxy. The fibers provide strength, and the resin provides corrosion resistance.

Pipe and connections are typically filament wound but can also be centrifugally cast. Joining methods are normally threaded, flanged, or adhesively bonded. Some joints include O-rings or gasket seals compressed by threads or lock rings.

In filament winding, the glass fibers are saturated with the resin material and wound on a circular, rotating mandrel at carefully controlled winding angles with relation to the axis of the pipe. The winding process continues until enough layers of glass fibers are wound to give a specified wall thickness. In centrifugally cast pipe, the glass fibers saturated with resin material are cast against the inside of a mandrel by centrifugal force.

Many categories of glass fibers can be used in fiberglass construction, including

- E: Electrical
- C: Chemical
- S: Strength

As this nomenclature suggests, each type is formulated for an end use. All oil field FRP products are made with E-glass, an alumino-borosilicate glass with less than 1% by weight alkali oxides. The fibers are wound at angles that depend on the application. For example, if the loading is along the axis, as in a rod-pumped well rod, the glass will be placed in that direction. However, pipe is under internal pressure and may see bending and other loads, so the glass is wound at an angle of 53.75° in a single or double wound helix.

Most pipe for flowlines is made without a chemically resistant internal C-glass liner. These liners are called the "resin rich layer" (90% resin 10% glass) and are usually found in chemical applications.

Several different resins with different properties are used in manufacturing FRP pipe and tanks. Most are epoxy, vinyl ester, or polyester resins. It is very important that the material matches the intended service. When ordering FRP pipe or tanks, the vendor needs to know what the item will be used for, or the product may fail. Just taking any FRP tank or pipe from the warehouse will not do. For example, no money is saved if the solvent in an oil field chemical dissolves the resin from the glass in a tank used to store the chemical (a true story). Thus, for specific applications, particularly for chemicals other than produced fluids, the chemical and FRP pipe- or vessel supplier should be consulted.

5.7.2 Designing an FRP Pipe System

Steel piping systems fail mainly after being in corrosive service for some period, whereas FRP piping systems can fail before commissioning. The following are ways to avoid early failures of FRP:

- Recognize the differences between nonmetallic and metallic materials and be able to work with both types.
- Use design factors that are unique to nonmetallic materials.
- Employ designers who are familiar with FRP piping system design, construction, and installation.
- Specify that the design must adhere to relevant API or other industry specifications (see Appendix 5.A).
- Purchase FRP from manufacturers that have qualified quality assurance and quality control plans.
- Qualify FRP pipe joiners for field joint makeup. (More about this issue follows.)
- Follow specifications for adequate burial depth and backfill requirements for onshore systems.
- Hydrotest the completed system. Unlike steel, FRP properties are time dependent, so leaks may not occur upon first application of pressure. Hydrotesting may require several hours, instead of the shorter times that are adequate for steel.

Just the same as requiring qualified welders to join steel and alloy pipe, nonmetallic pipe joiners should be qualified. Field installations that require joint makeup should be done by qualified joiners; steel pipe fitters are not necessarily qualified to join nonmetallic pipe. Training sessions conducted by the pipe manufacturer or other qualified specialists can be held to obtain qualified joiners.

There are several differences between steel and FRP that require attention during design:

- Pipe support spacing must be shorter than for steel.
- Buried pipe works best but requires backfill sand that is devoid of rocks, which can damage FRP pipe.
- FRP pipe laid on top of the ground without special supports is prone to failure due to soil erosion, unsupported span lengths, and point loads from rocks.
- Pipe anchors should avoid point contact with FRP pipe.
- The modulus of elasticity, Poisson's ratio, and coefficient of thermal expansion of FRP pipe depend on the winding angles. The manufacturer's data for that pipe should be used.

- Pressure is derated at much lower temperatures than for carbon steel or alloys.
- Ultraviolet light degrades the exterior surfaces of FRP. UV degradation of the resin penetrates a small distance into the pipe component and leaves a bleached mass of fibrous glass. Paint can prevent UV damage, but paint adhesion to FRP is poor. Fortunately, failures are not typically attributed to UV degradation.[31]
- Static electricity can build up and cause failure of an ungrounded FRP piping system, so FRP should be grounded.
- Buried FRP pipe cannot be detected using pipe locators used for metallic pipe. Some operators install a continuous length of copper wire with the FRP pipe to help locate it.
- Observe the manufacturer's recommended minimum bend radius to avoid resin cracking from excessive bending in pipe laying.
- Avoid cracking FRP flanges when bolted to metallic flanges, by using flat-face flanges or backup rings.
- If FRP pipe is threaded to metallic pipe or fittings, ensure that the FRP is the pin end of the pipe, and the metallic component is the box end.

The most common causes of failure of FRP line pipe are mechanical damage and joint leakage.[31-33] Many failures occur before commissioning or early in the life of the system due to mechanical external-force damage or hydraulic hammering, and to poor joining technique. In some cases, failure has been traced to poor manufacturing quality of the base pipe or fitting.

Sometimes FRP fittings are more difficult to obtain than metallic fittings made of carbon steel or stainless steel. Operators have resorted to inserting metallic components into FRP piping systems; however, because of significant differences in the modulus of elasticity, connecting a threaded steel or stainless steel pin end into an FRP threaded box results in cracking of the FRP box. When inserting a threaded metallic pipe into a threaded FRP piping system, an FRP pin end should be threaded into a metallic box end.

If a carbon steel flange is used to mate flanged steel pipe to a flanged FRP pipe, a steel backup ring should be inserted behind the FRP flange so that bolt torque does not cause excessive bending and subsequent cracking of the FRP flange, especially in raised face flanges. The radius of curvature of bent FRP pipe should not exceed the manufacturer's recommendations to avoid resin cracking.

5.7.3 FRP Pipe for Flowlines, Well Lines, Manifolds

FRP flowlines have operated for more than 30 years in petroleum production.[33] Designs adhering to API specifications 15LR and 15HR have improved the consistency of design and standardization of FRP products.[34-35] API RP15TL4 provides information for transporting, handling, installing, and reconditioning fiberglass tubulars in oil field usage.[36] The hydrostatic design basis is standardized in ASTM D2992.[37]

When API 15LR or 15HR are specified, Method A in ASTM D2992 is required. Unlike carbon steel, FRP "creeps" at ambient temperature. This means that its strength is time dependent, and over a period, will decrease. The ASTM D2992 method requires long-term testing, and a defined method to extrapolate test time to a 20-year design life. Plastic pipe designed using these standardized procedures has 20-year long-term hydrostatic strength. This method describes static and cyclic fatigue methods so that pressure cycling can be accounted for. Essentially, the result is an S-N fatigue curve for the pipe if Method A is used.

A significant difference between FRP pipe used in flowlines and manifolds, and FRP pipe used in chemical processing plants is the higher pressure encountered in the oil field. Operators have used threaded downhole tubing to make up high-pressure FRP line pipe if the diameter and pressure requirements can be met.

5.8 FRP Tanks

FRP tanks and vessels are used in many services in production related operations. Probably the most numerous are relatively small tanks used for oil field chemical and other chemical storage. However, many are used in water injection (water flood and salt water disposal) systems and as lease crude tankage (Figures 5.14 and 5.15). The consequences of exposing FRP tanks to high temperatures created by exposure fires should be considered. This material loses strength as the temperature increases. FRP tanks should be suitably protected against fire exposure or located so that any spills resulting from the failure of these materials could not unduly expose people, buildings, structures, or other equipment to the possible fire incident. The fire hazards and local regulations associated with storing flammable products in FRP tanks must be considered.

As was noted earlier, manufacturing FRP tanks uses similar techniques to FRP pipe manufacturing. FRP tanks are made using either a filament winding or spray-up process. The thermoset resins used in both filament-wound and spray-up fabrication commonly include polyesters, vinyl esters epoxies, and isophthalic reins. Spray-up is an open molding fabrication method used to make FRP tank bottoms, and dished tops and parts with large surface areas. Chopped fiberglass reinforcements and catalyzed resin are deposited onto a mold surface. Rollers or squeegees are used to remove entrapped air and work the resin into the reinforcements. Woven fabric is added to give the laminate greater strength.

Large diameter FRP tanks can be shop-fabricated in segments and shipped to the field site where the manufacturer's field technicians install the tank, assemble and attach the bottom and head, and install all fittings, manways, and internals. Large diameter, field installed fiberglass tanks are made using the same materials and fabrication methods as standard shop-fabricated tanks. The tank shell may be either chop glass/hoop wound or contact molded. The bottom, dished head, and all joints and secondary bonding are contact molded.

In many cases, FRP tanks are made up of multiple layers with different resins and possibly different glass thicknesses in each layer. In such cases, the inside layer is designed to be resistant to the substance stored in the tank, inner layers are structural, and the outer layer would be designed for atmospheric protection—including ultraviolet protection. The thermoplastic C glass liner, if required, is not considered to contribute to mechanical strength. The FRP liner is usually cured before winding or lay-up continues. API Spec 12P covers standards and specifications for materials, design, fabrication, and testing of FRP tanks.[38]

Because FRP is not a good conductor of electricity, static charge can build up in an FRP tank. Metal piping, thief hatches, bull plugs, vent valves, and level transmitters sit isolated on top of the tanks.

Lightning can induce voltages in these metal bodies, potentially resulting in arcing that can ignite vapors in the tank. It has been observed that FRP tanks can completely fail during lightning storms, and cause fires. FRP tanks require grounding to avoid this problem. API RP 2003 can be used as a guide.[39] In Figure 5.14A, a green grounding wire is found at the bottom of the middle tank (left side of photo).

A

B

Figure 5.14 FRP storage tanks used in the oil field: (A) crude oil and water service, with a tank capacity of 500 barrels (79.5 m^3); and (B) saltwater disposal tanks. *Photographs provided courtesy of LF Manufacturing.*

Figure 5.15 A vertical FRP water tank demonstrates the diversity of size and dimensions available in FRP. *Photograph provided courtesy of LF Manufacturing.*

5.9 FRP Sucker Rods

When FRP sucker rods first reached the market, they were considered a corrosion-resistant rod; the FRP rod body is corrosion resistant. However, the connections are steel and in some rod string designs it has been necessary to run steel sucker rods or weight bars to provide string weight for proper pumping action. Thus, corrosion inhibition programs are still required to protect the metal parts of the rod string and the steel tubing. FRP sucker rods do have their use; however, they need to be justified for mechanical reasons—not to eliminate or reduce inhibition costs.

FRP rods are recognized as viable additions to rod pumping equipment and API's Spec 11B and RP 11BR include FRP rods in their coverage.[40-41] Because mechanical loading is uniaxial, the direction of glass fibers in FRP rods is in the axial direction.

5.10 FRP Production and Injection Tubing

FRP downhole tubing can be considered for high-pressure water injection systems because of its superior resistance to corrosion. Water systems can become contaminated with dissolved oxygen and become corrosive to carbon steel. As with carbon steel tubing, threaded joints are used with FRP tubing. The design must account for all loads on the tubing string. The axial load imposed by the weight of the string can limit the depth that FRP tubing can be run downhole.

FRP tubing is filament wound. Epoxy is the most common resin used. Bisphenol A (BPA) and novolac epoxy are the two main epoxy resins used in petroleum production, with BPA being the most common, and novolac epoxy having the highest chemical resistance. Most manufacturers use E-Glass to minimize cost.

Thread makeup is critical to achieving a leak-tight system. A manufacturer's representative should be present to train the crew if they are not familiar with FRP. Chemicals to be used for future well activities should be confirmed they are compatible with FRP string. Acids containing HF (e.g., mud acid) should not contact FRP. If chemicals that are detrimental to FRP are required, the FRP string should be removed and replaced with a temporary work string.

Injection tubing (water and CO_2) for water-alternating gas injection has had successful application of an FRP liner that is grouted into a steel tubular. Steel provides pressure and tensile load resistance. Joints are made by adding a corrosion barrier ring, which provides a smooth, continuous joint; this ring does not allow any gas or liquid into the joint (refer to Chapter 6, Figure 6.22).

5.11 FRP Offshore Firewater Pipe

There are only a limited number of firewater systems that are fabricated from FRP located on offshore platforms. Note that most owners who installed FRP firewater systems first confirmed in fire tests that it would be a reliable system. Deluge firewater systems are an important type of fire protection used on wellhead and process areas of offshore platforms. Fire tests have demonstrated that FRP pipe can withstand a fire for a requisite amount of time in completely water-filled firewater systems with continuous flow, such as offshore ring mains. However, for incompletely filled or dry parts of a firewater system, other material choices, or fire-protective coatings on FRP must be used.[42-43] Most piping in these systems is maintained dry, with open spray nozzles and sprinkler heads. The dry systems are separated from pressurized wet systems by deluge valves, which open and create deluge upon detection of fire and/or high gas levels. Vast amounts of water are released through the sprinklers during deluge.

Some of the operators that have installed FRP firewater pipe originally installed internally coated carbon steel. Using FRP eliminated the corrosion associated with carbon steel offshore firewater piping. Internal coatings have not proved successful in eliminating sprinkler nozzle plugging from carbon steel corrosion products and disbonded coatings.

Pressure surges must be considered in the design of a firewater system, particularly one made of FRP. Surges occur when the system is on test or called into use in a fire. Field tests in one offshore location measured transient pressure surges in fiberglass firewater pipe up to 170% of the design pressure. This fiberglass system failed because of these pressure surges, poor-quality pipe (weeping), and poor pipe joining. It was replaced with 90Cu-10Ni pipe.

ASTM F1173 covers the classifications of, and basic requirements for, thermosetting fiberglass reinforced resin pipe systems with nominal pipe sizes (NPS) 25–1200 mm (1–48 in), which are to be used for all fluids approved by the authority having jurisdiction in marine piping systems.[44] The standard addresses specific requirements such as internal and external pressure, fire endurance, flame spread, smoke, and other toxic products of combustion, temperature, material compatibility, electrical resistance, static charge shielding, potable water usage, glass content, and wall thickness.

FRP competes with 90Cu-10Ni, titanium, and super austenitic stainless steels for offshore firewater applications using seawater. However, FRP has been selected only for a limited number of locations. This may have to do with unfamiliarity with composite materials, and a few horror stories about FRP failures in the industry that are associated with manufacturing or joint flaws. Oil companies that have considered using FRP have been discouraged by a lack of industry design and installation standards. Critical FRP applications, like firewater, require people who are knowledgeable about FRP properties and design, joining, and installation to be an important part of the design, specifications, and installation team.

5.12 Composite Repair Systems for Corroded Steel

The portability of FRP and its relative ease of application over pipe and pressure vessels make it suitable for use in repairing corroded or leaking metallic piping and pressure vessels. FRP has also been used to wrap a stainless steel pressure vessel that experienced chloride stress corrosion cracking (CSCC) at its coastal location (see the discussions on SCC and CSCC in Chapter 3, Sections 3.17 and 3.17.3). Once good bonding has been ensured, enough wrap is applied to withstand the loads. The design and application of composite repair systems is addressed in ASME and ISO standards in Appendix 5.A, and the references.[45-46] Operating companies normally have practices that address where composite repairs are classified as "temporary" or "permanent."

Although wraps can be applied to any configuration, the easiest is cylindrical shaped equipment. Nozzles and flanges are more difficult to get a leak-tight repair. Putting a wrapped nozzle or flange under vacuum after wrapping is completed can reduce porosity.

A composite repair is shown in Figure 5.16. The offshore piping experienced significant external corrosion. When corrosion is significant, the operator must ensure that abrasive blasting can be performed without perforating the pipe. The line was successfully abrasive blasted, and an epoxy resin was applied to wet out the carbon–glass hybrid composite wrap, which was in roll form. Total wrap thickness is 8 mm using four wraps, 2 mm each, with overlap. The repair system meets ASME PCC2 article 4.1.[45]

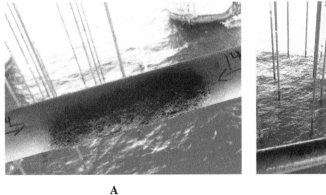

A B

Figure 5.16. A composite wrap of offshore piping with significant external corrosion: (A) externally corroded pipe before remediation; (B) pipe after remediation with 8 mm thick carbon–glass hybrid reinforced epoxy.
Photograph provided courtesy of Belzona International Ltd.

5.12.1 Cold Weld Compounds

Another use of reinforced epoxies to restore internally or externally corroded flange faces is "cold weld compounds," i.e., epoxy-based polymeric compounds filled with metallic or ceramic particles. These compounds are not fiber reinforced. After the required surface preparation, the polymer is applied and blended to restore the flange dimension. Consideration must be given to the temperature limits and flame spread rating of the compound in hydrocarbon service. Unlike composite repair systems, there are no industry standards for this application.

5.13 Elastomers

Another class of polymeric materials related to polymers is elastomers. The most common uses are as seals and packing. For example, most O-rings are made from elastomers. Another use of elastomers (in sheet form) is as tank linings. Although not directly involved in corrosion control, selecting the proper elastomer can have an effect on failures and downtime of equipment. Furthermore, the integrity of seals is the key to being able to operate systems oxygen free, as discussed in the Chapter 9 on controlling the corrosive environment.

5.13.1 Seals and Packing

Elastomers seal by deforming (extruding) into a confined space and filling the gap in that space. Elastomers are highly incompressible. Because of this incompressibility, they transmit pressure in the same way as a liquid. Rubber seals are chosen because they combine reasonable stiffness with an ability to conform to the shape of the sealing gap. Most seals used in petroleum production are made of nonmetallic elastomers (synthetic rubbers) or thermoplastics. In critical (high-pressure, sour) service where a leak can expose personnel or the environment to H_2S, metallic seals are often selected over nonmetallic seals—but, even in these services, elastomer seals are found.

"Rubber" and "elastomers" are terms that are often used interchangeably, especially if the term rubber is not limited to natural rubber and includes synthetic rubber. However, elastomers and thermoplastics are two entirely different classes of materials, so these terms should not be used interchangeably. Oil field equipment seals are manufactured from either synthetic rubber or thermoplastic materials. Thermoplastics are described earlier in this chapter. Depending on the specific thermoplastics selected, they offer a wide range of chemical resistance. Unlike elastomers, thermoplastics have low resilience, and may require springs or O-rings for seal energization. Thermoplastics creep under sustained load[3] and may be reinforced with glass fibers to improve wear properties and deformation under load[4]. Elastomers do not flow under sustained load.

Elastomers exist in two forms—glassy and rubbery—depending on the temperature. In amorphous (or glassy) and semicrystalline polymers, the glass transition is due to a reduction in motion of large segments of molecular chains with decreasing temperature. Upon cooling, the glass transition corresponds to the gradual transformation from a liquid to a rubbery material, and finally to a rigid

---

3. A sustained load is a constant, static force like internal pressure or tensile force. Even under a constant force, thermoplastics "creep" or change their elongation over time.
4. "Under load" is a sustained, static load—the force that is trying to break the seal apart (usually pressure).

solid. The temperature at which the polymer experiences the transition from rubbery to rigid state is termed the "glass transition temperature," T_g. The properties are quite different above and below the glass transition temperature. At temperatures below T_g, the polymer is rigid and glassy. Except below, at, or near T_g, elastomers are resilient materials, which recover quickly and fully after deformation. After stretching or compressing they rapidly return to their original shape. Melting and glass transition temperatures are important parameters relative to in-service application of polymers because they define the upper and lower temperature limits for many applications.[47]

5.13.1.1 O-Rings

A common elastomer seal is an O-ring, which is a circular cross-section torus molded from an elastomer that fits into a metallic groove called a "gland." O-rings may be used in static and dynamic seals; the latter involves maintaining a seal while the joint moves. Static seals are found on bolted covers and joints. Dynamic sealing is more difficult to achieve than a static seal and can be found in pump shafts and other moving parts.

5.13.1.2 Gaskets

Nonmetallic gaskets are used in stationary, flanged joints to provide a leak-tight seal in flat-faced and raised-face flanges. Using compressive forces applied by threaded fasteners, the gasket creates a seal between the two flange faces and contains the internal pressure at that joint. In spiral wound gaskets, the filler material between the inner and outer metallic ring includes graphite, ceramic, and fluorocarbon. Carbon and graphite-containing gaskets can cause crevice corrosion of stainless steel flange sealing surfaces because these materials are cathodic to stainless steel. Flat rubber gaskets are found in least hazardous and aggressive conditions such as low-pressure water service.

5.13.1.3 Other Seals

There are many other types of seals used in petroleum production. Centrifugal pumps often contain nonmetallic seal components. Valve stems may contain a series of stacked cup and cone cylinders. These seals may be constructed of graphite and/or or nonmetallic polymers such as PTFE, PEEK, perfluorocarbon (FFKM), or others depending on the severity of the service conditions. Refer to Section 5.13.2 for a discussion of these elastomers.

5.13.1.4 Flange Isolation

Electrical isolation is required to prevent galvanic corrosion between flanges of dissimilar metals (e.g., carbon steel to stainless steel flange in a corrosive fluid service) or to keep cathodic protection current from going past a joint and to isolate CP. The latter involves a change in custody, such as a pipeline going from offshore to onshore, or at a customer metering station. Full face insulating sets for flat face and raised face flanges include a high dielectric strength insulating gasket, insulating sleeves to be placed around the bolts, and insulating washers backed by steel washers (refer to Chapter 9, Figure 9.2). The nonmetallic insulator is often made from reinforced nylon, phenolic-based plastic with nitrile rubber facing, or an energized fluorocarbon seal. Phenolic-based products may contain reinforcing fibers to improve their tensile and compressive strength. They are brittle and prone to cracking. Glass-reinforced epoxy materials are used in buried lines, which are immersed in high moisture environments because of their lower water absorption than phenolic products. Bolt sleeve insulators are made from polyester.

Insulating gaskets must be checked periodically for electrical resistance and shorting. Their need can often be eliminated in the design stage, for example, increasing the wall thickness of the anode, coating the cathode, and moving the bimetallic joint into an oil-wetted zone that will not promote galvanic corrosion. Typical problems encountered include cracked insulating washers and/or bolt sleeves and missing insulating bolt sleeves. Shorting occurs where metallic connections occur from damage to the insulating gasket set, or because metallic connections that bridge the insulating gasket are made externally, such as instrumentation lines (refer to Chapter 9, Figure 9.3). Highly conductive fluid (such as brines) causes current to jump across the insulating gasket and requires a higher resistance path, such as an insulating pipe joint.

Monolithic isolating joints have been installed in locations where a flange is undesirable, such as transitions in high-pressure pipelines between offshore to onshore. These joints are not flanged and contain pipe stubs that can be welded into place. These monolithic insulating joints contain nonmetallic rings, so if the pipe weld requires postweld heat treatment, heat damage to the nonmetallic components should be considered.

5.13.2 Classification of Elastomers

Elastomers and thermoplastics play a vital role in providing a sealing function for numerous tools and equipment in the exploration for, and the production of, oil and gas. However, the choice of the correct seal material to suit the applications is difficult because of the wide variety of elastomers available and service conditions encountered. ASTM D1418 provides a classification system for elastomers in which 3- and 4-letter descriptors are used to generically categorize various elastomers.[48] Oil and gas producers tend to specify specific elastomers by trade name manufacturer instead of generic elastomer classes because of the wide range of properties within a given elastomer family.

The properties of elastomers and thermoplastics used for seals not only depend on the generic chemistry of the class of polymer, as described by the ASTM International classification, they also depend strongly on fillers that are used to make the final product. Fillers can be reinforcing or nonreinforcing, and include carbon black, talcum, clay, aramid, plasticizers, metal oxides, stabilizers, pigments, extending oils, lubricants (e.g., graphite, PTFE), and curatives. Fillers lower overall cost and improve certain properties. Curatives optimize cross-link density.

5.13.2.1 Classification of Commonly Used Seal Materials

Although there are many elastomers to choose from, the following classifications (3- and 4- letter designations) are the typical elastomers used for surface, subsurface, and subsea equipment:

- EPDM (Ethylene propylene diene monomer)
- NBR (Acrylonitrile butadiene, also known as nitrile)

- HNBR [(Hydrogenated acrylonitrile butadiene; also known as "HSN" (highly saturated nitrile)]
- FKM (Fluorocarbon)
- FEPM (Tetrafluoroethylene propylene)
- FFKM (Perfluorocarbon)

Following are the common thermoplastics classifications used for surface, subsurface, and subsea equipment:

- PPS (Polyphenylene sulfide)
- PEEK (Polyetheretherketone)
- PVDF (Polyvinylidene fluoride)
- PTFE (Polytetrafluoroethylene)

ASTM D4000 provides a standard classification for plastic materials with a 2–5 letter designation system.[49] This is a gateway standard to individual ASTM International standards for each plastic material.

5.13.3 Chemical Resistance of Elastomers and Thermoplastic Seals

Some seal materials perform satisfactorily in water, but not in oils; others may be used with oil if there is no H_2S; and many are sensitive to high-pressure gases. Table 5.2 presents only some the examples of possible environments and elastomers. As with other materials, for proper seal selection, it is essential that the equipment vendor knows the service environment—especially in critical situations. Detailed, up-to-date information on elastomer properties may be obtained from seal vendors. Basic elastomer property information may be found in the *NACE Corrosion Engineers Reference Book*, and in other reference books.[50–52] Downhole seals in valves, injection equipment, and subsurface safety valves are very critical, and the most resistant elastomers should be chosen for those applications. The fluoroelastomers (FKM, FEPM-ETP, and FEPM-TFE/P), and perfluoroelastomers (FFKM) are usually selected for maximum reliability.

5.13.3.1 Explosive Decompression and Chemical Degradation

When assessing a polymer's corrosion resistance, consideration must be given to the produced fluids and to chemicals that may contact the seal infrequently, or frequently in small concentration. Important produced fluids to consider are CO_2 and H_2S. H_2S can embrittle elastomers such as NBR and HNBR. CO_2 in high-pressure ($\geq$4000 kPa [600 psig]) gas can permeate into the polymer and result in explosive decompression damage whereby the polymer is internally blistered or cracked when pressure is released rapidly. NBR is particularly susceptible to explosive decompression, whereas fluorinated polymers have good resistance. Infrequent chemical contact compounds to consider include hydrate inhibitors (glycol, methanol) and cleaners made from aromatic compounds such as xylene and toluene. FKM is adversely affected by contact with 100% methanol but can withstand 90% or lower concentration. Aromatic compounds are detrimental to NBR, HNBR, and fluoroelastomers FEPM and TFE/P, which are copolymers of tetrafluoroethylene and propylene. Frequent contact with low concentration chemicals occurs when operators inject amine corrosion inhibitors, which adversely affect NBR and many varieties of FKM fluoroelastomers.

5.13.3.2 Maximum and Minimum Service Temperature

The maximum temperature for thermoplastics is greater than for elastomers. Examples are PPS to 205°C (400°F); PEEK to 260°C (500°F); and PTFE to 205°C (400°F). Thermoplastics can operate in colder temperatures than elastomers. For example, PPS and PEEK can be used to −60°C (−75°F), and PTFE to −200°C (−328°F). For elastomers, NBR can be used to −30°C (−22°F) if formulated for arctic service; HNBR to −23°C (−10°F); FFKM to −18°C (0°F); and EPDM to −51°C (−60°F). The cold service temperature limit for FKM varies with the type. For FKM 2, the cold temperature limit is −10°C (+14°F). The temperature limits included here allow for some safety factor. Similar but somewhat colder temperatures can be found in the references.[51]

Table 5.2 Examples of the selection of elastomeric seals for different environments.

Water use
EPDM—to 275°F (135°C) without oil present
HNBR, NBR, fluorocarbon—to 180°F (82°C) with oil

Crude oil use
Fluorocarbon, NBR—to 121°C (250°F) without H_2S
HNBR—to 121°C (250°F) with H_2S <0.3 kPa (0.05 psi)
H_2S partial pressures >0.3 kPa (0.05 psi) partial pressure require FKM, FFKM, or thermoplastics PPS, PEEK, or PTFE

Carbon dioxide (CO_2) use
EPDM—to 275°F (135°C)—qualification is recommended (NACE TM0192; NORSOK M-710. See Appendix 5.A)
Nitrile—to 275°F (135°C)—qualification is recommended (NACE TM0192; NORSOK M-710. See Appendix 5.A)

API, NACE, and NORSOK have standards that address testing elastomers and are described in Appendix 5.A:

- API Bull 6J[53]
- NORSOK M-710[54]
- NACE TM0187[55]
- NACE TM0192[56]
- NACE TM0296[57]

5.14 Offshore Riser Splash Zone Protection

Vulcanized chloroprene (neoprene) rubber has been used successfully for decades to protect import and export risers from seawater corrosion at the splash zone of fixed offshore platforms. The maximum temperature is approximately 80°C (180°F). The riser is shop-coated and vulcanized, then brought to the field for installation. For temperatures >80°C (180°F), hot risers can be sheathed with Monel 400 (UNS N04400) that is seal-welded to the riser (note: seal welds may adversely effect the corrosion-fatigue life of the riser).

Riser protection works best when it is brought high enough outside the splash zone up to the lower deck of the platform. A typical coverage is 4 m (13 ft) below mean sea level to 2 m (6.5 ft) above the sea deck. In this way, seawater mist does not enter into the interface, which can cause unseen corrosion, and the lower part of the riser is protected by the platform cathodic protection system.

There are many retrofit riser splash zone protection systems on the market. In addition to welded split-sleeve repairs, doubler plates, and clamps, nonmetallic materials include coatings (e.g., high-build epoxy), adhesively bonded FRP sleeves, sleeved petrolatum tapes, grouted composite sleeves, in situ vulcanized rubber, and many others. Users should consider testing any unfamiliar products before applying them in the field. Test methods are described in the references.[58] However, most test methods do not include testing under conditions of elevated temperature to test for the effect of disbondment. Shear stress created by differences in temperature and in coefficients of thermal expansion between a steel riser and the nonmetallic riser coating system, has resulted in disbondment of nonmetallic riser coating systems.

References

1. NACE SP0187 (latest edition), "Design Considerations for Corrosion Control of Reinforcing Steel in Concrete" (Houston, TX: NACE International, 2008).
2. NACE SP0395 (latest edition), "Fusion-Bonded Epoxy Coating of Steel Reinforcing Bars" (Houston, TX: NACE International, 2013).
3. NACE SP0290 (latest edition), "Cathodic Protection of Reinforcing Steel in Atmospherically Exposed Concrete Structures" (Houston, TX: NACE International, 2007).
4. NACE RP0390 (latest edition), "Maintenance and Rehabilitation Considerations for Corrosion Control of Atmospherically Exposed Existing Steel-Reinforced Concrete Structures" (Houston, TX: NACE International, 2009).
5. ASTM C150 (latest edition), "Standard Specification for Portland Cement" (West Conshohocken, PA: ASTM International, 2018).
6. IMO Resolution A.753(18) (latest edition), "Guidelines for the Application of Plastic Pipes on Ships" FTP Code (London, United Kingdom: International Maritime Organization, 2010).
7. Transport Canada TP14612 (latest edition), "Procedures for Approval of Life-Saving Appliances and Fire Safety Systems, Equipment and Products," https://www.tc.gc.ca/eng/marinesafety/tp-tp14612-menu-4477.html (Jan. 28, 2019).
8. ASME NM.1 (latest edition), "Thermoplastic Piping Systems" (New York, NY: ASME International, 2018).
9. API 15LE (latest edition), "Specification for Polyethylene (PE) Line Pipe" (Washington, DC: American Petroleum Institute, 2018).
10. ISO 24033 (latest edition), "Polyethylene of Raised Temperature Resistance (PE-RT) Pipes—Effect of Time and Temperature on the Expected Strength (Geneva, Switzerland: International Standards Organization, 2009).
11. ASTM D1598 (latest edition), "Standard Test Method for Time-to-Failure of Plastic Pipe Under Constant Internal Pressure" (West Conshohocken, PA: ASTM International, 2015).
12. ISO 12176 (latest editions Parts 1 through 4) "Plastics Pipes and Fittings, Equipment for Fusion Jointing Polyethylene Systems (Geneva, Switzerland: International Standards Organization, 2017).
13. ASME B31.3 (latest edition), "Process Piping" Chapter VII Non-Metallic Piping (New York, NY: ASME International, 2016).

14. API 15S (latest edition), "Spoolable Reinforced Plastic Line Pipe" (Washington, DC: American Petroleum Institute, 2016).
15. Hamner, N.E., Ed., *Corrosion Data Survey–Nonmetals Section*, 5th ed. (Houston, TX: NACE International, 1975).
16. Graver, D.L., Ed., *Corrosion Data Survey–Metals Section*, 6th ed. (Houston, TX: NACE International, 1985).
17. API 17J (latest edition), "Specification for Unbonded Flexible Pipe" (Washington, DC: American Petroleum Institute, 2017).
18. DNVGL ST-F119 (latest edition), "Thermoplastic Composite Pipes" (HÐvik, Norway: DNVGL, 2017).
19. API 17K (latest edition), "Specification for Bonded Flexible Pipe" (Washington, DC: American Petroleum Institute, 2017).
20. NACE SP0304 (latest edition), "Design, Installation, and Operation of Thermoplastic Liners for Oil Field Pipelines" (Houston, TX: NACE International, 2016).
21. Lebsack, D., R. Egner, and D. Hawn, "Extending Liner Installation Lengths," CORROSION 2004, paper no. 04714 (Houston, TX: NACE International, 2004).
22. Groves, S., and D. Hawn, "Connector Technologies for Joining Lined Pipeline Segments," *Materials Performance*, 39, 9 (Houston, TX: NACE International, 2000).
23. Franco, R.J., "Materials Selection for Produced Water Injection Piping," *Materials Performance*, 34, 1 (Houston, TX: NACE International, 1995).
24. NACE Publication 35101-2001, "Plastic Liners for Oil Field Pipelines" (Houston, TX: NACE International, 2001).
25. Goerz, K, L. Simon, J. Little, and H. Fear, "A Review of Methods for Confirming Integrity of Thermoplastic Liners—Field Experiences," CORROSION 2004, paper no. 04703 (Houston, TX: NACE International, 2004).
26. Lebsack, D., and D. Hawn, "Internal Pipeline Rehabilitation Using Polyamide Liners," CORROSION'97, paper no. 560 (Houston, TX: NACE International, 1997).
27. Mason, J.F., "Pipe Liners for Corrosive High Temperature Oil and Gas Production Application," CORROSION'97, paper no. 80 (Houston, TX: NACE International, 1997).
28. Wolodko, J., B. Fotty, and T. Perras, "Applications of Nonmetallic Materials in Oil Sands Operations," CORROSION 2016, paper no. 7298 (Houston, TX: NACE International, 2016).
29. NACE SP0298 (latest edition), "Sheet Rubber Linings for Abrasion and Corrosion Service" (Houston, TX: NACE International, 2007).
30. British Standard BS 6374-5 (latest edition), "Lining of Equipment with Polymeric Materials for the Process Industries—Part 5 Lining with Rubbers" (London, United Kingdom: British Standards Institution, 2017).
31. Chiu, A., and R.J. Franco, "FRP Linepipe for Oil and Gas Production," *Materials Performance* 29, 6 (Houston, TX: NACE International, 1990).
32. Oswald, K.J., "The Effect of 25 years of Oil Field Flow Line Service on Fiberglass Pipe," *Materials Performance* 27, 8 (Houston, TX: NACE International, 1988).
33. Oswald, K.J., "Thirty Years of Fiberglass Pipe in Oil Field Operations: A Historical Perspective," *Materials Performance* 35, 5 (Houston, TX: NACE International, 1996).
34. API 15LR (latest edition), "Specification for Low Pressure Fiberglass Line Pipe" (Washington, DC: American Petroleum Institute, 2018).
35. API 15HR (latest edition), "High-Pressure Fiberglass Line Pipe" (Washington, DC: American Petroleum Institute, 2016).
36. API RP15TL4 (latest edition), "Recommended Practice for Care and Use of Fiberglass Tubulars" (Washington, DC: American Petroleum Institute, 2018).

37. ASTM D2992 (latest edition), "Standard Practice for Obtaining Hydrostatic or Pressure Design Basis for "Fiberglass" (Glass-Fiber-Reinforced Thermosetting-Resin) Pipe and Fittings" (West Conshohocken, PA: ASTM International, 2018).
38. API 12P (latest edition), "Specification for Fiberglass Reinforced Plastic Tanks" (Washington, DC: American Petroleum Institute, 2016).
39. API RP2003 (latest edition), "Protection Against Ignitions Arising Out of Static, Lightning, and Stray Currents" (Washington, DC: American Petroleum Institute, 2015).
40. API Spec 11B (latest edition), "Specification for Sucker Rods" (Washington, DC: American Petroleum Institute, 2010).
41. API RP 11BR (latest edition), "Recommended Practice for the Care and Handling of Sucker Rods" (Washington, DC: American Petroleum Institute, 2015).
42. Thon, H., H. Haanes, R. Stokke, and S.H. Hoydal, "Use of Glass Fibre Reinforced Plastic (GRP) Pipes in the Fire Water System Offshore," 1991 ASME International Conference on Ocean, Offshore & Arctic Engineering, Volume III-B, Materials Engineering (New York, NY: ASME International 1991).
43. Fredriksen, A., D. Taberner, J.E. Ramstad, O. Steensland, and E. Funnemark, "Reliability of GRP Seawater Piping Systems," Veritec Report No. 90-3520 (Hövik, Norway: Veritas Offshore Technology and Services A/S, 1990).
44. ASTM F1173 (latest edition), "Standard Specification for Thermosetting Resin Fiberglass Pipe Systems to Be Used for Marine Applications" (West Conshohocken, PA: ASTM International, 2018).
45. ASME PCC-2 (latest edition), "Repair of Pressure Equipment and Piping," Article 4.1 "Nonmetallic Composite Repair Systems (High Risk Applications)," and Article 4.2 "High and Low Risk Applications (Low Risk Applications)" (New York, NY: ASME International, 2018).
46. ISO 24817 (latest edition), "Petroleum and Natural Gas Industries—Composite Repairs for Pipework—Qualification and Design, Installation, Testing and Inspection" (Geneva, Switzerland: International Standards Organization, 2017).
47. Callister, W.D., Jr., *Fundamentals of Materials Science and Engineering an Integrated Approach*, 2nd ed. (Hoboken, NJ: John Wiley and Sons, 2005): pp. 462–463.
48. ASTM D1418 (latest edition), "Standard Practice for Rubber and Rubber Latices—Nomenclature" (West Conshohocken, PA: ASTM International, 2017).
49. ASTM D4000, "Standard Classification System for Specifying Plastic Materials" (West Conshohocken, PA: ASTM International, 2016).
50. Baboian, R., *NACE Corrosion Engineer's Reference Book*, 4th ed. (Houston, TX: NACE International, 2016).
51. Schweitzer, P.A., *Corrosion Resistance of Elastomers* (New York, NY: Marcel Dekker, 1990).
52. Schweitzer, P.A., *Mechanical and Corrosion Resistant Properties of Plastics and Elastomers*, (Boca Raton, FL: CRC Press, 2000).
53. API Bull 6J (inactive but available for purchase), "Bulletin on Testing of Oil Field Elastomers—A Tutorial (Washington, DC: American Petroleum Institute, 1992).
54. NORSOK M-710 (withdrawn but available for purchase), "Qualification of Nonmetallic Materials and Manufacturers—Polymers (Oslo, Norway: NORSOK, 2001).
55. NACE TM0187 (latest edition), "Evaluating Elastomeric Materials in Sour Gas Environments" (Houston, TX: NACE International, 2011).
56. NACE TM0192 (latest edition), "Evaluating Elastomeric Materials in Carbon Dioxide Decompression Environments" (Houston, TX: NACE International, 2012).
57. NACE TM0296 (latest edition), "Evaluating Elastomeric Materials in Sour Liquid Environments" (Houston, TX: NACE International, 2014).
58. Semerad, V.A.W., F.A. Corsiglia, D. Weaver, and C. Cox, "Testing of Epoxy Adhesives for a Splash-Zone Coating Retrofit System for Marine Pipeline Riser Applications," CORROSION 2003, paper no. 03042 (Houston, TX: NACE International, 2003).

Appendix 5.A: Compilation of Standards Concerning Nonmetallic Materials Used in Petroleum Production

NACE International (Houston, TX)

SP0298 (latest edition), "Sheet Rubber Linings for Abrasion and Corrosion Service"

SP0304, "Design, Installation, and Operation of Thermoplastic Liners for Oil Field Pipelines"

TM0187 (latest edition), "Evaluating Elastomeric Materials in Sour Gas Environments"

TM0192, "Evaluating Elastomeric Materials in Carbon Dioxide Decompression Environments" (latest edition)

This TM provides procedures to measure the effect on elastomeric materials subjected to rapid depressurization from elevated pressures in dry carbon dioxide environments and is designed for testing O-rings or other specimens of elastomeric vulcanites.

TM0296, "Evaluating Elastomeric Materials in Sour Liquid Environments"

This TM describes an accelerated aging procedure with additional information on sour environment testing under pressures greater than atmospheric pressure, allowing data from separate laboratories to be compared if specified test conditions are used. Designed for testing O-rings or specimens of elastomeric vulcanizates cut from standard sheets.

TM0297, "Effects of High-Temperature, High-Pressure Carbon Dioxide Decompression on Elastomeric Materials"

This TM presents test procedures to measure the effect of rapid depressurization from elevated pressures and temperatures in dry CO_2 environments on elastomeric materials.

TM0298, "Evaluating the Compatibility of FRP Pipe and Tubulars with Oil Field Environments"

This TM provides a means to evaluate the relative resistance of most fiber-reinforced plastic (FRP) pipe and tubular products to specific oil field environments by comparison of apparent tensile strength before and after exposure.

American Petroleum Institute (API, Washington, DC)

API Bull 6J, "Bulletin on Testing of Oil Field Elastomers–A Tutorial"

This bulletin provides a tutorial for the evaluation of elastomer test samples of actual elastomeric seal members intended for use in the oil and gas industry. It also reviews testing criteria, evaluation procedures, guidelines for comparisons, and effects of other considerations on the evaluation of elastomeric seal materials and members.

API SPEC 12P, "Specification for Fiberglass Reinforced Plastic Tanks"

This specification covers material, design, fabrication, and testing requirements for FRP tanks. Only shop-fabricated, vertical, cylindrical tanks are covered. Tanks covered by this specification are intended for aboveground and atmospheric pressure service.

API Spec 12R, "Specification for Fiberglass-Reinforced Plastic Tanks" (latest edition)

This specification covers minimum requirements for material, design, fabrication, and testing of FRP tanks.

API Spec 15HR, "Specification for High Pressure Fiberglass Line Pipe"

This specification covers fiberglass line pipe and fittings rated for operating pressures greater than 1000 psi (6895 kPa). Quality control tests, dimensions, and performance requirements are included. Some spoolable pipe is manufactured to this specification (see API 15S).

API Spec 15LE, "Specification for Polyethylene (PE) Line Pipe"

This specification provides standards for PE line pipe suitable for use in conveying gas, oil and non-potable water in underground service for the oil and gas producing industries. Dimensions, materials, physical properties, and service factors are included. This specification does not address PE-lined steel pipe.

API Spec 15LR, "Specification for Low Pressure Fiberglass Line Pipe"

This specification covers glass fiber reinforced thermosetting resin line pipe suitable for use in conveying gas, oil, or nonpotable water in the oil and gas producing industries at a maximum rating of 1000 psi (6895 kPa).

API Spec 15LT, "Specification for PVC Lined Steel Tubular Goods"

This specification provides standards for PVC lined steel pipe or tubing suitable for use in conveying water and/or oil in the petroleum industry.

API Spec 15S, "Spoolable Reinforced Plastic Line Pipe"

This specification provides requirements for the manufacture and qualification of spoolable reinforced plastic line pipe in oil field and energy applications including transport of multiphase fluids, hydrocarbon gases, hydrocarbon liquids, oil field production chemicals, and nonpotable water, and performance requirements for materials, pipe, and fittings.

API RP 15TL4, "Recommended Practice for Care and Use of Fiberglass Tubulars"

This RP provides information on the transporting, handling, installing, and reconditioning of fiberglass tubulars in oil field usage. Appendices are also included to cover adhesive bonding, repair procedures, and inspection practices.

API Spec 17J, "Specification for Unbonded Flexible Pipe"

This specification applies to both static and dynamic flexible pipes used as subsea flowlines, risers, and jumpers.

API Spec 17K, "Specification for Bonded Flexible Pipe"

This specification applies to both static and dynamic flexible pipes used as flowlines, risers, jumpers and offshore loading and discharge hoses and constructed of bonded flexible pipe.

ASME International (New York, NY, USA)

NM.1, "Thermoplastic Piping Systems"

This ASME standard has been issued to address nonmetallic pressure piping systems. ASME NM.1 provides requirements for design, materials, manufacture, fabrication, installation, inspection, examination, and testing of thermoplastic pressure piping systems. This Standard address both pipe and piping components that are produced as standard products, and custom products that are designed for a specific application.

NM.2, "Glass Fiber-Reinforced Thermosetting Resin Piping"

The NM.2 Standard provides requirements for design, materials, manufacture, fabrication, installation, inspection, examination, and testing of glass-fiber-reinforced thermosetting-resin piping systems. This Standard address both pipe and piping components that are produced as standard products and custom products that are designed for a specific application. Glass-fiber-reinforced thermosetting-resin pipe and piping components manufactured by contact molding, centrifugal casting, filament winding, and other methods are covered.

RTP-1, "Reinforced Thermoset Plastic (RTP) Corrosion Resistant Equipment"

This Standard applies to reinforced thermoset plastic corrosion resistant vessels used for the storage, accumulation, or processing of corrosive or other substances at pressures not exceeding 0.1 kPa (15 psig).

Section X, "Boiler and Pressure Vessel Code Section X Fiber-Reinforced Pressure Vessels"

Section X includes three Classes of vessel design: Class I and Class III—qualification through the destructive test of a prototype; and Class II—mandatory design rules and acceptance testing by nondestructive methods.

PCC-2, "Repair of Pressure Equipment and Piping"

The standard addresses welded as well as nonwelded repairs and includes training and installation requirements for engineered composite repair systems.

ASME B31.3 "Process Piping, Chapter VII Nonmetallic Piping"

ASTM International (West Conshohocken, PA, USA)

ASTM C150 "Standard Specification for Portland Cement" (West Conshohocken, PA: ASTM International, 2018).

ASTM D1598, "Standard Test Method for Time-to-Failure of Plastic Pipe Under Constant Internal Pressure for Thermoplastic and Reinforced Thermosetting Resin Pipe"

ASTM D1418 "Standard Practice for Rubber and Rubber Latices–Nomenclature"

ASTM D4000, "Standard Classification System for Specifying Plastic Materials"

ASTM D2992, "Standard Practice for Obtaining Hydrostatic or Pressure Design Basis for "Fiberglass" (Glass-Fiber-Reinforced Thermosetting-Resin) Pipe and Fittings"

ASTM F1173, "Standard Specification for Thermosetting Resin Fiberglass Pipe Systems to Be Used for Marine Applications"

International Standards Organization (ISO, Geneva, Switzerland)

ISO 14692, "Petroleum and Natural Gas Industries—Glass-Reinforced Polymer (GRP) Piping"

   Part 1: Vocabulary, Symbols, Applications and Materials.
   Part 2: Qualification and Manufacture
   Part 3: System Design
   Part 4: Fabrication, Installation and Operation

ISO 24817, "Petroleum and Natural Gas Industries–Composite Repairs for Pipework–Qualification and Design, Installation, Testing and Inspection"

ISO 12176, "Plastics Pipes and Fittings, Equipment for Fusion Jointing Polyethylene Systems"

   Part 1: Butt Fusion
   Part 2: Electrofusion
   Part 3: Operator's Badge
   Part 4: Traceability Coding

ISO 24033, "Polyethylene of Raised Temperature Resistance (PE-RT) Pipes–Effect of Time and Temperature on the Expected Strength"

International Maritime Organization (IMO, London, United Kingdom)

   The International Maritime Organization is the United Nations specialized agency with responsibility for the safety and security of shipping and the prevention of marine and atmospheric pollution by ships.

IMO Resolution A.753(18), "Guidelines for the Application of Plastic Pipes on Ships"

Resolution provides guidelines covering acceptance criteria for plastic materials in piping systems, appropriate design and installation requirements and fire test performance criteria for assuring ship safety (including floating offshore drilling and production structures).

Norwegian Offshore Standards (NORSOK: Oslo, Norway)

M-622, "Fabrication and Installation of GRP Piping Systems"

The standard is based upon ISO 14692 (all parts) but extended with sections on quality control and NDT including two annexes describing ultrasonic and radiography testing, respectively.

M-710, "Qualification of Nonmetallic Materials and Manufacturers–Polymers"

The standard describes a methodology for establishing long term chemical compatibility for polymers based on an accelerated testing scheme for components wetted by production and injection hydrocarbon fluids. Further, for polymers (mainly elastomers) that can be exposed to sudden pressure drops in a gaseous environment, the standard describes a methodology for testing rapid gas decompression resistance.

Canadian Standards Association (CSA, Toronto, Ontario, Canada):

CSA Z662, Oil & Gas Pipeline Systems Section 13, "Fiberglass Pipelines" (latest edition)

CSA Z662 provides guidance on the safe design, construction and maintenance of fiberglass pipeline systems used for petroleum production. This standard also covers HDPE line pipe, spoolable reinforced thermoplastic pipe, and plastic liners for oil field pipelines.

Transport Canada (Ottawa, Ontario, Canada)

T14612, "Procedures for Approval of Life-Saving Appliances and Fire Safety Systems, Equipment and Products for Ships and Floating Offshore Structures"

The approval procedures for life-saving appliances and fire safety systems, equipment and products may be conducted by recognized organizations or product certification bodies under specific Canadian requirements.

DNVGL [Det Norske Veritas (Norway) and Germanischer Lloyd (Germany)] (Høvik, Norway)

DNVGL-OS-C501, "Composite Components"

The standard provides requirements and recommendations for structural design and structural analysis procedures for composite components. Emphasis with respect to loads and environmental conditions is put on applications in the offshore and oil and gas processing industry.

DNVGL-CP-0258, "Filament Wound Fibre Reinforced Thermosetting Resin Tubes for Machine Components and Special Pressure System Components"

The standard addresses the approval of manufacturers intending to supply these products.

DNVGL-ST-F119, "Thermoplastic Composite Pipes"

The standard describes requirements for flexible thermoplastic composite pipes for offshore applications.

American Bureau of Shipping (ABS, Spring, TX, USA)

ABS is the primary class society for the mobile offshore vessels and floating platforms for use in the United States and worldwide. It approves both installation and properties of the various products, for example those that follow the ABS Part 4 "Steel Vessels Rules for Building and Classing Vessel Systems and Machinery."

United States Coast Guard (USCG, Washington, DC, USA)

The USCG approves and inspects all U.S. flag vessels and foreign flag vessels entering U.S. waters to the requirements of Port State Inspection.

Floating vessels and platforms in U.S. waters carrying the U.S. flag are required to use USCG-approved equipment—including polymer piping that has been tested to meet the IMO Fire Testing Procedures (FTP Code) and comply with Title 46, Shipping Parts 41–69 of the *United States Code of Federal Regulations* for its fire and toxicity properties.

Protective Coatings

Timothy Bieri

6.1 Introduction

Chapter 1 covered the requirements for a corrosion cell, i.e., anode, cathode, metallic path, and electrolyte. Given all metals have three of the four requirements inherent, controlling access to the electrolyte is a primary method of controlling corrosion. As the name implies, a "protective coating" is a coating that is applied to a surface to protect it from the surrounding environment by establishing a barrier. Coatings that are applied for decorative or aesthetic purposes are usually referred to as "paints." There is some overlap—the coating used to protect the outside of a steel tank can be decorative, because it improves the appearance of the tank. Likewise, one of the functions of the paint selected for the wood trim on your house is to protect the wood from deterioration. In addition to controlling corrosion, internal coatings may also be applied to increase the efficiency of pipelines by reducing friction, to minimize deposition, or to prevent product contamination. For the purposes of this chapter, the word "coating" will refer to liquid or powdered applied protective coatings, linings, and/or coating systems, unless otherwise noted. Figure 6.1 and Figure 6.2 show the coated exteriors and interiors of tanks.

In general, a coating is a relatively thin film or layer, usually applied as a liquid or powder, which on solidification, is firmly adhered to the surface to be protected, and serves as a barrier keeping the environment from contacting the substrate (i.e., the metallic surface). Key characteristics of the coating to protect against corrosion for a practical period of time include flexibility, resistance against impact, chemical resistance to the environment to which it is exposed, resistance to permeation by moisture, good adhesion[1] and cohesion,[2] and resistance to the temperature to which they are exposed; refer to Figure 6.3.

---

1. Adhesion relates to the strength of the bonds between two different materials, i.e., coating molecules to substrate molecules.
2. Cohesion relates to the strength of the bonds within the same material. i.e., coating molecules to coating molecules.

Figure 6.1 Externally coated tanks at fluid handling facility.

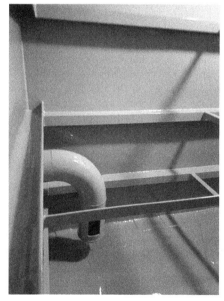

Figure 6.2 Newly coated tank.

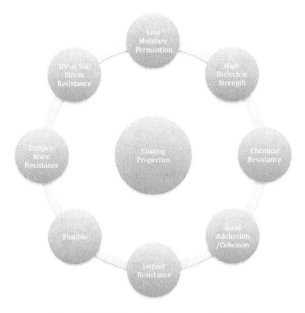

Figure 6.3 Coating performance considerations.

In addition to the coating performance, other factors need to be considered: design life, availability, and ease of application can affect the economics. Economics has to be considered when developing a coating program. For example, it may be more economical to invest in a premium (more expensive–long life) coating system, than to select a less expensive system, which will require periodic maintenance and recoating. The costs associated with the materials, surface preparation, application, and maintenance over the service life can be combined to create the lifecycle cost. This cost can be stated as "cost per mil per square foot (cost per micrometer per square meter) per year of service life."[1] The various economic decisions a company must make were discussed in Chapter 4, Section 4.1.

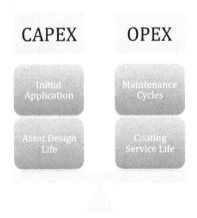

Figure 6.4 Balancing Capital Cost (CAPEX) against Operational Cost (OPEX).

CHAPTER 6: Protective Coatings

The first part of this chapter will discuss coatings topics such as classifications, chemistry, properties, surface preparation, application, and inspection in general. The latter part will discuss coating specific production equipment.

6.2 Coating Formulation

Usually available in cans, the substance called a "coating" is not just a single material—it is not just an epoxy or polyurethane or vinyl resin. Rather, it is a combination of materials, each of which is important to achieve whatever the desired properties are from that formulation. The formulation is key to enable the reaction from a liquid to a solid film, which will be resistant to the service environment. The basic components in a coating are the resin, pigment, vehicle, and additives. In the most general terms, these are the solid and the liquid portions of the coating. A coating "family" name is typically based on the resin; refer to Table 6.1. Not all coatings are available in all geographic areas, and some (e.g., coal tar epoxy) may not be permitted by local environmental regulations.

Table 6.1 Typical liquid-applied protective coatings.

Primary Resin (Binder)	Typical Use (Service)
Acrylics	External
Alkyds and modified alkyds	External—weathering
Coal-tar epoxy	Immersion
Epoxy—catalyzed (two part epoxy)	Immersion
Epoxy ester (one part epoxy)	External—Interior
Phenolic (baked)	Internal—vessel, tubing
Epoxy modified (baked)	Internal—vessel, tubing
Oil modified	External—marine
Polyesters and fiber reinforced polyesters	Immersion—tank/vessel
Silicones	High temperatures
Urethane	
Oil modified (Urethane Alkyd)	External—weathering
Moisture cured	External—concrete
Converted (catalyzed) Vinyl	External—internal
Vinyl acrylic	External—weathering
Zinc—rich	
Inorganic, post-cure, water-based	External—primer
Inorganic, self-cure, water-based	External—primer
Inorganic, self-cure, solvent-based	External—primer
Organic, one package	External—primer
Organic, two package	External—primer
Organic, three package	External—primer

6.2.1 Resin or Binder

The resin, also referred to as the "binder," has a number of functions including binding the pigments together to make a homogeneous film. The binder also serves to wet the surface (i.e., metal or previous coat), to promote adhesion, serves as the primary barrier to the environment, and maintains its integrity in a corrosive environment. While the binder in many coatings is composed of only one resin, many others have more than one compatible resin. The combination imparts specific properties.[2] Many different generic resins are available and are used in coatings for production equipment.

6.2.2 Pigment

The pigment is often thought of as just the coloring agents in the coating. However, they are much more than that. Table 6.2 lists reasons that pigments are used in coatings. To accomplish these tasks, many formulations will contain more than one type of pigment, e.g., inhibitive pigments, inert and reinforcing pigments, and color pigments. Each pigment has its own function in the final coating and is intended to improve the coating performance.

Table 6.2 Example pigment properties used in protective coatings.

Provide corrosion inhibiting characteristics
Decrease the permeability of the coating film
Hide the surface underneath
Provide color and the desired gloss
Protect the film from the effects of ultraviolet light and weather
Perform self-cleaning and controlled chalking
Provide specific surface finishes
Increase the coating consistency to allow thicker film application

The inert and reinforcing pigments are often added to improve the density, chemical resistance, toughness, adhesion, and bonding, as well as to increase the dry film thickness. Reinforcing pigments include glass flake, fiber roving, or mica flake. Additionally, pigments may be used to contribute to the viscosity, thixotropy, and overall workability of the coating in its liquid stage. Color pigments are used to impart opacity, to hide the underlying surface, to protect from ultraviolet (UV) ray deterioration, as well as provide color.[1]

6.2.3 Vehicle

The vehicle's purpose is to essentially help get the coating on the substrate. The vehicle is primarily made up of solvents, which as the name implies, are the fluids that dissolve the binder and create a usable liquid. As the coating cures, or dries, the solvents evaporate and are not part of the dry film. Traditionally, most solvents in protective coating formulations have been mixtures of various organic materials, such as aliphatic and aromatic hydrocarbons, ketones, esters, alcohols, and ethers. Water is used as the solvent in some formulations but, until recently, most water-based coatings have been for commercial rather than industrial use (with the exception of the inorganic zinc silicates). With the worldwide concern for air quality, many locations have tight limits on the release of volatile organic compounds (VOC). These restrictions have affected the use of many coating formulations. Coatings industry research is continually reformulating coatings to reduce or eliminate the VOC content. Water-based formulations, as well as solvent-free powder or liquid coatings, are receiving much of the attention.

6.2.4 Additives

Plasticizers are a type of additive that can be in the formula to provide flexibility, extendibility, and toughness.

6.3 Coating Types

6.3.1 Coating Classification or Type

Coatings may be classified or categorized in many ways:

- Service environment: atmospheric, immersion, or buried. Each of these environments has different requirements for the coating and each can be further subdivided. For example, atmospheric environments may be interior or exterior, where the same coating system could require different surface preparation.
- Equipment type: internal vessel coating and internal tubing coating—which may be quite different—yet both call for immersion coatings.
- Film thickness: "thick film" or "thin film" based on the thickness of the coating film after it has cured.
- More recently, coatings have been categorized as "original construction coatings" and "maintenance coatings."
- Classed by chemical composition.
- Curing method: air dry, forced cured, baked on, or fusion bonded.

While the above is not an all-inclusive list, it does show the need to be clear about what "coating" is in the conversation or specification.

6.3.2 Coating Thickness

The dry film thickness of a coating is important from several standpoints. The optimum thickness range for a specific coating depends on the coating type, substrate, and service conditions. Most of the coatings have a range for the allowable thickness. The required coating thickness will also determine the desirable anchor pattern to be obtained by surface preparation blasting. Coating performance can be adversely affected if it is applied too thick or too thin.

- Too Thick—The cohesive forces within the coating may be stronger than the adhesive forces that hold the coating on the surface. The result is that the too-thick coating can disbond or crack.
- Too Thin—A coating will not provide a proper barrier, and the metal surface will corrode.

6.3.3 Chemical Composition

Classification of coatings, based on chemical composition, include organic and inorganic coatings. Both types of coatings applied on vessels, tanks, piping, and tubing are basically liquid-applied coatings that are sprayed, brushed, or rolled onto the surface and allowed to cure. Table 6.1 lists many of the chemical types that are, or have been, used in production facilities. Organic coatings are the most common materials used to protect oil and gas production facilities and equipment. The basic components in organic coatings are organic resins, including the various epoxies, vinyls, phenolics, and urethanes.

Inorganic coating is a term used to classify coatings that are not organic. Liquid-applied inorganic coatings usually have a silicon-based vehicle, which may carry a pigment (e.g., the inorganic zinc silicate primers). This broad category also includes metallic coatings applied to protect carbon steel. Metallic coatings may be either sacrificial, such as galvanizing with zinc, or a corrosion-resistant barrier, such as electroless nickel or chromium plating used on valve, pump, and related equipment components. Other inorganic coatings are placed on the surface by plating, thermal spraying, or casting techniques. For example, thermally sprayed aluminum or cast cement lining for pipe products.

6.3.4 Curing Methods

The terms "curing" or "drying" are often used interchangeably, but in fact mean two different processes. Combined, these two processes convert the wet as-sprayed film to the hardened final protective film.

- Drying is essentially the evaporation of the vehicle (solvents).
- Curing refers to the process whereby a chemical reaction takes place to form the film. This reaction can be internal to the coating itself, a reaction to the environment, or a combination of both.

Different coatings have different drying and curing requirements.

"Air dried" is the most common form of curing, particularly for atmospheric coatings. Solvents or reaction products evaporate from the coating at ambient temperature. The curing is due to the reaction of the binder with the atmosphere. The drying time is unique to a specific formulation, and the application specification should emphasize the importance of following the coating manufacturer's instructions.

"Force" curing is related to air dried, except the equipment is heated (usually with hot air) to reduce the curing time. This approach is not applicable to all coatings because if the surface is dried too fast, the drying will not be complete and the coating will not be properly cured. Force cured is most likely to be used for internal tank or vessel coating. It is necessary to ensure coating compatibility with forced curing processes because not all coatings are amenable to forced curing.

"Baked" coatings require higher temperatures to cure and are actually baked in an oven at temperatures around 205 °C (400 °F). The epoxies, phenolics, and similar internal tubing coatings are baked coatings as are some vessel coatings. These materials are usually applied in very thin, multiple coats and receive intermediate bakes at lower temperatures after each coat.

"Fusion-bonded" coatings are a special type of material, which are applied as a solid powder to a heated metal surface. The melting particles adhere to the hot metal and fuse together to form the coating film. The heat also supports the curing of the coating. The most common material of this type is fusion-bonded epoxies (FBE) used as pipe coatings. Fusion bonded coatings have very low VOC, which makes them attractive to original equipment manufacturers.

6.3.5 Reinforced Coatings

Polyester or epoxy coatings reinforced with a matt fiberglass or other materials (carbon fiber, geotextile fabric, etc.) are used to provide thicker, stronger linings for new equipment, or for rehabilitating and restoring damaged or leaking tanks or vessel bottoms.

6.4 Coating Systems

The term "coating systems" is typically used to define protective coatings in oil and gas. The common perception is a coating system consists of a primer, intermediate, and topcoat—each with its own purpose. However, the number of coating layers can vary from a single coat to five (or more) coats. When the terms "two coat" or "three coat" are used, it typically refers to the different types of coatings with different functions—*not the actual number of times the structure was sprayed with coating.* However, some coating systems use the same material for more than one category of coat (e.g., the primer coat could be used as both primer and intermediate).

Figure 6.5 illustrates the three main ways coatings protect the substrate from corrosion. Different coating layers can be combined based on the coating selection process. Figure 6.6 shows a five-coat system and indicates the purpose of each of the three different kinds of coats.[3]

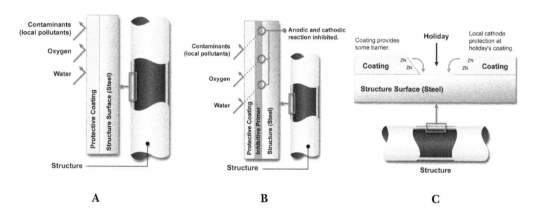

Figure 6.5 How coatings protect (A) barrier coating, (B) inhibitive coating, and (C) sacrificial coating.
Photograph provided courtesy of NACE International.

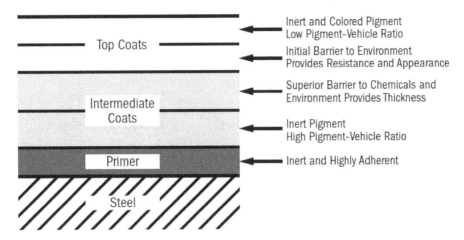

Figure 6.6 Five-coat impervious coating system.

6.4.1 Primers

The primer is required for most protective coatings and is considered to be a most important part of the coating system. Table 6.3 summarizes the primary purposes of a primer.

Table 6.3 Purposes of the primer in a coating system.

Primer Coat Purpose:
Provides an effective bond to the metal substrate
Provides a tie (intercoat bond) to the subsequent coat
May provide a strong resistance to corrosion and chemicals as a barrier
May provide corrosion protection as an inhibitor (inhibitive primer)
May provide corrosion protection with sacrificial metal (zinc-rich primer)

When equipment or facilities are fabricated, the surfaces may be prepared, and the primer coat applied before the equipment is shipped. It may be weeks or even months before the final coat(s) are applied. Thus, the primer may have to prevent atmospheric rusting for an extended period. There are three basic types of primers—barrier, inhibitive, and sacrificial (or galvanic or cathodic). All three types provide the bonds to the metal substrate and the next coat.

Barrier primers are often variations on the formulations of the other coats (i.e., the resin may be the same). This is the usual type of primer for immersion coating systems. Barrier primers rely on inert characteristics and very strong adhesion.

Inhibitive primers are formulated with inhibitive pigments that will suppress rust in atmospheric service. Historically, the most widely used inhibitive pigment was red lead; however, health and safety regulations required alternative materials. Pigments containing phosphates or similar nontoxic compounds are among those in use today.

Sacrificial primers contain a high concentration of zinc dust. The zinc acts as a sacrificial anode to protect the bare steel when the coating is damaged. Available as both inorganic zinc primers and organic zinc primers, the choice depends on the service and the topcoat material. The zinc-rich primers are sometimes used without a topcoat. Although similar in action to galvanizing, the zinc-rich primers can be easily topcoated with most resins, whereas galvanizing requires special treatment to be successfully topcoated. Zinc-rich primers are normally used with atmospheric coatings. Many marine/offshore coating systems use zinc-rich primers.

6.4.2 Intermediate Coats

The intermediate coat(s) in three or more coat systems is also called the "barrier coat" or "body coat." Table 6.4 summarizes the primary purposes of the intermediate coats.[4]

Table 6.4 Purposes of the intermediate coat in a coating system.

Intermediate Coat Purpose:
Provides thickness for the total coating
Provides increased chemical resistance
Provides increased resistance to moisture vapor transfer
Provides increased coating electrical resistance
Provides a strong bond to the primer and to the topcoat

The intermediate coat is a very critical coat in many aggressive environments. That is why many offshore facilities are coated with three-coat systems.

6.4.3 Topcoats

The topcoat, sometimes referred to as the "finish coat," is the first line of defense against chemicals, water, and the environment. Table 6.5 summarizes the typical functions of the topcoat.[4] The topcoat can also serve other functions in addition to, or in lieu of, a barrier, e.g., nonskid surface.

Table 6.5 Purposes of the topcoat in a coating system.

Topcoat Purpose:
Provides a resistant seal for the coating system
Forms the initial barrier to the environment
Provides resistance to chemicals, water, and weather
Provides a touch- and wear-resistant surface
Provides an aesthetic appearance

Many of the topcoat's functions are similar to the intermediate coat. Often the topcoat will be a different color than the intermediate coat, so the coater and inspector can judge the amount of coverage.

6.5 Coating Selection

Coating selection is not as simple as just picking a product. It is more appropriate to think of it as a system; the right product for the environment applied correctly to meet the functional requirements. These three elements are interrelated and failure of one can lead to failure of the system. There is an adage in the coating industry: "A poor coating material properly applied will outperform the very best coating material improperly applied." While that still may be true today, it is likely that neither scenario will achieve the desired service life.

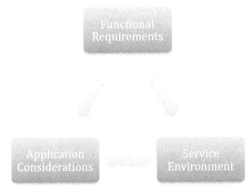

Figure 6.7 Coating selection dependencies.

6.5.1 Service Environment

Because the first thing to be decided when selecting a coating is the service, it is a convenient way to classify coatings. Atmospheric, buried, and immersed environments have their own considerations and requirements for coatings. ISO 12944-2[5] provides a classification of environments related to coating of steel structures, Table 6.6.

Table 6.6 Environment corrosivity categories based on atmospheric corrosivity (C) and the types of immersed (Im) water and soil. (The two categories are independent of each other.)

Atmospheric Corrosivity	Immersed Water & Soil
C1, Very Low	Im1 - Fresh Water
C2, Low	Im2 - Sea or Brackish Water
C3, Medium	Im3 - Soil
C4, High	–
C5, I Very High (Industrial)	–
C5, M Very High (Marine)	–

CHAPTER 6: Protective Coatings

Atmospheric coatings are those used on the external surfaces of facilities. The exposure may be the open atmosphere (exterior and uncontrolled) or may be inside a building (interior and controlled). Coating systems for open atmospheric service are designed to handle ever-changing environments and must be resistant to sunlight (UV), alternate wet/dry conditions, temperature changes, physical damage, and atmospheric contaminants. The selection of the specific coating system will depend on the severity of the atmospheric environment. These are usually air-dried coatings that cure because of the evaporation of solvents or reaction products of two component coatings. The surface preparation is more critical for coatings exposed to the weather than for those coatings in housed conditions.

Immersion coatings are coatings that are immersed in liquid all (or at least most) of the time. These are the coatings used inside vessels, tanks, pipes, and oil country tubular goods (OCTG), as well as external coatings for subsea equipment. Immersion in fluids service can be a very harsh environment for coatings, depending on the fluids in the tank, vessel, pipes, or tubulars. Air dry, force cure, baking, and fusion-bonded materials are all available as immersion coatings. Proper surface preparation and application techniques are critical for immersion coating performance. CUI (Section 6.7.1.2) creates an environment that is similar to immersion conditions and coatings used for corrosion protection under insulation should be selected accordingly.

Coatings for use underground, or buried, must being resistant to corrosive soil environments and soil stresses, and be resistant to damage by rocks and by "cathodic disbondment" from cathodic protection systems.

6.5.2 Functional Requirements

At a high level, the functional requirements can be defined as the primary purpose for the coating combined with the service life. While the primary purpose in this case is corrosion protection, there are several additional considerations:

- Is the coating going to be applied to new surfaces under construction, or are there preexisting coatings and/or corrosion present?
- What type of surface preparation can be performed?
- Are there constraints such as the time available for coating application and curing?
- Which generic coating systems are suitable for the service environment?
- How long does the coating need to last with or without maintenance, and how does that compare to the remaining life of the equipment or field?

6.5.3 Application Considerations

Regardless of the structure to be coated—offshore structure, vessel internals, piping internally and externally, etc.—the design must allow the surfaces to be properly cleaned and prepared, coated, and inspected.

Features as sharp edges, skip welds, blind areas, crevices, and similar design faults spell "coating failure" even before the coating job starts. Table 6.7[6] provides a list of items to consider and include in design specifications to ensure that the structure or equipment is amenable to coating. Additional considerations can be found in NACE SP0178 and ISO 12944-3.[7-8]

Figures 6.8–6.10 illustrate the results of overlooking some of these design items. Other design considerations will be mentioned as appropriate in the discussion on coating of specific production equipment.

Figure 6.8 Complex geometry. Note the rust breakthrough along the edges of beams and stiffeners. *Photograph provided courtesy of Deepwater Corrosion Services.*

A   B

Figure 6.9 Shop-applied coating showing (A) excessive and (B) inadequate application.

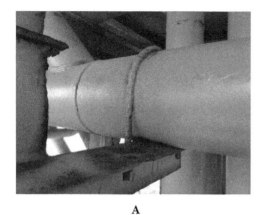 

Figure 6.10 Touch point corrosion: (A) unmitigated metal to metal contact and (B) mitigated with half-round nonmetallic support and coated U-bolt.

Table 6.7 Designing for protective coating (adapted from *Fundamentals of Designing for Corrosion Control*[6]).

Design Considerations to Facilitate Coating
Eliminate sharp corners
Use butt welding instead of lap welding
Remove weld splatter
Specify that welds are to be continuous, no skip welding
Avoid designs that will collect or hold water and debris
Provide drainage in recessed zones
Specify the removal of roughness and surfaces defects by grinding
Specify the rounding of all corners
Eliminate hard to reach places
Specify that surfaces be easily accessible
Provide a continuous and even surface to allow complete bonding of the coating to the metal surface
Eliminate intricate construction that would be difficult to coat; use pipe construction or other simple design
Specify that irregular surfaces, such as threads, be cleaned and caulked
When a thick plate is to be joined to a thinner plate, design the surface to be coated to be flat
Specify that baffles and other vessel internals be removable so the internal surface of the shell can be properly coated
Do not incorporate crevices or pockets in the design; where crevices cannot be avoided, specify that they be continuously welded and welds be ground prior to the application of the coating
Ascertain that mill scale be removed from the steel surface
Provide tanks, vessels or piping, with easy access for coating and inspection
When dissimilar metals are required for vessel internals they should be electrically isolated from the coated shell; where this is not possible, coat both the cathode and anode
All nozzles on internally coated vessels should be 2 in or greater internal diameter (ID) for proper surface preparation and coating

6.6 Coating Application

The term "coating application" is an all-inclusive term meaning the process of coating application—surface preparation, coating the structure, and inspection.

6.6.1 Surface Preparation

Surface preparation involves two areas—preparing a clean, smooth surface and preparing an anchor pattern. Each is important as even the best coating applied over poorly prepared surface can fail. Proper surface preparation maximizes coating adhesion and, all other things being equal, will result in longer service life. The importance of proper surface preparation is illustrated in Figure 6.11.[9]

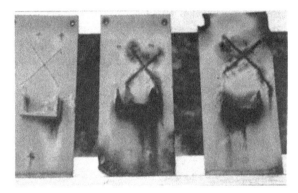

Figure 6.11 The effects of surface condition on coating life—corrosion at scribes in the same coating (three-coat vinyl system) applied over abrasive blasted steel (left), rusted steel (middle), and intact mill scale (right), exposed for 9 years in a marine environment.

Cleaning the surface may be as simple as washing the surface with solvent to remove dirt, oil, and grease, or it may involve grinding to remove weld splatter, to grind out defects in the surface, and to smooth rough welds. Visual comparators exist for weld condition, Figure 6.12. This plastic weld replica was molded from actual welds with various degrees of surface finish and illustrates the written requirements in NACE SP0178.[7]

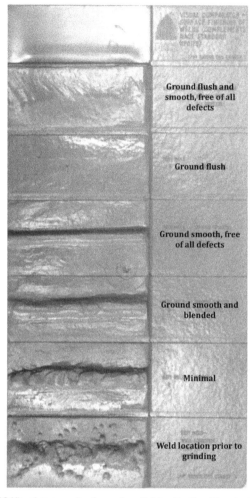

Figure 6.12 Visual comparator for surface finish of butt welds prior to coating.
Photograph provided courtesy of KTA-Tator, Inc.

There are several methods used for surface preparation, and the optimum method will depend on the coating and functional requirements. Blast cleaning is the most common method of surface preparation used for production equipment and facilities. Commonly referred to as "abrasive blasting," blast cleaning may use a number of different materials. Health and safety concerns have brought on regulations in many places concerning the use of sand. Currently, blast material selection varies from garnet, slag, and shot, to agricultural materials such as walnut shells. The choice depends on the cleaning and/or anchor pattern required, cost, and local regulations. Thin components and low-strength materials (e.g., aluminum) require less-aggressive blast media.

There are several different standards that can apply to surface preparation, Table 6.8 cross references the most common global standards. A complete definition of the cleanliness requirements is given in each standard. The standards also cover procedures before and after blasting (immediately prior to coating and lining), blast cleaning abrasives, blast cleaning methods and operation, and inspection requirements. Appropriate comments and explanatory notes are included so that the document can easily be part of the specification for a coating job.

Table 6.8 Surface preparation standards.

NACE	SSPC	ISO
No. 1 White Metal[10]	SP5 White Metal Blast	Sa 3 Visually Clean Metal
No. 2 Near White Metal[11]	SP10 Near White Metal	Sa 2½ Thoroughly Clean Metal
	SP11 Power Tool to Bare Metal	
No. 3 Commercial[12]	SP6 Commercial Blast	Sa 2 Thorough Blast Cleaning
No. 8 Industrial	SP14 Industrial Blast	
No. 4 Brush Blast[13]	SP7 Brush Blast	Sa 1 Light Blast Cleaning
	SP3 Power Tool	St 3 Very Thorough Hand/Power
	SP2 Hand Tool	St2 Thorough Hand Power
	SP1 Solvent Wipe	

Visual standards and comparators are available for use in evaluating the cleanliness on the job. For example, SSPC-VIS 1[14] provides color photographs for the various grades of cleanliness as a function of the preblast condition of the steel.

"Anchor pattern" is the term used to describe the profile of the metal's surface roughness. It is a measure of an average distance between the peaks and valleys created by blast cleaning. Having the proper anchor pattern is a necessary step in achieving satisfactory coating performance. Too small an anchor pattern (i.e., too low a profile) can lead to unsatisfactory adhesion between the coating and the metal, while too large an anchor pattern (i.e., too high a profile) and the coating will not cover the peaks.

Anchor patterns vary depending on many factors including the type and size of the blast material. The anchor pattern requirement is primarily a function of the coating thickness and will usually be specified by the coating manufacturer for a specific formulation. Table 6.9 can be used as a general guide to the relationship between coating thickness and anchor pattern on mild steel.[15]

Table 6.9 Relationship between coating thickness and anchor pattern.

Dry Film Thickness	Anchor Pattern
125-200 µm (5-8 mils)	20-25 µm (1-2 mils)
200-500 µm (9-20 mils)	50-75 µm (2-3 mils)
500 µm or more (over 20 mils)	75-125 µm (3-5 mils)

6.6.2 Coating the Structure

When the surface preparation is complete, the next step is the actual application of the coating. Spray coating is the most common method and the most practical with today's anticorrosion coating technology. Two types of spray equipment are common—conventional air spray and airless spray. The former atomizes the coating with compressed air. The latter forces the coating through the nozzle under high pressure, achieving atomization by the pressure drop at the nozzle orifice. Brush or roller application is also common where access is limited or if overspray can be an issue.

It is very important that the applicator follow the manufacturer's instructions for each specific coating material in the system. Items such as coating viscosity, thinners, mixing (two component coatings), time between coats, and application temperature limits are all critical for a successful coating job. Table 6.10 lists some examples of the application problems that can occur if specifications and instructions are not followed closely.

Table 6.10 Potential coating application problems.

The Coating Job Will Not Be Successful If The Coating Is:	
Too Viscous in Spray Pot:	
	May not spray properly
	The cure time will be improper
	The solvent may be trapped in the film
Too Fluid in Spray Pot:	
	May not spray properly
	May not build required wet film thickness
	Extra coats may be required to achieve desired dry film thickness
Too Thick on Substrate:	
	The cure time will be improper
	Runs, sags, cracks may occur
	The solvent may be trapped in the film
	May disbond from previous coat or substrate
Too Thin on Substrate:	
	May have rust bleed-through
	May allow premature failure
Applied Too Long After Abrasive Blast Cleaning:	
	Surface rusting may have started
	The coating may not properly bond to the surface
	Coating will prematurely fail

Table 6.10 (cont.) Potential coating application problems.

The Coating Job Will Not Be Successful If The Coating Is:	
Applied Too Soon After Previous Coat:	
	May result in solvent entrapment and improper cure
	Can lead to defects—wrinkling, blistering, delamination
Applied Too Long After Previous Coat:	
	May not bond properly if previous coat completely cured
	May not adhere to previous coat due to surface contamination
Too Cold Substrate:	
	May slow or prevent cure
	May not have adequate anticorrosive properties
Wrong Component or Ratio (Two Component Coating):	
	Will not cure or will over-cure and fail
Wrong Thinner:	
	May react and affect spraying
	May prevent cure

The spraying technique is an important aspect of coating application. Factors that can make the difference between a good coating job and a poor coating job include spray gun pressures, spray pattern, spray gun-to-structure distance, and coverage area.

6.6.3 Inspection

The process of properly applying coating has several stages. Quality assurance (QA) and quality control (QC) play important parts in the proper execution of each stage. Many coating project specifications will require a certified coating inspector as an integral part of the QA/QC process.[3]

The role of the coating inspector is a continuing one throughout a coating project. The inspector is there to witness and document the coating process. Ideally, the inspector will start at the prejob conference with the applicator and will continue until the final cure inspections are complete. Large coating jobs might be handled by third party inspectors, whereas smaller ones might be done by user-company personnel.

The primary duty of the inspector is to verify that appropriate specifications are met throughout the project. The inspector must have the authority to enforce the specifications, and if necessary, to see that any deviations are satisfactorily resolved. The inspector will run tests required by the speci-

3. The NACE Coating Inspector Program is delivered through NACE International-Education. NACE Coating Inspector Certification is administered by NACE International Institute. The Institute of Corrosion (ICorr) administers ICorr Coating Inspector courses and certification. SSPC administers the Protective Coatings Inspector Program courses and certification. FROSIO administers a Surface Treatment Inspector certificate program.

fication to evaluate and verify conditions and results. NACE SP0288 summarizes inspection requirements, equipment, and reference standards.[16] Although the document scope is aimed at immersion service, the outline can be adapted for atmospheric coatings as well. The coating specification should include the details of inspection and other quality control issues.

6.6.3.1 Inspection Timing

Initial inspections should take place before the surface preparation begins to ascertain the condition of the structure. Environmental conditions such as metal temperature, ambient weather, and atmospheric contaminants will be important concerns at each step in the project. Surface preparation efforts include verifying that the specifications for cleanliness and anchor pattern requirements are met. During application, the inspector will assure that the various manufacturer's recommendations are met to avoid the items listed in Table 6.10, as well as determine the wet and dry film thicknesses. As appropriate, the dried coats will be inspected for voids or damaged locations. Table 6.11 lists suggested inspection checkpoints. Although this table was adapted from an internal coating standard (NACE SP0181[15]), the same points apply to atmospheric coating projects as well.

Table 6.11 Coating inspection check points.

Typical Inspection Check Points
Preblast cleaning (e.g., ambient conditions, surface temperature, compressor and/or air cleanliness)
After blast cleaning (e.g., cleanliness, profile)
After each coat is applied (e.g., application equipment, mixing, application techniques, wet film thickness)
After final cure (e.g., holiday testing, adhesion tests,)
Final completion inspection (Includes damage inspection after scaffolds removed)

6.6.3.2 Inspection Tools

In addition to routine inspection equipment (mirrors, magnifiers, depth gages, etc.) and comparators previously mentioned, there is a whole host of coating inspection tools available. A selection of common items is presented in this section.

Film thickness gages are used to evaluate both the wet film thickness before curing and the dry film after curing. Figure 6.13 shows a notch-type wet film thickness gage. By knowing the amount of shrinkage during cure, the manufacturer can specify the wet film thickness to obtain the specified dry film. Thus, the applicator or inspector can tell if the correct amount of coating has been applied using a notch-type gauge, because the wet film thickness will be the largest notch that is completely liquid-filled.

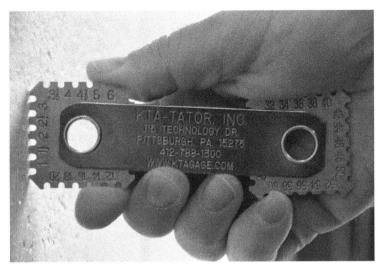

Figure 6.13 Notch-type wet film thickness gage.
Photograph provided courtesy of KTA-Tator, Inc.

Figure 6.14 shows an example of a magnetic pull-off type gage used to measure dry film. Thickness is determined by the amount of "pull" required to lift a magnet off the coating. Other types of dry film thickness include electronic and electromagnetic styles (Figure 6.15).

Figure 6.14 Magnetic coating thickness gage.
Photograph provided courtesy of KTA-Tator, Inc.

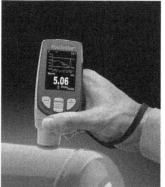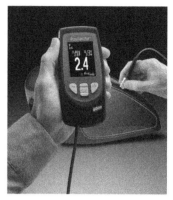

Figure 6.15 Electronic dry film thickness gages.
Photographs provided courtesy of DeFelsko, Inc.

The anchor pattern can be verified on the job by the replica tape method. NACE SP0287 covers the details of this method.[17] Figure 6.16 illustrates a method of measuring a surface profile using replica tape. Visual comparators may also be used to confirm the depth of the anchor pattern (Figure 6.17).

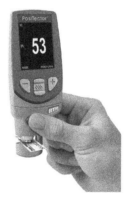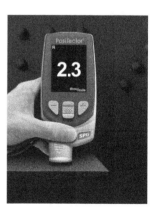

A B C

Figure 6.16 Surface profile (A) and (B) with replica tape, and (C) without replica tape.
Photographs provided courtesy of DeFelsko, Inc.

Figure 6.17 Surface profile visual comparator.
Photographs provided courtesy of KTA-Tator, Inc.

"Holiday detectors" are used to inspect a coating to locate discontinuities in a coating film. A holiday, is a discontinuity in the coating that exposes the unprotected surface to the environment.[18] There are other types of coating discontinuities such as voids, cracks, thin spots, foreign inclusions, or contamination in the coating. Any discontinuity may lower the dielectric strength of the coating in that area. Therefore, electrical techniques can be used to locate the discontinuities. Holiday testers are available in two principal types—low-voltage wet sponge type and high-voltage spark type.

The low-voltage wet sponge holiday detector is applicable for thin-film coatings (DFT < 500 μm [20 mils]) (Figure 6.19). The unit that supplies the low-voltage DC is grounded to the metal structure and connected to the wet sponge. As the wet sponge is passed over the coating, the water will penetrate the holidays and damaged places, completing the circuit and sounding an audible alarm. The water used in the sponge should be tap water with a low-sudsing wetting agent. The sponge speed should be slow enough to detect the holidays (0.3 m/s [1 ft/s] is recommended).[18]

Figure 6.18 Coating inspection with the wet sponge holiday detector.
Photograph provided courtesy of DeFelsko, Inc.

For thick-film coatings (DFT > 500 μm [20 mils]), a high-voltage spark type holiday detector should be used. The high-voltage unit uses a metal wire whisk or spring assembly (exploring electrode) to contact the coating, Figure 6.19. When the electrode reaches a holiday, a spark will jump the air gap to complete the circuit and trigger the alarm. To avoid burning holes in the coating, voltage and current must be controlled for the film thickness involved. The manufacturer can supply the figures for the maximum voltage for the specific coating. NACE SP0188-2006 gives suggested voltages based on dry film thickness.[18]

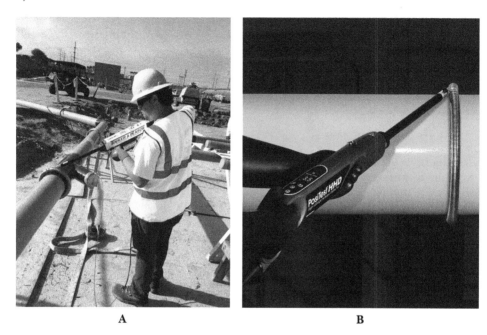

Figure 6.19 High-voltage holiday detection on piping.
Photograph A provided courtesy of Tinker-Rasor. Photograph B provided courtesy of DeFelsko, Inc.

6.7 Coating of Production Facilities and Equipment

This section will look at considerations or issues in the use of coatings for specific equipment and facilities. While coatings are not the total answer to oil and gas corrosion problems, they, by themselves and combined with other control methods, are viable, economical approaches to corrosion control.

6.7.1 Atmospheric Protection

As previously covered, there are five distinct atmospheric corrosivity categories (Table 6.6). Except for appearance, exterior protection of structural steel, tanks, vessels, and piping are essential in the C4 and C5 environments. In moist atmospheres—especially in marine and coastal regions (C5)—a coating system using a sacrificial primer, such as the organic or inorganic zinc-rich coatings, is common. In other areas, it is quite common to use galvanized structural steel and tanks without a topcoat where the thickness of the galvanized layer can meet the desired service life of the equipment. As mentioned previously, if the galvanizing is to be topcoated, there are special considerations regarding the surface preparation and type of primer used. Interior facilities may be externally coated to protect equipment, steel decks, or concrete floors from attack due to spills.

6.7.1.1 Offshore Platforms

Chapter 1, Section 1.9.2.1 discussed the various corrosion zones of an offshore platform: submerged zone, splash zone, and atmospheric zone. The environmental differences in each zone have caused the industry to develop different control approaches for each zone.[19] Cathodic protection in the submerged zone is discussed in Chapter 7.

Splash zone areas are considered by many as the most severe external environment and typically has the highest external corrosion rates on a platform's structure. The splash zone is alternately wet-dry; therefore, cathodic protection is not a practical method of corrosion control. Corrosion-resistant nickel-based and copper-based alloys may be welded over the steel. Other materials such as high-build fiberglass or glass flake coatings, rubbers, heat shrink sleeves, flame sprayed aluminum, and a number of proprietary materials have been used to protect this critical area. Many of these are described further in Chapter 5.

Atmospheric zone environments on offshore platforms are considered by many as the most severe of the atmospheric environments. The coating of platforms has evolved from the original painting schemes adopted from the shipping industry (including frequent "chipping and repainting") to today's sophisticated multicoat systems.

Platform coating is often broken into two categories: original construction and maintenance. Original construction coating is done onshore at the fabrication yard and involves the latest techniques of surface preparation, application, and inspection. Usually, the three to five coat systems are used.

Maintenance coating, as the name implies, is a program of periodic inspection and repair or touch-up. ISO 4628-3[20] (Table 6.12) and ASTM D610[21] (Table 6.13) and provide similar rating systems for evaluating the degree of coating degradation.

Table 6.12 ISO 4628-3 degree of rusting and rusted area.

Degree of Rusting (Ri)	Rusted Area, %
Ri0	0
Ri1	0.05
Ri2	0.5
Ri3	1
Ri4	8
Ri5	40 to 50

Table 6.13 ASTM D610 degree of rusting and rusted area.

Rust Grade	Rusted Area, %
10	0.01
9	0.03
8	0.10
7	0.30
6	1.00
5	3.00
4	10.0
3	16.0
2	33.0
1	50.0
0	<50

The objective is to have a continuous program of preventative maintenance, rather than waiting until the coating approaches total failure, then attempting a costly repainting in the marine atmosphere.[22] When to perform maintenance is an economic and risk-based decision. Some offshore operators use a schedule (e.g., 10 years) or rust rating (e.g., ASTM 7) to determine scheduling. Risk assessment can be considered in modifying the schedule, accelerating recoating of equipment that could result in higher consequence failures, and deferring maintenance coating for lower consequence equipment.

Different operators have different preferences and different specifications for their offshore coating systems. Table 6.14 provides examples of nine different multilayer coating systems used for atmospheric zone coating.

Table 6.14 Typical coating systems used in the Atmospheric Zone.
The groupings indicate the different coating types used in multilayer coating systems.

Coating	Thickness	
	µm	mils
Wash primer Vinyl, intermediate and topcoats (three to four coats)	13 200-250	0.5 8-10
Wash primer Chlorinated rubber, intermediate and topcoats (three to four coats)	13 200-250	0.5 8-10
Inorganic zinc-rich self-cured primer Epoxy intermediate coat Vinyl acrylic or polyurethane topcoat	75 125 50	3 5 2
Inorganic zinc-rich self-cured primer Epoxy intermediate and topcoat (two coats)	75 250	3 10
Inorganic zinc-rich self-cured primer Vinyl high-build intermediate coat Vinyl topcoat (two coats)	75 100-150 50	3 4-6 2
Inorganic zinc-rich post-cured primer Epoxy intermediate coat Vinyl acrylic or polyurethane topcoat	75 125 50	3 5 2
Inorganic zinc-rich post-cured primer Epoxy tie-coat Epoxy intermediate coat Vinyl acrylic or polyurethane topcoat	75 50 100-150 50	3 2 4-6 2
Inorganic zinc-rich postcured primer Copolymer tie-coat Vinyl high-build topcoat	75 50 150-250	3 2 6-10
Inorganic zinc-rich self-cured primer Epoxy tie-coat High-build polyurethane	75 50 150-200	3 2 6-8

Note: The specified number of coats and the thickness may vary among operators and manufacturers.

The criteria for selecting a coating system is outlined in NACE SP0108,[19] NORSOK M-501,[23] or ISO 12944-5.[24] Because real-time testing takes a number of years of onsite exposure, ISO 12944-6,[25] NACE TM0304,[26] and TM0404[27] accelerated test methods are referenced.

6.7.1.2 Corrosion Under Insulation (CUI)

Insulation can be applied to process equipment for one of three main purposes: personnel protection, noise abatement, or process purposes (i.e., keep hot things hot or cold things cold). The problem occurs when moisture enters the insulation system and becomes trapped against the steel surface. Figure 6.20 shows an example of CUI on a carbon steel piping segment.

Figure 6.20 Corrosion under insulation on carbon steel piping.

Given it is better to plan for water to enter the insulation system, many operators are requiring a protective coating on the steel surface to serve as a barrier to CUI. The common types of coatings include thin-film liquid, fusion bonded, thermal spray aluminum, and wax-tape. Selection of the coating must consider the immersion type environment and operating temperature. NACE SP0198[28] provides additional guidance for control of CUI.

6.7.2 Vessels and Tanks

Internal coatings are used in all types of field vessels and tanks, as well as anywhere that free-water (liquid water) can collect or be condensed from gas. Separators, free-water knockouts, heater treaters, water handling vessels, lease tankage, or stock tanks may be coated where corrosive conditions merit protection. Types of coatings include baked phenolics, epoxy-modified phenolics, or ambient temperature-cured and force-cured-catalyzed coatings such as amine, amine adduct or polyamide-cured epoxies, amine-cured epoxy-modified phenolics, amine- or polyamide-cured coal-tar epoxies, hot-applied asphalt and coal-tar pitch enamels, or solvent release polyesters and epoxies. Depending on the coating material, these are applied either in an applicator's shop or by an applicator after erection in the field. The coating choice depends on the service requirements, economics, and coating applicators available. If cathodic protection is to be used in addition to coating, a coating system resistant to cathodic disbonding will be necessary.

Regardless of the coating system selected, the vessel or tank must be designed for coating. The requirements in NACE Standards for immersion coatings—SP0178[7] and SP0181[15]—may be used as guidelines for planning coating projects. They also can be incorporated in whole or part in the coating specifications for a particular project.

Although in most production vessels, the only areas to require coating are the free-water section or along the bottom and the gas section, many times, it is more economical to coat the entire shell than to selectively coat portions. Large storage tanks, however, are more likely to have coating on the underside of the deck and down the sides to the working oil level, and on the bottom and sides up 1 m (3 ft).

The best time to coat a vessel or tank is before it is placed in service; however, used vessels and tanks are often coated to extend their lives. Surface preparation is critical because all the oil, grease, scale, soluble chlorides, and corrosion product must be removed. Shallow pits will blast clean. Sharp edged or deep pits should be ground smooth before blasting if they are to be properly cleaned. If the tank bottoms are corroding through from internal or external corrosion, fiberglass reinforced resin linings or concrete may be used to repair and to minimize further attack.

Bolted tanks present a coating challenge because of the design. The bolts and the gasket make it difficult to obtain a uniform, continuous coating film. Consequently, special treatment is required, such as in NACE SP0181,[15] which presents one approach—undercutting the gaskets, and caulking both the gaskets and bolt heads. Covering the seams and bolts with fiberglass and resin also has been used effectively. Note the resin and the coating must be compatible to assure proper adhesion and sealing between the two materials.

For maximum benefit from a coating system, once the vessel has been placed in service, care must be taken not to damage the coating. Operations and maintenance personnel should be aware that a vessel has internal coating. Labeling the exterior of the vessel is one approach to keeping everyone informed.

6.7.3 Tubular Goods

Internal coating of tubing and other downhole equipment is an established method for controlling corrosion in high-pressure gas condensate wells, flowing oil wells, gas lift oil wells, and water injection wells. To be successful downhole, internal coatings and linings need the following properties:

- Be capable of providing long term corrosion protection in corrosive, high temperature, high-pressure environments
- Have resistance to acid gases (H_2S and CO_2)
- Resist explosive decompression
- Resist the long-term effects of condensate and chemical treatments
- Resist the effect of abrasion/erosion due to sand entrainment
- Resist mechanical damage during completion of the tubing string and during service (wireline operations to run logs and pull valves, plugs, etc.)
- Resist cracking caused by normal flexing during transportation and handling of the tubular

The most common coatings are organic resin coatings very similar to the immersion coating used in vessels and tanks. Cement-lined tubing and cement plus glass reinforced epoxy-lined tubing also may be used in water injection and water alternating gas (WAG) service, subject to their temperature limits. The term "internally plastic coated" (IPC), is often used to designate tubular goods (e.g., tubing, casing, line pipe, and drill pipe) and accessories (e.g., couplings, valves, packers, etc.) that have been internally coated. NACE has several standards covering IPC tubulars including SP0191[29] and SP0291,[30] which respectively cover the application of IPC and its care, handling, and installation.

Several materials are available for downhole use. Basically, all are the so-called "high bake" materials (oven baked in the 204 °C [400 °F] range to achieve the proper cure). The coatings for hydrocarbon service are usually classed as "thin film coatings," less than 250 μm (10 mils) dry film thickness.[29] Water service IPCs are more likely to be "thick film coatings," 250–760 μm (10–30 mils) DFT.[29] Coatings can be selected by laboratory testing.[31] Figure 6.21 shows a coating failure of an epoxy-phenolic coated tubular under test conditions during an autoclave test of a complete ring specimen.

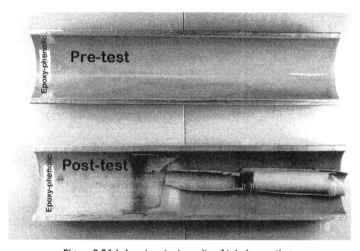

Figure 6.21 Laboratory test results of tubular coating.

When the use of IPC tubing is being evaluated, the tubing connection must be considered. It is important to have a continuous coating system throughout the tubing. Many connections are considered coatable. For example, EUE 8RD, 8RD long casing threads, connections that use a "corrosion barrier" ring—often referred to as a CB ring—and tubing buttress threads are all considered coatable. These connections allow the coating to extend around the nose of the pin to mate with the coating or CB ring in coupling (Figure 6.22).

 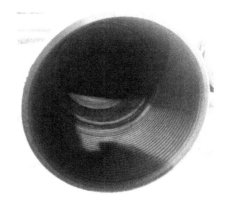

Figure 6.22 Pin and Box end of a connector with glass reinforced epoxy liner and CB rings.

Internal plastic coatings require the same attention to specifications for surface preparation, application, care, and inspection as other coating services. Holiday inspection of coatings is to be carried out to ensure satisfactory quality. Another important consideration is the proper care and handling of coated materials—during transportation, during installation, and while in service. Any pipe can be damaged by improper handling; coated pipe takes just a little more care. Such items such as the use of slings rather than hooks in the ends, leaving thread protectors in place until the joint is in the derrick, use of stabbing guides, etc., will reduce the odds for a premature failure.

Once an internally plastic coated tubing has successfully made it to the well site and has been properly installed, the most likely time for coating damage in a well is when wire-lining. Wireline operations can damage coatings, but when the use of IPC tubing is justified, the possibility of future wireline work should not affect the decision to use IPC. Figure 6.23 shows a sample of coated tubing from a well that had numerous wireline operations. Note that, although the coating is worn, bare metal is not showing, nor is there corrosion. This example is typical of the condition of the coated tubing from this well. No special care had been exercised during wireline operations because the well record did not show coated tubing. When the well was pulled, it was discovered that the lower portion was coated and the upper was bare pipe. When it is known that the tubing string is internally coated, wireline speed should be controlled to minimize damage, typically held to less than 30 m (100 ft) per minute.

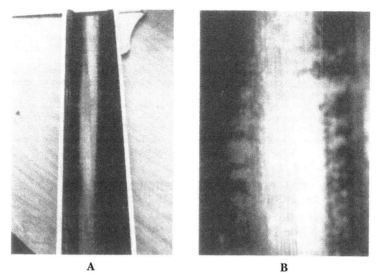

Figure 6.23 Internally plastic-coated tubing that had numerous wireline runs: (A) overall view of the split joint, (B) closeup of the worn area.

As an additional barrier, some companies have justified the use of inhibitors to supplement the coating for critical, high cost/high risk areas (such as high pressure, offshore, remote, urban, and high corrosion rate areas). This is simply a technique to approach complete corrosion control. Furthermore, if relying on inhibitors alone, and the program fails for some reason, massive corrosion would weaken the pipe, requiring well intervention to repair the damage. If the pipe is coated and the supplementary inhibition fails, there may be a hole in the pipe—but structural integrity is maintained—and pulling and replacement costs are held to a minimum.

6.7.4 Flowlines, Gathering Systems, Injection Lines, and Piping

From the standpoint of external corrosion control, the various surface piping systems (whether produced fluids, oil, gas, or water) are treated much alike. That is, aboveground atmospheric protection follows the same guidelines discussed previously. Buried and subsea (submerged) systems rely on pipeline coatings, which are reviewed in the next section. Typically, coatings are supplemented with cathodic protection, which is discussed in Chapter 7.

Internally, most oil well and gas well flowlines and gathering systems depend on inhibitors rather than internal coatings. In extremely corrosive fluids, corrosion resistant alloys are also used. Immersion coatings would be more competitive except for the joining problems. Because most flowlines and gathering lines are high-pressure systems, they are welded. Welding destroys the coating at each joint. A number of proprietary joining systems have been developed, but most have had limited specialized use.

One approach that allows the use of liquid-applied organic immersion coatings to welded lines is "in-place" coating—also called "in situ" coating. The cleaning and surface preparation is the greatest challenge for in-place coating. This highly specialized technique has been most successfully used to coat new lines. New, uncorroded pipe is relatively clean, and stands the best chance of a successful coating application. In spite of the challenges in successfully coating existing systems, in-place coatings have successfully reduced the number of failures and extended the life of some severely corroded piping systems.

Rehabilitating lines in-place with prefabricated liners is more widely used than in situ coating. As mentioned in the section of Chapter 5, Section 5.6.5, on nonmetallics, both thermoplastic and thermosetting pipe have been, and are being, used as liners to rehabilitate older corroded lines. As with other materials, the liners must be matched with the service environment for satisfactory performance.

On the other hand, the problem of internally coating long lengths of welded joints is not a concern in facilities piping systems, because they can be flanged. Flanged piping can be coated if it is designed for coating. The key to coatable piping is accessibility—accessibility for grinding of welds, surface preparation, coating application, and inspection. A rule of thumb is "the pipe piece should have only one change of direction (elbow)—all surfaces must be visible."

The same coating materials and techniques as discussed in the sections dealing with immersion service are used inside piping. Most piping systems can be coated at an applicator's shop where conditions are easier to control than in the field.

Water system gathering, injection, and plant piping is often internally protected. Cement linings and plastic liners have been the most widespread for new construction. Cement linings are usually centrifugally cast in the pipe joints. Techniques involving heat-resistant gaskets that allow welding cement-lined pipe were developed over 30 years ago.

6.7.5 External Pipeline Coatings

Historically, pipelines were welded together and then coated "over the ditch." In contrast, today's oil and gas production flowlines, gathering, and injection systems are primarily made from shop-coated pipe. The only field coating is at the joints or to repair coating damage incurred during transport, handling, and installation.

NACE SP0169[32] is the basic document on the subject. Although usually thought of as a cathodic protection document, NACE SP0169 covers external coatings as well. Much of the information is by reference to other documents that detail specific tests, materials, or procedures.

With a few exceptions, the materials used for external coatings on buried and submerged pipelines are different materials than those discussed previously. The common pipeline coatings are the hot applied coal tars (enamels and tapes), wax coatings (hot and cold applied and tapes),[33] prefabricated films (tapes), fusion-bonded epoxy coatings (FBE),[34-35] and extruded polyolefin coatings (polyethylene and polypropylene jacketed).[36-37]

Multilayer coatings are also commonly installed. The 2-layer system has a layer of about 250 μm (10 mils) mastic, covered and compressed by a 875–1,000 μm (53–40 mils) polyethylene or polypropylene jacket. The extruded poly coatings are used both onshore and offshore. The 3-layer system begins with a layer of FBE over the bare steel, Figure 6.24. The 3-layer coatings are used in higher temperature service and are more resistant to external damage than FBE alone.

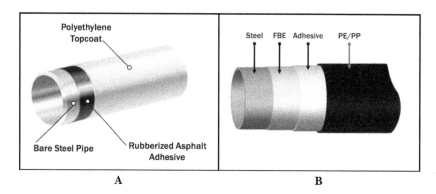

Figure 6.24 Polyolefin pipeline coatings: (A) 2-layer and (B) 3-layer.
Photograph provided courtesy of NACE International.

Figure 6.25 Tape being applied using a tension machine.
Photograph provided courtesy of NACE International.

Figure 6.26 FBE pipeline coating.
Photograph provided courtesy of NACE International.

Figure 6.27 Extruded polyethylene coating.
Photograph provided courtesy of NACE International.

Field joints can be coated with a variety of options[38] including shrink sleeves, liquid coatings,[39] or one of the self-adhesive tape coatings.[40] They also may be coated with field-applied FBE with techniques using an induction heater.[41]

A   B   C

Figure 6.28 Field joint coatings: (A) heat shrink sleeve, (B) tape, and (C) field-applied FBE.
Photograph provided courtesy of NACE International.

Field-applied plastic tapes must be used with care; that is, unless the tape is properly applied with the proper tension, overlap, and mastic or adhesive, it will fail to protect the pipe. Sometimes hand (manual) wrapping will be attempted, usually with poor results, because the tension will not be constant. Manual application will be necessary for some of the materials used as protective coatings for valves, fittings, and joints.

Pipeline coatings are typically supplemented with cathodic protection. The high dielectric strength of these coatings reduces the amount of cathodic protection required by limiting the exposed steel to coating holidays. However, this coating property can also create a shield in certain circumstances. If the coating is disbonded from the pipeline adjacent to a holiday, moisture from the environment can access the steel surface and enable a variety of corrosion mechanisms. The cathodic current beneath the disbonded coating will not be sufficient to provide adequate levels of protection.

External pipeline coatings require the same attention to specifications for surface preparation, application, care, and inspection as other coating services. Holiday inspection of pipeline coatings is covered in SP0274[42] (for general pipeline applications), and SP0490[43] (specifically for FBE pipeline coatings). Figure 6.29 shows the use of a ring type holiday detector scanning a pipeline.

Figure 6.29 High-voltage holiday detection on pipeline.
Photograph provided courtesy of Tinker-Rasor.

6.7.5.1 Subsea

Subsea pipelines or flowlines have similar considerations as buried pipelines, they also present some unique challenges. For example, the method of construction may mean the pipeline is welded and girth welds coated in an assembly line fashion as the pipeline is installed from a lay barge. The girth weld coating needs to be applied, cured (if applicable), and inspected in a continuous process. Another challenge is almost no coating maintenance is possible without bringing the component to the surface.

Current subsea coating systems are the same as the multilayer systems used for onshore-buried pipelines. Additional considerations apply if the subsea pipeline needs to be concrete weight coated or thermally insulated. Figure 6.30 shows a 3-layer polypropylene coating with several layers of syntactic thermal insulation.

Figure 6.30 Three-layer polypropylene coating with multiple layers of syntactic thermal insulation.

6.7.6 Misconceptions

Discussions about coated tubing and damages to the coating often bring up statements such as

> "Corrosion will be accelerated at holidays and damaged places in coating; therefore, we are better off with bare pipe!"

or

> "Wireline damaged coating is worse than no coating at all!"

Neither statement is true! Certainly, it is not when talking about coated tubing in oil field operations.

The reason these are not true has to do with the geometry of the corrosion cell—the cathode/anode areas. As discussed in Chapter 3, the rate of corrosion in a particular environment is related to the area of the anode (which is corroding) and the area of the cathode. Refer back to Figure 3.44 in Chapter 3 where the wireline cut bare pipe illustrates "wear-corrosion." The wireline wore a bright spot in the metal, which was anodic to the rest of the steel and corrosion was accelerated because of the large cathode area (majority of the surface area) versus the small anode area (wireline cut location).

Now coat that same joint and cut through the coating with the wireline exposing the same line of bright steel to serve as an anode—but what about a cathode? The large cathode area from the bare pipe is now isolated with coating. All anodes and cathodes will have to form on the bare metal exposed by the wireline wear—thus, there is only a small anode area–small cathode area relationship. The corrosion rate will lower. There could be less corrosion on damaged coated tubing than on bare tubing.

Even though the coating is tough, it can be damaged by wirelining. Too high wireline speeds can cause the tools to "rattle" and pound on the coating—particularly as they pass through couplings and connections. To minimize damage, tools should have rounded corners, and whenever possible, be coated. Wireline speeds should be held to less than 30 m (100 ft) per minute.

References

1. Vincent, L.D., *The Protective Coating User's Handbook* (Houston, TX: NACE International, 2016).
2. Munger, C.G., and L.D. Vincent., *Corrosion Prevention by Protective Coating* (Houston, TX: NACE International, 2014): p. 73.
3. Munger, C.G., and L.D. Vincent, *Corrosion Prevention by Protective Coating*, Figure 4.8 (Houston, TX: NACE International, 2014): p. 66.
4. Munger, C.G., and L.D. Vincent, *Corrosion Prevention by Protective Coating* (Houston, TX: NACE International, 2014): p. 66.
5. ISO 12944-2:1998, "Paints and varnishes—Corrosion protection of steel structures by protective paint systems—Part 2: Classification of environments" (Geneva, Switzerland: International Organization for Standardization, 1998).
6. Landrum, R.J., *Fundamentals of Designing for Corrosion Control* (Houston, TX: NACE International, 1984): p. 317.

7. NACE SP0178-2007, "Fabrication Details, Surface Finish Requirements, and Proper Design for Tanks and Vessels to be Lined for Immersion Service" (Houston, TX: NACE, 2007).
8. ISO 12944-3:1998, "Paints and varnishes—Corrosion protection of steel structures by protective paint systems—Part 3 Design Considerations" (International Organization for Standardization, Geneva, Switzerland, 1998).
9. Munger, C.G., and L.D. Vincent, *Corrosion Prevention by Protective Coating* (Houston, TX: NACE International, 2014): p. 203.
10. NACE Joint Surface Preparation Standard, "NACE No. 1/SSPC-SP 5 White Metal Blast Cleaning" (Houston, TX: NACE International, 2007).
11. NACE Joint Surface Preparation Standard, "NACE No. 2/SSPC-SP 10 Near-White Metal Blast Cleaning" (Houston, TX: NACE International, 2007).
12. NACE Joint Surface Preparation Standard, "NACE No. 3/SSPC-SP 6 Commercial Blast Cleaning" (Houston, TX: NACE International, 2007).
13. NACE Joint Surface Preparation Standard, "NACE No. 4/SSPC-SP 7 Brush-Off Blast Cleaning" (Houston, TX: NACE International, 2007).
14. SSPC VIS 1, "Guide and References Photographs for Steel Surfaces Prepared by Dry Abrasive Blast Cleaning" (Pittsburgh, PA: SSPC, 2002).
15. NACE SP0181-2006, "Liquid-Applied Internal Protective Coatings for Oil Field Production Equipment" (Houston, TX: NACE International, 2006).
16. NACE SP0288-2011, "Inspection of Lining Application in Steel and Concrete Equipment" (Houston, TX: NACE International, 2011).
17. NACE SP0287-2016, "Field Measurement of Surface Profile of Abrasive Blast-Cleaned Steel Surfaces Using a Replica Tape" (Houston, TX: NACE International, 2016).
18. NACE SP0188-2006, "Discontinuity (Holiday) Testing of New Protective Coatings on Conductive Substrates" (Houston, TX: NACE International, 2006).
19. NACE SP0108-2008, "Corrosion Control of Offshore Structures by Protective Coatings" (Houston, TX: NACE International, 2008).
20. ISO 4628-3:2016, "Paints and Varnishes—Evaluation of Degradation of Coatings—Designation of Quantity and Size of Defects, and of Intensity of Uniform Changes in Appearance—Part 3: Assessment of Degree of Rusting" (Geneva, Switzerland: International Organization for Standardization, 2016).
21. ASTM D610-08, "Standard Practice for Evaluating Degree of Rusting on Painted Steel Surfaces" (West Conshohocken, PA; ASTM, 2015).
22. Coke, J.R., "Protective Coatings for Offshore Equipment and Structures," *Materials Performance* 29, 5 (1990): pp. 35–38.
23. NORSOK M501-2012, "Surface Preparation and Protective Coating" (Oslo, Norway: Standards Norway, 2012).
24. ISO 12944-5:2007, "Paints and Varnishes—Corrosion Protection of Steel Structures by Protective Paint Systems—Part 5: Protective Paint Systems" (Geneva, Switzerland: International Organization for Standardization, 2007).
25. ISO 12944-6:1998, "Paints and Varnishes—Corrosion Protection of Steel Structures by Protective Paint Systems—Part 6: Laboratory Performance Test Methods" (Geneva, Switzerland: International Organization for Standardization, 2007).
26. NACE TM0304-2004, "Offshore Platform Atmospheric and Splash Zone Maintenance Coating System Evaluation" (Houston, TX: NACE International, 2004).
27. NACE TM0404-2004, "Offshore Platform Atmospheric and Splash Zone New Construction Coating System Evaluation" (Houston, TX: NACE International, 2004).
28. NACE SP0198-2017, "Control of Corrosion Under Thermal Insulation and Fireproofing Materials—A Systems Approach" (Houston, TX: NACE International, 2017).
29. NACE SP0191-2017, "The Application of Internal Plastic Coatings for Oil Field Tubular Goods and Accessories" (Houston, TX: NACE International, 2017).

30. NACE SP0291-2017, "Care, Handling, and Installation of Internally Plastic-Coated Oil Field Tubular Goods and Accessories" (Houston, TX: NACE International, 2017).
31. NACE TM0185-2006, "Evaluation of Internal Plastic Coatings for Corrosion Control of Tubular Goods by Autoclave Testing" (Houston, TX: NACE International, 2006).
32. NACE SP0169-2013, "Control of External Corrosion on Underground or Submerged Piping Systems" (Houston, TX: NACE International, 2013).
33. NACE SP0375-2018, "Field Applied Underground Wax Coating Systems for Underground Piping: Application, Performance, and Quality Control" (Houston, TX: NACE International, 2006).
34. NACE SP0394-2013, "Application, Performance, and Quality Control of Plant-Applied, Single Layer Fusion-Bonded Epoxy External Pipe Coating" (Houston, TX: NACE International, 2013).
35. ISO 21809-2:2014, "Petroleum and Natural Gas Industries—External Coatings for Buried or Submerged Pipelines Used in Pipeline Transportation Systems—Part 2: Single Layer Fusion-Bonded Epoxy Coatings" (Geneva, Switzerland: International Organization for Standardization, 2014).
36. NACE SP0185-2007, "Extruded Polyolefin Resin Coating Systems with Soft Adhesives for Underground or Submerged Pipe" (Houston, TX: NACE International, 2007).
37. ISO 21809-1:2011, "Petroleum and Natural Gas Industries—External Coatings for Buried or Submerged Pipelines Used in Pipeline Transportation Systems—Part 1: Polyolefin Coatings (3-layer PE and 3-layer PP)" (Geneva, Switzerland: International Organization for Standardization, 2011).
38. ISO 21809-2:2014, "ISO 21809-1:2011, "Petroleum and Natural Gas Industries—External Coatings for Buried or Submerged Pipelines Used in Pipeline Transportation Systems—Part 3: Field Joint Coatings" (Geneva, Switzerland: International Organization for Standardization, 2014).
39. NACE RP0105-2005, "Liquid-Epoxy Coatings for External Repair, Rehabilitation, and Weld Joints on Buried Steel Pipelines" (Houston, TX: NACE International, 2005).
40. NACE SP0109-2009, "Field Application of Bonded Tape Coatings for External Repair, Rehabilitation, and Weld Joints on Buried Metal Pipelines" (Houston, TX: NACE International, 2009).
41. NACE RP0402-2002, "Field-Applied Fusion-Bonded Epoxy (FBE) Pipe Coating Systems for Girth Weld Joints: Application, Performance, and Quality Control" (Houston, TX: NACE International, 2002).
42. NACE SP0274-2011, "High-Voltage Electrical Inspection of Pipeline Coatings" (Houston, TX: NACE International, 2011).
43. NACE SP0490-2007, "Holiday Detection of Fusion-Bonded Epoxy External Pipeline Coatings of 250 to 760 µm (10 to 30 mils)" (Houston, TX: NACE International, 2007).

Cathodic Protection

Timothy Bieri

7.1 Introduction

Cathodic protection (CP) is defined as a technique to control the corrosion of a metal surface by making that surface the cathode of an electrochemical cell[1]. Because corrosion takes place at the anode of an electrochemical cell, the cathodic surface is protected. In other terms, CP is the use of direct current from an external anode to make the corroding structure cathodic, thus stifling the corrosion. With CP, an electrochemical cell still exists, but the structure no longer corrodes—the structure is the cathode in a cell with an external anode. Of course, the cathode and anode must be in the same electrolyte (water or soil) and electrically connected (with a wire). The direct current may be supplied by a sacrificial (galvanic) anode or from a power source (impressed current).

CP is used to protect a wide variety of production, process, and pipeline equipment both onshore and offshore. CP is applied to the external surfaces of buried and submerged piping, well casings, tank bottoms, offshore fixed or floating structures, the internal surfaces of water handling tanks and vessels, the freewater sections in separation vessels (such as three phase separators and emulsion treaters), and tank bottoms (with a water layer). This chapter will briefly review the basics of CP, the types of CP, and finish with a discussion of the use of CP in different types of production situations.

7.2 Cathodic Protection Principles

The proper application of CP changes the flow of current in the electrochemical corrosion cell. Rather than the current flowing between the anodic (corroding) areas and cathodic (noncorroding) areas on a given structure, CP makes the entire structure cathodic. Direct current impressed on the structure from an external, more powerful, anode overrides the naturally occurring anodic areas. Figure 7.1 illustrates schematically a basic cathodic protection circuit. Note that the electrochemical corrosion current from anode to cathode (A) is replaced, or "overridden," by the current from the auxiliary (external) anode (B).

CP does not eliminate the corrosion reaction. It does, however, transfer corrosion from the corroding structure and moves it to an anode or anodes, which can be designed for long life and easy replacement. CP is only applicable to the surface of the metal that is exposed to the same electrolyte as the anode. For example, CP applied to the exterior surface of a saltwater tank bottom has no effect on the corrosion on the internal surface of the tank bottom.

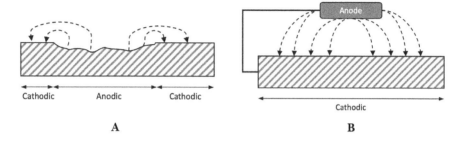

Figure 7.1 Schematic drawing illustrating that CP protects by making the metal structure the cathode in the electrochemical corrosion cell: (A) corroding, (B) with cathodic protection.

7.3 Cathodic Protection Systems

There are two approaches or systems for CP based on different ways of supplying the necessary cathodic protection current:

1. Galvanic anode systems in which the current is due to a galvanic couple set up between galvanic anodes and the structure to be protected.
2. Impressed current systems in which current is supplied (impressed) from an external direct current (DC) power source (commonly a rectifier) via relatively inert anodes.

The selection of the type of system for a particular application, as usual, is another technical as well as economic decision. For example, in some instances a higher driving force (voltage) might be required than is available from galvanic anodes and an impressed system would be necessary. In that case, the comparisons would be to select the most cost effective impressed current design. Table 7.1 lists guidelines that may be used when selecting a CP system.

Table 7.1 Sacrificial anode vs. impressed current cathodic protection systems.

When Galvanic Anode Systems Are Used	When Impressed Current Systems Are Used
Current requirements are low	Current requirements are high
Soil resistively is low (<10,000 ohm-cm)	Soil resistively is high
Electrical power is not available	Electrical power is readily available
Interference problems are prevalent	Long life protection is required

7.3.1 Galvanic Anode Cathodic Protection Systems

When a galvanic anode is attached to a steel structure an intentional corrosion cell is created. As mentioned previously, corrosion is not eliminated but there is a transfer of corrosion from the corroding structure to an anode specifically manufactured for the purpose.

7.3.1.1 Galvanic Anodes

A galvanic anode by definition is "a metal which, because of its relative position in the galvanic series, provides protection to metal or metals that are more noble in the series, when coupled in an electrolyte."[1] Table 7.2 shows the galvanic series for metals in seawater.

Galvanic anodes used in oil field CP are made of special magnesium, aluminum, or zinc alloys, which when connected to steel, exhibit sufficiently high potentials to develop sufficient current flow through the electrolyte to protect the structure. The principle is the same as that of galvanic corrosion in Chapter 3, Section 3.6. Table 7.3 provides a comparison of common galvanic anodes performance data, and Table 7.4 presents a summary of advantages and disadvantages. Galvanic anodes come in a wide variety of shapes, sizes, and mass.

Table 7.2 Galvanic series of various metals exposed to seawater

Table 7.3 Comparison of common galvanic anode performance.

Anode Alloy	Theoretical Capacity	Approximate Consumption Rate	Typical Efficiency
Zinc	780 A h/kg (350 A h/lb)	12 kg/A yr (26 lb/A yr)	95%
Magnesium	2,200 A h/kg (1000 A h/lb)	8 kg/A yr (17 lb/A yr)	50%
Aluminum	2,420 A h/kg (1100 A h/lb)	3 kg/A yr (7 lb/A yr)	85%

Magnesium anodes are widely used in soil applications because of their high driving potential. This allows them to be used in higher resistivity environments. Magnesium is the least efficient of the galvanic anodes, at approximately 50%. For soil applications, magnesium anodes are frequently prepackaged in a special backfill designed to enhance the anode performance. Magnesium anodes can also be used in fresh or condensed waters in vessels. There are several applicable standards for magnesium anodes.[2-3]

Aluminum anodes are primarily used in seawater and brines and have the highest energy capability per pound of anode. The alloy of aluminum, zinc, and indium (Al-Zn-In) is the most common anode used offshore today. There are several applicable standards for aluminum anodes.[4-6]

Zinc anodes are not as common in oil field systems as magnesium and aluminum. Zinc finds its greatest application in low resistivity soils and some waters. However, zinc has a disadvantage—at elevated temperatures, zinc can reverse polarity and become cathodic to steel. Thus, care must be exercised when selecting zinc for cathodic protection application. The reported reversals occur above 60°C (140°F) to 76.7°C (170°F) depending on the water composition. Even if the zinc anode polarity does not reverse, it may passivate at temperatures above 49°C (120°F).[7] Consequently, zinc anodes are not a good choice for fired vessels such as emulsion heaters (heater treaters). There are several applicable standards for zinc anodes.[8-9]

Table 7.4 Some advantages and disadvantages of galvanic CP systems.

Factor	Advantages	Disadvantages
Environment	- Depending on alloy, can be used onshore and offshore	- Not practical for use in high resistivity conditions - Restricted to well-coated structures because of the limited current available
Installation	- Straightforward installation	- Often bulky - Large quantity of anodes required for uncoated structures
Power source	- Independent of any power source	- Hydrodynamic loadings may be high - Anodes may restrict water flow in water system (e.g., for pump casing systems offshore) - Anodes may be required at a large number of positions
Interference	- Less likely to affect neighboring structures, because current output at any point is low	
Control	- Tendency for current to be self-adjusting	- Lifespan varies with local conditions - Replacements may be required at different times
Maintenance	- Low maintenance	- No ability to measure current flow to determine remaining life
Damage	- Robust, not very susceptible to mechanical damage	- Anode weight may not be supported by structure to be protected
Connection	- Designed mechanical connection or welded directly to surface of structure to be protected - Connections are cathodically protected	- Not able to determine if connection is solidly made
Hazards	- Magnesium anodes can be used in potable water tanks	- Because of sparking concerns, magnesium should only be used in confined spaces that are well-vented and not in areas containing hydrocarbons - Aluminium and zinc anodes should not be used in potable water tanks - Zinc anodes should not be used in diesel storage tanks that supply dual fuel gas turbines

7.3.1.2 Design and Installation of Sacrificial Anode Systems

Galvanic anodes are available in many shapes and sizes depending on the application and required life. Anodes for mounting in the water handling sections of vessels and tanks may be in the shape of blocks, slabs, or spherical. Anodes for burial are usually cylindrical and may be obtained prepackaged, or bare. Figure 7.2 is a schematic showing typical galvanic anode installations for pipelines—for tank bottoms or other structures in contact with soil.

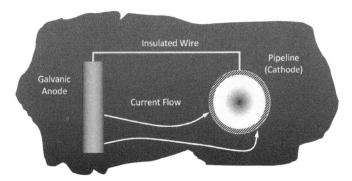

Figure 7.2 Example installation of galvanic anode adjacent to a pipeline.

There are basically three components in a galvanic anode installation in soil: the anode, the backfill, and the connection wire. The anode size and number are determined by the design life of the system and current requirement calculations. The purpose of the backfill mixture surrounding the anode is to provide a low resistance contact between the anode and the soil. Smaller anodes may be purchased with the anode and backfill in a permeable package ready to bury. These packaged anodes make it very easy to quickly and economically install anodes for localized ("hot spot") protection. The connection wire is attached to the pipe or structure and must be installed so it will remain mechanically sound and electrically conductive. Common methods of attachment are to use a thermite welding technique or pin brazing technique. Furthermore, buried or submerged connections should be coated with an electrically insulating material that is compatible with the pipe's coating and the wire's insulation.

7.3.2 Impressed Current Cathodic Protection Systems

Larger quantities of protective current usually require some type of impressed current system. Figure 7.3 shows a typical rectifier installation on a buried pipeline. The transformer rectifier takes the high voltage alternating current (AC) power and changes it into low voltage direct current (DC), which in turn is impressed on a "groundbed" of anodes. The anode bed is connected to the positive (+) side of the rectifier, while the pipeline is connected to the negative (−) side, to complete the circuit. Such installations normally generate 10–100 A or more of protective current at one location. Table 7.5 presents some advantages and disadvantages of impressed current systems.

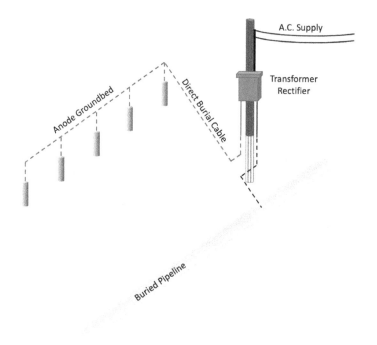

Figure 7.3 Impressed current groundbed installation along a pipeline.

The watchword for installing impressed current cathodic protection systems is "Think Positive." This refers, to the requirement to connect the groundbed to the positive terminal of the rectifier. Hooking the rectifier up backwards reverses the circuit, the pipeline or other supposedly protected structure becomes the anode, the groundbed becomes the cathode, and the structure corrodes to protect the groundbed! Any mistakes such as this will hopefully be caught before commissioning, or at least before severe damage has occurred. Unfortunately, that has not always been true and the first warning of trouble is a corrosion failure. So please—Always THINK POSITIVE!

Table 7.5 Some advantages and disadvantages of impressed current CP systems.

Factor	Advantages	Disadvantages
Environment	– Not restricted by high resistivity conditions	– May cause overprotection, coating disbondment, or hydrogen induced cracking of high tensile steels
Installation	– Good flexibility; can be applied to a wide range of structures	– Requires high level of detail design and installation expertise
Power source	– Controllable current output requiring fewer anodes because of higher current output; controlled current output caters for changing conditions	– External power source necessary with continuous power supply – DC polarity needs to be checked during commissioning, because connection with reversed polarity will accelerate corrosion on structure
Interference	– Generally requires a small number of anodes. – Lighter and has the least effect on water flow	– Effects on other structures that are near anode locations of protected structures should be assessed (but any interaction may be readily corrected)
Control	– Straightforward controls, which can be made automatic to maintain potential within close limits	– Monitoring and control required at regular intervals – Ability to switch current off to measure IR-free potentials
Maintenance	– Inspection can be maintained at relatively few points of structure – Large capacity, long life systems	– Although designed for long life, requires regular inspection, monitoring, and control
Damage	– Anodes can be located in a single location	– Lighter anodes less resistant to mechanical damage and, therefore, loss of anodes more critical – History of use on offshore structures shows that anodes are susceptible to damage during installation
Connection	– Fewer connections required – Can be flush mounted to structure, preventing turbulence or water flow restriction	– Connection more complex and requires high integrity insulation – Susceptible to water ingress at anode termination points, resulting in premature failure – Requires high integrity insulation on connection to positive side of transformer rectifier, which is in contact with soil or water; otherwise, connection will corrode
Hazards		– Diver risk from electric shocks – Impressed current anodes need to be switched off if divers are in vicinity – Overprotection can result in coating disbondment and hydrogen induced cracking of high strength steels (SMYS >550 MPa [80 ksi])

7.3.2.1 Impressed Current Anodes

Impressed current anode materials include graphite, high silicon cast iron, platinum over various substrates such as titanium or niobium, mixed metal oxides, or scrap steel such as pipe, abandoned well casing, or railroad rails. Table 7.6 presents performance characteristics for some of the common impressed current anodes.

Table 7.6 Comparison of common impressed current anode performance.

Anode Material	Current Density	Consumption Rate
Steel	<1.0 A/m² (<0.1 A/ft²)	9 kg/A yr (20 lb/A yr)
High Silicon Cast Iron	20-30 A/m² (2-3 A/ft²)	0.23 kg/A yr (0.5 lb/A yr)
Graphite	15-25 A/m² (1.5-2.5 A/ft²)	0.23 kg/A yr (0.5 lb/A yr)
Platinized Anodes	500 A/m² (50 A/ft²)	8-10 mg/A yr
Mixed Metal Oxide	500 A/m² (50 A/ft²)	1-2 mg/A yr

In all cathodic protection installations for buried structures, it is important that anodes be properly installed so that there is minimum electrical resistance between the anode and the surrounding soil. A low resistance material is usually packed around the anode to serve as backfill. Figure 7.4 shows a typical graphite (or silicon-iron) anode installation with coke breeze backfill.

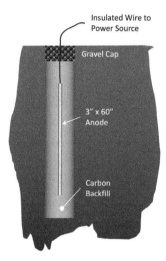

Figure 7.4 Vertical installation of impressed current anode.

Most impressed current groundbeds are relatively shallow. By definition, a shallow groundbed has one or more anodes installed either vertically or horizontally at a depth of less than 15 m (50 ft). Commonly, there are zones of low resistance soils near the surface, and the groundbeds will be much shallower. Often, the holes for vertical anodes can be drilled with an auger. Horizontal groundbeds are simply trenched.

In certain instances, however, deeper placement is required to reach low resistance stratum or to be able to get the anodes remote from the pipe or structure. Deep anode beds are defined as those with one or more anodes placed vertically in drilled holes at depths greater than 15 m (50 ft)—often they will be several hundred feet deep. Deep anode beds are very important to the economic success of many CP systems for pipelines, well casings, and surface facilities. NACE RP0572 outlines the requirements, considerations, and procedures for deep anode beds.[10] Figure 7.5 is a schematic of one design for deep anode beds. Figure 7.6 illustrates the positive distribution box for a deep anode bed.

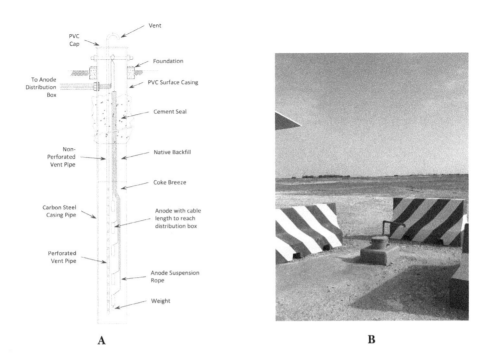

Figure 7.5 Deep anode bed (A) schematic, and (B) photograph of a completed installation.

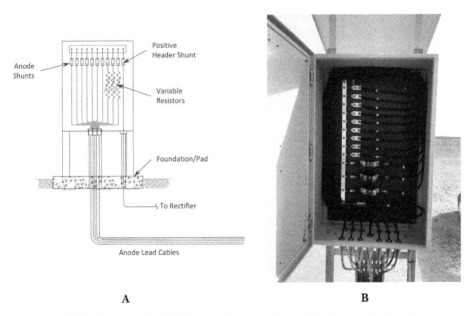

Figure 7.6 Positive (anode) distribution box (A) schematic, and (B) photograph of an actual box.

7.3.2.2 Impressed Current Cathodic Protection Power Sources

Although AC/DC rectifiers are the most common power supply for impressed current cathodic protection (Figure 7.7), other sources have been and are being used. Where AC power is not available, DC generators can be used for systems with heavy loads. Solar or thermoelectric generators may be used where lower currents are needed.

Figure 7.7 Examples of air-cooled rectifiers.

7.3.2.3 Cathodic Protection Interference

One possible problem when using impressed current CP is that of interference on unprotected lines or structure. Figure 7.8 illustrates the case where interference between a cathodically protected pipeline and foreign pipeline is aggravating corrosion of the foreign line. Current from the anode bed is picked up by both the pipeline being protected and the foreign pipeline crossing under it. To complete the circuit, the current picked up by the foreign pipeline must leave the pipeline, pass through the soil, be picked up by the protected pipeline, and return to the rectifier. The area where current leaves the foreign line is anodic and accelerated corrosion occurs where the current leaves. This is an example of static interference, or stray current, because the current remains relatively constant and normal monitoring survey procedures can be used to detect it.

If a proper metallic bond is placed between the foreign and protected pipelines, this problem does not occur. Of course, this means both structures are then receiving CP current, and the current required for adequate protection is correspondingly increased. Frequently, the metallic bond will have a variable resistor to minimize the amount of current going to the foreign pipeline.

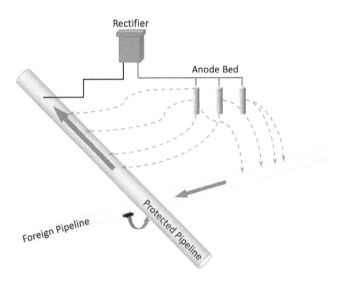

Figure 7.8 A representation of CP interference.

Figure 7.9 illustrates a dynamic stray current condition caused by telluric currents[1] overlayed with the system response to tidal influence. Dynamic stray currents require more advanced survey methods to detect and understand the source and magnitude of the problem.

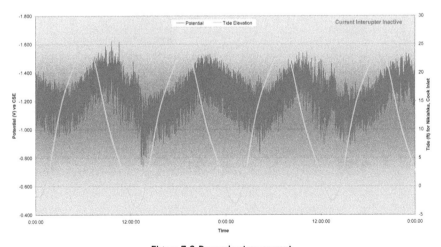

Figure 7.9 Dynamic stray current.

CP installations should be designed in such a way that interference with other structures will be minimal. Once CP is installed, it is necessary to determine whether interference exists and, if so, to take steps to mitigate it. This requires conscientious effort and cooperation among the several parties who may be involved. Detection and mitigation of CP interference can be a complicated matter,

1, "Telluric Current" is defined as current in the Earth as a result of geomagnetic fluctuations.

and NACE SP0169[1] discusses the mechanism of interference, detection of interference currents, and methods for mitigating interference corrosion problems.[1]

7.4 Criteria for Cathodic Protection

Criteria are needed to determine if the amount of cathodic protection available is adequate to mitigate corrosion.

7.4.1 Know Your Reference

Knowing which reference electrode is used is very important when determining adequate levels of CP. An analogy would be reporting the temperature without the reference, 35° can require a coat or short-sleeves depending if the reference is Fahrenheit or Celsius. For CP, criteria have been developed to establish the effectiveness of CP on various structures. The more common criteria involve measurements of voltage (differences in potential) between the protected structure and the electrolyte. Probably the most widely used criterion for buried pipes or external tank bottoms involves the use of a copper-copper sulfate electrode (CSE) as a reference electrode. This electrode consists simply of a copper rod immersed in saturated copper sulfate solution, both being housed in a plastic cylinder with a porous plug on the bottom end (for contact with the electrolyte) and the copper rod extending out of the top for connection to the measuring high-impedance voltmeter (Figure 7.10).

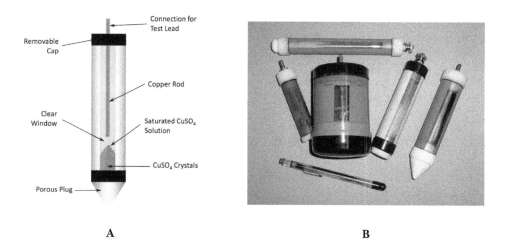

Figure 7.10 Reference electrodes: (A) schematic of a copper sulfate reference electrode; (B) different sizes of various reference electrodes.

Other reference electrodes may be substituted for the copper-copper sulfate reference electrode to meet specific needs. For example, the silver-silver chloride reference electrode is commonly used in seawater because the CSE may be inaccurate if the seawater penetrates the porous plug and contaminates the solution. Table 7.7 lists the equivalents for the commonly used reference electrodes.

Table 7.7 Common reference electrode equivalent voltages.[1]

Reference Electrode	Potential at 25°C/77°F (V_{SHE})	Potential at 25°C/77°F (V_{CSE})	Temperature Coefficient, k (mV/°C)/(mV/°F)	Equivalent Value to 0.850 V_{CSE} (V)	Typical Usage
Copper-Copper Sulfate Electrode (CSE)	+0.316	0.000	0.90/0.50	—	Soil, under tank bottoms, fresh water
Silver-Silver Chloride (SSC) Electrode	+0.256	-0.060	-0.33/0.18	-0.790	Seawater, brackish water[(4)]
Standard Hydrogen Electrode (SHE)	0.000	-0.320	—	-0.530	Laboratory
Saturated Calomel Electrode (SCE)	+0.244	-0.072	-0.70/0.39	-0.778	Laboratory
Zinc Reference Electrode (ZRE)	-0.800	-1.116	—	+0.266	Under tank bottoms, seawater

There are several factors that can affect the reference potential when using CSE or SSC.[11] The temperature coefficients in Table 7.7 can be used to correct the error using Equation (7.1a) or (7.1b).

$$E_{T°C}(mV) = E_{25°C}(mV) + k(T - 25) \tag{7.1a}$$

$$E_{T°F}(mV) = E_{77°F}(mV) + k(T - 77) \tag{7.1b}$$

7.4.2 Important Terms

The following are some of the important terms used when discussing CP criteria.[1] Figure 7.11 illustrates these terms on a potential vs. time graph.

- Polarization: The change in corrosion potential as a result of current flow across the structure-electrolyte interface. As more current is applied, the level of polarization increases.
- Polarized Potential: Specific to CP, it is the potential across the structure-electrolyte interface. This is the sum of the corrosion potential and the cathodic polarization. and is also sometimes referred to as the Instant Off potential.
- Voltage (IR) Drop: This is simply stated as the voltage across a resistance when current is applied. Voltage drop can make the measured potential appear more negative than the polarized potential. There are several resistances in the measuring circuit including the meter, leads, connections, electrolyte resistivity, and coating resistance.

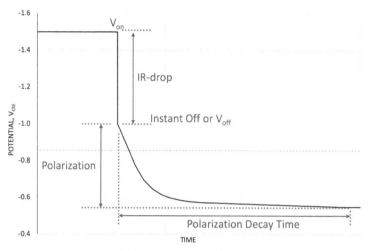

Figure 7.11 Representation of important terms.

7.4.3 Criteria

There are three primary criteria used for determining adequate levels of CP for low-alloy carbon steel.

1. -0.850 V_{CSE} potential: Negative (cathodic) potential of at least 0.850 V_{CSE} with the cathodic protection applied after IR-drop is considered.
2. -0.850 V_{CSE} polarized potential: Negative polarized potential of at least 0.850 V_{CSE}.
3. 100 mV polarization: Minimum of 100 mV of cathodic polarization.

Each of these criteria has benefits and limitations[1,12] that require consideration to determine their applicability to a given system. For example, 100 mV polarization is generally not applicable in a mixed-metal system such as where a copper earthing system is connected to the piping system. The 100 mV polarization needs to be applied to the most active/anodic metal in the mixed-metal system (steel) and it is difficult to know the starting potential if the two metals cannot be isolated from each other. There are additional considerations that can alter the criteria including higher operating temperature or the presence of bacteria. There are also criteria for other common metallics that are used in oil and gas production such as copper, aluminum, and corrosion resistant alloys.

For offshore and subsea structures protected by galvanic anodes, it is virtually impossible to interrupt the CP current. Fortunately, the environment resistivity is low (compared to soil or freshwater) and it is easier to place the reference electrode close to the structure. Consequently, IR-drop can be minimized when making structure-to-electrolyte potential measurements.

7.5 Survey and Test Methods in Cathodic Protection

Determining the need for, and applicability of, CP requires special instrumentation and judgment. Because both corrosion and CP are electrochemical in nature, the corrosion-CP survey consists essentially of a well-organized and correlated series of electrical measurements. The most frequent of these are

- Structure-to-electrolyte potential measurements
- Current flow (IR drop) measurements, either in the structure itself or in the surrounding electrolyte
- Electrolyte resistivity measurements
- Current requirement tests

These measurements in various combinations are used to determine the need for CP, to design CP systems, to commission and evaluate a new system, to monitor a system's performance, and to troubleshoot problems. A high level discussion of these methods is included in the following sections. More details are available in CP reference books including *Control of Pipeline Corrosion*, by A.W. Peabody[12] and *Cathodic Protection Survey Procedures* by Brian Holtsbaum.[13]

7.5.1 Structure-to-Electrolyte Potential Measurements

Structure-to-electrolyte potential measurements are probably the most common CP-related measurements. They can be used for testing isolation and/or continuity, demonstrating adequate levels of CP, and troubleshooting when things are not going as planned. Potential readings are customarily given in millivolts or volts. Because a voltage is really a difference in two potentials, the "potential" as reported is a voltage reading between the structure being investigated and a convenient reference electrode placed in the electrolyte near the structure. Typical potential values for different metals in neutral soil or water, measured with respect to a copper-copper sulfate reference electrode, are presented in Table 7.8. "Structures" can be anything immersed in, or in contact with, soils and waters or containing waters.

Table 7.8 Practical galvanic series.

Metal	Volts$_{CSE}$*
Commercially pure magnesium	–1.75
Magnesium alloy (6% Al, 3% Zn, 0.15% Mn)	–1.6
Zinc	–1.1
Aluminum alloy (5% zinc)	–1.05
Commercially pure aluminum	–0.8
Mild steel (clean and shiny)	–0.5 to –0.8
Mild steel (rusted)	–0.2 to –0.5
Cast iron (not graphitized)	–0.5
Lead	–0.5
Mild steel in concrete	–0.2
Copper, brass, bronze	–0.2
High silicon cast iron	–0.2
Mill scale on steel	–0.2
Carbon, graphite, coke	+0.3

*Typical potential normally observed in neutral soils and water, measured with respect to copper-copper sulfate reference electrode.

It is easier, for purposes of this particular section, to select for discussion one type of structure—a buried steel pipeline—with the understanding that the discussion is applicable to any structure in its particular environment.

The potentials encountered usually range from a few millivolts to several volts, and in order to obtain accurate potential measurements against a copper-copper sulfate or other reference electrode, a meter of high sensitivity is required. The meter, plus two reference electrodes, and necessary test leads and contact devices, are the minimum equipment required for a potential survey.

In making the survey, the structure is contacted at risers, through valve boxes, test stations, or by use of a device such as a contact bar. It is extremely important that a good, low-resistance electrical contact be established with the buried structure. The positive terminal of the voltmeter is connected through suitable leads to the pipeline contact, and the negative terminal is connected to the reference electrode. If a copper-copper sulfate electrode is used, the pipe will exhibit a negative polarity (Figure 7.12).

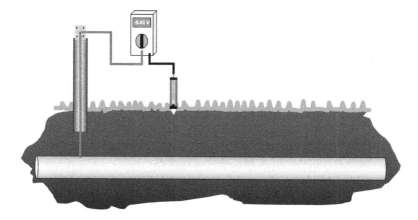

Figure 7.12 Pipe-to-soil potential measurement.

The potential of the structure is taken with the electrode firmly embedded in the earth directly over the structure. The structure-to-electrolyte potential is recorded. This procedure is repeated at suitable intervals throughout the entire system under study. When the whole pipeline can be surveyed in this fashion, it is called a "close-interval potential survey."[14] Figure 7.13A shows a close interval survey crew. Typically, the lead crew member would be locating the buried pipeline to ensure the survey data is collected over the pipeline. The CP crew member carries a thin gauge wire spool connected to the pipeline at a test station. Each of the poles has a reference electrode at the end, and the pipe-to-soil potential is measured and recorded as the crew member walks the length of the pipeline. Figure 7.13B shows an example of how the data from multiple locations can be plotted. If there is no CP applied, then areas of more negative potential indicate anodic or corroding areas.

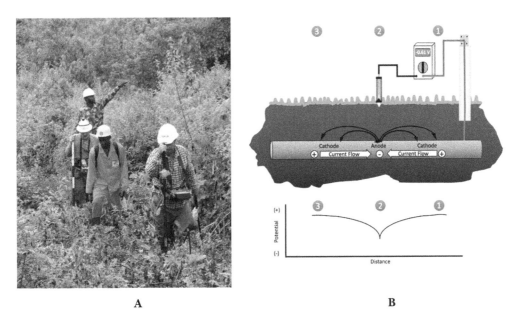

Figure 7.13 Pipe-to-soil potentials: (A) close interval survey crew; (B) schematic and plotting of the data.
Photograph A provided courtesy of Brown Corrosion Services.

7.5.2 Current Flow (IR Drop)

The direction of direct current flow in the structure (pipeline) can be determined using a direct reading millivolt meter or potentiometer. The area or location where current flows from the structure to the electrolyte is where corrosion is occurring. This is detectable by noting the direction and magnitude of current flow in the pipe. The point at which it reaches a maximum value and reverses its direction of flow is the point at which corrosion is occurring. The magnitude of current flow, which can be calculated or estimated from the millivolt readings and the pipe resistance between contact points, can be used to provide a rough approximation of how much metal loss per year is occurring.

Corrosion may still be occurring in the absence of measurable current flow in the structure due to local cell action between the contact points. For this reason, the IR drop measurement of current in the structure is useful only to locate gross anodic areas. The method is illustrated in Figure 7.14. Because accurate readings as low as 1 mV may be required, the resistance of the test leads and contacts to the structure must be low—on the order of 1%—compared to the internal resistance of the voltmeter. The potential difference is measured between contact leads A and B by a low resistance voltmeter. Current flow in the pipe is from (+) to (−); thus, anodic locations will be at the most negative points on the pipe.

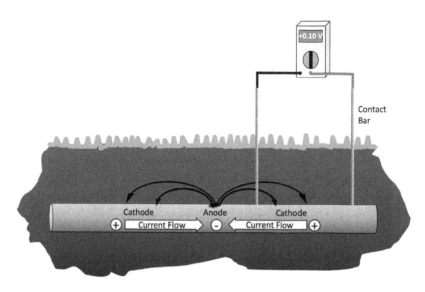

Figure 7.14 Diagram showing measurement of IR drop to determine direction and of current flow.

Knowing the resistance of the pipe, current flow can be calculated from Ohm's Law:

$$I = E/R \qquad (7.2)$$

where I is the current in amperes, E is the potential in volts, and R is the pipe resistance in ohms (between contact points). If the resistance of the pipe under investigation is not known, it may be estimated from published tables for various sizes of steel pipes, or calibrated on the spot by impressing a known current through a test section of the pipe and noting the IR drop produced.

An alternate method, called the "surface potential" or "two-electrode" survey method can be used when there no contact points are available. Because the same current flowing in the pipe must also be flowing in the adjacent soil, where the much greater soil resistance produces millivolt readings of much larger magnitude than in the pipe, a modification of the technique is widely used on buried bare pipe.

The surface potential survey requires a high impedance millivolt meter, two copper-copper sulfate electrodes, and suitable test leads. A survey crew will also need a pipe locator and a soil resistivity meter, because it is essential to stay immediately over the pipe and to know the soil resistivity at anodic areas. Figure 7.15 illustrates two reference electrodes connected through a voltmeter. Potential is measured between the two electrodes. When the two electrodes are moved to position (2), the reading will be more negative, indicating an anodic or corroding area. When the electrodes are moved to position (3), the readings will be more positive, thus helping to locate the anodic area.

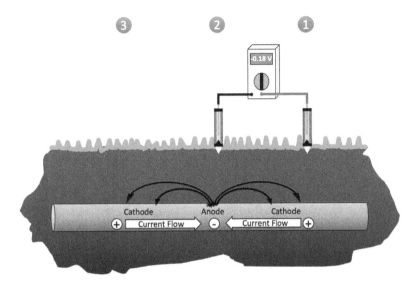

Figure 7.15 Surface potential method for determining corrosion.

The two copper-copper sulfate electrodes, which should be kept in good condition to read within 5 mV of each other, are placed over the pipe at a suitable distance apart, usually 6 m (20 ft), and the potential difference (in millivolts) and electrode polarity is read on the meter and recorded. The electrodes are then moved down the line 6 m (20 ft), maintaining the set spacing. In this manner, millivolt readings and electrode polarity are recorded over the full length of the line being surveyed. Anodic and cathodic areas are identified as points of potential reversal.

Because the anodic areas are of greatest significance, these are identified during the course of the survey. Soil resistivity measurements, usually coupled with "side drain" millivolt readings, also are obtained at that time. Side-drain measurements consist of placing one reference electrode directly above the pipe, and a second electrode is placed at a 90° angle to the pipe at a distance approximately equal to 2.5 times the pipe depth. Measurements are typically made on both sides of the pipe. These measurements will be needed later to properly interpret the severity of the anodic areas and to select the number and size of anodes needed in the "hot spots."

7.5.3 Soil (or Water) Resistivity Measurements

Understanding the resistivity of the electrolyte is important for two reasons: knowing how corrosive is the electrolyte, and the resistivity is a key input into the design of CP systems. Because current flow to and from the structure will generally occur at the areas of lowest earth resistivity, it is logical to assume that low resistivity areas could be anodic. Essentially, resistivity determines only the opportunity for corrosion and the location where such corrosion would be expected. It does not indicate whether or not corrosion is actually occurring, or how fast.

The common resistivity may be determined by using a four-pin instrument[15] (Figure 7.16) or single-point probes. This technique helps to determine the resistivity of the earth, expressed most correctly as ohm-centimeters, at preselected areas along the pipeline. Actual resistance values alone usually have little meaning; greater significance of the survey data lies in the differences in soil resistivity values along the line. Resistivity values also vary with depth and can be useful in planning depth of installation for anodes.

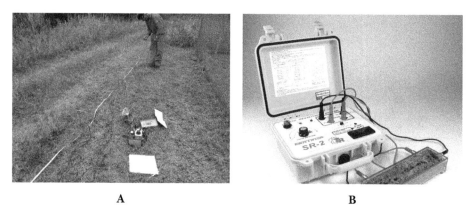

Figure 7.16 Wenner 4-pin soil resistivity testing: (A) field test, and (B) soil box test.
Photograph A is provided courtesy of Brown Corrosion Services. Photograph B is provided courtesy of Tinker-Rasor.

7.5.4 Current Requirement Surveys

The actual amount of current required for CP of a given structure can be estimated in several ways. One such method for existing structures involves a field test where a temporary current source is applied to the structure and the response is measured. For this discussion, CP is meant to imply complete corrosion control as evidenced by compliance with some selected criteria, such as -0.85 V_{CSE}. It is easier for this discussion, to select one type of structure—a buried steel pipeline—with the understanding that the discussion is applicable to any structure in its particular environment.

If a bare pipeline, or other structure, lies in soil or water of known general characteristics, the CP system may be designed based on applying a selected current density (in milliamps per square foot [mA/ft^2] or milliamps per square meters [mA/m^2]) to the structure, provided the current is properly distributed.

On large bare structures, it is seldom practical to temporarily apply the amount of current needed to achieve protective potentials; therefore, CP design is usually approached on the current density basis described earlier. In areas where previous experience may be lacking (e.g., produced waters, polluted rivers, estuaries, etc.), the use of test "coupons" can be helpful in arriving at the proper current density on which to base the full-scale design. A coupon is a small piece of steel, which can be placed in the same environment to measure how it responds to CP. In this case, a range of applied current densities over a period of several weeks or months. For example, if it is set at 1 A for 1 week, then 1.5 A for 1 week, etc., the size of the coupon is known, and the current density (in amps per square meter or square feet) can be calculated. Based on that coupon test, the system can be scaled for the entire structure.

With coated structures, taking a pipeline again as an example, it is possible to set up a temporary CP drain point and determine how much current will be needed to protect either the entire line, if it is relatively short and is insulated from other structures, or to protect a given portion of the line. On long pipelines, several such tests may be necessary, especially if substantial differences in coating condition and/or soil resistivity are anticipated in different areas. Temporary test currents of up to about 10 A can be applied by a storage battery, and as much as 100 A can be applied via a generator.

The temporary "groundbed" used to discharge the test current into the soil may be any existing non-critical structure that is not electrically connected to the pipeline under test, such as an abandoned section of line or well casing. In many cases, a suitable temporary groundbed for test purposes must be constructed of driven steel rods, aluminum foil, or actual anodes that can be left in place for a later permanent installation, or installed in such a manner that they can be removed for reuse. It is desirable to locate the test beds, at least for the larger current drains, at a distance from the pipeline similar to that which might be expected for the later, permanent groundbed installation.

7.6 Application of Cathodic Protection

CP, either by itself or in combination with another control method (such as coatings), is a viable method of corrosion control for most oil field corrosion problems, except inside pipes and oil country tubular goods (OCTG). As a generality, CP is not a sound approach inside piping or wells. The spacing is so close that it would be difficult, if not impossible, to get satisfactory current distribution. However, external well casing; buried lines (flow, gathering, distribution, and injection); surface vessels; and tanks that are set on the ground, as well as the internals of vessels and tanks that handle or store water (including those with water holding sections or water standing in the bottom), are all candidates for CP.

7.6.1 Cathodic Protection Design

There are numerous standards available related to the design of CP systems for various structures. Table 7.9 lists some of the common standards for different structures.

Table 7.9 Common CP design standards

Onshore Pipelines and Structures	Aboveground Storage Tanks	Underground Storage Tanks	Internal CP for Storage Tanks and Vessels	Well Casings	Offshore Pipelines and Structures
EN 12473[16]	API RP 651[17]	NACE SP0285[18]	EN 12499[19]	NACE SP0186[20]	DNVGL-RP-B401[21]
ISO 15589-1[22]	NACE SP0193[23]	API RP 1632[24]	NACE SP0196[25]		DNVGL-RP-F103[26]
NACE SP0169[1]			NACE SP0388[27]		DNVGL-RP-B101[28]
NACE SP0286[29]			NACE SP0575[30]		EN 12474
NACE SP0572[10]					EN 13173
					ISO 15589-2[31]
					NACE SP0176[32]
					NORSOK M-503[33]

Direct visual observation of the effectiveness of CP may be possible in many cases (offshore structures, inside tanks, etc.). In addition, coupons[34] of the same metal may be installed on the protected structure to periodically check the degree of effectiveness of the applied protective current. Finally, where experience has shown that a certain current density has been effective in protecting steel, in a given relatively uniform environment, this current density, when uniformly applied, may be considered an indirect criterion of protection. Current densities of 10.8 mA/m² (1 mA/ft²) of bare steel pipe will give the desired potential response in many soils.

In seawater, an initial current density of 64.6–86.1 mA/m² (6–8 mA/ft²) is required to protect steel structures. In severe environments (low temperature and/or high flow rates), requirements can be much larger. Table 7.10 presents some typical design current densities for protection of offshore platforms in different producing areas around the world. Polarization and calcareous deposit build-up tend to reduce the current density required to maintain protective potentials to one-half or even lower than the initial value.

Table 7.10 Design criteria for offshore cathodic protection systems.[32]

Production Area	Water Resistivity[B] (ohm-m)	Water Temp. (°C)	Environmental Factors[A]		Typical Design Current Density[C]			Typical Design Slope
			Turbulence Factor (Wave Action)	Lateral Water Flow	mA/m² (mA/ft²)			ohm-m² (ohm-ft²)
					Initial[E]	Mean[F]	Final[G]	
Gulf of Mexico	0.20	22	Moderate	Moderate	110 (10)	54 (5)	75 (7)	4.1 (44)
Deep water GOM	0.29	4-12	Low	Varies	194 (18)	75 (7)	86 (8)	
U.S. West Coast	0.24	15	Moderate	Moderate	150 (14)	86 (8)	100 (9)	3.0 (32)
Cook Inlet	0.50	2	Low	High	430 (40)	380 (35)	380 (35)	1.0 (11)
Northern North Sea[D]	0.26-0.33	0-12	High	Moderate	180 (17)	86 (8)	120 (11)	2.5 (27)
Southern North Sea[D]	0.26-0.33	0-12	High	Moderate	150 (14)	86 (8)	100 (9)	3.0 (32)
Arabian Gulf	0.15	30	Moderate	Low	130 (12)	65 (6)	86 (8)	3.5 (37)
Southeast Australia	0.23-0.30	12-18	High	Moderate	130 (12)	86 (8)	86 (8)	3.5 (37)
Northwest Australia					120 (11)	65 (6)	65 (6)	
Brazil	0.20	15-20	Moderate	High	180 (17)	65 (6)	86 (8)	2.5 (27)
West Africa	0.20-0.30	5-21	Low	Low	130 (12)	65 (6)	86 (8)	3.5 (37)
Indonesia	0.19	24	Moderate	Moderate	110 (10)	54 (5)	75 (7)	4.1 (44)
South China Sea	0.18	30	Moderate	Low	100 (9)	32 (3)	32 (3)	

[A] Typical values and ratings based on average conditions, remote from river discharge.
[B] Water resistivities are a function of both chlorinity and temperature. In the reissued *Corrosion Handbook*[14] by H.H. Uhlig, the following resistivities are given for chlorinities of 19 and 20 parts per thousand (ppt):

Resistivities (ohm-m)
Temperature (°C [°F])

Chlorinity (ppt)	0 (32)	5 (41)	10 (50)	15 (59)	20 (68)	25 (77)
19	0.351	0.304	0.267	0.237	0.213	0.192
20	0.335	0.290	0.255	0.227	0.203	0.183

[C] In ordinary seawater, a current density less than the design value suffices to hold the structure at protective potential once polarization has been accomplished and calcareous coatings are built up by the design current density. CAUTION: Depolarization can result from storm action.
[D] Conditions in the North Sea can vary greatly from the northern to the southern area, from winter to summer, and during storm periods.
[E] Initial and final current densities are calculated using Ohm's Law and a resistance equation such as Dwight's[15] or McCoy's[16] equation with the original dimensions of the anode. An example of this calculation is given in Appendix D (nonmandatory), which uses an assumed cathode potential of -0.80 V (Ag/AgCl [sw]).
[F] Mean current densities are used to calculate the total weight of anodes required to maintain the protective current to the structure over the design life. Examples of these calculations are given in Appendixes D and E (nonmandatory).
[G] Final current densities are calculated in a manner similar to the initial current density, except that the depleted anode dimensions are used. An example of this calculation is given in Appendix D (nonmandatory).

(This table is from NACE SP0176-2007. Note that the appendix notations and reference numbers given in the footnote sections are those from the actual NACE standard, i.e., not this chapter.)

7.6.2 Well Casings

External corrosion of well casing is not a universal problem (i.e., well casings do not fail from corrosion in every well during its producing life). When they do fail, a number of the failures are due to internal, not external, corrosion. However, enough wells have failed from external corrosion in most operating areas, that control procedures have been developed. The review of oil field corrosion and its control in Chapter 1, Section 1.5.1.3 mentioned that cathodic protection is an accepted approach to controlling external casing corrosion. Depending on the circumstances, both sacrificial and impressed current systems are in use. NACE SP0186 covers many of the details of design, operation, and maintenance of both type systems.[20]

Current requirements for well casings, with few exceptions, fall in the range of 1–25 A. The smaller requirements can frequently be met using galvanic anodes. In many instances, even for such small currents, soil resistance is too high for galvanic anodes, and an impressed current system becomes necessary. For economy, one rectifier groundbed is frequently installed to take care of several well casings, either by running separate negative connections to the wells, or by using the flowlines as current conductors to the wells. In either case, the well casing must be insulated from the flowline. A current-controlling resistor can be placed across this insulating fitting to drain a little current from the flowline to the well (to afford some cathodic protection to the flowline, while eliminating possible cathodic interference). Alternately, current can be drained from the well to the flowline where the latter has been used as the current conductor back to the rectifier.

In many areas, the soil resistivity is such that even with rectifiers, design of an economical groundbed to produce the desired current at a reasonable voltage becomes a problem. Deeper strata are often lower in resistance. In such cases, it becomes economical to use deep anode groundbeds, as described earlier and illustrated in Figure 7.6. These deep anode groundbeds have all the anodes placed one above the other in a vertically drilled 152 or 203 mm (6 or 8 in) diameter hole.

The problem of rectifier adjustment arises when a large number of well casings are protected in a field where the surface groundbed resistance to earth may change radically with soil moisture variations. This has led to an increasing usage of "constant current" type rectifiers in well casing CP installations. These units are designed so that, over a reasonable range of groundbed resistance, the rectifier automatically compensates for increased circuit resistance and tends to maintain the preselected current output of the unit. This feature is not normally required with deep well groundbeds, which are little affected by surface moisture conditions.

Interference with well casings by nearby CP installations is a greater problem than that with pipelines because both detection and alleviation are difficult. Surface measurements, backed up by a casing potential profile when in doubt, must be used in any case of suspected interference with a neighboring well casing. Field wide application of CP to well casings—together with balancing of individual well potentials—minimizes interference problems from well to well. However, there has been a somewhat greater problem with interference by pipeline cathodic protection systems on well casings.

7.6.3 Flowlines, Gathering Lines, Distribution Lines, Injection Lines, and Pipelines

All of the buried field piping made of steel, whether it is used for flow, gathering, distribution, or injection lines may be subject to external corrosion. Coatings or cathodic protection or both are the usual methods of corrosion control. Older lines are often bare steel that is repaired or replaced when failures occur.

If the line is well coated and not very long, the current requirements will usually fall in the range where installation of galvanic anodes, singly or in groups, is the most practical means for achieving CP. Figure 7.2 shows typical magnesium anode installations.

Bare field lines may not justify CP except under very corrosive conditions. They may be reconditioned, coated, and placed under complete CP, as mentioned above, or they may be surveyed by the surface potential profile (two electrode) technique and subsequently placed under "hot spot" protection with galvanic anodes. Known leak areas, or "hot spots," may be satisfactorily protected by a limited number of galvanic anodes.

Critical trunk lines and transmission lines are normally surveyed, spot protected if bare, or placed under complete CP if coated externally. Most newly installed lines of any importance will be well coated, making the task of complete CP very economical and relatively simple.

Lines under cathodic protection should be electrically isolated (using insulating unions, flanges, couplings, or nipples) from unprotected structures so that the CP current will be confined to the structure for which it was intended. Even one metallic contact (e.g., to a well casing not under CP) can cause substantial loss of protection on the pipeline. This is because the bare well casing adds a large amount of steel surface to the CP circuit; probably much more surface than was used in the system design. Therefore, an inspection and maintenance program for insulating fittings is important to ensure continuing isolation. Insulating fittings have other uses besides isolating CP systems. They are used to break up corrosion currents between pieces of equipment and to isolate different metals from one another. Electrical isolation and insulating fittings are discussed in greater detail in Chapter 9, Section 9.4.

7.6.4 Surface Equipment and Vessels

Equipment such as production and saltwater disposal tanks, heater-treaters, accumulators, separators, and filters are prone to corrosion from produced brines. Elevated temperatures, coupled with the presence of sulfate-reducing bacteria (SRB) in many instances, leads to short service life unless CP is applied. This is particularly true, as might be expected, for the fire tube, emulsion, and freewater sections in heater-treaters. In one South Texas field, corrosion failures of the fire tubes occurred in about three months without protection.

An adage states that an anode can protect only what it can "see"—it cannot see around corners into the next compartment. Figure 7.17 compares the coverage of an anode in a vessel compartment to that of a light bulb in a similar situation. Multiple anodes are required because baffles separate the sections of the treater, and the close confines limit how much steel surface each anode can "see." Furthermore, in the short distances within any vessel, the coverage of any one anode is limited.

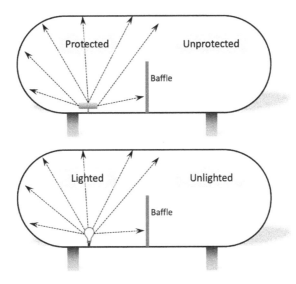

Figure 7.17 Sketch showing how an anode 'lights' only some of the surface.

Current density requirements can range from 50–400 mA/m² (5–40 mA/ft²) of bare water-immersed steel.[29] In the absence of specific data, 100 mA/m² (10 mA/ft²) is commonly used for design estimates. The choice of a current source is primarily a matter of the availability of AC power.

Where AC power is available, rectifiers may be used to protect one or more vessels, using graphite or other anode material installed in special through-wall mounts. Some of the newer anodes are very small and may be conveniently lubricated in and out of the mount while the vessel remains in operation. While graphite anodes can suffer some reduction in output from production chemicals in produced fluids, the anodes' large surface area allows development of adequate current flow in all but very rare situations. Chlorine-resistant, high-silicon, cast iron anodes may be used where graphite is adversely affected by oil saturation.

Where AC power is not present, adequate protective current can be generated by either magnesium or aluminum anodes mounted through the wall in the same fashion as the graphite anodes. The use of jumper wires allows current flow to be measured and adjusted for each individual anode at its point of mounting. Because magnesium is very active, its current output to the vessel must be highly restricted to avoid wasteful overprotection. However, this results in greatly reduced anode efficiency because the anode will continue to corrode at almost the same rate and only a fraction of its corrosion current is being used to protect the vessel. The aluminum anode offers much lower, but still adequate, output in oil field brines, while maintaining current efficiencies on the order of 60–70%. Some operators have resisted penetrating the pressure boundary of a pressure vessel to install jumper wires and shunts to measure current. Instead, they install sacrificial anodes bolted to blind flanges or manways without means of monitoring their performance. Unfortunately, when the vessel is entered years after anode installation, the anodes are often found to be missing, broken, or corroded, with no knowledge available when this damage occurred. Some acid jobs have been allowed to flow into separators and heater treaters, and these fluids are highly corrosive to galvanic anodes.

Zinc anodes should not be used in produced water above ambient temperature unless tests in the water under service conditions indicate that they are satisfactory. As previously mentioned, zinc can reverse polarity with steel in many electrolytes at elevated temperatures and cause corrosion of steel, rather than protect against it.

Brine storage tanks are usually protected from internal corrosion by CP or a combination of a coating and CP. Where coatings are used, care must be exercised to restrict the potential across the coating film; too high a potential can result in blistering and severe coating damage. Coating damage is particularly easy to do with improper adjustment of a rectifier system, but even magnesium anodes can develop injurious potentials unless current output is properly restricted. Once again, aluminum—with its lower potential—offers a safe and more than adequate source of protective current for coated vessels.

Impressed current and galvanic anodes are often suspended from the tank deck via special deck mounts. Anodes can be inspected and replaced conveniently with such an arrangement. Some aluminum and magnesium anodes are designed to mount on a threaded stud welded to the vessel shell. Figure 7.18 shows these 'condenser' style anodes mounted in an internally coated vessel. These shell mount anodes are available in several sizes and shapes, and a number may be stacked for increased life. Vessel entry is required to inspect and replace this type of anode.

Figure 7.18 An example of galvanic anodes mounted in a vessel.

While "fresh" waters may not be as corrosive as brines, equipment handling surface waters for water flood projects must be protected from corrosion. Once again, a combination of coatings and CP can approach 100% protection against corrosion of the submerged surfaces of such equipment. Unless the water is quite conductive, magnesium anodes or an impressed current system will be required to provide sufficient protective current. Care must be exercised in locating the anodes to give good current distribution to all submerged equipment surfaces, especially in vessels containing internal compartments or baffles.

7.6.5 Storage Tanks

Present environmental concerns and regulations have focused attention on leaking, aboveground storage tanks (ASTs). Emphasis has been placed on avoiding leaks, detecting leaks, and secondary containment, should a leak occur. Inspection and verification of tank soundness is very important. Construction practices and installation designs to allow water to drain from beneath the bottoms, has helped to minimize the effects of corrosion. Goals of "zero tolerance" for leaks have resulted in a continuing evolution of inspection and corrosion control approaches and procedures.

7.6.5.1 Internal Protection

In oil storage tanks, where the layer of conducting basic sediment and water (BS&W) is usually quite shallow, CP of an internal tank bottom is much more difficult than if the tank was partially filled with water. Internal coating of the bottom and up to 1 m (39 in) up the side, have been more common than CP. However, on larger tanks in centralized production facilities, in terminals, and in refineries, galvanic anodes distributed over the bottom on insulating pads or blocks have been used to give good results. Because the anodes will not operate properly if covered with oil, it is necessary to maintain a water level that is higher than the anodes. Anodes of magnesium, zinc, and aluminum have been used for this purpose, with aluminum again offering the advantage of longer life. Figure 7.19 shows two different configurations for larger AST.

 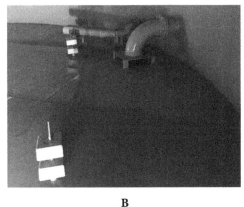

A　　　　　　　　　　　　　　　　　　B

Figure 7.19 Examples of different galvanic anodes mounted in a tank: (A) cylindrical anodes suspended from a ceiling, and (B) block anodes mounted on threaded rod to tank floor.

7.6.5.2 External Protection

With the emphasis on eliminating tank leakage mentioned above, external tank bottom protection techniques are continuing to be developed. CP designs have changed to match the changing requirements. NACE RP0193[23] presents guidelines for the design, installation, and maintenance of these CP systems.

For example, with containment designs that place an impermeable membrane in the tank's sand or gravel foundation, there may be only inches of room between the membrane and the tank bottom. This places quite a restriction on past methods of anode positions to get current properly distributed under the tank. The tight spaces have led to numerous designs utilizing anodes made as "mats" or looped wire anodes (Figure 7.20). When installing new tanks, the corrosion control design must be worked out to match the containment and other environmental concerns.

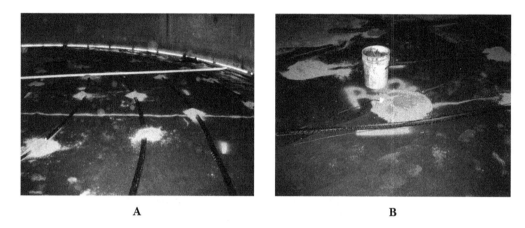

Figure 7.20 Placement of mixed-metal oxide anode loops and reference electrode for a new slotted tank bottom: (A) concentric rings held in place with sand, and (B) location of fixed reference electrode.

Existing tanks can be retrofitted with CP where and when required. Smaller tanks, or groups of tanks, can be successfully protected with galvanic anodes. Impressed current systems may have large rectifier units with distributed beds of graphite or silicon iron anodes. A centrally located deep well groundbed may be used to distribute the current needed for protection of large areas of steel tank bottoms. A current density of 10.76 mA/m² (1 mA/ft²) will usually prove adequate, if the current is well distributed.

A problem occurs in measuring the effectiveness of the applied CP because the tank potential at the periphery will usually not represent the true potential at the center of the tank. For this reason, some operators carry a potential 0.05–0.10 V higher (more negative) than the minimum –0.85 V_{CSE} at the rim, to be reasonably assured that the center area will be at or near –0.85 V_{CSE}.

Where new tankage is being constructed, a "permanent" reference electrode can be placed below the tank bottom at representative locations. Both silver-silver chloride and high purity zinc encapsulated in an appropriate backfill are used as permanent reference electrodes. As shown in Table

7.7, the equivalent voltages to the copper-copper sulfate electrode's –0.850 V_{CSE} are 0.790 V_{SSC} and +0.266 V_{ZRE}, respectively. A lead wire is brought out to a test box at the tank rim, where tank-to-reference cell potentials can be measured once CP is applied. Techniques have also been developed for placing reference electrodes under existing tanks.

7.6.6 Offshore Structures

The oil industry has a tremendous investment in offshore drilling and production structures, both fixed and floating (movable), including marine terminals of various types. Corrosion of these structures has been an expensive problem, and a great deal has been learned about corrosion rates, structure design to minimize corrosion damage, and methods of combating corrosion.

NACE SP0176[31] presents the various corrosion control approaches. Figure 1.8 in Chapter 1 illustrates the corrosion zones, relative corrosion rates, and corrosion control approaches for each zone. Also listed in Chapter 1, Section 1.9.2.1 is NACE SP0176's definitions of the three corrosion zones: atmospheric, splash, and submerged. The submerged zone is the underwater portion of the platform below the splash zone including the portion below the mud line. It is also the zone where CP is applicable and is the preferred approach to corrosion control. When properly designed, installed, and conscientiously maintained, CP can be very effective in controlling corrosion in the submerged zone. CP systems for marine structures may be impressed current or sacrificial, depending on the platform design, AC power availability, and economics (including a consideration of the relative ease of maintenance between the two systems).

7.6.6.1 Platform Design

The design of the jacket and the shape, size, and configuration of the structural elements affects the effectiveness of the CP system. NACE SP0176 discusses design features to simplify the application of CP, such as the need to use tubular members for the submerged structure, the need to avoid skip and tack welding (all welded joints should be continuous), and the need to avoid having underwater lines positioned in a way that they shield the structure from CP.

7.6.6.2 Criteria for Protection

The basic criteria for CP, i.e., using a silver-silver chloride seawater reference electrode for the potential measurements, is a negative voltage of at least –0.790 V_{SSC} between the reference and the structure. The most common method for measuring potentials involves suspending the electrode from the deck to various water depths. This approach gives the general condition of the CP, but it does not usually define problem areas. A more precise approach is for the reference electrode to be handled by a diver or a remotely operated vehicle (ROV), which can result in much more detailed and more meaningful data. Visual inspection, including ROV videos, can reveal the actual condition of the structure, such as, any pitting or cracks as well as the presence of calcareous deposits as a result of CP.

Calcareous deposits develop over time on cathodically protected structures in seawater. These deposits have some beneficial effects in that they perform like a coating, albeit a poor coating in the conventional sense. They can ultimately reduce current requirements by 50% or more, and they are very helpful in extending current spread along a cathodically protected structure. As one would imagine, water chemistries around the world's oceans can vary by location and depth. As a consequence, the ability to form stable calcareous deposits varies as well.

7.6.6.3 Galvanic Anodes

Like anodes for other applications, anodes for offshore platforms may be alloys of magnesium, zinc, or aluminum. These alloys are available in a variety of shapes and sizes. Thus, they can be selected to provide protective current to specific offshore platforms with optimum current distribution for the design life of the structure. Different methods may be used to attach the anodes to the structure depending on their type and application, but most importantly, a low resistance electrical contact must be maintained throughout the operating life of the anodes. The performance of galvanic anodes in seawater depends on the alloy's composition (particularly true for zinc and aluminum). The aluminum anode galvanic system is the most common CP method for offshore structures (Figure 7.21). The complexity of the structure makes the design and control of current density much more difficult with impressed current systems than with galvanic systems.

Figure 7.21 Galvanic anodes installed on an offshore structural steel (or "jacket").
Photograph provided courtesy of NACE International.

Anodes are selected to provide a specific life, often a 20- or 25-year life. Fixed structures will have the anodes attached in the fabrication yard. It is important that the total weight of all anodes be included in the structural design calculations for the platform. Cases have been reported where an impressed current system was selected for an offshore platform because the required total weight of galvanic anodes exceeded the design loading for the platform.

Galvanic anodes suspended from the deck find an application where the platform is in relatively shallow water, where supplemental protection is needed, or as a retrofit. Because of well recompletions and secondary recovery projects, the service life of many platforms has been extended beyond the design life of the cathodic protection system. Consequently, considerable work has been done developing economical techniques for effectively retrofitting anodes on platforms.[35]

Figure 7.22 Galvanic anode retrofit system for an offshore installation. *Photograph provided courtesy of Deepwater Corrosion Services.*

7.6.6.4 Impressed Current Systems

Typically, impressed current systems employ rectifiers with large current output ratings to minimize the number and space occupied by such units. Historically, a widely used anode material was lead-silver anodes. However, concerns over the use of lead means these anodes are rarely used in new or retrofit systems.[36] Other anode materials have been used with some success; the platinized and mixed-metal oxide anodes are most common today. The impressed current anodes may either be suspended or placed in special holders for rigid attachment to the underwater platform members. Suspended systems are somewhat more susceptible to mechanical damage, but they are simple to install and relatively easy to maintain. Impressed current anodes also may be mounted on concrete sleds that lay on the bottom with the cables installed through "J" tubes. The sleds may be placed under the platform, or set at a remote distance, or both as required for current distribution.

Impressed current systems are capable of long-term protection, but are less tolerant of design, installation, and maintenance shortcomings than sacrificial anode systems. Good service can be expected if proper attention is paid to mechanical strength, connections, cable protection, choice of anode type, and integrity of power source.

7.6.6.5 Subsea Structures

Subsea completion wellheads, flowlines, and oil storage units are widely used as oil production moves into deeper water offshore and into shipping lanes. The concerns, problems, and solutions are essentially the same as for offshore platforms with some additional considerations.[37] Coatings supplemented with CP is the usual approach to corrosion control. However, corrosion-resistant metals are seeing widespread use in some offshore areas (e.g., the North Sea). Guidelines for preventing hydrogen embrittlement damage to duplex stainless steel that is cathodically protected are available.[38] External CP may be the impressed current or galvanic anode type, depending on the particular circumstances.

Figure 7.23 Subsea CP monitoring using an ROV.
Photograph provided courtesy of Deepwater Corrosion Services.

7.7 Role of Protective Coatings

As discussed in Chapter 6, protective coating systems are one of the basic approaches to corrosion control. However, because no approach is 100% effective 100% of the time, many critical situations use the combination of coatings and cathodic protection. In fact, when it comes to buried or submerged structures, the combination is often the most economical approach. For example, pipeline coatings supplemented with CP can be a more economical approach to control corrosion on a pipeline than either technique would be by itself. Not only can the combination provide near 100% protection, it can also allow the use of a lower-cost coating system and a much simpler CP system. In many cases, an uncoated line would require an impressed current system for adequate corrosion control, yet corrosion on the same line could be completely controlled with coating supplemented with a sacrificial anode system to protect holidays (voids) and damaged places in the coating. Coatings that are used with CP need to resist cathodic disbondment.[39]

The use of sacrificial anode systems to supplement internal vessel and tank coatings is an accepted technique for controlling corrosion in water handling vessels and water sections of multiphase vessels.

Coatings have only been used in limited cases on the exterior of well casings. However, some operators cement all the way to the surface to provide a cement sheath on the casing. Even though the cement sheath is not complete, the bare surface area is drastically reduced, and again, the demands on the CP system are decreased. In fact, the submerged areas of steel offshore platforms are probably the only place where coatings have not been economically justified as a means of reducing the cost of corrosion control by CP.

References

1. NACE SP0169-2013, "Control of External Corrosion on Underground or Submerged Metallic Piping Systems" (Houston, TX: NACE International, 2013).
2. ASTM G97-97(2013), "Standard Test Method for Laboratory Evaluation of Magnesium Sacrificial Anode Test Specimens for Underground Applications" (West Conshohocken, PA: ASTM International, 2013).
3. ASTM B843-13, "Standard Specification for Magnesium Alloy Anodes for Cathodic Protection" (West Conshohocken, PA: ASTM International, 2013).
4. NACE SP0387-2014, "Metallurgical and Inspection Requirements for Cast Galvanic Anodes for Offshore Applications" (Houston, TX: NACE International, 2014).
5. NACE TM0190-2017, "Impressed Current Laboratory Testing of Aluminum and Zinc Alloy Anodes" (Houston, TX: NACE International, 2017).
6. NACE SP0492-2016, "Metallurgical and Inspection Requirements for Offshore Pipeline Bracelet Anodes" (Houston, TX: NACE International, 2016).
7. NACE SP0193-2016, "Application of Cathodic Protection to Control External Corrosion of Carbon Steel On-Grade Storage Tank Bottoms" (Houston, TX: NACE International, 2016).
8. ASTM B418-16a, "Standard Specification for Cast and Wrought Galvanic Zinc Anodes" (West Conshohocken, PA: ASTM International, 2016).
9. ASTM F1182-07(2013), "Standard Specification for Anodes, Sacrificial Zinc Alloy" (West Conshohocken, PA: ASTM International, 2013).
10. NACE SP0572-2007, "Design, Installation, Operation, and Maintenance of Impressed Current Deep Anode Beds" (Houston, TX: NACE International, 2007).
11. Ansuini, F.J., and J.R. Dimond, "Factors Affecting the Accuracy of Reference Electrodes," Materials Performance 33, 11 (1994): pp. 14–17
12. Peabody, A.W., *Peabody's Control of Pipeline Corrosion*, 2nd ed. (Houston, TX: NACE International, 2001).
13. Holtsbaum, B., *Cathodic Protection Survey Procedures*, 3rd ed. (Houston, TX: NACE International, 2016).
14. NACE SP0207-2007, "Performing Close-Interval Potential Surveys and DC Surface Potential Gradient Surveys on Buried or Submerged Metallic Pipelines" (Houston, TX: NACE International, 2007).
15. ASTM G57-06(2012), "Standard Test Method for Field Measurement of Soil Resistivity Using the Wenner Four-Electrode Method" (West Conshohocken, PA: ASTM International, 2013).
16. EN 12473-2014, "General Principles of Cathodic Protection in Seawater" (Brussels, Belgium: Comité Européen de Normalisation [European Committee for Standardization], 2014).
17. API RP 651-2014, "Cathodic Protection of Aboveground Petroleum Storage Tanks" (Washington, DC: American Petroleum Institute, 2014).
18. NACE SP0285-2011, "Corrosion Control of Underground Storage Tank Systems by Cathodic Protection" (Houston, TX: NACE International, 2011).
19. EN 12499-2003, "Internal cathodic protection of metallic structures" (Brussels, Belgium: Comité Européen de Normalisation [European Committee for Standardization], 2003).
20. NACE SP0186-2007, "Application of Cathodic Protection for External Surfaces of Steel Well Casings" (Houston, TX: NACE International, 2007).
21. DNVGL-RP-B401-2010, "Cathodic Protection Design" (Oslo, Norway: DNV 2010).

22. ISO 15589-1:2015, "Petroleum and Natural Gas Industries—Cathodic Protection of Pipeline Transportation Systems—Part 1: On-Land Pipelines" (Geneva, Switzerland: International Organization for Standardization, 2015).
23. NACE SP0193-2016, "External Cathodic Protection of On-Grade Carbon Steel Storage Tank Bottoms" (Houston, TX: NACE International, 2016).
24. API RP 1632-2010, "Cathodic Protection of Underground Petroleum Storage Tanks and Piping Systems" (Washington, DC: American Petroleum Institute, 2010).
25. NACE SP0196-2015, "Galvanic Anode Cathodic Protection of Internal Submerged Surfaces of Steel Water Storage Tanks" (Houston, TX: NACE International, 2015).
26. DNVGL-RP-F103-2016, "Cathodic Protection of Submarine Pipelines By Galvanic Anodes" (Oslo, Norway: DNV, 2016).
27. NACE SP0388-2014, "Impressed Current Cathodic Protection of Internal Submerged Surfaces of Carbon Steel Water Storage Tanks" (Houston, TX: NACE International, 2014).
28. DNVGL-RP-B101-2016, "Cathodic Protection of Submarine Pipelines By Galvanic Anodes" (Oslo, Norway: DNV 2016).
29. NACE SP0286-2007, "Electrical Isolation of Cathodically Protected Pipelines" (Houston, TX: NACE International, 2007).
30. NACE SP0575-2007, "Internal Cathodic Protection Systems in Oil-Treating Vessels" (Houston, TX: NACE International, 2007).
31. ISO 15589-2:2015, "Petroleum and Natural Gas Industries—Cathodic Protection of Pipeline Transportation Systems—Part 2: Offshore Pipelines" (Geneva, Switzerland: International Organization for Standardization, 2015).
32. NACE SP0176-2007, "Corrosion Control of Steel Fixed Offshore Platforms Associated with Petroleum Production" (Houston, TX: NACE International, 2007).
33. NORSOK Standard M-503-2007, "Cathodic Protection" (Oslo, Norway: Standards Norway, 2007).
34. NACE SP0104-2014, "The Use of Coupons for Cathodic Protection Monitoring Applications" (Houston, TX: NACE International, 2014).
35. Britton, J., "Ageing Subsea Pipelines External Corrosion Management," CORROSION 2017, paper no. 9642 (Houston, TX: NACE International, 2017).
36. Britton, J., "Impressed Current Retrofits on Offshore Platforms The Good, The Bad, and The Ugly," CORROSION 2001, paper no. 01505 (Houston, TX: NACE International, 2001).
37. NACE Technical Committee Report 7L192-2009, "Cathodic Protection Design Considerations for Deep Water Projects" (Houston, TX: NACE International, 2009).
38. DNVGL-RP-F112-2017, "Design of Duplex Stainless Steel Subsea Equipment Exposed to Cathodic Protection" (Oslo, Norway: DNV, 2017).
39. NACE TM0115-2015, "Cathodic Disbondment Test for Coated Steel Structures Under Cathodic Disbondment" (Houston, TX: NACE International, 2015).

Chemical Treatment

Robert J. Franco

8.1 Introduction

The phrase "chemical treatment" covers many activities in oil field operations (drilling, production, and plant facilities), including control of acid gas corrosion (CO_2, H_2S, brine), oxygen corrosion (oxygen scavengers), microbiological corrosion (biocides), and water corrosion (cooling and boiler water). Many chemicals are used for purposes other than corrosion. These include additives for drilling mud, clay stabilization, scale inhibition, flow assurance (e.g., paraffin inhibitors, hydrate inhibitors, drag reducers), demulsification, defoaming, and others.[1] Although these additives are not used for corrosion control, corrosion inhibitors must be compatible with them if they are in the system together because interference between two or more chemicals can accelerate corrosion or result in other problems such as scaling. Some chemicals are used to transport a corrosion inhibitor into the fluid stream, for example, a glycol hydrate inhibitor is often used to carry a corrosion inhibitor into a gas stream.

These chemicals are referred to as "oil field chemicals." This chapter will deal with the chemicals that are used directly or indirectly in the control of corrosion. Some materials, such as those used to scavenge oxygen or control microbiological activity are discussed in Chapter 9, Sections 9.5.3 and 9.8, respectively.

By NACE International definition, a "corrosion inhibitor" is a chemical substance or combination of substances that, when present in the proper concentration and forms in the environment, reduces the corrosion rate. This is not a detailed definition because it addresses so many types of corrosion inhibitors. What is important to understand is the potency of corrosion inhibitors: a small concentration of tens-to-hundreds of parts per million (ppm) can enable carbon steel construction to perform for decades, whereas in the absence of a corrosion inhibitor the expected service life could be only a few years. Effective inhibitors reduce steel corrosion rate by more than 90% at an economical cost. Corrosion inhibitors—and indeed many oil field chemicals—are truly miracle workers.

8.2 Fundamentals of Corrosion Inhibitors

A successful inhibitor program involves selecting the proper inhibitor, applying it with a technique that ensures the inhibitor is transported where it needs to go and has a chance to form a protective film, and monitoring and performing periodic reviews to confirm control and identify changes in conditions that require treatment modification. Therefore, inhibitor programs must be designed for specific wells, pipelines, or other systems, their mechanical features, and operating procedures.

Corrosion inhibition is accomplished by one or more of three fundamental mechanisms:

1. Organic corrosion inhibitors retard corrosion by adsorption to form an invisibly thin film (monomolecular) or thicker film (macromolecular) on the metal surface.
2. Inorganic corrosion inhibitors form thicker precipitates than those formed by an organic inhibitor monomolecular film. These precipitates coat the metal and protect it from attack. Included in this mechanism are chemicals that cause the metal to corrode in such a way that a combination of adsorption and corrosion product forms a passive layer. Inorganic corrosion inhibitors function by affecting the anodic, or cathodic, or both half-cell corrosion reactions.
3. Substances that, when added to an environment, retard corrosion, but do not interact directly with the metal surface. This type of inhibitor causes conditions in the environment to be more favorable for forming protective precipitates (e.g., oxygen removal), or it neutralizes an acidic component in the system by adjusting the pH.

There are many types of corrosion inhibitors for various applications. Generally, they can be grouped into two broad categories—inorganic and organic (Table 8.1). The inorganic inhibitors are most commonly used in cooling waters and heat mediums, e.g., closed loop or open, recirculating cooling water systems, as well as in gas sweetening processes. The organic film formers are used in oil, gas, and produced water systems, and are the most common inhibitors in oil field applications.

Most oil field inhibitors are organic inhibitors that cover local cathodes and anodes. The first layer formed by adsorption of an organic molecule onto a metal surface may be bonded, perhaps by an electrical charge exchange analogous to chemical bonding. Such bonding is called "chemisorption." Weaker physical bonding forces are also involved, especially during deposition of subsequent layers of the film.

Table 8.1 Corrosion inhibitor classification and use.

Service Fluid	Inorganic Corrosion Inhibitors	Organic Corrosion Inhibitors	Refer to Section Number
Cooling tower water	X		8.10.2.6
Closed loop heat/cooling mediums	X		8.10.2.7
Glycol dehydration	X	X	8.10.2.3
Sweetening system amine solutions	Corrosion control is primarily by controlling operating parameters		8.10.2.5
Oil, gas, and water wells		X	8.9
Flowlines and pipelines		X	8.10.1
Gas systems		X	8.10.1.1, 8.10.2.4
Produced and condensed water		X	8.11

8.3 Inorganic Corrosion Inhibitors

Inorganic corrosion inhibitors are usually metal salts, which will passivate the surface of a metal that normally would not be passivated in the environment. These are water-soluble chemicals that react with the metal surface to produce a layer having protective qualities. A constant (or fixed) concentration is required; therefore, close control is necessary. Care must be taken with some materials because too high or too low concentration can cause accelerated corrosion. Furthermore, they are usually pH sensitive (i.e., the pH must be maintained within certain limits to ensure passivation). Another reason they are not used in production streams is that the inorganic inhibitors usually will not work well in the presence of a large quantity of chloride ions and most of the waters in production operations contain chlorides in the brine. The inorganic inhibitors fall into two classes: anodic and cathodic.

8.3.1 Anodic Inorganic Inhibitors

As the name indicates, the anodic inhibitors react mainly at the anodic areas—where corrosion is taking place—and the film formed is a reaction product that is essentially a corrosion product (a protective iron oxide). The net effect is to slow down or stop the corrosion reaction by eliminating the anodic areas. Anodic inhibitors can be divided into two categories: passivators and nonpassivators.

The passivators inhibit in the absence of oxygen. Examples of anodic passivators include soluble salts of chromates, nitrites, and molybdates. At one time, chromates were the inhibitors of choice in the open cooling water systems in plants; however, because of their toxicity, environmental regulations have eliminated their use.

The nonpassivators, conversely, require oxygen to be present for their reaction. A common example would be the polyphosphates.

Some sources refer to the anodic inhibitors as "dangerous inhibitors." An insufficient dosage will result in incomplete coverage of anodic areas. These small, bare anodic sites will be coupled to large cathodic areas. As discussed in Section 3.6, the large cathode–small anode relationship can lead to greatly accelerated pitting at the anodic sites.

The film created by an anodic inorganic corrosion inhibitor is persistent only if an adequate concentration of the inhibitor is present in the liquid. When concentration falls below a critical value, the inhibitor not only fails to protect steel, it accelerates localized corrosion. Therefore, anodic corrosion inhibitors are used at high concentrations to prevent this.

8.3.2 Cathodic Inorganic Inhibitors

If compared to anodic inhibitors, cathodic inhibitors are active at the cathodic sites and thus reduce the cathode/anode relationship and slow down the corrosion reaction. Examples of the cathodic inhibitors are calcium salts, magnesium salts, and zinc salts. While these may be effective inhibitors in fresh water, most oil field brines contain dissolved salts or gases that would react and cause the inhibitors to precipitate. For example, zinc salt inhibitors will precipitate as zinc sulfide in the presence of hydrogen sulfide, or calcium can be precipitated as calcium sulfate scale by dissolved sulfates.

The film created by cathodic inorganic corrosion inhibitors is persistent only if an adequate concentration exists in the liquid. The effectiveness of the film decreases as concentration diminishes, and lower dosages increase corrosion rate, but—unlike anodic inorganic corrosion inhibitors—this does not lead to accelerated localized corrosion.

8.4 Organic Corrosion Inhibitors

The most common corrosion inhibitors in use in oil field systems are the so-called "organic film forming inhibitors." Therefore, unless stated otherwise, the remainder of this chapter will be dealing with the organic inhibitors. These are polar organic materials consisting of molecules that have a charge on each end. The polar molecule can be depicted as if it were a tadpole, with a head and a tail (Figure 8.1A), the head being positively charged at one end (typically by nitrogen-containing N^+) and hydrophilic (water attracting), and the tail being nonpolar, hydrophobic (water repelling), oily molecules. Because of the charges, these polar organics are surface active and are attracted to, and tend to cling to, solids (Figure 8.1B). Thus, they can form a protective barrier (film) on a metal surface. The inhibitor adsorbs onto the corrosion byproduct layer rather than a clean metal surface. If this were not the case, inhibitors would rarely function because the metal surface is never free of corrosion product. The long-chain hydrocarbon tail on the inhibitor molecule is oleophilic (i.e., it has an affinity for oil and incorporates oil into the tail). The net result in oil systems (or produced water systems with traces of oil carryover) is a monomolecular, hydrophobic inhibitor–oil film on the metal surface. Mass transfer through the inhibitor film decreases as film thickness increases. In effect, the corrosion inhibitor establishes an oil-wet condition on the metal surfaces (Figure 8.1C). Oil-wet surfaces repel water and thus break up the corrosion cell by effectively placing a barrier between the electrolyte and the metal surface (local anodes and cathodes).

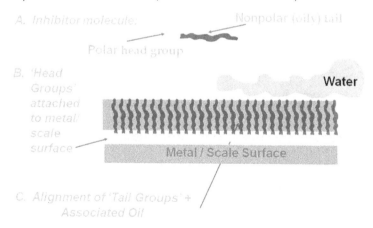

Figure 8.1 Artist rendering of a polar organic corrosion inhibitor: (A) inhibitor molecule with polar head and nonpolar tail; (B) inhibitor heads attached to metal-scale surface forming a film (monomolecular layer); and (C) inhibitor film plus associated produced oil form a macrofilm, which keeps water away from metal-scale surface.

8.5 Corrosion Inhibitor Formulations

Most inhibitors used in oil field applications are liquids. Inhibitors are formulated to contain many chemicals besides the active inhibitor molecule. It is a misconception to talk about inhibitors as if the delivered product contains 100% of the molecules that form the film. The inhibitor is a mixture of materials compounded to give certain properties (both physical and chemical) to the formulation. Even though these other chemicals are not part of the active ingredient, they are essential for the proper functioning of the product.

The purchased drums contain a material called an "inhibitor," which is a mixture of inhibitor's active ingredient (the positively charged head and long-chain hydrocarbon tail) and additives in a diluent (sometimes called the "carrier" or the "solvent") (Figure 8.2).

Figure 8.2 Components of a corrosion inhibitor (not drawn to scale).

8.5.1 Introductory Chemistry of Inhibitor Molecules

The corrosion engineer does not need to be an organic chemist to understand oil field inhibitors and how to properly use them. For many oil field inhibitors, the positively charged head is achieved by functional groups containing nitrogen. The simplest form of the nitrogen group is the organic analog of the ammonia (NH_3) molecule. This organic equivalent of ammonia is written in shorthand as R-NH_2 and is known as an "amine." Here, "R" stands for the long chain hydrocarbon tail containing CH_3-(CH_2)11 (i.e., CH_2 is repeated 11 times in the chain). Typically, organic molecules do not begin to exhibit surface activity until the hydrocarbon chain is around 12 carbon atoms long, and many oil field corrosion inhibitors contain at least a 12-carbon atom chain, but oftentimes this "carbon backbone" will extend to 18 or more carbon atoms, $\geq C_{18}$).

Other forms of organic corrosion inhibitors include

- Quaternary amines R–N$^+$–(CH$_3$)
- Amine salts R–NH$_3^+$–R1–COO$^-$, where R and R1 are two hydrocarbons with different carbon chain lengths; these are salts of nitrogen-containing molecules with carboxylic acids (fatty acids, naphthenic acids)
- Amides –N–C=O
- Imidazolines (Figure 8.3)

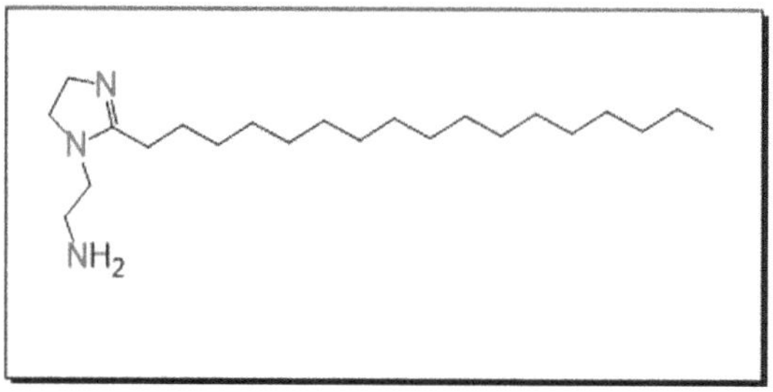

Figure 8.3 Imidazolines: the zig-zag line represents the carbon backbone.

The inhibitor bases are compounded of various chemicals that will form films and have specific properties. Most of the bases are organic amines and/or amine salts. They may be long-chain molecules (with 16 or 18 carbon atoms) or ring compounds. Generically, they may be classified as diamines, quaternary ammonium chlorides, poly-amido-amines, or imidazolines. Often the base materials are amine salts, which means that they have been reacted with other organic materials such as organic acids (e.g., dimer-trimer acids, acetic acid, etc.). Several different base chemicals are used, and the same base chemical may be in many different formulations designed to give various specific properties of solubility, dispersibility, and film persistence for specific applications (refer to Section 8.6 for definitions of these terms).

The additives are used to help these properties, to stabilize the mixture, and to prevent emulsion formations.

8.5.2 Solvents and Co-solvents

Solvents are used to control physical characteristics, to assist inhibition, and to maintain a reasonable cost per unit volume. Most inhibitors in their concentrated form are extremely viscous and without suitable dilution with solvent would not be pumpable. The solvent may be many materials (from diesel oil to heavy aromatic naphtha, kerosene, water or alcohol). Solvents allow the inhibitor to be soluble or dispersible in the crude or water system. Because the solvent is the lowest-cost component in the delivered chemical, cost "competitive" inhibitors are often high in solvent content and low in active inhibitor components. In cold climates, there are valid reasons to increase the solvent content in wintertime in order to protect the chemical from freezing in outdoor storage. The corrosion engineer should not be concerned about the price per gallon (liter). It is the "use" cost (volume per day) that is the most important.

Often, a small amount of another solvent is added to improve the overall stability of the final product. These are known as co-solvents or coupling agents and are usually highly aromatic solvents or alcohols. Co-solvents also provide an inexpensive means to completely dissolve the inhibitor molecule if commercial hydrocarbon solvents do not.

8.5.3 Additives

No inhibitor can control corrosion unless it is transported to the metal surface. Surfactants and dispersants are used to modify the overall behavior of corrosion inhibitors, imparting water solubility to an oily inhibitor molecule, making oil soluble inhibitors water dispersible, or transporting an insoluble inhibitor throughout the produced fluid.

By their very nature, corrosion inhibitors are powerful surface-active molecules, which can promote severe emulsion problems. Usually, demulsifiers are incorporated into the product, especially those products that are used in high concentration for batch treating wells and lines.

8.6 Corrosion Inhibitor Properties

Selection of a specific inhibitor formulation for a specific program requires matching the inhibitor properties with the system's fluids, its environment, and application technique to provide cost-effective corrosion control, while at the same time avoiding introducing other operating problems. For example, the ideal inhibitor should be soluble enough in a system's fluids to have the mobility to transport to the metal surface, yet insoluble enough to have good film persistency, but it must not form "gunk" that plugs injection wells or surface equipment. It must have sufficient surface-active properties to clean the surface and work its way through and under deposits, but it cannot form unbreakable emulsions of the well fluids. A film-persistent inhibitor (also referred to as a gunking inhibitor or a "gunker") is used to batch treat oil wells. Its lengthy film persistency allows occasional batch treatment. However, its properties must be controlled to prevent plugging and emulsion problems.

The need for different properties is the reason each oil field chemical supplier has so many different formulations available. The properties needed for treater truck inhibition of a rod pumping well are not the same as those for a gas-lift well; in fact, the properties required by all rod pump wells are not the same because of differences in standing fluid levels, production rates, corrosivity (CO_2 and H_2S concentration), water cut, gas-oil ratio and other differences. An inhibitor to be used in a gas well flowline will have different properties than the inhibitor used downhole in the same well. Table 8.2 lists some of the properties and characteristics that should be considered when selecting a corrosion inhibitor.

TABLE 8.2 Important considerations for corrosion inhibitor selection.

Desired Property	During Chemical Storage and Handling	In Diluent and Carrier Fluid	In Produced Fluids (Oil, Water, Gas)	Refer to Section No.
Inhibits corrosion	Note 1		X	8.6.1
Develops persistent film			X	8.6.2
Good film efficiency			X	8.6.3
Rapid film time			X	8.6.4
Soluble or dispersible		X	X	8.6.5
Partitions between oil and water		X	X	8.66
Compatible with other chemicals			X	8.67
Does not form stable emulsions		X	X	8.6.8
Is pumpable (pour point & viscosity)	X	X		8.6.9
Does not freeze	X	X		8.6.10
Stable after freeze/thaw cycles	X	X		8.6.11
Stable at maximum system temperature			X	8.6.12
Is not corrosive	Note 1	X	X	8.6.13
Mobility of individual components		X	X	8.6.14
Does not promote foaming	X	X	X	8.6.15
Compatible with downstream processes			X	8.6.16
Meets environmental regulations	X	X	X	8.6.17

Note 1: Some inhibitors and other chemicals, such as chemical oxygen scavengers, hydrate inhibitors and biocides, are corrosive in their neat or concentrated form, and pumps (including seals) and storage tanks should be made of compatible metallic and nonmetallic materials.

8.6.1 Inhibits Corrosion in the Well or System

Inhibitor effectiveness is defined as the percent reduction in uninhibited corrosion rate, as defined in Equation (8.1). Effectiveness is determined by laboratory testing and monitoring field trials. All corrosion inhibitors reduce corrosion rate, but their effectiveness is system-dependent, and variables such as temperature, flow rate, acid gas partial pressure, and the presence of deposits influence effectiveness.

Corrosion inhibitors can be demonstrated in laboratory testing to reduce corrosion by 98% or even more. The effectiveness (E) of the corrosion inhibitor is determined by subtracting the inhibited corrosion rate (R_{IC}) from the uninhibited corrosion rate (R_{UIC}) and dividing the difference by the uninhibited corrosion rate as in

$$E = \frac{R_{UIC} - R_{IC}}{R_{UIC}} \times 100 \qquad (8.1)$$

Criteria used for gauging effectiveness varies with the operator. Values of $E \geq 90\%$ or 95% are commonly specified. Some operators specify an inhibited corrosion rate (R_{IC}) ≤ 0.1 mm/y (4 mpy).

Chemicals should be injected into flowlines, pipelines, and piping using quills to avoid impinging them in their concentrated form directly against the pipe wall. If not well distributed into the flowing stream, localized corrosion called "injection point corrosion" can occur.

8.6.2 Film Persistency

The inhibitor film is not permanent. It can be removed by desorption into the hydrocarbon stream, high wall shear stress (high flow velocity) and entrained solids. As the concentration of an organic corrosion inhibitor decreases, its effectiveness decreases and corrosion rate increases. However, after a protective film has adsorbed onto the metal, organic corrosion inhibitors can be effective for some time even if the concentration of the corrosion inhibitor falls to zero in the produced fluid. This is known as film persistency, which is a measure of how long an inhibitor film will remain intact (i.e., how long will it take to wear away or break down the inhibitor film to the point where it is not sufficiently protective). During this time, the inhibitor can maintain a film and repair holidays (breaks or voids) in the film. The extent that an inhibitor film persists controls whether the inhibitor must be introduced periodically or continuously. Generally, the inhibitors used to batch treat oil and gas well systems have good film persistency and re-treatments can be infrequent (days to weeks, even months).

8.6.3 Filming Efficiency (Percent Protection)

Filming efficiency is another inhibitor characteristic; it is a measure of the completeness of the film and the protection it provides. Usually called "percent protection," it is determined by measuring the metal loss with and without the inhibitor in laboratory and/or field test procedures [Equation (8.1)]. Efficiency is connected to corrosion inhibition effectiveness. Desirable values $\geq 90\%$ are typical in laboratory tests.

8.6.4 Filming Time

This is an important inhibitor characteristic that is often overlooked and very seldom measured. It is the time it takes for an inhibitor film to form. It is common to assume that all that is necessary is to expose a metal surface to an inhibitor solution and instantly the inhibitor molecules will adsorb onto the metal forming the protective barrier. Adsorption takes a finite amount of time. As would be expected, the filming time is a function of inhibitor concentration, solubility, temperature, and other variables of the specific inhibitor and system being treated. The filming time is very important in most of the batch treatments discussed later in this chapter. Because inhibitor injection is stopped after the batch treatment has been completed, film formation should be rapid and remain persistent.

In many laboratory tests, a corrosion coupon or probe is dipped into the inhibitor and diluent, then removed quickly and inserted into a test chamber containing the simulated corrosion environment but without any additional inhibitor. An inhibitor with a short filming time will perform well in this test if the inhibitor film remains persistent on the metal surface.

8.6.5 Solubility or Dispersibility

Solubility or dispersibility is another way of classifying inhibitors. There are three categories of inhibitor solubility:

1. Soluble
2. Insoluble
3. Dispersible

A product is soluble in a produced fluid when it forms a clear solution that does not separate. An insoluble product will form a separate, identifiable layer after mixing. An inhibitor product is dispersible in a produced fluid (e.g., produced water) if it forms a mixture by gentle mixing, is not clear (hazy), and separates slowly, if at all. A dispersible inhibitor forms a suspension of the inhibitor molecule of colloidal size or larger in brine. Many inhibitor molecules can be made to appear soluble in any fluid through the choice of solvent or surfactant, although they are truly highly dispersible. Most water-soluble corrosion inhibitors are mixtures of inhibitor, solvents, and surfactants designed to produce a clear solution when mixed with water or brine. Except for a few specific molecules, very few inhibitors are truly water soluble, but true water solubility is not a crucial factor in inhibitor effectiveness.

Manufacturers classify their corrosion inhibitors as to their water and oil solubility and dispersibility characteristics. Generally, the manufacturer's product data sheets give solubility in oil, fresh water, and brine. Inhibitors may be described as "oil soluble-water dispersible," "water soluble-oil dispersible," "oil soluble-water insoluble," "water soluble-oil insoluble," or some other descriptor that indicates their solubility or dispersibility characteristics. These characteristics are important for several reasons.

Because the inhibitor must get to the metal surface if it is to form the film, its solubility or dispersibility in the produced fluids must be considered. The equipment being inhibited, and/or the treating method, are also considered in the selection process. For example, when treating a flowline where water collects and flows along the bottom, the water solubility of the inhibitor may control the selection of the inhibitor even if most of the fluids are hydrocarbons. Yet, in a gas well being treated by tubing displacement batch treatment, the most cost-effective inhibitor will most likely be a heavy film-former, which is basically water insoluble. Water or brine solubility is important for packer fluid inhibitors because it is important to prevent separation over a long period of time at bottomhole temperature. Brine packer fluids are therefore typically inhibited with water-soluble corrosion inhibitors (refer to Chapter 1, Section 1.5.1.3). To manage corrosion associated with produced water, water-soluble inhibitors are often used with water injection systems, gas transmission lines, and lines with high water content.

8.6.5.1 Effect of Temperature

Because the solubility or dispersibility at system temperatures can be quite different than at room temperature, the temperature effects on these properties become an important consideration. It is often thought that as temperatures are increased, materials become more soluble. This is not necessarily true. Some inhibitors that are water soluble at room temperature will separate at well temperature. While that might be an advantage inside the tubing string (there might not be as much tendency for the inhibitor to desorb), it would be a disadvantage for the inhibitor to float to the top if the inhibitor was in the packer brine in the tubing-casing annulus.

8.6.5.2 Solubility or Dispersibility in the Carrier Fluid

In addition to considering the solubility or dispersibility of the inhibitor product, consideration must also be given to these properties in the carrier fluid, often referred to as a diluent. A carrier fluid is added in the field to transport the inhibitor to the metal surface; the inhibitor solvent is what is delivered in the purchased inhibitor, provided by the chemical supplier. Many treating techniques call for diluting the inhibitor in a proper solvent (carrier fluid) before application. Dilution in the carrier fluid allows for better wetting of the metal surface and transport of the inhibitor molecules to the metal. Without the use of a carrier fluid, the treatment volume would be small and incapable of reaching all the metal surfaces. Depending on the inhibitor formulation, water, methanol, glycol, field crude, diesel, or kerosene may be selected.

The carrier fluid and the inhibitor's solvent system must be compatible to avoid problems. For example, an inhibitor formulated in an aromatic solvent may require an aromatic diluent, or else it could come out of solution if the diluent is not an aromatic oil. The inhibitor supplier should be consulted for the proper choice of a field diluent. An inhibitor is generally considered *soluble* in a diluent if the inhibitor–diluent mixture remains clear. An inhibitor is considered *dispersible* if it can be evenly dispersed in the diluent by moderate hand shaking. The quantity of carrier fluid is almost always equal to or greater than the quantity of inhibitor. The dispersion of the inhibitor in the carrier fluid may break rapidly (i.e., in less than a minute). This is a temporary dispersion. An inhibitor that remains uniformly dispersed in the diluent is considered a "dispersible inhibitor."

In a system that is dry of liquid hydrocarbons, the inhibitor may require a liquid hydrocarbon carrier fluid to develop a protective film. The carrier fluid must remain as a liquid to be effective. This is a consideration when treating high temperature gas wells (refer to Section 8.6.12.1).

8.6.6 Partitioning Between Oil and Water

This is an indication of the proportion of the inhibitor that will be in the oil phase and the water phase. This can be important in some applications. For instance, will the inhibitor used to treat a rod pumping well also perform satisfactorily in the flowline where free water runs along the bottom of the line? Also, will enough inhibitor be in the water leaving the heater treater to provide inhibition in the water disposal system?

The partition coefficient between water and oil provides information about the disposition of the inhibitor in both the oil and water phases and allows the corrosion engineer to estimate what the concentration of the inhibitor is in the water phase. The utility of this knowledge is being able to use data from laboratory corrosion testing in water phase to approximate how much inhibitor concentration is needed in the field. Going direct from lab to field is actually not this straightforward, but it provides a first order approximation.

The partitioning equation is shown in Equation (8.2):

$$C_w = \frac{C_t \times P}{V_o + P \times V_w} \qquad (8.2)$$

where

C_w = concentration (parts per million) of inhibitor to be expected in the water phase;

C_t = concentration (parts per million) of inhibitor fed to system (based on total fluid volume);

P = water-to-oil partition coefficient (available from the manufacturer or laboratory study). This parameter is temperature dependent.

V_o = Oil volume fraction in the produced liquid; and

V_w = Water volume fraction in the produced fluid.

Example:

100 ppm of inhibitor is fed to a produced liquid that is 75% oil and 25% water, and the inhibitor water-to-oil partition coefficient is 60%.

Concentration expected in water phase = 100(0.6)/(0.75 + 0.6 × 0.25) = 67 ppm.

8.6.7 Compatibility of Corrosion Inhibitors with Other Chemicals

Compatibility must be considered when the inhibitor and other chemicals are present, even in use dosages (concentrations of a few parts per million). Compatibility includes physical properties such as solubility and performance properties that include interferences that reduce performance. In some cases, two or more chemicals will react with each other, nullifying their effectiveness. In addition, in the concentrated form (as they are supplied) many oil field chemicals are not compatible with corrosion inhibitors (or each other) because of variations in solvent system, type of chemical (cationic [positive] versus anionic [negative] charge), etc. Many corrosion inhibitors are not compatible with each other. Therefore, chemicals should not be mixed before injection, nor should they be injected at the same point in a line or vessel. Each chemical should be injected with its own tank, pump, line, and injection fitting. Furthermore, the separate injection points should be far enough apart so that the chemicals are well diluted before they meet.

In many fields that produce wet sour gas in the Middle East, it was found that the corrosion inhibitor interfered with the kinetic hydrate inhibitor. Because the criticality of using the kinetic hydrate inhibitor was so great, the corrosion inhibitor was reformulated to be compatible.

8.6.8 Emulsification Properties

Emulsification properties of a corrosion inhibitor can be a concern because their surface-active nature can cause emulsions in water-oil systems. Some of these emulsions break quite readily, while others are extremely stable and practically impossible to break. The inclusion of an emulsion breaker in a corrosion inhibitor is no guarantee against formation of stable emulsions. To minimize the chance of forming stable emulsions, the specific inhibitor being considered should be tested for emulsion-forming tendency with actual produced liquids samples.

8.6.9 Pour Point

Pour point is the temperature where the inhibitor will no longer pour. The pour point, freeze point, and freeze-thaw stability are very important properties in cold climates. A low pour point only guarantees that the product will flow; it does not address whether that product separates into two or three phases upon cooling.

8.6.10 Freeze Point

Freeze point is the temperature where the inhibitor freezes—a very important value in cold climates, particularly when the inhibitor drums (or other containers) are stored outdoors.

8.6.11 Freeze-Thaw Stability

Freeze-thaw stable means that after a frozen inhibitor is thawed, it will continue to be (or return to) a homogeneous material with the same properties as before freezing.

8.6.12 Thermal Stability

Thermal stability indicates that the inhibitor will retain its characteristics at the highest and/or its coldest operating temperature in the specified environment. In many cases, the coldest temperature comes from the ambient temperature of the stored inhibitor. It is a potential problem in arctic areas. The high temperature usually comes from the process conditions. For example, temperatures can get quite hot at the outlet of gas compressors for an inhibitor that is injected into the discharge piping of a gas compressor in a gas compression system. Downhole inhibitor placement near or at the bottomhole location also exposes the inhibitor to high temperature. In continuous downhole treatment (Section 8.9.13.7), the inhibitor may be stored in the tubing-casing annulus and is made available for continuous well treatment through chemical injection valves. The annulus is hot, and the inhibitor is exposed for an extended time period. Tubing displacement treatment (Section 8.9.2) transports the inhibitor to the bottom of the well. The physical cold-temperature properties are addressed in Sections 8.6.9–8.6.11, above. Rarely is inhibitor performance required at cold temperatures because corrosion rates are low. However, both physical and performance properties are required at elevated temperatures.

Because many inhibitor bases are made at reaction temperatures above 204 °C (400 °F), degradation of the molecule is not as much of a concern as physical changes in inhibitor properties, for instance, solubility or separation of the components. However, degradation can occur in the 204–260 °C (400–500 °F) temperature range in high-pressure gas wells. Inhibitor performance at elevated temperature is typically degraded, requiring either an increased dosage or a change in inhibitor formulation.

8.6.12.1 Phase Behavior

In addition to thermal stability, phase behavior of the produced hydrocarbons is important in high-pressure and high-temperature gas wells that are continuously treated with corrosion inhibitor. If the produced hydrocarbon phase is supercritical, liquid hydrocarbons are absent from the flow stream entering the tubing, and a liquid corrosion inhibitor injected at that location will evaporate and not function unless it remains in the liquid state. The phase behavior of an inhibitor carrier–well fluid system must be such that at bottomhole conditions the inhibitor is filmed out of the liquid phase. Where this is not possible, the well may be batch treated during shut-in (if conditions can maintain a liquid phase), or a corrosion-resistant alloy (CRA) tubing string may be required.

8.6.13 Materials Compatibility

Materials compatibility as a property of a corrosion inhibitor may sound counterintuitive, but some inhibitors in their "as delivered condition" may be quite corrosive, even though they inhibit corrosion in their at-use concentration in the system. Corrosion inhibitors are often delivered in lined or nonmetallic drums and are injected through CRA equipment. Information on the corrosion characteristics of a specific corrosion inhibitor is important when selecting materials for chemical injection facilities. Because they might be changing formulations during a project's life, some operators install all chemical injection facilities with corrosion-resistant components (often 316 [UNS S31600] stainless steel is the material of choice). Elastomer components in both the injection and producing equipment are susceptible to degradation during exposure to many inhibitor formulations, and compatibility should be confirmed in laboratory testing. Elastomers such as nitriles (nitrile butadiene rubber [NBR]) and fluoroelastomers (such as FKM and FFKM), are particularly susceptible to attack by amine inhibitors. Refer to Chapter 5, Section 5.13.2.1 for a description of these elastomers.

8.6.14 Mobility

Mobility of the corrosion inhibitor base, when the solvents are exposed to temperature and pressure, is critical in situations where the inhibitor is injected into dry gas (such as gas-lift injection gas). An inhibitor should be formulated with a suitable solvent that remains liquid under the temperature and pressure of the application.

8.6.15 Foaming Properties

Foaming properties of an inhibitor are not usually a concern at the normal low use dosages. However, in some batch treating situations where high concentrations can enter separation facilities or plants, foaming of the produced fluids could create operating problems. Thus, the foaming tendencies of the inhibitor in the specific field fluids should be determined. Many water system corrosion inhibitors are "heavy foamers" and an overdose (such as might be caused by a runaway pump or a broken check valve) can upset separators and water plants. Reports of water plants "filled with soap suds" are not uncommon.

8.6.16 Compatibility with Downstream Processes

Compatibility is important to avoid the situation where the inhibitor would upset or foul a process or operation. For example, inhibitors could cause foaming and glycol loss in glycol dehydration units, upset gas sweetening processes, foul dry bed (molecular sieve) units, or upset refinery catalysts. Crude sale to a refinery may have restrictions imposed on the concentration of certain additives.

"MSDS—It's the Law"

The Material Safety Data Sheet (MSDS) is just what the name implies—a form that supplies information on a material for use in an emergency or accident.

In the United States, each chemical manufacturer or vendor is required to furnish an MSDS for any material supplied to a location—including samples for testing. Other countries refer to it as a Safety Data Sheet (SDS) or Product Safety Data Sheet (PSDS). The user is required to have the MSDS (or equivalent) available to company and emergency personnel. These regulations were designed to minimize risks at the time of spills, fires, or other accidents.

MSDS forms have 16 sections (this varies by country), covering a variety of topics, including such items as product identification, composition or information on ingredients, hazards identification, first aid, firefighting, accidental release measures, handling and storage requirements, and ensuring personnel safety—including minimizing personnel exposure and required personal protective equipment. Typical physical and chemical properties, and stability and reactivity data are given. Toxicological, ecological, disposal, transportation, and regulatory information is listed. The last section is for information that may be required by a specific state or local authority.

The amount of detail and thoroughness of the information may vary depending on the chemical, its hazard level, and the supplier. Depending on the chemical, the composition may only be very generic, but it gives a clue to the chemistry and to the solvent system used. Except for biocides and other materials that must be registered (labeled) the exact formula (percentage composition) is not required. The physical and chemical properties section is another that gives very helpful information in case of a spill or accidental contact. It is well worth the effort to get familiar with the sheets for chemicals in use.

8.6.17 Environmental Concerns

Environmental concerns include not only the implications of an inhibitor spill, but also, in normal operation, where does the inhibitor ultimately end up, and could that be a problem? Inhibitors that are discharged to open bodies of water are a concern.

Many of the above properties and information can be found on the inhibitor supplier's product data sheets. Other product information may be found on the Material Safety Data Sheet (MSDS) that must be supplied for each chemical delivered to a user's location. (See the sidebar, "MSDS—It's the Law.") Together these two sources supply a lot of information on production chemicals.

8.7 Selecting the Inhibitor

There are three basic questions that must be answered in selecting an inhibitor:

1. What product?
2. What concentration?
3. What application method?

The first step in inhibitor selection is to review the system, its physical layout, mechanical considerations, and fluids being handled (locating any special or unique factors). It is important to try to understand what the most likely corrosion mechanism is if failures have occurred or corrosion rates are higher than expected. If failures are due to oxygen contamination, filming inhibitors will not solve the problem. The same is true for microbiologically influenced corrosion (MIC). Look for open valves, contamination sources, etc., before changing or starting an inhibitor program. If failures are occurring in areas of high turbulence, an inhibitor that has high film persistency at high-wall shear stress should be considered. Are failures occurring in dead legs (i.e., nonflow piping segments)? (Dead legs are described in Chapter 1, Section 1.5.5.3) Can these dead legs be eliminated? What are the conditions at the points of failure, i.e., CO_2, H_2S, brine, water concentration, flow rate, temperature? Are there fluid sources present that continually or intermittently introduce fresh bacteria? Are there sources of oxygen contamination?

Identifying the location of injection points and corrosion monitoring (and fluid sample points) that are already installed must also be included in the system review. Are injection quills installed or only fittings? What injection and monitoring facilities are required for optimum corrosion control and monitoring? Is the existing injection equipment adequately corrosion resistant? What nonmetallic seals are used in it, and can they be deteriorated by the corrosion inhibitor?

What materials of construction are in the production system? Besides carbon steel, are there elastomers, other nonmetallic materials, and/or corrosion resistant alloys present? Most corrosion inhibitors require 316 (UNS S31600)[1] stainless steel injection equipment; however, even higher alloyed materials may be required for a limited number of chemicals.

The concentration of the inhibitor required depends on whether the inhibitor is batch or continuously treated, as well as the inhibitor effectiveness under the service conditions. Inhibitors adsorb onto the metal-scale surface. A low concentration results in low adsorption (insufficient surface coverage and thickness), and poor effectiveness. Effectiveness increases with inhibitor concentration until it reaches a point where it levels out. At that point, adding more inhibitor does not result in a significantly lower corrosion rate but does increase the cost of the treatment. Selecting the inhibitor concentration should take the effectiveness beyond the point of leveling out, but not too far beyond. Selecting concentration should be based on field experience, laboratory results (Section 8.7.1), the

---
1. See Chapter 4, Section 4.6.4.3.

recommendation of the chemical supplier, and economics. The production rate and flow rate of hydrocarbons and water plays an important role in deciding the required inhibitor concentration because of inhibitor desorption.

The third step is to select the application method(s) that ensures the chemical is transported to *where* it is needed *when* it is needed. This third step may be the step that controls the other two. Sometimes there is only one choice, and other times several techniques could be used. As an example, are failures occurring in downhole tubing nearest the wellhead or closer to bottomhole conditions? The answer may swing the decision to tubing displacement for bottomhole-located failures rather than a batch and fall technique for near-surface failures. Other examples: Is a pipeline experiencing corrosion at the top or bottom of the line? Does the pipeline have facilities for maintenance pigging?[(2)] The inhibitor must reach the points of failure or highest corrosion rate, and the application method depends on the answers. The final decision is based on both technical and economic factors.

The corrosion engineer should review the properties that will be required for the application method (other than it must be a corrosion inhibitor). How will the inhibitor be transported to where it is needed? What diluent should be used? Does the system need a water-soluble, oil-dispersible inhibitor or a heavy filming amine (a "gunker")? What other chemicals are in the system that may cause incompatibility? Is an emulsion breaker required?

Finally, after the system is reviewed, the application methods selected, and the property requirements outlined, an inhibitor may be selected. Oftentimes, the specific inhibitor is picked based on past experiences in similar situations (e.g., a company's experience, or experiences from a neighboring field) and a field trial is started. In other cases—particularly for large programs, or when unique properties are required—screening tests are performed to compare several inhibitors. The most effective is then selected as the one to go to field trial. In such cases, important physical characteristics must be compared as well as performance tests.

8.7.1 Laboratory Inhibitor Testing

The chemical supplier-service company will recommend an appropriate corrosion inhibitor product and the dosage rate. The operator needs to play a key role in determining the appropriate application method—either batch or continuous—inhibitor injection locations, and other details. Even though the chemical supplier-service company has performed screening tests, often, the user needs to conduct laboratory testing of competitor inhibitors.

There are many reviews of corrosion inhibitor laboratory screening methods and test requirements.[2-5] Papavinasam et al. extensively compared laboratory methods to evaluate corrosion inhibitors for oil and gas pipelines.[6] In addition, the authors provide a list of references and a method to isolate the study of pressure and flow on corrosion rate. They found that the rotating cage (ASTM G170) provided the best correlation of laboratory and field worst-case general and pitting corrosion rates.[7] The wheel test—a common screening test used by chemical suppliers—can differentiate a good inhibitor from a poor inhibitor, but not a better inhibitor among good inhibitors.[8] Papavinasam et al. caution

---

2. A pig is a device with blades or brushes inserted in a pipeline for cleaning purposes. These devices are also called "scrapers." Pigs require launchers and receiver facilities to operate. (Modified from the Schlumberger Oil Field Glossary <https://www.glossary.oilfield.slb.com/Terms/p/pig.aspx>.) Refer to these sections for an in-depth discussion of pigging: Sections 8.10.1.1–8.10.1.3; Chapter 1, Appendix 1.A; and Chapter 9, Section 9.8.1.

that the absence of standards, and the absence of agreements among the inhibitor supplier, the user, and any third-party laboratory makes it difficult to evaluate the merits of a corrosion inhibitor because the ranking of inhibitor efficiency varies among different laboratories.[5] The lowest-ranked methodologies—the wheel test, bubble test, and static test—are commonly used by inhibitor suppliers for screening inhibitor selection. The corrosion engineer must insist that more comprehensive inhibitor testing be performed when warranted.

It is generally accepted that inhibitor qualification for major projects that require pipeline, downhole, and subsea inhibitor applications requires a systematic approach that addresses several issues before laboratory testing commences. A systematic approach includes a number of steps:[9-11]

- Complete a flow assurance model that develops temperature profiles along the line, identifies the flow regimes during steady state flow and upsets, and the determines the composition and phase behavior changes along the length of a large diameter pipeline or downhole tubing string. A flow model can also examine the distribution of a corrosion inhibitor during batch pigging of a pipeline. Chapter 10, Figure 10.3 depicts different flow regimes.
- Identify corrosion mechanisms that include corrosion location (bottom and/or top of a pipeline, wellhead or bottomhole for tubing), and sweet versus sour mechanisms. Other potential mechanisms are considered where appropriate (MIC, O_2, etc.).
- Use corrosion models to identify likely corrosion mechanisms and likely corrosion locations. Corrosion modeling using proprietary corrosion models have been largely successful for CO_2 corrosion, but sour corrosion models are in development and are available.
- Develop a laboratory testing program to evaluate corrosion mitigation. Test conditions include test duration, wall shear stress, and many details. Discussions should be held with chemical suppliers regarding their test capabilities.
- Develop laboratory tests to determine the compatibility of the corrosion inhibitor with metallic and nonmetallic materials used in chemical storage tanks and injection pumps (refer to Section 8.6.13).
- Establish a program to evaluate the field effectiveness of the inhibitor, including corrosion monitoring and equipment inspection.
- Optimize the inhibitor dosage based on the results of the corrosion monitoring and equipment inspection. Corrosion monitoring includes corrosion coupons and other methods, which are described in Chapter 10. When first applying a corrosion inhibitor, commence with a higher dosage rate than anticipated from laboratory testing, monitor the results, and optimize downward over time (months to years). Inspection methods provide long-term data on equipment condition and include intelligent pipeline inspection tools ("smart pigs") and downhole calipers for tubing inspection.

It is important to have standard test protocols; however, just as important are the myriad ways to test the consequential variables that have not yet been introduced into test standards. There are standards that address flow rate (or wall shear stress), pressure, temperature, gas composition, and water chemistry to simulate field brine composition. There are, however, many nuances that are not easily defined. For example, in high-pressure gas, should the test simulate the partial pressure of CO_2 and H_2S, or the fugacity[3] of these gases? Flow regimes change over time, as does temperature, pressure, production rate, and water chemistry. Should the test simulate early production, end production, or both? Until more widely accepted standards are developed, the user must have a systematic approach

---

3. Fugacity is a function used an analogue of the partial pressure in applying thermodynamics to nonideal (real) systems.

and use consistent test methods in inhibitor qualification. Given this situation, it is likely that different laboratories—who have developed and use their own variety of protocols—may obtain different results, and perhaps a different rank order of inhibitors. Despite the difficulties in correlating laboratory and field corrosion data, laboratory methods are invaluable as a first step to evaluate corrosion inhibitors for further testing in the field.

8.8 Inhibitor Application

No matter what the chemical formulation or how good an inhibitor is, it will not do the job unless it is properly applied. Therefore, inhibitor application is quite important. The inhibitor must be transported to the metal surface, if it is going to lay down its film. Many so-called "inhibitor failures" were "application failures." There are many methods of introducing the inhibitor into a well or surface system. Some of the methods involve periodic (batch) treatments, and others require continuous injection.

8.9 Inhibition of Producing Wells

This section describes the common methods used to apply a corrosion inhibitor to a well. It also describes the treatment methods with limited applications that may be useful when the common methods cannot be applied.

8.9.1 Overview of Treatment Methods

Inhibiting any component of a petroleum producing facility (downhole to pipeline) is as much art as science because the corrosion engineer relies on the expertise of a few service companies without necessarily understanding the product chemistry, and because the application techniques are field-specific. Inhibitor concentration, fluid pressure and temperature, and many other quantitative values shown in the following sections and in Appendix 8.A are based on experience and lab data. However, no number is absolute, and readers will find many exceptions to the guidance provided. To some degree, this is due to different practices among major oil and gas producers, improvements in inhibitor molecules, and different products that are available from several competing chemical manufacturers and service companies. Although the concentration, treating frequency, etc., may vary among users, the variables affecting these guidelines, such as production rate, fluid levels, etc., are similar among chemical service companies and major producing companies.

There are many approaches to inhibitor applications when it comes to producing wells. Table 8.3 lists many methods for both wells with "packers" and those with open annuli. A packer is a subsurface tool that provides a seal between the tubing and the casing. "Packered well" or "packered annulus" is shorthand for a well that contains a packer (Chapter 1, Figure 1.3). Conversely, a well that does not have a packer is called "open annulus" also often known as "unpackered" or "packerless," (Chapter 1, Figure 1.4). As noted in the table, several of these methods have very limited application; however, they were included because they may be used for unusual cases. Some of the more common or more widely used applications are briefly reviewed in the following paragraphs. The details of the various methods are presented in Appendix 8.A. The description of each application technique includes the type of well(s) where it is used, basic approach and mechanism, treating procedure, treating frequency, monitoring, and comments (pros and cons).

Table 8.3 Corrosion inhibitor application methods for well treatment.

Type of Inhibitor Treatment (B = Batch, C=Continuous)	For Wells with Packered Annuli	For Wells with Open Annuli	Section No.
Tubing Displacement (B)	X		8.9.2
Nitrogen Squeeze or Nitrogen Displacement (B)	X		8.9.3
Batch and Fall (B)	X		8.9.4
Yo-Yo Treatment (B)	X		8.9.5
Treating Strings (C)	X		8.9.6
Formation Squeeze (B)	X		8.9.11
Gas Lift Gas Addition (C)	X		8.9.8
Annular Batch (B)		X	8.9.13
Continuous (C)		X	8.9.13.7
Weighted Liquids (B)	Limited use		8.9.13.8
Dump Bailers (B)	Limited use		8.9.13.8
Wash Bailers (B)	Limited use		8.9.13.8
Inhibitor Sticks (B)	Limited use		8.9.13.8

8.9.2 Tubing Displacement Treatment

Tubing displacement is commonly used to treat highly corrosive, packered gas wells, and—less frequently—packered gas-lift oil wells[4] and packered flowing oil wells. This treatment requires pumping the inhibitor plus diluent to reach bottomhole, but not into the formation. Basically, the procedure displaces the inhibitor or inhibitor mix to the end (bottom) of the tubing string, ensuring complete coverage from wellhead to bottomhole. It inhibits the entire tubing string, even below shut-in fluid levels. However, the plan is to avoid pumping the inhibitor mixture through the perforations.

Film-persistent corrosion inhibitors are used, and the treatment provides a lengthy film life. A theory to explain the extended film life is that the multimolecular film (macrofilm) provided by the heavy film-forming inhibitors is, in effect, self-repairing. That is, as inhibitor molecules desorb from the thick film, they are available to (and do) replace molecules that have desorbed further up the tubing. Retreatment frequency depends on the production rate and the corrosivity of the well fluids. Monthly batch treatment is typical for highly corrosive wells.

Tubing displacement is a higher-cost per treatment than several other treatment methods in use. It ensures that the inhibitor reaches the bottom of the tubing string, however, it requires higher treatment volumes. Tubing displacement also incurs a higher risk of killing the well when the downhole pressure cannot unload a full column of liquid liquid after the well is restarted. Displacement with nitrogen can be used instead of liquids to avoid killing the well (see Section 8.9.3, below). The key

---
4. See Section 8.9.8 for a discussion of gas-lift oil wells.

advantage of both tubing displacement treatment methods is in treating highly corrosive gas wells. The key disadvantage of tubing displacement is it can damage the formation if the chemicals enter the perforations.

8.9.3 Nitrogen Squeeze and Nitrogen Displacement Treatments

To "squeeze" a chemical is to displace it into the reservoir rock. Using liquids, and their associated densities, can result in production problems in gas wells if the production pressure cannot unload the treatment's liquid column. As the names imply, these treatments are variations where the "squeeze" or displacement fluid is nitrogen rather than a liquid hydrocarbon. These treatments have been used in gas wells where the bottomhole pressure is insufficient to unload the column of liquid. Nitrogen displacement is a type of Tubing Displacement treatment. Nitrogen squeeze is a type of Formation Squeeze treatment described in Section 8.9.11.

8.9.4 Batch and Fall

This method is applied to gas and gas condensate wells, mainly with low bottomhole pressure. It requires shutting in the well, pumping in a mixture of corrosion inhibitor and carrier fluid (diluent) through the crown valve by means of a pump truck, allowing sufficient time for the inhibitor to fall down the tubing by gravity drainage, then restarting production. The treatment is periodically repeated at a frequency that is dependent on the corrosivity of the well fluids. These retreating frequencies vary greatly from weekly to once every three months.

The inhibitor volume is estimated by assuming that the treatment (inhibitor + diluent) will be several percent of the tubing volume. An estimation of the treatment volume is made assuming it will coat the inside tubing surface area with a certain film thickness. Some operators use a film thickness of 0.75 mm (30 mil). Once the total fluid volume is calculated, the assumption is the inhibitor-to-diluent ratio is 4:1 (inhibitor occupies 20% of the fluid volume, the diluent occupies 80%). Depending on the inhibitor selected, the diluent can be field produced water, lease crude or condensate, or diesel. Field practice varies greatly, and operators have reported the percentage of inhibitor in the total inhibitor + diluent fluid to be 2–20%.

When a gas well is shut in, a standing column of liquid forms from the bottomhole up to some height. Gas resides above the fluid column. Gas wells with low standing fluid levels (liquid level lies below the corroding areas of the tubing) can be treated effectively with the batch and fall method.

As the corrosion inhibitor mixture enters the tubing string, it falls faster through gas than through liquid. The rate of fall is also a function of the shut-in wellhead pressure, as shown in Table 8.4. Note that the approximate inhibitor fall rate through liquid phase is too long to be cost effective so it is not included in the table.

Table 8.4 Batch-inhibitor fall times in a gas well.

Shut-In Wellhead Pressure	Approximate Inhibitor Fall Rate Through Gas Phase [1]
<20.7 MPa (3000 psi)	600–900 m/h (2000–3000 ft/hr)
20.7–41.4 MPa (3,000–6,000 psi)	300–600 m/h (1000–2000 ft/h)
>41.4 MPa (6000 psi)	Very slow, if at all. A tubing displacement treatment or continuous treatment is necessary.

Note 1: Fall rates in highly deviated wells (>30°) are about 60% of the values shown above. Inhibitor will not fall to the horizontal section of a horizontal well (Section 8.14.4).

The inhibitor and diluent do not form a cylindrical slug. Instead, as the inhibitor mixture is pumped, the column has a leading edge that falls faster than liquid column slug. If the pumping rate is too slow, a slug will not form, and the mixture preferentially flows down one side of the tubing string. A minimum pumping rate of about 20 L/m (5 gpm) is required. Actual pumping rates are much greater (90–160 L/min or 21–42 gpm).

The Batch and Fall method does not generally apply to crude producing wells because of higher standing liquid levels.

A potential disadvantage of the Batch and Fall method is deferred production during well shut-in, and the potential for killing a well when it is shut-in. The method's best advantage is treating sweet gas wells that produce minimal amounts of formation water. Water condenses near the top of the tubing string and the ensuing corrosion locations can be reached with the Batch and Fall method.

8.9.5 Partial Tubing Displacement and Yo-Yo Treatments

This is a variation of the Batch and Fall method for treating packered gas and gas condensate wells, and packered oil wells. It increases the depth of tubing covered by adding more displacement fluid (diluent) versus Batch and Fall, and the Yo-Yo method circulates the inhibitor mixture multiple times to improve film formation. The percentage of the tubing volume displaced is greater than a Batch and Fall, approximately 1/3 or 1/2 of the tubing volume. A volume of inhibitor mixture plus displacement fluid (diluent), less than the volume of the tubing, is pumped into the tubing, the well is shut-in long enough for the liquids to drain to the bottom, and then the well is brought back on. When the liquid reaches the wellhead, the well is shut-in, and the fluids fall to the bottom once more. The sequence of liquid to surface—shut-in to fall may be repeated several times. Thus, the name "yo-yo," because the inhibitor is "yo-yoed" up and down, providing multiple passes to allow more opportunity for film formation. The advantage of this method compared with Batch and Fall is the extended depth of tubing inhibited and improved film formation due to repeated exposure to inhibited fluid. The disadvantage, however, is the extended amount of time the well is not producing.

8.9.6 Treating Strings

This treatment application can be used with any oil or gas well with a packer. The treating string approach is an offshoot of the earlier practice of running a dual string of tubing—one the producing string, the other the "kill string"—in higher-pressure gas wells. The kill string was to provide a way to pump drilling fluid to kill the well if problems developed. Both parallel and concentric configurations were used. When these wells needed corrosion inhibitor treatment, the kill string became the treating string. Although the kill string approach has generally been abandoned, treating string use has become more widespread with the advent of small diameter (9.53 or 12.70 mm [0.38 or 0.50 in, respectively]) stainless steel tubing. The treating string approach is the most versatile method for inhibitor application in many situations where other methods have not been successful or are extremely costly. For example, the deep, hot, high-pressure gas wells—which require continuous treatment, and in situations where packered gas or oil wells cannot be economically shut-in for treatment—are candidates for treating strings. For batch treatment, a coiled tubing unit can be brought in and run to the bottom without killing the well, and the inhibitor placed at the bottom without forcing wellbore fluids back into the formation. The well can be quickly restored to production following treatment. There is also a potential application for some of the new small diameter (9.53 mm [0.38 in] or less) strings in gas-lift wells that have a serious corrosion problem.

8.9.7 Batch Treating Frequency for Gas Wells

Frequency depends on corrosivity of the well fluids, the gas production rate, fluid velocity, water production rate, presence of gas condensate, and the presence of particulates, especially produced sand. In other words, there is no standard treating frequency. The batching frequency must be determined based on field experience, well caliper results, the frequency of tubing failure, laboratory data, and the risk associated with tubing failure (the operator must determine this risk). Typical batch treating frequencies range from once per month to once every three months. Produced sand requires increasing the frequency, as often as weekly.

8.9.8 Gas-Lift Oil Wells

A gas-lift oil well is an artificially lifted well that injects gas into the tubing string to lighten the liquid column and allow the well to flow. The "lift gas" is often used as a carrier for the inhibitor in oil wells produced by gas lift. The "lift gas" is also known as the "gas-lift injection gas" and "gas-lift gas." The main drawbacks to this approach are twofold:

1. the interior of the well tubing will only "see" inhibitor from the lowest working gas-lift gas injection valve up, and
2. an inhibitor designed for lift-gas service may not be the most effective inhibitor for the well's fluids.

For successful well inhibition, the inhibitor must be injected into each individual well's lift gas stream. Inhibitor injection at a central location in a trunk-and-lateral lift gas system looks very cost effective on paper. However, this approach has not been successful, even with attempts to atomize the inhibitor so it will be carried with the lift gas. Tests to determine "inhibitor carry through" have shown that the inhibitor droplets fall from the gas in a relatively short distance (even within the short distances on an offshore platform). The liquid stream running along the bottom of the line is not evenly distributed to the wells. A central injection point might not provide the correct amount of corrosion inhibitor to each well. The well that needs the highest lift gas volumes may not be the one that needs the most inhibitor and vice versa.

Inhibition of gas lift wells in the lift gas is feasible if tubing corrosion is occurring above the lowest working gas lift valve mandrel. When inhibitor is introduced with the lift gas, it enters the tubing at the working gas injection valve and thus, there is no treatment below that valve. Furthermore, inhibitor effectiveness may be reduced because of compromises that must be made in inhibitor selection. Most of the commonly used solvents flash out of the inhibitor formulation when it is injected into a dry lift gas, so selection is limited to those materials that remain as mobile liquids by using an appropriate solvent. In addition, corrosion inhibitors injected into the gas-lift injection gas are specially filtered to ensure they do not contain particulates that can plug the injection valves.

One solution has been to internally run plastic-coated tubing and use the lift gas-introduced inhibitor to protect the damaged areas in the coating above the working valve—that is where the most wireline damage should be. It is difficult to obtain good coating quality at the injection valve and mandrel. At least one operator has had success using an immersion grade epoxy coating system, which worked well in separators and tanks, and performing meticulous surface preparation, coating application, and quality control. In many cases, CRA tubing such as 13Cr (refer to Chapter 4, Section 4.6.4.1) is run below the working valve. One operator has reported long-term success with installing API 5CT grade J55 tubing below the lowest gas lift gas injection valve because its pearlite-ferrite microstructure is more resistant to CO_2 corrosion than quenched and tempered grade L80.[12] Note that J55 is a lower strength grade than L80 and may not be suitable in many wells designed for L80 tubing.

Another alternative to treating the gas lift gas is to perform tubing displacements or formation squeeze treatments on gas lift wells. Gas lift wells usually have a standing fluid level in the tubing when the well is shut-in. If tubing displacement is selected, sufficient fluid must be pumped at each treatment to move the inhibitor to the bottom of the tubing string.

8.9.9 Hydraulic Pump Oil Wells

Hydraulic pump oil wells use a power fluid (power oil or water) to artificially lift the oil. Power fluid is often used as a carrier for the inhibitor in oil wells produced by subsurface hydraulic pumps. The appropriate inhibitor can be *continuously injected* into the power fluid. The dosage is based on total production (power fluid plus formation fluid) just as if it were a rod pumped well and is usually at least 20–25 ppm based on total fluid volume. Care must be taken to select an inhibitor that will not foul the power cylinder.

When power water is used, the power water tank should be gas blanketed to maintain air (oxygen)-free conditions, just as if it were a water flood facility to avoid oxygen entry into the well. (Air-free operation is reviewed in Chapter 9, Appendix 9.A) A highly water dispersible—or preferably a water-soluble inhibitor—should be selected for power water.

If power oil is used, the corrosion inhibitor should be oil soluble or highly oil dispersible. If the power oil has been degassed so that no formation gas is evolving in the power oil tank, it may be desirable to gas blanket that tank also. Because the inhibitor is in the power fluid, the well only will be treated from the pump's engine-end up to the surface. Other means of corrosion control may be necessary below that point.

8.9.10 Electrical Submersible Pump (ESP) Oil Wells

These wells are usually prolific fluid producers with high water cuts. The resulting tubing fluid velocity is relatively high so that film persistency is limited. For this reason, *continuous treatment* is often the preferred method. The inhibitor is pumped from the surface down a capillary tube, which is strapped to the outside of the production tubing. The inhibitor is introduced below the fluid level into the suction of the submersible pump. Inhibitor concentration is estimated to provide 20–25 ppm based on total fluid production. Inhibitor is introduced into the wellhead annulus. The inhibitor is diluted with an appropriate flush fluid to provide good distribution and prevent the inhibitor from remaining in the upper annulus and drying out. The inhibitor should be injected during well pump operation.

The inhibitor should be noncorrosive to both the electrical cable armoring (which is usually galvanized steel), as well as with the cable insulation material.

8.9.11 Formation Squeeze Treatment

Squeeze treatment into the formation is usually applied to gas wells but has been used in gas lift oil and flowing oil wells. As the name implies, for the formation squeeze the inhibitor is squeezed (pumped) into the formation by displacing it with liquid (e.g., crude, diesel, condensate, or other appropriate fluid) into the perforations. Often, the inhibitor selected is a heavy film former that has superior film persistency. The inhibitor adsorbs onto the formation rock surfaces and gradually desorbs into the produced fluid as production ensues. The inhibitor then adsorbs onto the tubing string as the produced fluid flows up the tubing string. The bulk of the squeezed fluid will be produced back when the well is put back on production. Thus, more inhibitor is filmed onto the tubing as it is produced. The theory for the long film life is based on the inhibitor being desorbed from the formation rock over time, thus providing inhibitor molecules to repair the film on the walls of the tubing. Depending on the specific application, treatment may provide inhibition for 3–6 months, sometimes longer. Thus, retreatments are very infrequent.

The amount of corrosion inhibitor required to be squeezed into the formation must be estimated. The production rate, the time of effectiveness (e.g., 3 months), and the effective inhibitor dosage (in parts per million) are used to determine total inhibitor volume. However, a certain amount of the inhibitor is lost permanently by adsorbing onto formation rock, and more is "lost" when a portion of it flows back immediately when the well is returned to production. Therefore, the total inhibitor volume may be three times the amount first calculated.

The main concerns with using formation squeezes are the cost of individual treatments and the possibility of "formation damage." The higher cost of individual treatments is usually offset by the long time between treatments. Formation damage is a possibility in some reservoirs (those where the formation is natively water wet) because the inhibitor will form its inhibitor-oil film and will, in effect, oil-wet the reservoir rock. Relative permeability characteristics will change because of the reversal in wettability, which could result in lower well productivity. Depending on the formation characteristics and the actual production rate, the oil-wet film theoretically could restrict flow. Other potential detrimental effects are emulsion forming and clay swelling. Prior to selecting a formation squeeze treatment, laboratory tests should be run on emulsion-forming characteristics, and—if core sample are available—clay swelling and wettability (relative permeability) tests should also be performed. Inhibitor pump pressure should be limited to values below the formation fracture pressure.

As described earlier (Section 8.9.3), nitrogen can be used to squeeze the inhibitor into the formation if a liquid column cannot be unloaded by production pressure.

8.9.12 Continuous Treatment

As previously mentioned, continuous downhole treatment may also be used in submersible pumped wells, hydraulic pumped wells, gas lift wells (via the lift gas), naturally flowing, and packered oil wells. Packered gas wells that are aggressively corrosive may also be treated continuously by different methods: through treating strings or small diameter capillary tubing that extend to the bottom of the tubing string; and injecting the inhibitor and diluent into the annulus and through a subsurface injection valve located immediately above the packer. If the annulus is dirty with rust or corrosion products, these products may break loose when contacted by a surface-active corrosion inhibitor and plug the downhole injection valve. Some operators have resorted to cleaning the annulus with a detergent solution in seawater, followed by inhibited acid wash and then neutralization. Because the tubing-casing annulus is used to store the inhibitor mix, the mix must remain thermally stable under downhole conditions. Thermal stability can be confirmed in laboratory tests.

Subsurface injection valves have experienced plugging problems, especially in sour gas wells. They can stick closed (no inhibition) or open and empty the inhibitor inventory in the annulus. Valves can be inadvertently damaged during a well work over. When they function properly, downhole injection valves provide good inhibition above the chemical injection valve. Corrosion resistant alloy tubing is required below the chemical injection valve.

In high-pressure, high-temperature gas wells, other problems with continuous downhole treatment have occurred with phase behavior, as described in Section 8.6.12.1 "Phase Behavior."

8.9.13 Treatments Down the Tubing-Casing Annulus for Wells with an Open Annulus (Without Packers)

Batch treatment down the casing annulus is the usual method to treat wells without packers. Most sucker-rod pumping wells are treated by some version of this approach.

8.9.13.1 Fluid Level

Many guidelines are available to approximate the amount of inhibitor required and the treatment frequency for oil wells without packers. Appendix 8.A provides one of several approximations. The fluid level in the well is a determining factor in treating oil wells with an open annulus. Fluid levels affect which batch treating method to use, the circulation time in the Extended Batch (Section 8.9.13.3), and Batch and Circulate (Section 8.9.13.5) methods, respectively). Fluid level is measured as the distance from the pump to the top of fluid level in the annulus.

In wells with a high gas-oil ratio, the accuracy of determining the fluid level may be suspect because of foaming. This results in errors in calculating the circulation time. When in doubt, the corrosion engineer should consider conducting chemical tracer studies to determine an accurate circulation time. A high gas-oil ratio is approximately 2000 cubic feet of gas per barrel oil or 350 cubic meters of gas per cubic meter of oil (at standard conditions). In addition, wells with high gas-oil ratios may oppose the fall of the treating chemical because of the rising gas.

8.9.13.2 Batch-and-Flush Treatment

The Batch and Flush method is typically applied to open-annulus oil wells with a fluid level less than about 150 m (500 ft). The well is taken off production. A small quantity of inhibitor is lubricated or pumped down the annulus followed by 1–2 barrels of fluid to flush it to the fluid level. The inhibitor concentration is estimated at 25 ppm (based on total fluid volume oil plus water production rate) for a well with low corrosivity, and 50 ppm for a well that has experienced failures. However, the inhibitor concentration in the batch treatment at the well pump is much higher, about 1000 ppm, and contact time is at least one hour. The most common treatments are those done with treater trucks. A truck equipped with tanks, meters, pumps, and quick-connect hoses hooks up to a well, injects the prescribed amount of chemical followed by several barrels of flush fluid (oil or water), disconnects, and drives to the next location. There are, however, still many wells treated manually by the pumper or operator who will fill a "chemical pot" (after first bleeding off pressure) on the casing wing valve and divert a portion of the well stream down the annulus for a few minutes to flush down the well. Typical treating volumes and dosages are presented in Appendix 8.A.

Other variations of treatment down the annulus of open-annulus wells include the Batch-and-Circulate treatment, Automated (Semibatch) treatment, and Continuous treatment with bypass flush.

8.9.13.3 Extended Batch Treatment

This is sometimes referred to as "circulate and park" and is another variation of the standard batch treatment developed for wells with high pumping fluid levels in the annulus. A large amount of inhibitor (as much as one drum or 200 L) is placed (parked) in the annulus of the well. The well is put on complete circulation and circulated until the inhibitor goes down the annulus, up the tubing, and back into the annulus. The well is then put back into production leaving the inhibitor in the annulus. As the annular fluid level fluctuates, small quantities of inhibitor are carried in the oil into the tubing, thus periodically treating the tubing and pump for weeks or months after the chemical was added. A variation on this type of treatment recirculates in the well every month or so, each time leaving the bulk of the inhibitor in the annulus. It must be remembered that this technique depends on very careful calculation or measurement of the circulated volumes to ensure that the inhibitor is stored in the annular fluids. If volumes are missed and the inhibitor slug is transported beyond the well, into the well line, and possibly farther into the separator, treatment will cease without the operator's knowledge, and emulsion problems may occur downstream.

8.9.13.4 Automated (Semibatch) Treatment

This uses treating equipment designed for automatic injection of inhibitor followed by produced fluid flush or circulation. Times can be set to vary the amount of inhibitor, flush, and frequency.

8.9.13.5 Batch-and-Circulate Treatment

This approach is generally reserved for treating problem wells when the batch-and-flush treatments are not effective. It is used when fluid levels exceed 150 m (500 ft). The procedure is the same except rather than a simple flush, the well's production may be diverted to the annulus so that the inhibitor–fluid in the annulus is circulated around one or more times. The circulation time is calculated as the time it takes for the chemical to fall through the gas zone plus the time required to fall through the fluid zone plus a percentage (typically 50%) of the time for the chemical to come up the tubing. Equations to calculate circulation time are provided in Appendix 8.B.

8.9.13.6 Batch Treating Frequency for Open-Annulus Oil Wells

Treating frequency depends on fluid level above the pump and the total production rate. Batch frequencies between twice weekly to once monthly are typical (see Appendix 8.A). Frequency can be optimized by monitoring corrosion rates (iron counts, coupons, etc.) and rod and tubing failure frequency.

8.9.13.7 Continuous Treatment

Rod pump wells may also be treated continuously by providing an injection pump or lubricator box on each well. However, continuous treatment has high maintenance costs resulting from operating many chemical injectors in the field, so batch treatment is more commonly practiced. A bypass flush involves a chemical feed pump to continuously inject inhibitor into a small stream of produced fluids bypassed into the annulus. Several types of chemical pumps may be used including mechanical pumps operated by a push rod from the walking beam or small gas-powered pneumatic pumps. The volume of the bypass-produced fluids is not critical if a positive flow is maintained down the annulus.[12]

8.9.13.8 Infrequently Used Downhole Treating Methods

Infrequently used treatment methods for packered and open annuli wells include weighted inhibitors, wash bailer, dump bailer, and inhibitor sticks. The effectiveness of these methods is limited and generally poor, but they have been successfully used and may be a last resort treatment for marginal, low-pressure gas or oil wells. Appendix 8.A provides example procedures, treating frequency, and comments.

- Weighted inhibitors: Inhibitors formulated into a specific gravity higher than well fluids. They are injected "neat" (i.e., without dilution) into the wellhead then allowed to fall to the bottom of the well. After the well is returned to service, the chemical slowly disperses into the production fluid.

- Dump bailer: This method can be used after performing a caliper survey on a well that is not corrosion inhibited and minimizes corrosion at the caliper tracks left by the feeler gauges. At the completion of the caliper survey, the caliper company uses the caliper wireline to add inhibitor at the bottom of the well before the well is turned on.
- Wash bailer: A wash or spray type hydraulic bailer is used to place corrosion inhibitor throughout the tubing string.
- Inhibitor sticks: An inhibitor formulated with a waxy compound and furnished into a stick form. The inhibitor is dropped through the crown valve, and upon reaching bottom the wax melts, releasing the inhibitor.

8.10 Inhibition of Surface Facilities

For this section, surface facilities include flowlines, gathering lines, injection lines, pipelines, separation equipment, piping, heat exchangers and coolers, vessels and tanks, gas handling facilities, glycol dehydration equipment, gas sweetening equipment, heating and cooling media, water injection facilities, and gasoline plant facilities, i.e., everything in the oil field except the wells themselves. Surface facilities often require different corrosion inhibitor formulations than those used in producing wells to match the various application schemes.

8.10.1 Flowlines and Gathering Systems

The downhole inhibitor will often provide protection for the flowline. That is the usual case with artificial lift well lines. However, many corrosive streams from gas wells will require additional inhibition for flowlines and gathering systems.

In multiphase flowlines, understanding flow behavior and regimes in the surface line helps to understand how the inhibitor will be transported to the metal surface. For example, in stratified flow (smooth and wavy), even with production from low water-cut wells, as the water flows through the horizontal line, water separates out and flows along the bottom of the line, water-wetting the bottom of the line. Exceptions exist, such as those situations where the water forms an oil-external emulsion, and no free-water phase forms. When a separate water phase forms, a liquid corrosion inhibitor treatment will reach the entire inner surface of the line in slug flow, annular mist flow, and dispersed bubble flow. Multiphase flow regimes are shown schematically in Chapter 10, Figure 10.3. NACE SP0110 describes how flow regime and other factors (water content, liquid holdup, flow velocity, temperature, and pressure changes) affect internal corrosion.[13] Another good reference explaining how flow regime affects corrosion can be found in the bibliography—Palacios (2016).

When flowline inhibition is required, it is injected continuously at the wellhead. A water-soluble or highly water-dispersible inhibitor is used. Thus, the inhibitor will be carried in the water along the bottom and in low places in the line. The dosage is usually based on water production.

Lines from naturally flowing oil wells that are not being treated may likely require inhibition. These lines will have low water cut. As in sales crude lines, however, low points in the line can accumulate water. Even low water-oil ratio fluids can be corrosive, and the lack of downhole treatment may be because the wells are not producing much water. An oil-soluble water-dispersible inhibitor can be batched into the pipeline. Periodic pigging can remove standing pools of water; however, a pig launcher and receiver must be installed to accomplish this, and often they are not.

Even in wells producing a lot of water, there may be no downhole treatment because the downhole tubing is made from corrosion resistant alloy, so a carbon steel flowline and the gathering system require treatment. A high water-cut crude oil line water-wets the interior surfaces. It can be batch treated with film-persistent inhibitor. Treatment dosage should ensure a high concentration of inhibitor in the slug (example 1000 ppm based on total fluid) with a contact time of at least one hour, but this practice varies based on many factors.

Some of the larger diameter gathering lines likely will require additional inhibition from that arriving from the downhole treatment or the flowlines. For severely corrosive service involving full wellstream and high water-cut oil, a combination of continuous and periodic batch treatments has been successful. A high water-cut oil combined with a sweet gas had resulted in full well-stream pipeline failures until this comprehensive treatment was initiated. This combination of continuous plus batch treatment is more effective than continuous treatment alone because of improved ability to repair a damaged inhibitor film. Continuous treatment by itself is more effective than batch treatment alone. Think of the following to compare the effectiveness of treating a large diameter flowline and pipelines transporting full well stream fluids:

$$\text{Continuous + batch} > \text{continuous} \gg \text{batch}$$

If batch treatments are considered to supplement continuous treatment, any unwanted consequences such as foaming and emulsion forming should be addressed.

8.10.1.1 Top of the Line Corrosion

If the upper part of the pipeline is cooled (such as on a subsea pipeline transporting wet gas), water may condense along the top of the line. If the flow rates and liquid volumes are high enough to have mist flow or slug flow, then the inhibitor will be transported to the top of the line. On the other hand, if the rates are low, the inhibitor will only be in the liquids along the bottom, and another approach will be needed (e.g., periodic pigging or modification of piping to increase the velocity).

If there are no pigging facilities, and flow regime is such that liquids do not get transported to the top of the line, more assessment is required to determine if top-of-the-line corrosion could be an important integrity threat. If it is, then an inhibitor formulation containing both a volatile component and a filming inhibitor could be used. It is difficult to determine the dosage requirements for a volatile inhibitor, and several operators prefer inhibition with pigs if pig launchers and receivers are installed.

Chapter 9, Section 9.10.2.2 describes pH stabilization as an alternative to using organic corrosion inhibitors to control corrosion in wet sweet gas. The chemicals listed in the chapter are neutralizers and not film formers. However, they do not control top-of-the-line corrosion. Internal CRA cladding or full-thickness CRA may be required.

8.10.1.2 Batch Treating Lines Using Pigs

If a line (flowline, pipeline, transmission line, water injection line, etc.) is equipped for pigging, an inhibitor slug can be periodically pushed through it between two pigs. By using two pigs, the inhibitor will be wiped on all the internal surfaces. In such cases, a heavy film forming inhibitor (such as those often used for tubing displacement treatments) is used. The heavy film formers have a highly persistent film and relatively infrequent treatments are required. Thus, the line would only need to be retreated when it is pigged for liquids removal, typically once a month.

When a chemical such as a corrosion inhibitor (or biocide) is batched into a flowline, gathering line, or pipeline, it is usually most effective if a slug of fluid containing a high concentration of the chemical traverses the pipeline length. Many flowlines, gathering lines, and pipelines do not have facilities for maintenance pigging. With no pig, if chemicals are injected into the line, the produced fluids will mix into the chemical slug and dilute it, even an oil soluble inhibitor. With one pig behind the batch pushing the chemical slug down the pipeline, the chemical slug stays intact for a longer distance than with no pig, but then it mixes into the produced water and eventually into the produced oil. With the batched chemical and its diluent between two pigs (as shown in Figure 8.4), the chemical slug can stay intact the length of the line and remain undiluted. However, the time between launching the first and second pig must be short enough to keep the slug intact. Some operators mistakenly launch the second pig a day or days after the first, and this has no positive effect.

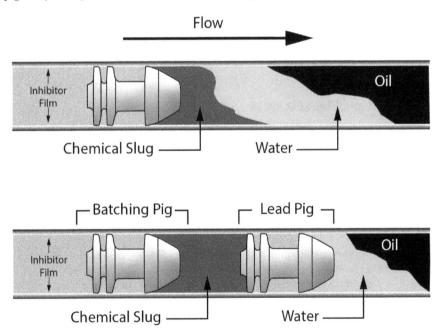

Figure 8.4 Chemical batching with one pig (top) and two pigs (bottom).

CHAPTER 8: Chemical Treatment

To estimate the amount of inhibitor required per batch, use the equations (8.3a and 8.3b) below. The equations are based on "painting" a thin (≈3 mil or 0.08 mm) film thickness onto the inner circumference of the pipeline. Add 25–50% excess to allow for pig bypass.

$$V \text{ (gallons)} = \text{Pipeline ID (feet)} \times \text{pipeline length (miles)} \times 31 \quad (8.3a)$$

$$V \text{ (liters)} = \text{Pipeline ID (millimeters)} \times \text{pipeline length (kilometers)} \times 0.239 \quad (8.3b)$$

where V equals the solution volume (inhibitor plus diluent).

8.10.1.3 Batching Pigs

There are two types of pigs that are used to batch treatment chemicals:

- A typical batching pig consists of a two (polyurethane)-cup pig, a multidisc pig (bi-directional discs on either end of the mandrel); or, a combination of a disc and cup arrangement, with one or more discs at the front end, and at least one cup on each end of the mandrel.
- A commercial pig that sprays a chemical to the top of the multiphase gas gathering or gas transmission pipeline is available. It is designed to have higher pressure flow through its body and spray head. By-pass flow provides the motive force to transfer fluid while creating a low-pressure area in the spray nozzles as it traverses the pipeline. This pressure drop creates a vacuum at the pig's front inlet ports, which draws the inhibitor from the bottom of the pipeline to the top.

Pig selection and design are discussed in Chapter 9, Section 9.8.1.

8.10.2 Production and Plant Facilities

When it comes to production facilities and plants, the use of corrosion inhibitors is a viable method of corrosion control. However, inhibitors do have their limitations. For example, do not depend on inhibitors for corrosion control in valves, pumps, compressors, or similar equipment where velocities and turbulence may be high. Use corrosion-resistant metals where corrosion control is needed.

8.10.2.1 Vessels and Tanks

Vessels and tanks are indirectly inhibited if the fluid carried into them contains sufficient corrosion inhibitor. A separator vessel requires a corrosion allowance—which can be estimated as the inhibited corrosion rate over the design life of the equipment—accounting for lapses in inhibitor treatment. Causes of lapses are discussed in Section 8.15 "Chemical Delivery and Reliability." Relying on corrosion inhibition to provide pressure vessel integrity depends on many factors. Vessels accumulate deposits (scale, corrosion product, sand), which drop out of the fluid because of lower velocity in the vessel compared to the inlet piping. Inhibitors are surface active and are attracted to deposits, diminishing their effectiveness below the deposits. There is the possibility that accelerated corrosion occurs beneath deposits in part to galvanic corrosion between well-inhibited steel and poorly inhibited steel beneath deposits. Most filming amine inhibitors remain in the oil and/or water phases. Even if the inhibitor added upstream of the vessel would provide protection in the liquid portions of the vessel, the gas space will not see any inhibitor and may be subject to corrosion if the metal temperature is colder than the water dew point, allowing water to condense. Condensation depends on gas composition, pressure, and temperature.

Although inhibitors may work in some vessels, the use of immersion-grade internal coatings—supplemented with internal cathodic protection in free-water sections—and corrosion-resistant materials are standard practices. Protective coatings are discussed in Chapter 6, and cathodic protection in Chapter 7. CP of vessels is also described in the references.[14-16]

8.10.2.2 Piping Systems

Corrosion in plant piping, often can be controlled with chemical treatment. Inhibitors have applications in liquid piping and many auxiliary systems.

Manifolds, however, can present problems—distribution of the inhibitor is the cause of the difficulty. If areas of the manifold do not "see" the inhibited fluids, the inhibitor film will not be formed or maintained. Poor distribution of an inhibitor can occur when the chemical is fed to a common point and the line splits into parallel passes of slightly different pressure drop. Some passes are starved of inhibitor. This is especially true in multiphase flow regimes such as stratified and wavy, where each pass may see a different content of gas and liquid due to differences in pressure drop and flow rates. When the manifold corrosion is to be controlled with inhibitors, the manifold needs to be designed and operated for good inhibitor distribution. This may require making certain that the entire header is active. That is, there should be flow throughout the header. To accomplish this requires planning in the design and in the day-to-day operation. Alternatively, separate inhibitor injection points can be installed into each piping segment.

Other important design considerations are out-of-service branch connections and other nonflow piping segments ("dead legs") in manifolds. To minimize the dead area, valves should be as closely coupled to the header as possible. In some cases, dead legs in headers can be minimized by having block valves in the header itself. This technique can be useful when some of the header is designated "for future use." Extra valves do add to the cost of construction, but they can be the most economical approach when corrosive fluids are being handled. The purchase of an extra valve or two may be more economical than building the manifold from corrosion-resistant materials—particularly if the fluids contain inhibitor to protect other piping.

If for some reason it is impossible or impractical to have the entire manifold active on a routine basis, then it may be necessary to periodically flow through normally "dead" areas.

Dead leg problems and solutions are further reviewed in Chapter 1, Section 1.5.5.3 and Chapter 9, Sections 9.8.1 and 9.9.1.

8.10.2.3 Gas Handling Systems and Plants

Corrosion inhibitors have several applications in gas handling systems. In many systems, the first piece of equipment the produced gas, condensate, and water will see is a flowline heater. The flowline inhibitor should provide satisfactory failure control in the heater tubes (unless the velocity through the tubes is extremely high and erosion-corrosion is the cause of failures—inhibitors cannot be expected to control particulate erosion—inhibition has been very successful in controlling erosion-corrosion).

After the gas stream is heated and choked, it usually goes to a separator—either at the well site or at a central facility. As mentioned earlier, the inhibitor will be carried with the liquids; thus, the flowline inhibitor can provide inhibition to the liquid hydrocarbon condensate and water streams off the separator, depending how it partitions. In some cases, it may be necessary to add inhibitor to the water to provide sufficient corrosion control in the water system.

The gas from the separator will be wet, and thus may be corrosive wherever any water condenses on the wall of the gas outlet piping. In some systems, inhibitor may be added to the wet gas piping, while in others, resistant materials may present the most economical approach, and still in others, monitoring, inspection, and planned replacement may be best. If an inhibitor is used, it will need to be either water soluble or a vapor phase neutralizing type. The dosage will have to be worked out for the specific case, but an estimate of 25 ppm based on liquid volume and extent of corrosivity (0.5 pint/MSCF[5] gas or 8 L/Mm^3 continuous injection). When injecting a neutralizer, the quantity of inhibitor that is required should be enough to control the pH of any condensed water to about pH 6-7. Care must be exercised if the gas is going to a glycol dehydrator. Excessive inhibitor concentration can cause foaming of the glycol and excessive glycol losses.

As discussed in Chapter 9, Section 9.7, dehydration is the best method for controlling internal corrosion in pipelines transporting gas streams containing CO_2 and H_2S, and as such, is the preferred method of control in gas pipelines, distribution, and injection systems. Chapter 9, Section 9.7.4 also discusses corrosion control for the glycol system itself, including pH control with chemical addition.

8.10.2.4 Gas Compression Systems

Wet gas coolers present a challenge for inhibition. This is true whether they are air-cooled, or shell and tube. Achieving proper distribution of the inhibitor into each tube or throughout the shell interior and on all the tube surfaces can be a challenge. When it is known at the design stage that a cooler will be handling a corrosive gas, the wet gas will normally be on the tube side of a shell and tube exchanger, and the tubes will be made of a corrosion-resistant alloy. On the other hand, many coolers are installed with carbon steel tubes. Later, if it is discovered that the stream is corrosive, the decision must be made to retube with an alloy or to try to control the corrosion with inhibitor. Many operators have selected to upgrade air-cooled exchanger tubing and the downstream piping from the heat exchanger to CRA (a grade of stainless steel that is adequately resistant to internal corrosion and to external chloride stress corrosion cracking from the marine atmosphere, such as Alloy 825). This avoids the need to transport inhibitor offshore and frees personnel from maintaining the chemical inventory and injection equipment.

The inhibitor is injected continuously upstream of the cooler but should always be injected downstream of the gas compressor. If injected into the compressor suction, the inhibitor must be qualified for the compressor discharge temperature and it must not cause compressor fouling. Piping downstream of the cooler is protected with continuous inhibition like those used in gas handling systems, above. One difference is the higher temperature gas exiting the compressor and entering

---

5. MSCF stands for millions of standard cubic feet per day.

the cooler. This may require a higher inhibitor dose (16 L/Mm³ [1 pint/MSCF]) and qualification of the inhibitor in lab testing. Also, injection point corrosion is a concern, especially at elevated temperature. Some inhibitors become more corrosive at gas compressor discharge temperature. An injection quill is advised to distribute the inhibitor into the flowing gas stream without allowing it to impinge onto the pipe wall. If there are two heat exchanger trains and piping in parallel, each train should be separately treated.

If inhibition is the choice, a vapor phase neutralizing amine can also be considered. Depending on the injection point and flow characteristics, the usual liquid film former may not enter all the tubes in parallel passes of the heat exchanger equally. If the exchanger is made from CRA, this is not a concern. A vapor phase neutralizing inhibitor will be carried in the gas stream and dissolve in the water when and where it condenses. The inhibitors are usually formulated from low molecular weight neutralizing amines (e.g., morpholine, cyclohexylamine, others)—but the lightest and cheapest is ammonia, usually in the form of ammonium hydroxide liquid. However, ammonia is the most difficult to use (small variations in dosage can create large swings in pH), and handling is difficult. Morpholine is more capable than ammonia in condensing and neutralizing the first droplets of condensed acid gas because it distributes more in the condensed liquid than in the uncondensed gas. Morpholine is not nearly as strong of a base as ammonia, so more of the chemical is required to neutralize any acid.

The main problem with the use of the vapor phase neutralizing materials is determining the proper dosage. Because they neutralize the acid gases, monitoring the pH of water sampled from the knock-out drum located downstream of the cooler is the best way to make sure that the dosage is in the right range. As mentioned earlier, the objective of the use of the neutralizing amines is not to neutralize all the acid gas in the gas stream, only the amount that dissolves in any condensed water. Corrosion coupons or other corrosion monitoring devices are used to verify that corrosion is being controlled. Maintaining the proper pH is important, because too low a pH will not control the corrosion and too high a pH is a waste. Extreme overdoses of inhibitor could cause other problems (e.g., incompatibilities with downstream processes, for instance, glycol dehydration). Do not use the vapor phase amines (particularly ammonia) if copper alloys (e.g., brasses or bronzes) or aluminum alloys are present in the system.

Despite all these restrictions, there are instances where inhibitor can be used to control corrosion in gas compression systems.

8.10.2.5 Gas Sweetening Systems

Most sweetening systems employ one of a variety of chemical solvents to remove CO_2 and/or H_2S from gas streams. In upstream facilities, these include alkanolamines (such as methyl diethanolamine, MDEA) and include physical solvents such as Selexol®[6] and Sulfinol®[7] (a mixture of sulfolane, which is a physical solvent, and diisopropanolamine, a chemical solvent). Gas containing CO_2 without any H_2S is more corrosive because no iron sulfide film forms to limit corrosion.

Gas sweetening systems do not normally require inhibition; corrosion can be adequately controlled by maintaining the proper amine solution strength, the proper acid gas loading, and preventing heat

---

6. Selexol is a registered trademark of Honeywell UOP, Houston, TX
7. Sulfinol is a registered trademark of Shell Catalysts & Technologies, Houston, TX

stable salts and amine degradation and decomposition products (such as thiosulfates and bicine) from exceeding acceptable concentration limits. Minimizing amine degradation and decomposition products includes controlling regeneration temperature, operating a reclaimer, and preventing air contamination. One source of air entry is the amine makeup tank, which may be open to the atmosphere. New systems are often constructed with corrosion-resistant alloy solid, or as an alloy cladding, in selected areas. However, corrosion of carbon steel can still be a problem in several areas, notably around the regenerator (stripping) column reboiler. Like gas compression systems, the decision is to upgrade to a resistant alloy or to inject a corrosion inhibitor.

Inhibitors in gas sweetening systems can promote foaming, so antifoam chemicals are required additives. Some companies market inhibitors for refinery units to prevent cracking problems associated with hydrogen from wet H_2S (refer to Chapter 3, Sections 3.10–3.13 inclusive, for hydrogen cracking mechanisms). There is limited experience in oil and gas production facilities with these inhibitors.

Oxidizing passivators, such as sodium metavanadate ($NaVO_3$), act as inorganic, oxidizing anodic inhibitors, which can work only in the absence of H_2S. Sodium metavanadate is the primary corrosion inhibitor used in hot potassium carbonate solvent solutions used to remove CO_2 from gas streams free of H_2S. Oxidizing passivators work when they are properly maintained. They must be routinely analyzed for their concentration, and proper valency (oxidation state), and these parameters must be maintained within control limits. In addition, they must be protected against impurities that destroy them, including hydrogen sulfide. Iron corrosion products—both soluble and insoluble—can interfere with their functioning if in sufficient quantity. Note that when oxidizing passivators fail, they aggravate local attack. Many of them are toxic heavy metals and make it expensive to discard their solutions.

In general, corrosion inhibition is not a primary corrosion control method in gas sweetening units, especially those that use alkanolamines as their circulating solvent. Control of operating variables is the first line of defense against corrosion, and material upgrades (solid or clad austenitic stainless steel) are installed where needed.

8.10.2.6 Cooling Water Systems

There are two basic types of cooling water systems: the once-through systems, and the recirculating systems (which involve cooling towers). These systems and/or their source water are open to the atmosphere, so the water is saturated with oxygen.

Many of the once-through systems operated in the production industry use seawater and are located offshore or on the coast. Because the water is returned to the ocean, the discharge water with its chemical treatment may be regulated by government agencies. Corrosion in raw seawater is controlled by using appropriate alloys and other corrosion-resistant materials, such as nonmetallic materials, or by using carbon steel and removing dissolved oxygen in deaerator units and with chemical oxygen scavengers (Chapter 9, Section 9.6.3).

On the other hand, corrosion control with chemicals is the standard for the open recirculating systems. These systems are open to atmospheric air. Most systems require a corrosion inhibitor, chemicals for pH control, scale control (dispersants and antiscaling chemicals), and chemicals to control

MIC and fouling (e.g., from algae, sulfate reducing bacteria [SRB], acid producing bacteria [APB]; refer to Chapter 2, Section 2.5 and Chapter 3, Section 3.5.4). Each system is unique because of variations in the makeup waters, cycles of concentration, flow rates, temperature, and contamination from hydrocarbon leaks; therefore, they need to be handled on a case-by-case basis. In most cases, cooling water treatment to control corrosion, scale, and biofouling is handled by a cooling water treating vendor. Many of the chemicals used in cooling water systems are subject to rather strict environmental regulations, and these are continually evolving.

Most corrosion inhibitors in open recirculating water systems are cathodic or mixed anode–cathode inorganics such as phosphate-containing chemicals, zinc, and even organic phosphorous-containing (phosphonate) chemicals. These chemicals prevent scale and inhibit corrosion in cooling waters whose pH is maintained at more alkaline levels than the pH of saturation of many inorganic scales ($CaCO_3$, $CaSO_4$). Many ortho- and polyphosphate inorganic phosphate treatments require controlling pH within a specific range. Controlling pH often requires injection of acid, which is fed in low quantities into areas of the cooling tower that have good mixing; pH can be automatically controlled if control systems are added.

Many different chemicals are used to control algae and microorganisms. Oxidizing chemicals include chlorine, sodium hypochlorite, and chlorine dioxide. Bromine substitutes are available. Nonoxidizing biocides are also regularly used. In selecting the biocide, consideration must be given not only to its effectiveness in controlling microorganism growth, MIC, deposits, and Legionnaire's Disease, but also to personnel safety and environmental restrictions on discharge.

8.10.2.7 Heating and Cooling Media

These are closed recirculating heating and cooling systems usually circulating low dissolved solids water or glycol-water solutions. These systems are sealed from contacting atmospheric air. Ethylene or propylene glycol may be added to a closed system to freeze-proof the water. These systems are not open the atmosphere, but air ingress is possible through leaks, open tanks, etc. They are commonly used to cool gas engines and compressors and where atmospheric conditions do not favor using open, recirculating water.

Nitrite Inhibitors

Experience has demonstrated that inhibitors included with the glycol package need to be supplemented with inorganic corrosion inhibitors. The most common chemicals added are oxidizing anodic inhibitors such as nitrite and molybdate (see Section 8.3.1 "Anodic Inhibitors"). Nitrite residuals of 600–1200 ppm (as nitrite ion) are targeted. The concern about under-dosing an anodic inhibitor, mentioned earlier in this chapter, drives many operators to maintain the nitrite residual at the higher end of the range. The water system is closed so water losses are kept low and chemical cost relates to occasional batch treating. Nitrite-treated water may require select biocides to control nitrifying bacteria. When nitrite is used, pH 8–9 is maintained by incorporating sodium borate into the formulation. The maximum water temperature that can be treated with nitrite is about 120°C (250°F). Nitrite does not need the presence of dissolved oxygen to function. Oxygen reacts and forms nitrate ions, which are not corrosion inhibitors.

Molybdate Inhibitors

Molybdate inhibitors are known to function at higher temperature than nitrites and are therefore preferred to a nitrite-only treatment above 120°C, to a maximum of 180°C (350°F). Molybdate treatments are more expensive and may be subject to regulatory oversight in discharge water. Like nitrites, molybdates passivate steel surfaces. According to the literature, molybdate inhibitors require at least a limited amount of oxygen (e.g., 1 ppm) or an oxidizer to function properly in a closed water system. Although molybdates have been used alone, they are often used in combination with nitrites, which provide an oxidizer. Combination treatment programs are effective over a broad range of composition, 50:450 (ppm:ppm, i.e., both in parts per million) $MoO_4^{-2}:NO_2^{-2}$ to 350:200 (ppm:ppm) $MoO_4^{-2}:NO_2^{-2}$. Often, 250–500 ppm (as the molybdate MoO_4 ion) is targeted. As the chloride content of the water increases, molybdate concentration should be raised. Borate is added to maintain water pH 8.5–10.5.

Silicates

Silicates are more environmentally friendly but are less effective inhibitors. Silicates are anodic, filming inhibitors. They have a variable composition of $SiO_2:Na_2O$. Generally, a ratio of $SiO_2:Na_2O$ of 2.5–3.0 is effective. With a silicate treatment, the protective film develops slowly and may take weeks to form. The film is believed to consist of silica gel along with ferric hydroxide precipitates. They only function in water with low total dissolved solids content <500 ppm. A concentration of 100 ppm SiO_2 is used at a pH 8–9.5. Operators may be enticed to use silicates because of their environmental acceptability in discharge water. However, experience with silicates in oil and gas production closed recirculating cooling water systems is limited, and they are weak inhibitors.

Water Quality

To ensure that the inhibition package in closed cooling water systems works properly, the makeup water should have a low mineral content to avoid scaling at elevated temperature. Water quality is maintained according to the manufacturer's instructions. Most vendors specify a maximum chloride content of 100 ppm. Even at this relatively modest concentration, chloride is corrosive, but acceptable if nitrite levels exceed five times the sum of (chlorides plus sulfates) concentrations. If there is calcium in the water, part of the inhibitor formulation may precipitate out and cause scaling on exchanger tubes. Scaling can be a problem in heating systems. As would be expected, corrosion is more of a problem in heating systems than in cooling systems. These systems will have fewer problems when operated air-free, and with gas blankets on the tanks. Oxygen oxidizes nitrite to nitrate and reduces the effective inhibitor dosage. It also results in oxygen concentration cell corrosion.

Closed water systems that are heavily laden with corrosion product (black iron oxide if no oxygen is present, red if oxygen is present) require flushing to purge the system, otherwise, deposits will cause fouling and under-deposit corrosion.

8.11 Water Injection Systems

Water injection systems include both produced water disposal and water floods where source water and produced water are injected. It was mentioned previously that the major cause of corrosion in water is oxygen, and that in most water systems air-free operation is all that is required to control corrosion. However, some produced water and some subsurface source waters are corrosive even air-free (and they will be even more corrosive if air enters).

If an inhibitor is required, air-free conditions must still be maintained. Organic film-forming inhibitors will not control oxygen corrosion. The types of inhibitors that will control oxygen corrosion in fresh waters will not work in brines. Inhibitor selection often is based on the inhibitor's solubility in the injected water at the various temperatures in the system.

In most cases, a water-soluble filming inhibitor will be injected continuously at dosages of 10–15 ppm (by volume). Depending on the system, the inhibitor may be injected at one central location, or injection may be necessary at each water source point.

A filming corrosion inhibitor may restrict well injectivity by plugging or altering the wetting characteristics of the reservoir rock. Therefore, before commencing treatment, all necessary steps should be taken. Bacteria should be controlled with a biocide injection. Corrosion in the surface-water handling facility can be controlled by using coatings and nonmetallic materials (e.g., fiber-reinforced plastic [FRP], liners of high density polyethylene [HDPE]) where appropriate. This will minimize the amount of corrosion product entering the injection well. Oxygen must be eliminated thoroughly.

Although not a corrosion control measure, injected water should be clarified and filtered to a maximum of 2 ppm in sandstone or 4 ppm in dolomite (limestone) reservoirs. A typical maximum-acceptable oil content for water injection is 25 ppm under normal conditions. A surfactant may be helpful if oil concentration is causing plugging.

8.12 Gas Injection Wells

Injected gas is usually dehydrated to below the wellhead temperature dew point, so corrosion is not expected, even if the gas contains H_2S or CO_2. If undehydrated gas is to be injected, corrosion inhibition will control corrosion but will likely plug the injection well, as described in water injection systems.

CO_2 used in gas injection is normally dehydrated. In water-alternating CO_2 gas (WAG) injection, corrosion is controlled by proper materials selection. FRP-lined steel tubing is used (Chapter 6, Figure 6.22). Cement is used to grout the liner to the tubular. Wellheads are 316 stainless steel, and precipitation hardened nickel-based alloys are used for valve stems. Corrosion inhibitors are not used.

8.13 Other Types of Inhibitors

Inhibitors used for reducing the corrosion rate during well acidizing are a different family of corrosion inhibitors than previously described in this chapter. Volatile corrosion inhibitors are used in contained spaces, often under atmospheric conditions containing moisture and oxygen, to control corrosion in service, or during storage and shipping of equipment.

8.13.1 Well Acidizing Inhibitors

Well stimulation acid is used to increase production from wells by reacting with acid soluble portions of the formation rock, thus increasing the permeability of the formation. They are also used to remove acid soluble scales and deposits from the tubing and the face of the formation. Various acid solutions are used. The most common acid is 15% hydrochloric acid (HCl). Hydrofluoric acid (HF), HF-HCl mixtures, and many organic acids may be used. These acids are extremely corrosive to carbon steel at well-bore temperature and can cause corrosion of CRAs. Therefore, special inhibitors (commonly referred to as "acid inhibitors") are used. The acid inhibitors are generally proprietary materials provided as a package by the acidizing service company. The inhibitor package is designed to match the conditions of a specific well. Well temperature, acid volume, time of exposure of the tubing to the acid, presence of hydrogen sulfide (H_2S), and tubing metallurgy are important considerations. NACE SP0273 has guidelines for using acid inhibitors and operating considerations for minimizing acid corrosion.[17]

For years, inorganic passivating inhibitors were used successfully. Arsenate compounds were the most widely used; however, they fell out of favor. First, problems of "poisoning" refinery catalysts restricted their use. For most purposes, arsenates have been eliminated from oil field applications because of health, safety, and environmental concerns.

Today's stimulation acid inhibitors are organic compounds; however, by and large, they have different structures than the organics used in the typical oil/gas/water inhibitors discussed in most of this chapter. Most acid inhibitor formulations combine several different organic materials that act synergistically to provide the desired protection. Various acetylenic alcohols (e.g., propargyl alcohol), octynol, iodine salts or derivatives of formic acid, and similar materials make up the acid inhibitor.

When an acid job is to be performed, an inhibitor and dosage are selected based on the expected downhole temperature. Dosages vary from a range of 0.5–1% by volume (%v) at lower well temperatures (below 150°F [66°C]), to the range of 2–3%v at higher well temperatures (above 300°F [149°C]). For high-temperature applications, cooling down the well is recommended in NACE SP0273.[17]

A major corrosion concern involves the spent acid that returns when the well is brought back onto production. In most cases the acid will be only partially spent, with a pH of 4 or less (still on the acidic side and corrosive), and the corrosion inhibitor will have been left in the formation, adsorbed onto rock. Thus, the spent acid will be corrosive and should be removed from the well as quickly as possible, per the practice in SP0273.[17]

When stimulating horizontal wells with "mud acid," the exposure time of the downhole metals to acid could be 24 hours or even several days—much longer than in vertical wells. Horizontal wells may be completed with CRA and low-alloy steel in the same tubing string. The presence of CRAs favors the use of organic stimulation acids (acetic or formic acids, or both) with HF. Laboratory testing has demonstrated that acid inhibitors function well in organic acids for 24 hours; however, inhibition of inorganic acid formulations (HCl + HF) was not effective.[18] Section 8.14 discusses horizontal wells further.

When the use of CRA increased, it was discovered that the acid inhibitors that worked well with carbon steel tubulars did not adequately protect the CRA, particularly at elevated temperature. However, special inhibitors were developed. Since the mid-1980s, quite a bit has been published worldwide on the investigations of acid corrosion of CRAs and the effects of inhibitors. (Several articles on acid inhibition and CRA are listed in the bibliography at the end of this chapter. The books listed cover many facets of inhibition, including discussions on acid inhibitors.)

8.13.2 Volatile Corrosion Inhibitors

Another class of corrosion inhibitor that is different than the oil/gas/water system inhibitors discussed in the bulk of this chapter are called "volatile corrosion inhibitors" (VCIs). These are also vapor phase corrosion inhibitors and are sometimes listed as "vapor phase inhibitors" (VPIs). However, the chemistry as well as the usage is quite different than the vapor phase neutralizing amines that were discussed earlier in this chapter in Section 8.10.2.4 "Gas Compression Systems."

VCIs are used in contained spaces, often under atmospheric conditions with moisture and oxygen. In other words, VCIs are primarily inhibitors for metals in storage or transit, in pipeline casings and cased crossings, and in above-ground storage tanks. Wrapping papers, plastic film, envelopes, and bags impregnated with VCIs are used for protecting parts and apparatus until they are needed. Various powder or liquid formulations may be placed inside vessels, piping, engines, heat exchangers, and similar equipment for protection during mothballing (lay-up) of off-line facilities. One important application is their use in instrument, electronic device, and electrical enclosures to prevent corrosion or tarnishing of switches and contacts.

VCIs protect by the emission of vapors from a solid or liquid formulation. Vapors fill the confined space to reach the various surfaces to be protected. Their big advantage is probably that the inhibitive materials can reach inaccessible crevices and gaps in equipment or parts.[19]

The amount of VCI required is a function of the volume, and the length of time protection will be required. In instances where the VCI protected area is opened for service or operation, the VCI will have to be replaced periodically.

8.14 Horizontal Wells

The majority of Chapter 8 describes downhole inhibition practices for conventional wells. In the last decade, horizontal wells have become a valuable technique in oil and gas production. Horizontal drilling, also known as "directional drilling," is a recent way of obtaining oil and gas. While considered unconventional, this technique involves drilling at angles that surpass 80°. Horizontal wells have proven particularly useful as a component of the hydraulic fracturing ("fracking") process. Fracking is used in extracting natural gas ("shale gas") and oil ("shale oil") from huge shale reservoirs in the United States. These deposits tend to be inaccessible to traditional vertical drilling because of the impermeability of the shale formations.

Horizontal drilling, followed by hydraulic fracturing, is key to facilitating production of gas and/or oil from shale. Once the well has been drilled and casing installed, cement is added to fill the annular space between the casing and the hole, as in conventional wells. The casing and cement are perforated in the intended production zones (as in conventional wells). The well is then hydraulically fractured ("fracked") in a multistage process, with high-pressure water containing various additives, including a proppant (a material used to keep the fractures open) and various chemicals (acid, corrosion inhibitors, scaling inhibitors, etc.).[20]

8.14.1 Definition of Horizontal Well

A horizontal well is an oil or gas well drilled at an angle of at least 80° relative to a vertical wellbore. This technique has become increasingly common and productive in recent years. The horizontal well is a type of directional drilling technique.

Initially, the well is drilled in the same way, by drilling straight down. The tangent section of the well is drilled along a deviated well path to just above the reservoir section, to what is known as the "kick-off point." From the kickoff point, the well is drilled at an increasingly higher angle, arcing around toward an angle close to horizontal (i.e., close to 90°). The point at which the well enters (or lands on) the reservoir is called the "entry point." From there on, the well continues at a near-horizontal orientation with the intention of keeping it substantially within the reservoir target until the desired length of horizontal penetration is reached.

8.14.2 Unique Properties of Shale Oil

Compositionally—with its high wax content and low asphaltene fraction—shale oils are often closer to condensates than conventional black oil. Shale oils tend to poorly wet, and therefore poorly protect, steel surfaces, primarily due to the low concentration of polar compounds. The oil volatility can have an effect on the solubility of the acid gases in the hydrocarbon phase. Poor wetting and higher volatility make it more difficult for the corrosion engineer to predict potential threats based on past experience with conventional oil wells.[20]

8.14.3 Materials and Corrosion Concerns

Horizontal wells encounter several potential material and corrosion concerns, which are unique to their well design:[20]

- Erosion and corrosion damage occur during fracking due to the high frac pump rates.
- Corrosion damage occurs to the production casing throughout the completion and fracking process due to the presence of introduced fluids and subsequent flow up the casing (without downhole tubing installed) during early production.
- Internal corrosion of the liner section occurs in the production casing in the horizontal section due to uneven sections ("hills and valleys") from true horizontal. Gas bubbles accumulate in the hills, and water and solids collect in the valleys, resulting in corrosion of the long lateral liner and in the subsequent production of high iron concentration in the produced water.
- Internal corrosion or erosion-corrosion occurs during flowback of the large volume of water returning to the wellhead with produced gas. Flowback water contains entrained silica sand (proppant), returned stimulation acid (typically HCl), and many other chemicals such as scale inhibitors and surfactants.

- Environmental cracking of high strength (grade P110) casing and couplings occurs, even in wells that are absent of measurable quantities of H_2S. Hydrogen sources include corrosion from stimulation acid, conventional sulfide stress cracking (SSC) due to H_2S production, and downhole MIC activity.
- Pitting corrosion that initiated SSC has been associated with the presence of spent H_2S scavenger thiazine (refer to Chapter 9, Section 9.9.2). H_2S scavenger is used extensively in shale fields, and the spent reaction products may reside in equipment for a long time.
- Internal corrosion is associated with the large volume of fresh water injected. Mechanisms include MIC caused by using a contaminated source of fresh water, and localized corrosion caused by oxygen dissolved in the water used for fracking. MIC and CO_2 are major causes of internal corrosion.
- Internal corrosion of flowlines and facilities pressure vessels due to the low velocity develops 6 months to 2 years after initial production.

Horizontal wells share corrosion concerns with conventional wells in that internal corrosion depends on the bottomhole temperature, the presence of water, the pH of the water, and the concentrations in the water phase of

- H_2S
- CO_2
- Chloride
- Organic acids
- Oxygen
- Dissolved iron

In some wells, martensitic stainless steel[8] is used, but most wells are completed with carbon and low-alloy steel casing and tubing.

8.14.4 Corrosion Inhibition

In production, oils wells use artificial lift such as rod pump and ESP, installed in the vertical section of the well above the kickoff point. Gas wells may use plungers[9] to lift gas up the tubing string and dewater the well. Inhibitor delivery is by batch or continuous treatment, depending on the aggressive nature of the corrosion. Most lift devices are located above the kickoff point. This introduces an impossibility of getting a corrosion inhibitor to the points of interest in the horizontal section.

8.14.4.1 Inhibiting Shale Oil Wells

Shale oil wells are typically open-annulus design using rod pump or ESP. Some wells are packered and are artificially lifted using lift gas. Treatment consists of first treating the completion fluids for MIC and scale. Once the well is in production, inhibition relies on the original treatment of the completion fluids, especially in controlling MIC. Treatment methods follow those described in the

---

8. See Chapter 4, Section 4.6.4.1.
9. Plungers are an artificial-lift method principally used in gas wells to unload relatively small volumes of liquid. A plunger is dropped into the production tubing string during shut-in, and it and the column of liquid are carried up the tubing string when the well is returned to production.

relevant Sections 8.9.1–8.9.11, depending on the well design and treatment method. Capillary tubes are sometimes retrofitted to deliver corrosion inhibitor (Sections 8.9.10 and 8.9.12). Squeeze treatments are sometimes used to inject scale inhibitors; however, this is not a typical method to feed a corrosion inhibitor.

Inhibitor treatments reach the kickoff point of the vertical section, but do not reach the horizontal portion of the well. Wells treated these ways have only been in service about 6 years at the time of writing this book, and most still are in service. A few wells have had the horizontal section collapse because of corrosion. Corrosion of the horizontal section can be monitored by analyzing the produced water for manganese—a component of steel—but is rarely a natural component of produced water. If manganese concentration rises significantly, inhibition is not an option, and the well may require recompletion.

8.14.4.2 Inhibiting Shale Gas Wells

Most shale gas wells are packered. The completion fluids are treated similarly to shale oil wells. The gas is compressed, and a corrosion inhibitor is injected into the produced gas; and reinjected into the downhole tubing string where it inhibits the tubing string in a manner resembling an inhibited lift gas (Section 8.9.8). The inhibitor type is the same as used in gas-lift oil wells. The inhibitor is not transported below the gas injection point, nor in the horizontal section. Manganese concentration is used to monitor downhole corrosion.

8.15 Chemical Delivery and Reliability

The success of a corrosion inhibitor program is more dependent on how well the chemical injection volumes and feed rates meet the goals of the program than on the inhibitor effectiveness (refer to Section 8.6.1 for the definition of effectiveness).

Inhibitor deliverability or availability is a measure of success. Chemical feed systems that are well maintained can attain 95% uptime (deliverability). Field numbers range wildly from below 50%, up to 85–90%, but rarely above 95%. There are a number of causes of poor deliverability:

- Failure of the chemical feed equipment, particularly the injection pump—Diaphragm failure is the most common problem. Pressure control valves are also problematic.
- Limited surveillance of the injection equipment performance—No one is watching the trends in pumping rates, chemical tank inventory, inhibitor residual measurements, and corrosion rates. Often, an increase in corrosion rate is a sign of poor inhibitor deliverability.
- Mechanical failure—cranes and lifting equipment are not functioning offshore to lift the chemical tote tanks from work boat to platform.
- Insufficient spare parts inventory for injection equipment that break down.
- Lack of automation to control chemical injection—level gauges, flow meters, flow controllers, valves.
- Low criticality of the injection pump assigned by the maintenance and repair organization compared to the criticality of the equipment it is protecting.
- Pumps running dry due to poor chemical delivery.
- Adverse weather, especially offshore during hurricane or typhoon season.

The corrosion engineer should challenge any new project that requires >95% uptime of the chemical feed equipment. Uptime exceeding 90% should be challenged for brownfield projects (add-ons to existing operations), especially if the history of chemical feed uptime in the existing operation is not near this value. Crossland et al., describe control systems and provides schematics of a reliable chemical injection system.[21]

Several operators have formed chemical management teams consisting of corrosion staff, operations, the chemical supplier, and management. The team periodically reviews inhibitor usage, quantities injected (actual versus planned) and compares the results of corrosion monitoring with changes in operations and operating conditions to determine if chemical delivery and reliability, and the inhibitor and its dosage rate, should be adjusted. Teams provide a well-rounded viewpoint and are more effective than having individual oversight of an inhibitor program.

8.15.1 Injection Fittings

In many fields, a corrosion injection location consists of a 50 mm (2 in) nipple and a valve located at the top of the line to be inhibited. High-pressure fittings are commercially available. An injection quill improves the distribution of the corrosion inhibitor into the fluid flow. It consists of a small diameter tube with an end that is cut at a 45° angle, which introduces the inhibitor into the mid-diameter of the fluid line. The cut is placed in the back side of the flow direction. A quill disperses the inhibitor into the fluid better than a nipple connection.

The simplest design for the job should suffice. Spray nozzles with fine orifices can get plugged and fail to work. Any device that intrudes into the fluid flow may be subjected to vibration and fatigue failure, especially in large diameter lines that require long-length injection devices. The design can be checked against the guidelines recommended for thermowell design.[22]

8.16 Comparison with Other Corrosion Control Options

The main competitors to corrosion inhibition are CRAs, internal coatings (with or without cathodic protection), and nonmetallic materials.

CRAs are capital-intensive and can require long delivery times. The operator must weigh the "operating costs" (OPEX) of corrosion inhibition versus the "capital costs" (CAPEX) of CRAs. Carbon steel with inhibition will require more frequent workovers to replace downhole tubing than CRAs. Workovers frequently fall into the OPEX expense category, and can be expensive, especially offshore. The results of an economic analysis of inhibited carbon steel versus CRA depend on who is doing the economic comparisons and what assumptions they are making. For example, will carbon steel with inhibition last more than 10 years without a corrosion-related workover?

Considering the high initial cost of the CRA option, it is usually more economical to defer major costs, so higher OPEX often seems to be the economical choice. However, when the operator assesses the risk of loss of pressure containment and considers the likelihood too high or the consequences (safety, environmental and/or financial) too severe, CRAs may be their preference. CRAs can be solid or clad onto carbon steel to reduce cost. Typical service envelopes for many CRAs are described by alloy type in Chapter 4. CRAs can fail by localized corrosion or environmental cracking in a brief time if the alloy is poorly manufactured or if it is operated outside of its service envelopes.

Unless the corrosivity of the produced or processed fluid is benign, internal coatings should not be considered a substitute for inhibition. Many times, coatings complement the inhibitor program because they reduce the area of steel exposed to the environment. In addition, coatings can reduce workover costs. However, coatings are susceptible to damage (wireline damage—Chapter 6, Section 6.7.3; chemical damage—Chapter 6, Figure 6.21) requiring periodic inspection to determine their integrity, or they are run to failure. If an internally coated carbon steel downhole tube fails, it will typically leak through a corrosion hole. If a bare steel tube with inhibition leaks, internal corrosion over a great extent of the tubing can weaken the tubing enough to cause it to part when pulled during a workover. The internal coating saves a fishing job, which can be an expensive portion of the overall workover cost.

If the uninhibited corrosion rate is ≥6 mm/y (240 mpy), inhibition is reaching a borderline to control corrosion economically, and CRA should be considered.[23] At these extreme corrosion rates, the effectiveness of the corrosion inhibitor (efficiency and deliverability) is not likely to guarantee attaining a long service life.

Nonmetallic materials are discussed in Chapter 5. Fiberglass is cost competitive to inhibited carbon steel in water service and where it can be manufactured to withstand the operating temperature and pressure for the desired diameter pipe. Polyethylene-lined steel is cost competitive, especially if used to rejuvenate internally corroded steel pipe. Both materials are cost effective in mature fields with declining production volumes compared to continued inhibition.

Magnetic devices should never be considered as an alternative to inhibition. They have not been scientifically demonstrated to reduce scale, paraffin deposits, or corrosion rate.

8.17 Summary of Corrosion Inhibition

Corrosion inhibition can be summarized as follows:

- Corrosion inhibition is cost effective and allows for the use of carbon steel in corrosive services.
- There is a wide variety of corrosion inhibitor formulations available for use. They can be selected to handle most of the environments in oil and gas production systems, except where oxygen is present (oxygen exclusion is required). Filming amine corrosion inhibitors do not function well in the presence of oxygen.
- The application technique used must match the system's mechanical and process considerations to ensure that the inhibitor is transported the metal surface when and where it is needed. The application technique plays a significant role in selecting a corrosion inhibitor.
- The choice of the specific inhibitor program (chemical and application technique) is a combination of technical and economic considerations.
- Inhibition programs should be monitored and periodically reviewed, because process systems are continuously changing (refer to Chapter 10).
- Inhibitor programs need to be modified or adjusted periodically to optimize the program for cost effectiveness.

References

1. Fink, J.K., *Oil Field Chemicals* (Burlington, MA: Gulf Professional Publishing, 2003).
2. Pacheco, J.L., F.C. Ibrahim, and R.J. Franco, "Testing Requirements of Corrosion Inhibitor Applications for Pipeline Applications," CORROSION 2010, paper no. 10325 (Houston, TX: NACE International, 2010).
3. Papavinasam, S., "Inhibitor Selection for Corrosion Control of Pipelines: Comparison of Rates of General Corrosion and Pitting Corrosion Under Gassy-Oil Pipeline Conditions in the Laboratory and in the Field," CORROSION 2000, paper no. 00055 (Houston, TX: NACE International, 2000).
4. Papavinasam, S., R.W. Revie, and M. Bartos, "Testing Methods and Standards for Oil Field Corrosion Inhibitors," CORROSION 2004, paper no. 04424 (Houston, TX: NACE International 2004).
5. Papavinasam, S., R.W. Revie, T. Panneerselvam, and M. Bartos, "Standards for Laboratory Evaluation of Oil Field Corrosion Inhibitors," *Materials Performance* 46, 5 (Houston, TX: NACE International, 2007): pp. 46–51.
6. Papavinasam, S., R.W. Revie, M. Attard, A. Demoz, and K. Michaellian, "Comparison of Laboratory Methodologies to Evaluate Corrosion Inhibitors for Oil and Gas Pipelines," *Materials Performance* 59, 10 (Houston, TX: NACE International, 2003): pp. 897–912.
7. ASTM G170 (latest edition), "Standard Guide for Evaluating and Qualifying Oil Field and Refinery Corrosion Inhibitors in the Laboratory" (West Conshohocken, PA: ASTM International, 2012).
8. NACE Publication 1D182, "Wheel Test Method Used for Evaluation of Film-Persistent Corrosion Inhibitors for Oil Field Applications" (Houston, TX: NACE International, 2005).
9. Asperger, R., "A Systems Analytical Approach to Wet Gas Pipeline Inhibitor Selection and Application," CORROSION'92, paper no. 1 (Houston, TX: NACE International, 1992).
10. Skogsberg, L., B. Miglin, S. Ramachandran, and K. Bartrip, "Establishment of Corrosion Inhibitor Performance in Deepwater Conditions," CORROSION 2001, paper no. 01005 (Houston, TX: NACE International, 2001).
11. Pacheco, J., T. Martin, J. Regina, P. Abrams, S. Hickman, and L. Talley, "Corrosion Inhibitor Qualification Testing For Subsea Wet Gas Pipelines," CORROSION 2008, paper no. 08626 (Houston, TX: NACE International, 2008).
12. Russ, P.R., "Oilwell Batch Inhibition and Material Optimisation," SPE Asia Pacific Oil & Gas Conference, paper no. 28810 (Richardson, TX: Society of Petroleum Engineers, 1994).
13. NACE International Standard Practice SP0110 (latest edition), "Wet Gas Internal Corrosion Direct Assessment Methodologies for Pipelines," (Houston, TX: NACE International, 2010).
14. Lewis, R.E., and D.K. Barbin, "Mitigating Carbon Dioxide Corrosion in Production Vessels," *Materials Performance* 33, 7 (Houston, TX: NACE International, 1994): pp. 15–19.
15. Turnipseed, S.P., "Cathodic Protection in Oil Field Brine," *Materials Performance* 30, 12 (Houston, TX: NACE International, 1992): pp. 16–20.
16. John, G., T. Rosbrook, J. Robinson, and I. Munro, "Use of Impressed Current Cathodic Protection Systems for CO_2 Corrosion Control in Offshore Separators," CORROSION 2000, paper no. 00011 (Houston, TX: NACE International, 2000).
17. NACE International Standard Practice SP0273 (latest edition), "Handling and Proper Usage of Inhibited Oil Field Acids," (Houston, TX: NACE International 2016).
18. Bandeira, C.J., R.F. Brito, B.C. Barbosa, F.D. de Moraes, A.Z.I. Pereira, and L.C. do Carmo Marques, "Performance of Corrosion Inhibitors for Acidizing Jobs in Horizontal Wells Completed with CRA: Laboratory Test," CORROSION 2001, paper no. 01007 (Houston, TX: NACE International, 2001).

19. Fiaud, C., "Theory and Practice of Vapour Phase Inhibitors" in *A Working Party Report on Corrosion Inhibitors*, European Federation of Corrosion Publications, Number 11 (London, England: The Institute of Materials, 1994): p. 1.
20. Craig, B., D. Blumer, A. Huizinga, D. Young, and M. Singer, "Management of Corrosion in Shale Development," CORROSION 2019, paper no. 13189 (Houston, TX: NACE International, 2019).
21. Crossland, A., J. Vera, A. Fox, and G. Hickley, "Developing a Highly Reliable Approach to Corrosion Inhibitor Injection Systems," CORROSION 2010, paper no. 10324 (Houston, TX: NACE International, 2010).
22. ASME PTC 19.3 TW (latest edition), "Thermowells" (New York, NY: ASME International, 2016).
23. Hedges, B., and D. Paisley, "The Corrosion Availability Corrosion Model," CORROSION 2000, paper no. 00034 (Houston, TX: NACE International, 2000).

Bibliography

Books—Collections

European Federation of Corrosion, "A Working Party Report on Corrosion Inhibitors," European Federation of Corrosion Publications Number 11 (London, England: The Institute of Materials, 1994).

Hausler, R.H., Ed., "Corrosion Inhibition"—Proceedings of the International Conference on Corrosion Inhibitors, May 16–20, 1983. International Corrosion Conference Series, NACE-7 (Houston, TX: NACE International, 1988).

Nathan, C.C., Ed., *Corrosion Inhibitors* (Houston, TX: NACE International, 1973).[10]

Palacios Tenreiro, C.A., Corrosion and Asset Integrity Management for Upstream Installations in the Oil/Gas Industry, Chapter 6, Basenji Studio, LLC, 2016.

Papers and Articles on Acidizing

Cizek, A., "A Review of Corrosion Inhibitors Used in Acidizing," CORROSION'93, paper no. 92 (Houston, TX: NACE, 1993). Also published as "Corrosion Inhibitors Used in Acidizing" *Materials Performance* 33, 1 (Houston, TX: NACE International, 1994): pp. 56–61.

Hill, D.G., and A. Jones, "An Engineered Approach to Corrosion Control During Matrix Acidizing of HTHP Sour Carbonate Reservoir," CORROSION 2003, paper no. 03121 (Houston, TX: NACE International, 2003).

Kane, R.D., and S.M. Wilhelm, "Compatibility of Stainless and Nickel-Based Alloys in Acidizing Environments," CORROSION'89, paper no. 481 (Houston, TX: NACE International, 1989).

Scoppio, L., G. Mortali, E. Piccolo, J. Cassidy, P. Nice, L. Intiso, H. Nasvik, and H. Amaya, "Corrosion Testing of Stimulation Acidizing Packages on High Strength Corrosion Resistant Tubing Alloys for High Pressure Deep Water Wells," CORROSION 2014, paper no. 3945 (Houston, TX: NACE International, 2014).

Walker, M.L., J.M. Cassidy, K.R. Lancaster, and T.H. McCoy. "Acid Inhibition of CRA: A Review," CORROSION'94, paper no. 19 (Houston, TX: NACE International, 1994).

Walker, M.L., and T.H. McCoy, "Effect and Inhibition of Stimulation Acids on Corrosion-Resistant Alloys," CORROSION'86, paper no. 154 (Houston, TX: NACE International, 1986).

---

10. This book is outdated but several chapters of interest to oil and gas production are still relevant.

Appendix 8.A: Downhole Corrosion Inhibitor Application Techniques for Oil and Gas Wells

This appendix outlines examples of corrosion inhibitor treatments using various application techniques that could be used for downhole corrosion control in oil and/or gas producing wells. These examples may be used for estimating well, equipment, and personnel requirements, as well as costs for comparisons with other corrosion control approaches. However, they are not "final" designs. Details of specific treatments would need to be developed for actual treatment of specific wells. The main text of this chapter provides additional details.

Application techniques outlined include formation squeeze, tubing displacement, nitrogen squeeze or displacement, partial tubing displacement (and yo-yo), treating string, weighted liquids, dump bailer, wash bailer, inhibitor sticks, chemical injection valve, gas lift gas, and "batch and flush" or "batch and circulate" treatment down the annulus.

In many instances in this appendix, Imperial units precede metric units because chemical volumes and injection pump rate guidelines were developed in Imperial units as approximate values or ranges of values in round numbers.

8.A.1 Tubing Displacement

Section 8.9.1 describes this treatment method.

8.A.1.1 Type of Well

Usually gas wells with high shut-in standing fluid level, but it may be used for gas-lift and flowing oil wells.

8.A.1.2 Basic Approaches and Mechanism

1. Corrosion inhibitor mix is pumped into the tubing, pumped to the bottom with a flush fluid, the well is shut-in for a period, and then slowly returned to production. The inhibitor may be preceded by a slug of fluid to precondition the surface for the inhibitor film.
2. The inhibitor film is formed on the tubing during pump-in, distributed and strengthened during shut-in, and refilmed when the well is slowly returned to production.
3. The inhibitor used is a so-called "heavy film former" or "macrofilm former," which provides a multimolecular film. Extended film life is obtained because this type film tends to repair itself with molecules from the multimolecular layer.

8.A.1.3 Treating Procedure

1. Surface conditioning (for water removal and to prepare the surface for the inhibitor film): pump methanol or a diesel-surfactant mix into the tubing. The selection and exact formula for the pretreatment (prewetting) slug is determined for each well.

2. Inhibitor: pump in inhibitor mix in a 1:4 ratio of inhibitor to diluent (e.g., 1–2 drums, equal to 55–110 gal [208–416 L]) inhibitor concentrate diluted in 5–10 bbls (795–1590 L) of condensate or diesel. Use an oil-soluble film-forming type inhibitor that can form a macrofilm. A macrofilm is an inhibitor film that is adherent and measurable in size—approximately 0.1–3.0 mil (3–76 µm), as differentiated from a monomolecular inhibitor layer). In this step, the corrosion inhibitor is mixed into a diluent such as diesel for the first few barrels of the treatment. This maximizes the inhibitor concentration contacting the tubing. The remainder of the treatment consists of diluent to displace the inhibitor slug to the well bottom (step 3).
3. Displacement: follow inhibitor mix with sufficient diesel, lease condensate, or crude to displace mix to the bottom of the well.
4. Pumping rate: 0.5–1 bbl/min (80–160 L/min) (if higher pumping speeds are required, use larger amounts of inhibitor mix).
5. Shut-in: 4 hours.
6. Production: slowly return the well to production for the first 2–4 hours. This will allow additional filming as the inhibitor is brought to the surface.

8.A.1.4 Treating Frequency

Typically, retreatment is required every 1–2 months. Consider using the Batch and Fall method if the well requires less frequent treatment. Schedules will vary, depending on well production rates and fluid characteristics. The actual schedule would be worked out from monitoring results.

8.A.1.5 Monitoring

In addition to corrosion rate monitoring, the inhibitor slug return may be monitored for inhibitor concentration on early treatments. Wellhead coupons (or downhole, if available) should be exposed for about 90 days, or 1–2 treatments.

8.A.1.6 Comments

Pros:

- From just the standpoint of corrosion inhibition alone, this is a "preferred" method of inhibitor application for gas wells and packered oil wells.
- Applicable for use with existing completions.
- Positive placement of inhibitor to bottom of well.
- Good chemical distribution.
- Allows preparation of metal surfaces for filming.
- Inhibitor film repair based on "macrofilm."
- Relatively infrequent retreatments are required (1–2 months).
- Relatively low volume of chemical required on yearly basis.
- Less likelihood of emulsion by slugs of macrofilm inhibitor than with more oil soluble and/or dispersible compounds.
- Odds of achieving good corrosion inhibition (e.g., chemical placement, film formation, etc.) when treating schedule is followed—excellent.

Cons:

- Downtime required for pumping in and for shut-in period (less downtime than for formation squeeze).
- Tubing volume of produced fluids is displaced into formation.
- Relatively large volumes of fluid required (one tubing volume).

Supplemental Information:

- Surface equipment requirements are high-pressure pumps, chemical mix tanks or trucks, and flush fluid tanks or trucks.

8.A.2 Nitrogen Formation Squeeze or Tubing Displacement

Section 8.9.3 describes this treatment method.

8.A.2.1 Type of Well

Gas wells with bottomhole pressure insufficient to unload a full column of liquid.

8.A.2.2 Basic Approaches and Mechanisms

Same as formation squeeze or tubing displacement, except nitrogen is used to displace fluids to the bottom of the well.

8.A.2.3 Treating Procedure

Same as formation squeeze or tubing displacement, except for nitrogen displacement. Special service companies that handle nitrogen are required.

8.A.2.4 Treating Frequency

Same as formation squeeze or tubing displacement for a given well.

8.A.2.5 Monitoring

Same as formation squeeze or tubing displacement.

8.A.2.6 Comments

Pros:

- Allows positive placement of inhibitor in low-pressure wells.
- The use of nitrogen reduces the relative permeability effect, which would be expected with liquid flushes.
- Reduces possibility of formation damage compared to liquid tubing displacement (Section 8.9.2) or formation squeeze (Section 8.9.11).
- Odds of achieving good corrosion inhibition (e.g., chemical placement, film formation, film maintenance, etc.) when treating schedule is followed—very good.

Cons:

- Requires nitrogen service.

8.A.3 Batch and Fall, Partial Tubing Displacement, and Yo-Yo Treatments

Sections 8.9.4 and 8.9.5 describes these treatment methods.

8.A.3.1 Type of Well

Gas wells with bottomhole pressure insufficient to unload full column of liquid.

8.A.3.2 Basic Approaches and Mechanism

1. Corrosion inhibitor mix is pumped into the tubing, followed by a partial tubing volume of flush fluid, the well is shut-in to allow the liquid to fall to the bottom, and then slowly returned to production. The inhibitor may be preceded by a slug of fluid to precondition the surface.
2. For yo-yo treatments (also known as batch and circulate treatment), the well is shut-in for a second time as soon as liquid reaches the tree.
3. The inhibitor film is formed on the tubing during pump-in and falls to the bottom, and refilmed when the well is slowly returned to production. Two additional "passes" of the inhibitor are provided with each yo-yo.
4. The inhibitor used is a so-called "heavy film former" or "macrofilm former," which provides a multimolecular film.

8.A.3.3 Treating Procedures

Refer to the description in the main text to distinguish between displacement volumes used for Batch and Fall (Section 8.9.4), and Partial Tubing Displacement (Section 8.9.5).

1. Surface conditioning (for water removal and to prepare the surface for the inhibitor film): pump methanol or a surfactant diesel mix into tubing.
2. Inhibitor: pump in inhibitor mix in a 1:4 ratio of inhibitor to diluent (e.g., 1–2 drums, equal to 55–110 gal [208–416 L]) inhibitor diluted with 5–10 bbls (795–1590 L) of lease condensate or diesel. Use an oil-soluble film-forming type inhibitor that can form a macrofilm.
3. Displacement: follow inhibitor mix with partial tubing volume (approximately 1/3 to 1/2) of lease condensate or diesel. Note: use lower volume for Batch and Fall. See main text.
4. Pumping rate: 0.5–1.0 bbl/min (80–160 L/min).
5. Shut-in: a minimum of 8 hours (or until "normal" shut-in wellhead pressure is achieved) to allow mix to reach bottom.
6. Production: slowly return the well to production for the first 2–4 hours. This will allow filming as the inhibitor is brought to the surface.
7. For yo-yo treatment: when liquid hits the tree, shut-in the well again (repeat steps 5 and 6).

8.A.3.4 Treating Frequency

Typically, retreatments will be required every 1–2 months. The schedule will depend on well production rates and fluid characteristics, and the actual schedule would be worked out from monitoring results. Aggressively corrosive wells may require full tubing displacement, and higher treat frequencies.

8.A.3.5 Monitoring

Coupons should be changed during the treatment so that they are exposed to the inhibitor return.

8.A.3.6 Comments

Pros:

- From just the standpoint of corrosion inhibition alone, this is a "preferred" method of inhibitor application for low bottomhole pressure (BHP) gas wells.
- Applicable for use when BHP is too low for full tubing displacement.
- Partial tubing displacement and yo-yo treatments provide more positive placement of inhibitor than by Batch and Fall (Section 8.9.4).
- Yo-yo treatment ensures good film formation.
- Allows preparation of metal surfaces for filming.
- Inhibitor film repair based on "macrofilm."
- Relatively infrequent retreatments are required (1–2 months).
- Relatively low volume of chemical required on a yearly basis.
- Less likelihood of emulsions. Slugs of a macrofilm forming inhibitor reduce the likelihood of forming emulsions compared to batch treating with a more oil soluble-water dispersible inhibitor.
- Odds of achieving good corrosion inhibition (e.g., chemical placement, film formation, film maintenance, etc.) when treating schedule is followed—good.

Cons:

- Downtime required for pumping in and for shut-in period. Well may be down for a day.
- Relatively large volumes of liquid required (1/3 to 1/2 tubing volume).

Supplemental Information:

- Successful treatment depends on sufficient shut-in to get inhibitor to bottom.
- Good chemical distribution if the inhibitor reaches the bottom of the tubing string.
- Surface equipment requirements are high-pressure pumps, chemical mix tanks or trucks, and flush fluid tanks or trucks.

8.A.4 Treating String

Section 8.9.6 describes this treatment method.

8.A.4.1 Type of Well

Any gas or oil well with a packer.

8.A.4.2 Basic Approach and Mechanisms

A parallel or concentric treating string is run to the bottom of the well, preferably to below the packer. Thus, corrosion inhibitor may be introduced at the bottom of the hole by pumping down the treating string. Either batch or continuous treatment may be used.

8.A.4.3 Treating Procedure

1. Inhibitor: any type of corrosion inhibitor can be used and injected "neat" (without dilution) or mixed with carrier oil, depending on well requirements.
2. Treatment: can be batch or continuous, as required. Treater truck use is possible.

8.A.4.4 Treating Frequency

Would depend on the individual well's need.

8.A.4.5 Monitoring

Corrosion rate and inhibitor carry through (presence and residual).

8.A.4.6 Comments

Pros:

- From the standpoint of corrosion inhibition alone, this is one of the most versatile methods of inhibitor application, and for some operators, it is the preferred method for packered oil and gas wells (including high-pressure, high-temperature, and acid gas wells).
- Positive placement of inhibitor to the bottom of well if the string is run below the packer. Otherwise, inhibitor is introduced just above the packer through a chemical injection valve.
- Good chemical distribution.
- No downtime required for treatment.
- No fluids are displaced into the formation during treatment.
- Relatively small volumes of fluids required for flush or diluent.
- Low-to-medium volumes of chemical required on a yearly basis.
- Odds of achieving good corrosion inhibition (e.g., chemical placement, film formation, film maintenance, etc.) when treating schedule is followed—excellent.

Cons:

- Requires well workover to run treating string to existing wells. String should be run to bottom (i.e., through the packer).
- Fairly frequent retreatments may be required; however, the schedule can be optimized by matching treatment and chemical.
- Presence of treating string may interfere with future well workovers.

Supplemental Information:

- Inhibitor film is maintained by optimizing dosage and treating frequency.
- Pretreatment solution can be injected to prewet surfaces before initial treatment.
- Surface equipment requirements depend on treating approach. The approach could require a chemical injection pump and tank at each well, or it could use a treater truck for batch treatments.

8.A.5 Formation Squeeze

Section 8.9.11 describes this treatment method.

8.A.5.1 Type of Well

Usually used for gas wells but has been used in gas-lifted oil and flowing oil wells.

8.A.5.2 Basic Approach and Mechanism

1. Corrosion inhibitor is pumped into the tubing, displaced into the formation with an over-flush of fluid, the well is shut-in for a period, and then slowly returned to production. The inhibitor may be preceded by a slug of fluid to precondition the surface for the inhibitor film.

2. The inhibitor film is laid down on the tubing during pump-in and "reinforced" by the high concentration of inhibitor, which returns when the well is first returned to production.
3. A portion of the inhibitor remains in the formation and feeds back at low concentration over several weeks' time, thus providing inhibitor molecules to repair the film.

8.A.5.3 Treating Procedure

1. Surface conditioning (for water removal and to prewet the surface for the inhibitor film): pump methanol or a diesel-surfactant mix into the tubing. The selection and exact formula for the pretreatment slug is determined for each well.
2. Inhibition: pump in corrosion inhibitor "neat" (without dilution). Both oil soluble and heavy film former types of inhibitor are used. The volume will vary for the well. Two to 40 drums (110–2200 gal [416–8327 L]) have been used.
3. Displacement: follow inhibitor with lease condensate, crude, or diesel, and displace into formation with a sizable over-flush (50–200 bbls [8–32 m^3]). The size of the over-flush will depend on the formation volume and how much volume is required to move the inhibitor down the tubing string and into the formation.
4. Pumping rates: as slow as practical. Maximum 4–6 bbls/min (0.6–0.95 m^3/min) (the slower the pump-in, the more exposure of the inhibitor to the tubing).
5. Shut-in: 8–12 hours to allow inhibitor dispersion and filming in the reservoir.
6. Production: slowly return the well to production for the first 2–4 hours. This will allow additional filming by the returning inhibitor and reduce the possibility of treater upsets.

8.A.5.4 Treating Frequency

Typically, retreatments are required every 2–6 months. The actual schedule would be worked out from monitoring results. Usually, retreatments have a longer life than the first treatment on a well.

8.A.5.5 Monitoring

In addition to corrosion rate monitoring, the inhibitor slug return and residual chemical should be monitored (at least for initial treatment). Coupons are to be changed during each treatment so that coupons will be exposed to the inhibitor.

8.A.5.6 Comments

Pros:

- From the standpoint of corrosion inhibition alone, this is a "preferred" method of inhibitor application for gas wells and packered oil wells.
- Applicable for use with existing well designs, i.e., the well design does not need to be reconfigured; the treatment can be applied as-is without a workover.
- Provides positive placement of inhibitor to bottom of well.
- Good chemical distribution to tubing surfaces.

- Allows prewetting of metal surfaces for filming.
- Relatively infrequent retreatments required (2–6 months)
- Odds of achieving good corrosion inhibition (e.g., chemical placement, film formation, film maintenance, etc.) when treating schedule is followed—excellent

Cons:

- Does not provide control of inhibitor available to "repair" film; inhibitor film repair is dependent on rate of return of the inhibitor from the formation.
- Downtime is required for pump-in and shut-in periods.
- Possibility of damaging formation.
- Relatively large volumes of fluids are displaced into formation: produced fluid from wellbore, methanol or surfactant slug, inhibitor, and diesel.
- Large volume of fluid must be handled for each treatment.

Supplemental Information:

- Surface equipment requirements are high-volume, high-pressure pumps, displacement fluid tanks, or trucks.

8.A.6 Weighted Liquids

Section 8.9.13.8 describes this treatment method.

8.A.6.1 Type of Well

Low-pressure gas and oil wells.

8.A.6.2 Basic Approach and Mechanisms

1. Corrosion inhibitor base is formulated in a high-density, heavy, viscous solvent to make a weighted liquid corrosion inhibitor (≥9 lb/gal [≥1.1 kg/L]).
2. The required amount of inhibitor is batched into the tubing and the well is shut-in long enough to allow the inhibitor to fall to the bottom, preferably into the rat hole.[11] These weighted batches fall more rapidly to the bottom of the well.
3. When the well is returned to production, the chemical slowly disperses into the production fluids flowing into the wellbore. The inhibitor film is formed where the liquid contacts the tubing wall and when the well is brought back on production.

---

11. A rathole is a small amount of extra hole drilled to allow for junk, hole fill-in, and other conditions that may reduce the effective depth of the well. (https://www.glossary.oilfield.slb.com)

8.A.6.3 Treating Procedure

1. Inhibitor: use a weighted liquid corrosion inhibitor. If H_2S is present, the inhibitor must be formulated for use where H_2S is present. Weighted inhibitors up to 12 lb/gal (1.4 kg/L) are available for use where there is no H_2S.
2. Pump or lubricate chemical into the well:
    A. Initial treatment: use enough chemical to displace rat-hole volume with chemical (for example, if rat hole holds 4 bbl (168 gal [635 L]), use 4 bbl of inhibitor for initial treatment).
    B. Routine treatment: use 80–160 L of inhibitor.
3. Shut-in: allow chemical to fall to the bottom. Guideline for gas wells: allow 1 hour for each 1000 ft (305 m) of depth. For oil wells, time will vary depending on the weight of inhibitor. To be conservative, consider shut-in overnight.
4. Production: slowly return the well to production to allow filming from produced fluids and to reduce possibilities of treater upsets.

8.A.6.4 Treating Frequency

Varies. Could be as often as weekly or could be monthly. Schedules would be determined from monitoring results.

8.A.6.5 Monitoring

Corrosion rate and inhibitor return.

8.A.6.6 Comments

Pros:

- Does not require downhole equipment change.
- Chemical distribution should be relatively good if the chemical gets to the bottom. Chemical should run along the bottom of deviated portion of tubing where water resides.
- Little, if any, fluids displaced into formation.
- No flush or diluent required.

Cons:

- Odds of achieving good corrosion inhibition (e.g., chemical placement, film formation, film maintenance, etc.) when the treating is schedule followed–fair to poor.
- Long downtime required.
- Relatively high volume of chemical required on a yearly basis.
- Inhibitor film repair requires frequent retreatment (monthly).
- Provides no surface prewetting for filming. Thus, filming is dependent on inhibitor's surfactant and filming ability.
- Inhibitor selection is limited by the number of available weighting agents that may be used.
- Inhibitor will be very viscous at cold temperatures.

Supplemental Information:

- Surface equipment requirements are lubricator or high-pressure pump. The well can be constructed with necessary valves to be "truck treated."
- Inhibitor placement to the bottom of well will depend on the accuracy of estimate of shut-in time requirements (and wells being shut-in for required time).

8.A.7 Dump Bailer

Section 8.9.13.8 describes this treatment method.

8.A.7.1 Type of Well

Low-pressure gas or oil wells.

8.A.7.2 Basic Approach and Mechanism

1. The corrosion inhibitor is placed at the bottom of the tubing string with a wireline dump bailer. Sufficient runs are made to place the required dosage.
2. The tubing is exposed to the inhibitor as the well is brought back on production. Thus, contact time will be relatively short and insufficient to give long film life.

8.A.7.3 Treating Procedure

1. Inhibitor: use an inhibitor that is both oil soluble and water dispersible and that exhibits properties for quick filming
2. Dump inhibitor at the bottom of the well with dump bailer. Because of low volume capacity of the dump bailer, make as many runs with bailer as required to achieve dosage:
   A. Initial treatment: 10–20 gal (38–76 L).
   B. Routine treatment: 5–10 gal (19–38 L).
3. Production: return well to production at the lowest practical flow rate to give maximum contact time for film formation.

8.A.7.4 Treating Frequency

Once per week, although the actual treating frequency and dosages will be determined by monitoring results.

8.A.7.5 Monitoring

Corrosion rate and inhibitor residual monitoring.

8.A.7.6 Comments

Pros:

- Can be used with existing completions.
- Positive placement of inhibitor at bottom of tubing.
- Little, if any, fluids are displaced into formations.
- No flush or diluent fluids required.
- Medium amount of chemical is required on a yearly basis.

Cons:

- Odds of achieving good corrosion inhibition (e.g., chemical placement, film formation, film maintenance, chemical selection, etc.) when treating schedule followed—fair.
- Film maintenance depends on frequent retreatments (weekly) before the "old" film is destroyed (relatively frequent retreatments are required).
- Provides no surface prewetting for filming. Thus, initial filming is dependent on the inhibitor's surfactant and filming ability.
- Downtime required depends on wireline time requirements.

Supplemental Information:

- Surface equipment requirements are wireline unit with associated equipment, dump bailer, and pump for transferring inhibitors from drums to bailer.
- Chemical distribution—fair to good, depending on the ability to return well to production very slowly.
- Useful for inhibiting the tubing string after a caliper run. Because the wireline is already rigged, the added cost of the bailer run is nominal.

8.A.8 Wash Bailer

Section 8.9.13 describes this treatment method.

8.A.8.1 Type of Well

Low-pressure gas wells or low-pressure oil wells.

8.A.8.2 Basic Approach and Mechanism

1. A wash or spray type hydraulic bailer is used to place corrosion inhibitor throughout the tubing string. Multiple runs are required to provide complete coverage.
2. The tubing is thus exposed to a high concentration of inhibitor as it is sprayed from the bailer, while the well is shut-in for the bailer runs, and as the well is brought back on production.

8.A.8.3 Treating Procedure

1. Inhibitor: use a macrofilm forming-type inhibitor that has limited oil and water dispersibility.
2. Bailer: run bailer to the bottom, trip and unload inhibitor as the bailer is pulled out of hole. Repeat, tripping the bailer at higher levels for each run (overlap previous run to ensure coverage). Coverage and, thus, number of runs will depend on the capacity and design of the bailer.
3. Production: Slowly return the well to production until the tubing volume unloaded. This allows more filming time.

8.A.8.4 Treating Frequency

Estimate retreatments required each 1–3 months. This will vary depending on well production rates and fluid characteristics. Actual schedules will be based on monitoring results.

8.A.8.5 Monitoring

Corrosion rate. Coupons should be changed during treatment so that coupons will be exposed to inhibitor as the well is brought back on.

8.A.8.6 Comments

Pros:

- Applicable for use with existing downhole equipment.
- Reasonably positive chemical placement throughout tubing string.
- Little, if any, fluids are displaced into formation.
- Relatively infrequent retreatment required (1-3 months).
- No flush or diluent required.
- Relatively low volumes of chemical required on a yearly basis.

Cons:

- Odds of achieving good corrosion inhibition (e.g., chemical placement, film formation, film maintenance, etc.) if the treating schedule is followed—fair to good.
- Provides no surface prewetting for filming; thus, film life may be shorter than with tubing displacement (Section 8.9.2) when using the same macrofilming chemical.
- Downtime required depends on wireline time requirements.

Supplemental Information:

- Surface equipment requirements are wireline units with associated equipment, wash bailer, and pump to transfer inhibitor from drum to bailer.
- Chemical distribution should be good if the bailer operates properly.
- Inhibitor film repair based on macrofilm properties.

8.A.9 Inhibitor Sticks

Section 8.9.13.8 describes this treatment method.

8.A.9.1 Type of Well

Low-pressure gas.

8.A.9.2 Basic Approach and Mechanisms

1. Corrosion inhibitor base is combined with a wax and cast into solid sticks, 1.5 in (38 mm) diameter, 18 in (0.46 m) long. One stick is equivalent to approximately 1 qt (0.95 L) of field strength liquid inhibitor.
2. The required number of sticks are lubricated into the tubing and the well is shut-in sufficiently long to allow the sticks to fall to the bottom.
3. The sticks are designed to melt at bottomhole temperature, thus releasing the inhibitor to the produced fluids.

8.A.9.3 Treating Procedure

1. Inhibitor: corrosion inhibitor sticks. Dosage is usually based on 1 stick/MSCF[12] production per day.
2. Lubricate: lubricate sticks into tubing.
3. Shut-in: allow sticks to fall. (Guideline: shut-in 1 hr per 1000 ft (305 m) of depth in straight hole; fall rate in deviated hole would have to be determined.)
4. Production: slowly return the well to production to allow filming from produced fluids.

8.A.9.4 Treating Frequency

Weekly.

8.A.9.5 Monitoring

Corrosion rate monitoring and inhibitor residual.

8.A.9.6 Comments

Pros:

- Applicable for use with existing downhole equipment.

---

12. Million standard cubic feet.

Cons:

- *This method is not normally recommended.*
- Odds of achieving satisfactory corrosion control by this method—very low due to uncertainties in chemical placement, film formation, film maintenance.
- Odds are poor that sticks will fall to the bottom; they may hang up at the subsurface safety valve (SSSV) or in deviated holes, could "bridge"[13] across the tubing.
- Sticks must reach the bottom to melt and release inhibitor. If they are not exposed to near bottomhole temperature, they will soften, but may not release enough inhibitor, and at worst, the solid stick could be produced back and plug the SSSV or wellhead equipment.
- Odds of operational problems resulting from sticks not melting—high.
- Wax in sticks could create emulsion and/or foaming problems.

8.A.10 Chemical Injector Valve

Section 8.9.12 describes how chemical injection valves allow for continuous inhibitor treatment.

8.A.10.1 Type of Well

Packered oil or gas well.

8.A.10.2 Basic Approach and Mechanism

1. A chemical injection valve is run in the tubing string just above the packer and the tubing-casing annulus is filled with a corrosion inhibitor-diesel mix.
2. Additional inhibitor-diesel mix is pumped into the annulus to open the valve and inject inhibitor into the stream. Treatment may be by batch or continuous.

8.A.10.3 Treating Procedure

1. Inhibitor: load annulus with inhibitor-diesel mix:
   A. Use an inhibitor that is both oil-soluble and water-dispersible with excellent long-term solubility in diesel.
   B. Mix should be 10–20% inhibitor. The exact mixture will be based on the rate at which the chemical can be pumped through the chemical injection valve, as well as the producing volumes and expected treating requirements.
2. Inject inhibitor mix by pumping additional mix into the annulus. Volumes and pumping times should be developed based on treatment results:
   A. Batch treatments: inject inhibitor mix at a rate to achieve a dosage of approximately 300 ppm for at least 1 hr. For example: a 3000 barrels fluid per day (BFPD) (477 m^3/d) would require 16 gal (60.6 L) of a 10% mix injected for 1 hour for each treatment.

---

13. "Bridge" means to get stuck in the tubing string cross-wise so it makes a bridge, and does not fall to the bottom of the tubing string.

B. Continuous treatment: if only very low rates can be injected through the chemical injection valve, continuous treatment could be required. For oil wells, a dosage of 12–25 ppm (based on total fluid volume) would probably be necessary. This would be 0.5–1 gal (1.9–3.8 L) of inhibitor (5–10 gal [19–38 L] of a 10% mix) per 1000 bbls (159 m^3) of fluid treated. For gas wells, assume 1 qt (0.95 L) of inhibitor (2.5 gal [9.5 L] of a 10% mix) per MSCF (28,300 m^3) gas.

3. Initial Treatment:
    A. It is assumed that the tubing will be exposed to the inhibitor mix when that fluid is placed in the annulus. If the routine treatment is started within a week, no other prewetting treatment should be required.
    B. If routine treatment does not start immediately, an initial treatment to establish the film will be necessary. Details of such a treatment will have to be developed by the user and chemical supply company. Such a treatment might involve pumping an inhibitor mix at a relatively high rate for several hours with the well flowing at a low rate.

8.A.10.4 Treating Frequency

The optimum batch treating schedule would have to be developed based on well needs—for estimation purposes, assume that a 1 hour batch treatment would be required:

- Once per week for low-volume wells.
- Twice per week for mid-volume wells.
- Daily (essentially continuous) for high-volume wells.

8.A.10.5 Monitoring

Corrosion rate and inhibitor residual monitoring could be used.

8.A.10.6 Comments

Pros:

- Chemical distribution should be good above the chemical injection valve.
- Allows varying of dosage and treating frequency to optimize film maintenance.
- No downtime required for routine treatment.
- No fluids are displaced to formation during treatment.
- Provides significant flexibility for dosages and treating frequency.
- Odds of achieving good corrosion inhibition (e.g., chemical placement, film formation, film maintenance, etc.) when everything is working, and the treating schedule is followed—good.

Cons:

- Because of mechanical problems, the chemical injection valve method is not normally recommended. Valve failure is common (see Section 8.9.12).

- Odds of mechanical problems with chemical injection valve (plugging or cutting out)—high. Also, problems with the chemical valves could lead to dumping the annular inhibitor (i.e., the inhibitor mix stored in the tubing-casing annulus) mix while changing out valve.
- Requires workover to install chemical injection valve, cleaning of annulus of mud and solids, and placement of inhibitor-diesel mix in annulus of existing well.
- Relatively large volume of inhibitor and diesel are required (annular volume plus routine treatment).
- Provides placement of chemical at chemical injection valve location. Tubing below valve will be unprotected.
- Fluids displaced into formation only if the workover to install the injection valve requires killing the well.
- High-production volume wells could require daily treatment.
- Difficult to change inhibitor formulation.

Supplemental Information:

- Surface equipment requirements are chemical injection pump and chemical injection tank at each well. Could use a treater truck for batch treatments.
- Inhibitor film formation depends on initial filming procedure for surface preparation.

8.A.11 Gas-Lift Gas

Section 8.9.8 describes this treatment method.

8.A.11.1 Type of Well

Packered, gas-lifted oil wells

8.A.11.2 Basic Approach and Mechanisms

Inhibitor is introduced into the tubing through the working gas-lift valve to provide inhibition from that valve to the surface. Inhibitor may be injected continuously into the lift gas at the wellhead or batched in periodically. The inhibitor formulation must be designed for injection into dry gas (to prevent "gunking"[14] out of the inhibitor).

8.A.11.3 Treating Procedure

1. Inhibitor: an inhibitor formulated for introduction into dry gas (i.e., no evaporation of solvent or the active ingredient is a liquid after solvent evaporation) would be injected into the gas at each wellhead. Usually a corrosion inhibitor exhibiting both properties of oil solubility and water dispersibility would be used.
2. Initial treatment: a batch in 20 gal (76 L) inhibitor mixed with 80 gal (304 L) diesel.

---

14. Gunk is a thick, sludge-like deposit that is capable of plugging equipment.

3. Treatment could be batch or continuous with the volume and injection rate based on treatment results:
   A. Batch treatment: inject an inhibitor diesel mixture of up to 10 gal (39 L) inhibitor and 40 gal (151 L) diesel.
   B. Continuous treatment (common and most effective): base dosage on 12–25 ppmv (parts per million by volume) inhibitor in produced fluids (0.5–1 gal inhibitor/1000 bbl of fluid per day (BFPD) (1.9–3.8 L inhibitor/159 m^3 per day).

8.A.11.4 Treating Frequency

Batch treatment would vary with well production—typically once per week. The inhibitor can also be injected continuously in with the gas lift gas (preferred method).

8.A.11.5 Monitoring

Corrosion rate and inhibitor carry through (presence and residual).

8.A.11.6 Comments

Pros:

- Chemical distribution should be good above working gas-lift valve.
- Odds of achieving good corrosion inhibition (e.g., chemical placement, film formation, film maintenance, etc.) from the working valve up, when treating schedule followed—very good.
- Tubing string surface is prepared for inhibition by the initial inhibitor dosage. No separate prewetting is required.
- Inhibitor film is maintained by continuous treatment.
- No downtime is required for treatment.
- No fluids are displaced into the formation.
- A medium amount of diesel is required as diluent.
- Low-to-medium yearly volume of chemical required.

Cons:

- Treatment is effective only from working gas-lift valve to surface.
- No inhibition provided below the working gas-lift valve.
- Chemical selection somewhat limited by properties required for injection into dry gas.

Supplemental Information:

- Applicable for use when wells are being gas lifted.
- Surface equipment requirements are chemical injection pump and supply tanks required at each well. Batch treatment could be performed with treater truck.
- This treatment method can be used to supplement internally plastic-coated tubing by providing protection in the wireline-damaged areas.

8.A.12 Batch and Flush/Circulate Treatment Down Annulus

Section 8.9.13.5 describes this treatment method.

8.A.12.1 Type of Well

Rod pumped oil wells and other open-annulus wells.

8.A.12.2 Basic Approach and Mechanism

1. Inhibitor with a suitable volume of flush is injected into the open-annulus of a rod-pumped oil well after the annulus has been prewetted with flush fluids. Target concentration is 25–50 ppm based on total fluid production rate. The flush fluid may be well fluids, or deaerated water from a treater truck. In some cases, the treater truck may use lease crude for flush. Some wells may require circulation to achieve satisfactory treatment. In those instances, after the inhibitor is added to the well, the production of the well is then redirected down the annulus to circulate the inhibitor throughout the tubing and annulus.
2. The frequency of the treatment along with the type and amount of inhibitor and flush are process variables.
3. The treatment can be administrated by either production personnel or by truck treatment.

8.A.12.3 Treating Procedure

1. Prewet: to wet the annular volume (the volume in the tubing-casing annulus) so the inhibitor will run down the casing to the fluid level, produced fluids are flushed down the annulus for several minutes. If treatment is by treater truck, typically one barrel (160 L) of flush fluid is used for prewet. The prewet is particularly important in some wells to prevent a buildup of inhibitor at the top of the annulus.
2. Inhibitor: typical inhibitor for this use is an oil soluble or oil dispersible material with little or no water solubility or dispersibility. The volume of flush is typically 0.5 bbl for every 1000 ft of well depth.
3. Inject: volume of inhibitor followed by the flush into well annulus. Oil should be used as the flush in wells with high carbon dioxide (CO_2) contents. Water is acceptable in other cases, although not preferred. For circulation treatments, after the annulus is treated with the inhibitor-flush fluid, redirect the production from the tubing down the annulus to circulate the well. Circulation should continue long enough to distribute the inhibitor throughout the tubing and annulus.
4. The initial treatment: done with 5–25 gal (19–95 L) of inhibitor to lay down a good film. This should be done each time the well is pulled. Thereafter, the regular treatment should be performed.

8.A.12.4 Treating Frequency

Use Table 8.A.1 as a guide to develop the treating frequency and inhibitor volume requirement based on well production rate. Adjustments should be based on corrosion monitoring data.

Table 8.A.1 Batch treating frequency and inhibitor volume based on well production rate.

Well Production Rate	Treating Frequency	Inhibitor Volume
<25 BPD	Monthly	1 gal (3.8 L)
25–50 BPD	Bimonthly	1 gal (3.8 L)
50–100 BPD	Weekly	1 gal (3.8 L)
100–500 BPD	Weekly	1 gal/100 BPD (3.8 L/16 m^3/day)
500–1000 BPD	Weekly	5 gal (19 L)
>1000 BPD	2–3 times weekly	2–3 gal (7.6–11.4 L)

8.A.12.5 Monitoring

Corrosion rates and failure records can be used. Corrosion rates can be determined in the flowline. The frequency of rod failures can also be used to examine corrosion trends if the rod failures are associated with corrosion and/or corrosion-fatigue.

8.A.12.6 Comments

Pros:

- This is the preferred method of treating open-annulus rod-pumped oil wells.
- Odds of achieving good corrosion inhibition (e.g., chemical placement, film formation, etc.) when the treatment schedule is followed—very good.
- Chemical distribution should be excellent because the inhibitor is circulated.
- Adequate circulation of inhibited fluid greatly improves the inhibitor effectiveness.

Cons:

- Well is off production during circulation.

Supplemental Information:

- No special equipment is needed except a chemical pot and flush piping for operator treating and connections for a hose and recirculation piping for treater truck applications.
- The well must not have a packer.

Appendix 8.B: Circulation Time Required for Rod Pumped Well Batch and Circulate Treatment

These equations are based on those provided in the following paper:

Cameron, G., L.S. Lam, R.D. Myers, G.S. McNab, "Optimizing Downhole Corrosion Inhibition Treatments Circulation Program Based on Field Tracer Studies," NACE CORROSION'88 paper no. 100 (Houston, TX: NACE International, 1988).

Circulation time (in minutes) is calculated from three components:

1. Time for the chemical to fall through the gas zone
2. Time for the chemical to fall through the fluid zone
3. Time for the chemical to come up the tubing

1. Time for the chemical to fall through the gas zone

$$\frac{\text{Pump Depth} - \text{Fluid Level}}{28.08 + \dfrac{\text{Total Fluid Production Rate} \times 42}{1440 \times \text{Casing-Tubing Capacity}}}$$

2. Time for the chemical to fall through the fluid zone

$$\frac{\text{Fluid Level}}{13.91 + \dfrac{\text{Total Fluid Production Rate} \times 42}{1440 \times \text{Casing-Tubing Capacity}}}$$

3. Time for the chemical to come up the tubing

$$\frac{\text{Pump Depth}}{\dfrac{\text{Total Fluid Production Rate} \times 42}{1440 \times \text{Tubing Capacity}}}$$

where

- Pump Depth and Fluid Level are in feet
- Total Fluid Production (oil + water) is in barrels per day
- 28.08 and 13.91 are the gas and fluid fall velocity in feet per minute, respectively
- 1440 is the number of minutes in a day
- Tubing-Casing Capacity is the annular volume between the casing and tubing in barrels per foot
- Tubing Capacity is the volume of the tubing string in barrels per foot

Minimum Circulation Time

= Components (1) + (2) + [lesser of 25% of Component (3) or 1 hour]

Optimum Circulation Time

= Components (1) + (2) + [lesser of 50% of Component (3) or 2 hours]

Maximum Circulation Time

= Components (1) + (2) + [lesser of 75% of Component (3) or 3 hours]

Time to Circulate to Surface

= Components (1) + (2) + (3)

Internal Corrosive Environment Control

Robert J. Franco

9.1 Introduction

This chapter describes techniques that allow control of the internal corrosion environment (or at least modify it to reduce corrosion problems). Subjects such as system design to minimize or eliminate internal corrosion using oxygen exclusion, oxygen removal, dehydration, control of deposits, and control of microbiological growth are the main topics of discussion. While some of these techniques involve the original planning, design, and installation (or retrofit), most are operational and maintenance considerations.

The corrosive environment encountered in petroleum production consists of the internal environment (e.g., produced fluids, water) and external environment (e.g., marine environment, buried soil environment). In most circumstances, it is not possible to control the external atmospheric environment, so protective coatings are required to provide a barrier to corrosion. For buried service, a combination of cathodic protection and protective coatings provides a barrier against external corrosion. In small volume enclosures, such as void spaces in a floating production storage and offloading (FPSO) marine vessel, the corrosivity of the external environment can be controlled using dehumidification, but in general, corrosion barriers such as coatings (Chapter 6) and cathodic protection (Chapter 7) are necessary. Chapter 8 discusses corrosion inhibitors as a barrier to internal corrosion. Chapter 9 discusses chemicals other than corrosion inhibitors, such as oxygen scavengers, biocides and pH control, and nonchemical means to control internal corrosion.

9.2 Designing to Avoid Corrosion and Designing for Corrosion Control

The principle to remember is "design corrosion out—don't design corrosion in." The design of any well, pipeline, piping system, facility, or plant can affect the severity of corrosion at that facility and/or affect the costs of corrosion control efforts.

The first step is to determine where and when in the project corrosion problems are likely to occur. On large projects, this occurs prior to the front-end engineering design (FEED)—known as the "Pre-FEED" stage. Wells, lines, facility piping, vessels, and auxiliary equipment are all candidates for corrosion control either from design, startup, or sometime later in the project's life. As mentioned in earlier chapters, gas wells and gas handling facilities are most likely to present corrosion problems as soon as startup begins. Oil well projects, on the other hand, may not require corrosion control until later in their life—when water production increases. In sour oil production where hydrogen sulfide (H_2S) is present, sulfide stress cracking (SSC)-resistant materials will be needed from the beginning. The production from water floods using seawater as the injection water has traditionally experienced reservoir souring, which generally occurs later in the project life. Therefore, if seawater injection is planned, SSC-resistant materials should be considered during the design phase for the producing side. Water injection systems themselves require corrosion control planning and design. (Oxygen removal from waters and its exclusion from all systems are covered in Section 9.6).

When designing a facility, it is very important to consider operating and maintenance requirements—not just maintenance of the corrosion control system—but general facility maintenance. Many corrosion problems can be avoided with proper preventative maintenance programs. As will be discussed in Section 9.5, many of the steps to maintain an oxygen-free system are based on operating and maintenance procedures. Equipment should be selected that not only does the job intended, but is relatively simple to repair (replace seals, packing, etc.).

9.2.1 Process Layout

The location or position of equipment in a process can influence corrosion control requirements. For example, the farther upstream a corrosive gas is dehydrated, the less corrosion control will be required in the remainder of the system. With sufficient dehydration, bare carbon steel is all that is required for most downstream facilities. (The term "sufficient dehydration" is also discussed later in this chapter.). Even though sour gas is dehydrated, many operators choose to select SSC and hydrogen-induced cracking (HIC)-resistant steel to prevent hydrogen-induced failures in case of incorrect or abnormal operation of the dehydration unit. (Refer to Chapter 3, Sections 3.10–3.13 inclusive, for a discussion of these cracking mechanisms.)

Equipment located closest to the coastline or, if offshore, in the general wind direction, will experience more atmospheric corrosion than equipment located down wind. Locations selected for equipment inspection should account for this difference.

Control Corrosion Now or Later?

Oil production may start with a low water cut, which leaves carbon steel well components, flowlines, and gathering lines "oil-wetted" and protected from internal corrosion. The question facing the corrosion engineer is when to initiate inhibition? Crudes vary in corrosivity, and there is no industry-accepted water cut below which corrosion cannot occur. Critical water cuts vary from ≥1% to as high as ≥80%. Laboratory testing aids in determining the water cut that results in water-wetted equipment and requires inhibition. In general, heavier gravity crudes provide more protection than lighter crudes and condensates.

Simple laboratory kettle tests performed early in development of an offshore West Africa field suggested that a water cut of more than 40% was necessary to initiate corrosion, and early production was at a considerably lower water cut. The field was started with no corrosion inhibition. However, simple kettle testing does not simulate flow regimes and many other parameters that might affect corrosion. After a decade of operating without corrosion inhibition, the water cut slowly was rising, and the corrosion engineer recommended a safety factor of two (20% water cut) to initiate inhibition. By this time, operations management and field personnel were not accustomed to needing inhibition. The engineer had to justify why inhibition was a wise decision from a risk perspective and was successful in doing so before failures occurred.

If the corrosion engineer waits until water cut reaches a critical level, some corrosion damage may have already occurred because laboratory tests cannot simulate all relevant variables. Often, downhole tubing leaks occur before the corrosion engineer realizes that corrosion damage has already occurred. Once corrosion damage has initiated, it is harder to control.

During the project phase, it is easier to initiate a corrosion inhibition program. Tanks, pumps, and chemical injection facilities can be designed, planned, and accommodated before construction commences. After years in operations, the footprint required for chemical injection facilities may not be available, especially offshore (even though drawings show the location of the "future inhibitor injection skid"). Operators who are not accustomed to feeding corrosion inhibitors must be trained. The corrosion engineer must address the question whether it is wiser to spend money on inhibition on startup or wait. Present value economics favors waiting, but it does not address the intangible disadvantages mentioned here: convincing management, having operators change their procedures, and accommodating room on an already crowded platform that used the empty space for other purposes. Injecting inhibitor at a low dosage from the start of the project eliminates these disadvantages while controlling operating expenses. It is often easier to raise dosage than to initiate injection.

9.2.2 Design for Coating

It cannot be overemphasized that a vessel should be "designed for internal coating" if it will be coated new or in the future. The features of removable internals (piping, baffles, etc.), properly designed and executed welds, nozzle sizes, etc., must be included in the design specifications. Support beams that will be externally coated should have sharp corners and water traps eliminated. Vessel and tank internal coatings are discussed in Chapter 6, Section 6.7.2.

9.2.3 Design for Inspection

As discussed in Chapter 10, Sections 10.4.2 and 10.5, inspection is a very important part of any corrosion control program, but many oil field facilities are not designed for inspection.

Vessels and tanks often do not have manways (or sufficient manways) to enter and properly clean or inspect their condition. Many that do have adequate manways often do not have all the necessary valves to isolate the system for purging or inspection. Even when the valves were installed, they frequently have not been maintained and they will not seal leak tight. In those cases, the entire facility may require a partial or complete shutdown—a costly procedure in many facilities. To minimize the effect of valve problems, some operators use "double-block-and-bleed" valve arrangements. This uses two valves connected in tandem with a bleed valve in between. After the valves are closed, the pressure between them is bled off. The theory is that even if one of the valves is leaking, the other will hold and the vessel may be safely depressurized and "purged for entry," i.e., it is safe for a person to enter because the vessel has been drained, is gas-free, and safe for breathing.

Thermally insulated piping and pressure vessels are typically not designed for inspection of the underlying steel. Visual inspection is often limited to examining the condition of the weather jacketing and to locate areas where there are openings or new penetrations through it. Removing the weather jacketing and insulation for inspection is costly and can result in creating sites where water can enter the weather jacketing if re-insulation is not done with diligence. Rather than cutting holes for inspection purposes, or removing insulation, operators today are using inspection strategies based on advanced nondestructive testing methods, such as guided wave ultrasonic testing and digital radiography, whose selection depends on the equipment diameter, wall thickness, and temperature. Selecting long-life immersion-grade coatings for hot metal surfaces extends the life of thermally insulated equipment. These strategies should be developed during the design stage of a project. Eliminating unnecessary insulation is a successful strategy in operation.

Insulated or bare piping that rests directly on pipe supports is inaccessible for inspection at the support contact point. Piping support design is discussed later in the chapter, Section 9.3.5.

A frequently encountered problem in inspecting equipment is insufficient clearance for the inspector to gain access to areas deemed to be important for inspection. A minimum of 1 m (3.3 ft) clear working space is normally needed for inspection activities around equipment and associated piping. Piping that rests on or close to the ground may not be inspectable at the critical area along the bottom where water accumulates internally, and externally at the pipe supports. Piping that enters or exits from the bottom head of a vertical pressure vessel may not be inspectable inside the vessel support skirt. Inspection points should be located where corrosion damage is anticipated, *but these should be in accessible areas.* Difficult to access locations require personnel to climb ladders or scaffolding, rappel down ropes, or require a hydraulic crane with a platform (also known as a "man lift" or "cherry picker"). Difficult to access locations may have been selected from piping and instrumentation drawings, which do not show three-dimensional layout. At worst, the inspection locations are never accessible because they lie close to and behind adjacent equipment or cannot be reached from a platform or walkway, and alternative locations must be found.

9.2.4 Design for Monitoring

Types of monitoring techniques besides inspection are discussed in Chapter 10. In the early life of some facilities, corrosion will not be much of a problem, so monitoring requirements may be minimal. Once the facility is in service, however, control efforts, such as inhibition, may justify extensive monitoring. When coupon mounting locations, fluid sample connections, and the like are included in the initial construction, they are relatively inexpensive—certainly when compared to retrofitting costs (and downtime). The operability of new monitoring devices should be verified before startup begins. It is not unusual to find a high percentage of monitoring devices not functioning as designed. Some examples include the following scenarios:

- Installing a monitoring device at the bottom (6 o'clock) position of a pipe without having sufficient space below for a coupon retractor. Some high-pressure retractor tools for removing or installing coupons or corrosion probes require as much as 2 m (6.5 ft), although more compact retractors may be available. (See Chapter 10, Figure 10.10 for coupon retractors and Section 10.4.1.2 for more on coupons.)
- Excessive weld reinforcement on the access fitting reduces the internal diameter of the fitting enough to prevent a coupon or probe from passing through it.
- Locating monitoring points in inaccessible areas. (These inaccessible areas were discussed in the previous section.)

These deficiencies must be rectified before startup.

Typically, a minimum of 1 m (3.3 ft) clear working space for monitoring activities is required around equipment and associated piping. Permanent platforms should be provided for monitoring points that are 2 m (6.6 ft) or more above grade or deck level. If these activities are performed on five-year intervals or more frequently, then permanent access should be considered.

9.2.5 Design for Chemical Treatment

The cost of chemical treatment can be minimized by providing valves for chemical injection in the construction of new facilities, even if the treatment is not needed immediately (e.g., a new oil well system where initial water production is low). Valves are needed to allow isolation of the chemical injection piping from the main process piping. History has proven that some forms of chemical treatment—e.g., corrosion inhibitor, scale control, biocide, etc.—will be needed later in the field's life. The most likely points for future injection can be selected during the design stage and small (0.25–0.5 in [6.35–12.7 mm]) valves can be installed at minimum cost during construction. When system downtime is factored into the cost, adding chemical injection points later when they are urgently needed during operation will be more expensive than installing them during initial construction.

9.2.6 Design for Pigging

While most cross-county pipelines have pig launchers and receivers, most flowlines, gathering systems, and water or gas injection systems do not. Nevertheless, the ability to pig such lines can play a key role in controlling their internal corrosion. Pigs can be used to "wipe" heavy film corrosion inhibitors on the inside surface of wet-gas lines, clean deposits and debris from lines, apply biocide while removing biofilm, or inspect the line for corrosion and damage. These inspections are accomplished by using "smart" pigs, which are instrumented devices that gather data, such as wall thickness, along the length of the pipeline. (Refer to Chapter 10, Section 10.5.1.2 for a discussion of smart pigs.) These duties are in addition to the more common use of pigs to periodically remove water and hydrocarbon condensate from gas lines or waxes, and solids from crude oil lines.

Designing pigging facilities into piping systems is particularly important for corrosive gas systems and water systems where surface waters are the supply water, or where two or more waters are mixed. The ability to pig both field- and interplatform lines offshore is also very important because of the higher costs of operating a facility offshore. Even if the pigging facilities are not installed at the start, if the piping system is designed for pigging, the actual launchers and receivers can be added later much more economically than rebuilding the system. It must be mentioned that even in piping systems that were designed for maintenance pigging, the pigging facilities may not be able to accommodate the longer intelligent tools ("smart" pigs) without adding deck space or designing for vertical launching and receiving, which is very costly. Designing for smart pigging must also avoid wall thickness changes, dual diameter piping, and insufficient pipe bend radius that will not allow the smart pig to traverse the pipeline. These, too, should be addressed during the design phase.

9.2.7 Mixing Waters from Various Sources

It is not good practice to mix source and produced waters. Not only are there potential compatibility problems from the standpoint of scale and deposits, but there are some little-understood corrosion problems that can occur.

The industry has examples of water floods where two waters (produced and source waters) are essentially noncorrosive when air-free and kept separate. However, if they are mixed, the result is severe corrosion problems. Possible reasons for this include nutrient supply for microbiologically influenced corrosion (MIC), an increase in suspended solids leading to deposit formation, introducing oxygen, or causing a change in water chemistry that affects scale and corrosion product stability.

Ideally, therefore, the two water systems should be kept separated—going to separate pump and separate injection wells. Manifolds can be designed that will let wells be switched from source water to produced water as the produced water volumes increase. When this is not practical or possible, waters have been injected in tandem—that is, pump one water for a while, then switch to the other, back and forth. This can be accomplished by using sequencing diverter valves. The one thing to watch out for with such a system is leaking diverter valves. Valve maintenance is extremely important to ensure there is no mixing.

9.3 Piping Design

In the design of piping systems, three conditions can lead to corrosion problems: water traps, dead legs, and high velocities.

9.3.1 Water Traps

Water traps are low sections of lines that allow water to collect (and thus water-wet the pipe) even when very little water is present. Probably the most common water traps in plant installations are created when piping is brought down to ground level for a control loop. These will collect water and can lead to internal corrosion, leaks or failure to meet fitness for service criteria. Another common water trap occurs in plants and in field flowlines where lines drop down to go under other lines or roads. Where ground subsidence is a problem, a section of aboveground piping may sag relative to adjacent sections because of piping support subsidence and trap water if the flow velocity is insufficient to sweep the water phase out.

It is virtually impossible to avoid all water traps in piping. However, they should be kept to a minimum. Where practical, double-block-and-bleed-drain valves should be installed in the trap and periodically drained. *Water traps should be some of the first areas to inspect when an inspection program is initiated.* Figure 9.1 shows a photograph of a common sight in plants and large production facilities: almost every line drops to "working level" before rising to the next vessel, creating water traps.

Figure 9.1 Typical plant piping, forming water traps; note the U-shape of the control valve piping.

9.3.2 Dead Legs

Dead legs—which are common in most piping systems—are branch lines or the ends of piping that have no flow and the fluid is stagnant. They may be dead all the time (e.g., the end of piping systems or the end of extensions put in for future expansion); they may be dead during part of the project (sections of manifolds); or they may be "no-flow" lines that are used intermittently (relief lines, drain lines, recirculation lines, bypass lines, etc.). Dead legs may also be created if equipment (such as a heat exchanger) is no longer needed and is removed, but the piping to it is isolated by installing an end plate known as a "blind flange," which is bolted to the piping and prevents flow. Chapter 1, Figure 1.7 shows the typical locations of dead legs in a plant facility.

Dead legs can be particularly troublesome in water injection system piping. Not only does corrosion occur in dead legs, but stagnant water can serve as a breeding ground for bacteria. Dead legs are out of the process flow and miss any chemical treatment; they serve as a source of re-infecting the system after treatment.

It is virtually impossible to eliminate all dead legs in piping, but they can be minimized in the piping design by such things as

- Eliminating dead ends in manifolds
- Placement of valves to have the shortest practical dead legs, such as at the pressure vessel nozzle
- Providing double-block-and-bleed drains so the dead areas can be periodically flushed
- Using ells rather than tees
- Having branch lines come off the top rather than the side or bottom

9.3.3 Velocity Effects

The opposite of dead legs can also create problems, that is, areas of high velocity. Chapter 3 discussed erosion-corrosion where solids in liquid streams or liquid droplets in gas streams can remove corrosion products exposing the clean metal to more corrosion.

Up to now, about the only guideline available to determine erosional velocities (sometimes referred to as "critical velocities") is the formula in API RP 14E:[1]

$$V_e = \frac{C}{\sqrt{\rho_m}} \tag{9.1}$$

where

V_e = fluid erosional velocity, ft/s (m/s),

C = empirical constant (guidance for values to use is provided in RP14E, repeated below), expressed with English units herein, and

ρ_m = gas/liquid mixture density at flowing pressure and temperature, lb/ft³ (kg/m³).

This equation was first introduced in Chapter 1, Section.5.1, where its limitations are described.

There are indications that this formula for erosional velocity is conservative and that higher velocities are tolerable in many situations in the absence of sand production. The designer can adjust the empirical constant (the "C-Factor") upward to allow for higher velocity if there is evidence from that field or facility that higher velocity can be tolerated safely. Corrosion resistant alloys can operate at much higher C-Factors than carbon steel. The experience in many high production gas fields and in findings from lab experiments shows that 13Cr and duplex stainless steel can be operated at a C-Factor of 350–450.[2] At least one major oil company does not use the API RP 14E equation with corrosion resistant alloys (CRAs). As velocity increases, pressure drops, which becomes the practical limiting factor for design in the absence of a C-Factor.

9.3.4 Sand Effects

The relevance of the equation in API RP 14E to modeling sand erosion has been questioned, although the simple equation remains popular with oil field piping and subsurface tubing designers.

Technical papers, along with laboratory testing and/or computational fluid dynamics modeling, have demonstrated that C-Factors can be modified for sand production.[2-3] Several articles on erosional velocity are listed in the references and bibliography at the end of this chapter. Examples of software that can be licensed for modeling sand erosion are from the University of Tulsa (Tulsa SPPS Sand Production Pipe Saver)[4] and DNVGL-RP-O501 (refer to Bibliography) provides the basis for DNVGL licensed software, and guidelines for performing computational fluid dynamics modeling of sand erosion. The velocity of sand particles is important in estimating sand erosion because erosion is roughly proportional to the sand particle velocity to the cube power, i.e.,

$$SE \propto V_{SP}^{n} \tag{9.2}$$

where SE is the sand erosion, V is velocity, SP indicates the sand particles, and n depends on the material being exposed to sand erosion. If the pipeline or piping system is fabricated from carbon or low-alloy steel, the value of n is approximately 3 (cube power). Hence, sand production can be controlled by controlling the production rate.

Because erosion damage caused by sand production is not always predictable in extent and location, the safest thing to do is to periodically inspect piping that is operating at a high velocity to forecast the life of the piping. Points of momentum change (elbows, tees) are the likely areas to monitor wall thickness. Some operators install continuous wall thickness monitoring using ultrasonic measurements at these points, whereas others perform periodic inspections at these locations. Chapter 1, Figure 1.6 shows cushion tees installed between the choke and flowline to minimize failure caused by sand erosion. Sand erosion at the wellhead and choke is more prevalent in gas systems than in oil, because of higher velocity and the inability of a gas to significantly reduce the momentum of sand particles. Sand control should start before sand is produced and allowed to enter the facility. Where sand production is anticipated, sand screens are installed downhole. Using externally-clamped acoustic probes or intrusive sand probes helps the well operator limit well production rate to the point where sand breakout can be avoided. Having the signal from an acoustic sand probe wired into the distributed control system allows a rapid response when the well choke's valve position is being changed.

9.3.5 Pipe Support Design

Good piping design goes beyond controlling velocity and eliminating water traps and dead legs. Pipes resting directly on racks or fixed supports is subject to accelerated corrosion and failure at the contact points. Expansion and contraction of the pipe, combined with possible vibration, causes any paint to be scraped off, exposing the bare steel pipe surface. Water from the environment (e.g., precipitation, wind-blown sea spray, deluge system testing, wash downs, cooling tower drift) is trapped in the crevice created between the pipe and the surface on which it is resting, creating a local crevice corrosion cell and oxygen concentration cell (see Chapter 3, Figure 3.20). Such corrosion cells remain wet longer than adjacent surfaces and are at the temperature of the pipe contents, which is frequently elevated. Pipe failures have occurred on racked piping under these conditions, notable at coastal and offshore facilities. The preferred method to protect piping from external corrosion damage due to contact with pipe racks are "pipe shoes." These are devices that are either clamped or welded to the pipe, elevate the pipe, and prevent direct contact between the pipe and the pipe support beam. Pipe shoes allow the piping to thermally expand or contract, and they directly contact the support beam. Chapter 6, Figure 6.10 depicts another means of minimizing differential aeration cell corrosion at the contact point of offshore piping by inserting a nonmetallic half-round rod between the pipe and the pipe support beam. The half-round bar sheds water and prevents moisture being held against the pipe, and it is reinforced to resist sagging under the pipe weight.

9.3.6 Thermally Insulated Piping

Insulated piping systems are designed to protect against the entry of water; however, field experience indicates that at some point in the life of the equipment, the entry of water into the insulation should be expected, and corrosion under insulation (CUI) will inevitably occur.

CUI is described in Chapter 1, Section 1.10 and is shown in Figures 1.9–1.11, and in Chapter 6, Figure 6.20. Carbon steel operating with a skin temperature in the temperature range −4°C (25°F) to 175°C (350°F) has the greatest likelihood of CUI. Corrosion at colder and warmer temperatures still occurs if the temperature cycles in and out of the range. Temperature cycling accelerates the rate of corrosion. Corrosion of steel is rapid under the combined conditions of the presence of liquid water in or under the insulation, elevated temperature, and atmospheric oxygen.

Damage to the weather barriers and deterioration of the caulking sealants are frequently the cause of water intrusion. Failure to restore the water-tight integrity of the weather barriers after removal and reinstallation is also a cause. Applying an immersion-grade protective coating to insulated steel is essential to protect against CUI. Also essential is sealing openings in the insulation system, minimizing protrusions through the insulation, and following the design guidance in NACE SP0198.[5]

9.4 Electrical Isolation

The term "electrical isolation" (sometimes called "insulation") is a corrosion control method that truly breaks up the electrochemical cell. It is applied to internal and external corrosion. The objective of the installation of an isolation fitting is to break the metallic path, or at least to increase the resistance in the corroding circuit as much as possible. The use of electrical isolation requires a simple review of the system, keeping in mind the objective of breaking up possible current paths.

Electrical isolation is used in the following cases:

- Breakup the *external* corrosion circuit in buried facilities, such as the circuit between well casings and flowlines, or well casings and tank bottoms.
- Breakup bimetallic couples (galvanic couples) to prevent dissimilar metal *internal* corrosion, such as stainless steel piping connected to the nozzle of a coated vessel or stainless steel internals in an internally coated vessel.
- Isolate cathodic-protected structures to control *external* corrosion, such as isolating a cathodic-protected flowline from its unprotected well casing or a cathodic-protected buried line from a tank battery.

In the first two uses, electrical isolation is the corrosion control method, whereas in the third, it augments the primary method, cathodic protection (CP). Electrical isolation ensures that protective current flows to the intended structure (e.g., pipeline) and is not drained by inadvertently protecting other buried structures (e.g., tank batteries).

Chapter 5, Section 5.13.1.4 describes the polymers used in constructing electrical isolation. It also mentions design methods that can be considered to eliminate the need for such isolation.

9.4.1 Insulating Flange Kits and Monolithic Isolating Joints

Electrical isolation is discussed in Chapter 5, Section 5.13.1.4. There are several types of fittings for electrical isolation. Probably the most common is the insulated flange. Figure 9.2 shows a cut-away drawing of an insulating flange assembly and its components.[6] (The type of flange face is generic, and Figure 9.2 applies to different flange face designs.) A flange insulation kit contains insulating sleeves for the studs or bolts, the insulating washers for each end of each bolt, and the insulating gasket. Kits are available to match sizes, pressure ratings, and types of flanges. Insulating gaskets are available for flat-faced, raised face, and ring-groove flanges. Figure 9.3 shows a field installation of an insulating flange assembly.

Available types of isolation include insulating flanges, as well as insulating unions, hex nuts, and hammer unions. Monolithic insulating joints that are welded into pipelines isolate one section of the pipeline from receiving cathodic protection current. They have been installed in locations where a flange is undesirable, such as transitions in high-pressure pipelines between offshore to onshore. These joints contain pipe stubs that can be welded into place. Monolithic insulating joints contain nonmetallic rings, so if the pipe weld requires postweld heat treatment, avoiding heat damage to the nonmetallic components should be considered.

Insulating couplings, made like standard pipe couplings, but with nonmetallic threaded inserts, are manufactured for low-pressure service. Some operators use short, 0.3–0.6 m (1–2 ft) long, fiberglass nipples in flowlines at the wellhead to provide well-to-flowline isolation. Sketches of these and other isolation fittings, as well as detailed discussions on electrical isolation, are available in NACE SP0286.[7]

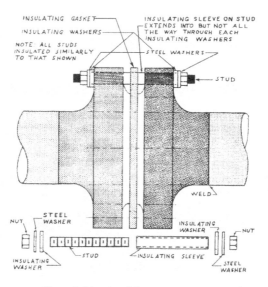

Figure 9.2 Insulated flange assembly details.[6]
Image provided courtesy of NACE International.

Figure 9.3 Photo of an insulating flange assembly. Note the use of nonmetallic washers and bolt sleeves.
Photograph provided courtesy of Advance Products & Systems.

When installing electrical isolation, it is very important that the isolation not be bypassed or shorted by another connection. Figure 9.4 is a photo of just such an installation where a metal chain short-circuits the insulated flange. Often, instrumentation tubing, metal straps, metal tags, and other metallic objects cause the isolation flange to be bypassed.

Figure 9.4 Photograph of insulated flanges in the field (painted brown). Note how the metallic chain contacts the pipe and the valve (hand wheel), thus electrically shorting the insulating flange.

9.4.2 Avoiding Galvanic Couples

Chapter 3, Section 3.6 discusses galvanic attack, corrosion due to the coupling of metals that are far apart in the galvanic series in an electrolyte that is corrosive to one of the metals in the couple. It is at the design stage where these couples can be minimized or avoided. The design needs to include electrical isolation if different metals are required to be connected because of other corrosion control considerations. For example, when stainless steel piping is used with a coated carbon steel vessel, the stainless steel must be isolated to avoid accelerating corrosion at "holidays" in the coated vessel nozzle. This is because of the large stainless steel cathode and small carbon steel anode. In coated equipment, small holidays are assumed to be present that allow corrosion current to contact the bare metal underneath the coating. In coated carbon steel equipment, the anode area is small (at the holidays) whereas the entire stainless steel area acts as the cathode. The unfavorable relationship of small anode-to-large cathode promotes localized corrosion and may require electrical isolation. The stainless steel tubing can be made to rest on nonconductive, nonmetallic spacers.

If a bimetallic line consists of an internally coated carbon steel pipe connected to a stainless steel pipe, and the line is handling low resistivity (high conductivity) produced water, flange insulation alone may not be satisfactory; a longer current path could be necessary to increase electrical resistance. In such cases, the first few feet of the stainless steel (cathode) could be internally coated. This will give a longer current path through the water. Furthermore, if there are holidays in the coating in the stainless steel spool, the galvanic couple (carbon steel anode to stainless steel cathode) will be a small anode and small cathode.

When a vessel and/or the above bimetallic line piping are mounted on a metal floor (such as on an offshore platform or in a building), the insulated flange can be electrically shorted through the vessel supports, the floor, and the pipe supports. In that case, two insulated fittings will be required—one at each end of the internally coated stainless steel spool. The pipe supports should be outside the spool. If these insulating fittings are necessary, the design should be simplified to eliminate as many of the isolation joints as possible.

Figure 9.5 illustrates another use of insulation to avoid a direct bimetallic couple. The stainless steel instrument flange is electrically isolated from the carbon steel nozzle flange. Isolation should be considered only in a corrosive aqueous fluid. Nonconductive fluids, such as sales crude, do not require electrical isolation to prevent galvanic corrosion. Dry gas systems do not require electrical isolation; neither does a wet gas system if there is no continuous water phase.

Figure 9.5 Flange insulation isolating stainless steel instrument (at left) from carbon steel piping. Note the use of nonmetallic bolt sleeves and washers.

Regardless of the purpose for installing the insulation, it is very important to periodically test the flange to verify it is still providing isolation. The wellhead-flowline insulation should always be tested whenever it has been removed for well work. NACE SP0286 lists several tests including checking the pipe-to-soil potential each side of the flange in a pipeline with one side under CP, audio frequency pipe locators, radio frequency meters, and a magnetometer system.[7]

9.5 Oxygen Exclusion

Oxygen is the greatest accelerator of corrosion, and its entry point is typically in water systems after the produced fluid has reached a surface treating facility. Therefore, oxygen exclusion is a primary method of corrosion control in all systems.

Because oxygen is not native to producing formations, the only oxygen in a system will be what is let in because of the design of the system, or the operating and maintenance practices. After primary separation, produced water flows into low-pressure storage facilities, and afterwards is pumped downhole under high pressure into injection wells. In low-pressure facilities, the partial pressure of CO_2 and H_2S are low, so acid gas corrosion is less of a primary corrosion threat than oxygen corrosion and MIC. Oxygen enters through seal leaks and tanks that are insufficiently vapor blanketed.

In many systems, oxygen exclusion is the only method of corrosion control needed (i.e., if oxygen is kept out, no significant corrosion will occur; if oxygen enters, severe corrosion problems will occur.)

When there are other corrosive agents, such as CO_2 and H_2S, the corrosion will be much more severe if oxygen enters the fluids. In most cases, organic, film-forming corrosion inhibitors used to control acid gas corrosion, will not satisfactorily control oxygen corrosion, and chemical oxygen scavengers may be required (Section 9.6.3).

9.5.1 Sources of Oxygen in Producing Wells

Because most flowing oil and gas wells operate with a positive pressure, oxygen seldom creates a problem. However, artificial lift wells may have low enough pressure to have oxygen enter under many conditions. For example, pumping wells that are produced on a cycle or that are shut-in periodically can draw in air as the annular fluid level is drawn down from the shut-in level to the pumping level.

Rod-pumping well-stuffing boxes[1] can also be oxygen entry points. When pumping wells are having corrosion problems in the pumping tee and the first few feet of flowline, oxygen may be the problem. Oxygen can be drawn in on the downstroke (even if oil and water do not leak out on the upstroke). Oxygen will corrode the steel in the pumping tee and maybe 3 m (10 ft) of flowline before it is consumed. From that point on, the system is oxygen-free, but the corrosion damage is done.

9.5.2 Sources of Oxygen in Water Systems

Most oxygen problems exist in surface facilities, particularly water injection systems. Reference 8 describes the problems of oxygen in water injection and disposal systems.

It is not difficult to design and operate a water system oxygen-free if a few simple steps are followed and the personnel involved understand what they are trying to accomplish (and why).

---

1. A stuffing box is the annular chamber provided to provide a leak-tight pressure seal around the polished rod (the uppermost joint in the string of reciprocating sucker rods used in a rod pump well).

The most common method of maintaining oxygen-free conditions is using gas blankets on supply wells and tanks, and to have vessels (e.g., filters) fluid packed. Maintenance of all seals and packing is important. Conditions allowing a vacuum to be created on wellheads, trees, or lines should be avoided. Most of the maintenance items are common sense and can become second nature to properly trained, experienced personnel.

9.5.2.1 Design of Gas Blanket Systems

A "gas blanket" is a gas phase maintained above a liquid in a vessel or tank to protect the liquid against air contamination. The gas source is located outside the vessel or tank.[2] The design of gas blankets is not complicated.[8] The details of a blanketing system will vary to fit the needs of each installation.

The objective of the gas blanket is to always maintain a positive pressure on the gas space so that atmospheric air cannot enter. Included along this line of thinking are the following thoughts:

- High blanket pressures are not required. A small positive pressure of 17 kPa (12.7 mm mercury or 0.5 in mercury) is sufficient on most tanks. Gauges should be installed so that the pressure can be routinely checked and adjusted when required.
- The regulators and gas lines should be sized to admit gas at a rate high enough to maintain the pressure when the fluid level drops. Some installations where tank drawdown rates vary tremendously have parallel regulators to ensure a sufficient gas supply under any conditions. The small volume regulator opens first, and usually this is adequate to handle normal pressure fluctuations. If the drawdown is too rapid for that regulator to keep the blanket pressure up, the second (larger) regulator opens.
- Regulators should be located as close to the gas entrance point as practical and should be installed so that condensation will not block off the gas flow (Figure 9.6). Drain valves should be installed on the blanket gas lines to allow removal of trapped water or hydrocarbons. This is particularly important in areas where the regulator is housed in a building to avoid freezing, and the low-pressure gas is piped to the top of the tank. Unless it is periodically drained from the line to the tank, the condensed liquids will build up and shut off the gas flow.
- When "field gas"[3] is used, regulators should be installed for easy removal for cleaning.
- The gas source should be oxygen-free. That may sound unnecessary to mention, but there has been at least one instance where gas that came off of a deaeration tower was used to blanket the tank of deaerated water. In another case, residue gas containing 2–3% oxygen was used for the blanket gas.

There are several ways to minimize the amount of gas required to maintain a gas blanket. The simplest method is to design the system for minimum fluid level fluctuations; this is done by sizing supply water well pumps to closely match water injection requirements. When the wells do not have to cycle often, their fluid levels remain constant, and gas is not pumped in and out. In many cases, minimum cycling has other advantages, such as longer pump life, less sand production, etc. Minimum cycling of supply wells also lets the supply tank operate with minimum fluctuations of water level. The tank gas-blanket requirements are also held to a minimum.

---

2. Modified from the definition given in the Schlumberger Oil Field Glossary, https://www.glossary.oilfield.slb.com.
3. Field gas is produced gas or fuel gas (gas obtained from the field and not purchased separately). It is not necessarily clean or dry and can foul the gas regulator.

Checking a Gas Blanket

The ideal way to check a gas blanket is to check it with an oxygen meter to ensure it is oxygen-free. However, in most cases in the field, a blanket is checked merely by opening the thief hatch[4] to see if there is an outward "blow" of gas.[5] This is quite satisfactory if the regulator can be heard dumping blanket gas to the tank when the hatch is opened, and if the tank is being pumped down or has a constant fluid level and is not filling.

Safety note: Operators require safety procedures be followed by all personnel before and during sending a person on the roof of a tank to open a thief hatch.

Figure 9.6 Gas-blanket pressure regulator mounted on top of a water tank. The short horizontal pipe on the tank side of the regulator effectively eliminates the possibility of condensation blocking off the gas flow.

Self-Blanketing

Some underground waters are quite "gassy," i.e., underground natural water may contain a high concentration of dissolved gas (e.g., CO_2) when under pressure such that gas is evolved in the low-pressure tank; this in turn can create an oxygen-free gas blanket (known as "self-blanketing"). Be very careful when depending on self-blanketing. Make certain that sufficient gas is evolved in the supply well annulus to replace the void when the well starts ("kicks on"). If the annular pressure drops to zero or below as the fluid level drops, supplementary gas will be required. The same holds true in supply water tanks.

If self-blanketing is used, blankets should be periodically checked to make certain conditions have not changed as the water supply zone is depleted. Experience has shown that self-blanketing does not reliably work on produced water tanks, and should not be depended upon.

---

4. A thief hatch is an opening in the top of the tank. The thief hatch allows tank access for a sampling or level measuring devices. It is used when to take or "steal" a sample. (Modified from the definition given in the Schlumberger Oil Field Glossary, https://www.glossary.oilfield.slb.com.)

5. A tank that is under a slight positive pressure because of the presence of a gas blanket, will expel (blow out) gas through the thief hatch when the latter is opened.

Oil Blanketing

Oil blankets were sometimes used on supply wells and tanks rather than gas blankets. However, oil blankets will not keep a system oxygen-free. Oxygen is soluble in oil. In fact, it may be 5–25 times as soluble in hydrocarbons as in oil field waters. At their best, oil blankets may slow down oxygen entry. At their worst, they oil coat solid particles in the water and create well-plugging problems, or the dump valve fails to close, and the oil blanket is pumped to the injection wells.

9.5.2.2 Other Sources of Oxygen Entry in Water Systems

As mentioned earlier, proper maintenance is important in maintaining oxygen-free conditions.

Pumps

An overlooked source of oxygen entry is the centrifugal pump. If the seals start leaking, air (oxygen) is sucked into the pump. Although it is counterintuitive to believe that high-pressure water can suck atmospheric pressure air in, it has been demonstrated that air can enter through a leaking stuffing-box seal.

Figure 9.7 is a reproduction of a chart from a recording oxygen meter located downstream of a centrifugal transfer pump.[8] Water from a deaeration tower was being pumped through filters. Water from the tower contained less than 20 parts per billion (ppb) dissolved oxygen. The dissolved oxygen levels downstream of the pump varied as high as 800 ppb (left portion of the chart). The seals in the pump were leaking. When the pump's seals were replaced, the oxygen dropped to the tower outlet level. The same phenomenon has been noted on centrifugal charge pumps for positive displacement injection pumps. Plumber pumps (sump pumps, submersible pumps) themselves can leak oxygen when not properly packed.

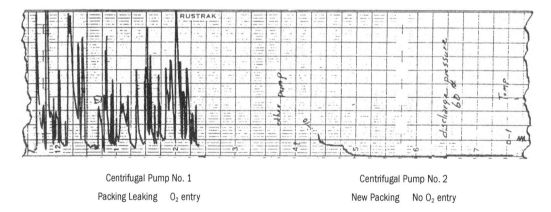

Centrifugal Pump No. 1   Centrifugal Pump No. 2
Packing Leaking   O₂ entry   New Packing   No O₂ entry

Figure 9.7 Oxygen meter recorder chart. Note the oxygen meter was calibrated in percentage oxygen saturation equals 0–10% full scale (equivalent to 0 = 0.79 mg/L dissolved oxygen in this specific water).[8]

Water Supply Wells

Other sources of oxygen entry in water systems include the supply wells themselves. The annulus should be gas blanketed unless there is sufficient gas for a self-blanket. The supply wells most likely to cause problems are those that operate cyclically. If there is insufficient blanket, air is drawn into the annulus each time the well kicks on and the fluid level drops.

A similar problem occurs when the foot valve[6] at the pump and/or a check valve fails when the pump shuts in. In this case, the water drains back out of the tubing and/or out of the supply line. A vacuum is created, and air is drawn into the line. When the well comes on, a tubing full (or line full) of air is pumped to the water supply tank. It may take several hours to work off the effects of such intermittent quantities of air.

For these reasons, it is good operating practice in systems with several supply wells for most wells to run continuously with only one well used as a "swing" well.[7] This is another reason why it is important to size supply well pumps close to the water requirements so that cycling can be kept at a minimum. Foot valve and check valve operation should be checked periodically to make certain they are holding, and the water does not drain back.

Water Injection Wells

Some water injection systems have injection wells that are fed the required amount of water under a vacuum (particularly in the early life of a water flood). Injection wellhead fittings will often allow atmospheric air to leak into the injection water if the fittings are under a vacuum, even though these fittings hold pressure without leaking. Thus, it is good practice to maintain a positive pressure on all water injection wellhead fittings and take the pressure drop at the last valve before the water goes downhole into the injection well. Pinching back on such a valve and holding a positive pressure of 100–170 kPa (15–25 psig) on all well site fittings has been a very effective method for controlling corrosion in injection well tubing in several floods.

Appendix 9.A is a checklist for troubleshooting oxygen-free water injection systems.

9.5.3 Sources of Oxygen in Gas Plants

Oxygen problems are not exclusively the concern of oil fields and water injection systems—plants have many oxygen problems.

Vacuum gathering systems are especially subject to accelerated corrosion due to the oxygen brought in from leaks in the incoming lines. The challenge of oxygen exclusion becomes one of first removing the oxygen.

A plant does not have to have a vacuum gathering system to have problems—any location where the pressure gets to or below atmospheric is a potential oxygen entry point.

---

6. A foot valve is a type of check valve. Both are used to prevent water flowing upward out of the water injection well to the water injection well pump.
7. A swing well is the supply water well an operator uses to adjust the water flow rate, allowing the other supply water wells to operate at near-constant flow rate.

Corrosion itself is not the only problem from air entry in a plant. At least one plant had to replace a major portion of the plant inlet piping. Not only was the oxygen corroding their system, but the nitrogen that was left from the air was diluting the hydrocarbon gas so much they could not meet the heat content (joules or British Thermal Unit [BTU]) requirement of their gas sales contract.

Oxygen contamination in amine plants accelerates the corrosion rate. Amines react with CO_2 and contaminants, including oxygen, to form organic acids (formic, oxalic, acetic, and others). These acids then react with the basic amine to form heat stable salts (such as bicine and thiosulfates), which accumulate in the amine solution and significantly increase the corrosion rate of carbon steel. They must be either removed by special distillation processes or the amine solution must be replaced.

Oxygen contamination in amine plants is typically a result of

- Using aerated make-up water to make lean amine solution
- Air contaminated lean amine and other liquids pumped into the system
- No, or insufficient, gas blanketing of the lean amine make-up tank
- Leaking seals in centrifugal pumps
- Air-contaminated inlet hydrocarbon gas

A plant reported that it was having corrosion and solids problems in their amine sweetening system. It turned out that much of their problem was due to oxygen. They did not have a vacuum gathering system, but the incoming gas carried some solids, and they were regularly having to change the elements in the inlet filter separators. There were no provisions to purge the air from the filter separators after each filter change, and air was carried right into the amine absorber tower. Apparently, the oxygen not only contributed to corrosion, but it oxidized dissolved iron, which in turn became insoluble iron sulfide and precipitated in the system.

Another plant solved some of their amine system problems by deoxygenating their make-up water for the amine system.

9.6 Oxygen Removal

Surface waters (e.g., river, lake, or seawater) used in some water floods as source water are aerated, and the oxygen must be removed to minimize corrosion problems. The most common methods of removing oxygen from aerated waters are mechanical deaeration (using counter-current gas stripping towers or vacuum towers) and chemical scavenging.

The choice of the removal method depends largely on the economics of the project. As a generality, stripping towers are used when copious quantities (levels in parts per million) of dissolved oxygen are to be removed and/or large volumes of water are to be handled. Chemical scavengers are used to remove small quantities (levels in parts per billion) of oxygen, and sometimes for removing the residual oxygen after tower deaeration. The final decision is an economic-technical one. Where large quantities of oxygen are present in the water source (e.g., 1–7 ppm), some operators have selected chemical oxygen scavengers to avoid the capital cost of adding a deaerator. However, the operating costs are high, and operators have a challenging task maintaining oxygen concentrations that are sufficiently low to avoid corrosion of carbon steel. The consensus is to maintain oxygen concentration ≤20 ppb to avoid corroding carbon steel.

9.6.1 Gas Stripping Towers

Gas stripping towers are the most frequent type used in oil field operations (they usually will be tray-type towers [Figure 9.8]). Natural gas is the most common stripping gas. However, nitrogen has been used when natural gas was not available, or when the available gas supply was unsatisfactory, for instance, high in H_2S. Gas stripping towers work by reducing the partial pressure of oxygen, which reduces oxygen solubility in the water. By increasing the partial pressure of the stripping gas, and decreasing the oxygen partial pressure, oxygen evolves from the water, or is "stripped out."

Gas stripping towers are (usually) properly designed if the manufacturer is supplied with sufficient information regarding the project. However, there are several practical considerations to remember when planning, installing, and operating towers:

- Flow meters should be installed on both the water and stripping gas streams. These are needed to evaluate tower performance, select the optimum gas-water ratio for the specific installation, and to allow varying the stripping gas feed with changing water and oxygen conditions. The latter is important because in practice, towers are seldom operated under design conditions, certainly not throughout the life of the project.
- The stripping-gas[8] source should be sweet gas. (Be aware that stripping gas high in CO_2 may contribute to sweet corrosion in the tower and downstream). Although sour gas (i.e., gas that contains H_2S) will perform the stripping function, H_2S and oxygen will chemically react to form elemental sulfur in the system (refer to Chapter 2, Section 2.4.3.4 for a discussion of the formation and corrosivity of elemental sulfur). Under severe conditions, elemental sulfur will be deposited on the tower's trays and eventually put it out of service. At best, only small quantities of elemental sulfur will be formed. However, they will be carried with the injection water and very effectively plug injection wells when combined with other possible plugging solids in the water. Removing elemental sulfur from the system and wellbore is difficult and must be done either mechanically or very selectively (and carefully) with a sulfur solvent treatment.
- The stripping gas should be free of oxygen, but as stated previously, some residue gas streams do contain oxygen.
- Low back pressure on the tower is critical. For example, the difference between operating a specific tower at 124 kPa (18 psig) rather that 345 kPa (50 psig) could mean a difference between 0.1 mg/L (100 ppb) and 1.0 mg/L (1000 ppb) oxygen in the deaerated outlet water. It is usually, therefore, more cost effective to operate a tower at a low pressure and pump water from it than to operate it at a high tower pressure without a transfer pump.
- Fluid level control in the tower is also critical for satisfactory tower operation. Varying levels can result in flooding the lower tray, entraining gas in the water, or "blowing" gas out of the water outlet. Some installations may require a surge tank for smooth operation with varying loads. If a surge tank is used, it should be designed and installed so that it does not become a stagnant area thus serving as an incubation chamber for microorganisms.
- A tower needs to be clean to operate at design efficiency. If the water contains solids that can settle out on the trays, periodic cleaning should be scheduled. Towers handling waters containing slime-forming bacteria will require periodic slug treatment with a biocide for microorganism control.
- Injection of chemical oxygen scavengers into the tower outlet is usually needed in most systems to meet the desired oxygen level ≤20 ppb. These chemicals are discussed in Section 9.6.3.

---

8. Stripping gas is the gas used to remove dissolved gas, such as O_2 or H_2S from water or oil. The stripping gas must be free of the gas it is removing, for example, natural gas, nitrogen, or even CO_2 can be used to remove O_2 or H_2S.

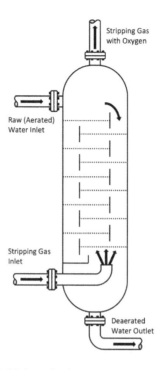

Figure 9.8 Schematic of tray-type gas stripping tower.

All gas stripping towers are not tray towers; sometimes packed column towers are used. However, they are usually harder to properly design and may be harder to operate efficiently than tray towers.

9.6.2 Vacuum Stripping Towers

Vacuum towers are used when there is an inadequate supply of stripping gas for a gas stripping tower, or when the gas supply is high in CO_2 or H_2S. Vacuum towers are continuously evacuated by vacuum systems, reducing the oxygen partial pressure to create a driving force for mass transfer of oxygen from the liquid to the gas phase. Water enters at the top of the deaeration tower and is distributed evenly across the vessel cross-section. The water trickles down over a bed of mass transfer packing, where it is broken up into thin films, thereby forming a large interfacial area between the water and the surrounding vapor phase. A vacuum system extracts all gases from the vapor phase, thereby lowering the partial pressure of oxygen in the vapor phase to near zero. This creates a driving force for oxygen molecules dissolved in the water to diffuse to the liquid surface and into the vapor phase, thereby reducing the concentration of oxygen in the water.

Vacuum towers are not as popular as gas stripping towers because they require packed columns rather than trays. Figure 9.9 is a drawing of a three-stage, packed-column, vacuum stripping tower using gas eductors on each stage. The eductors in turn feed a vacuum pump.

Historically, vacuum towers have had high maintenance costs because of the vacuum pump requirements. They also tend to quickly lose efficiency if the packing is fouled.

Most vacuum tower installations require the addition of a chemical scavenger after the tower to remove the last traces of oxygen to achieve specification water ≤20 ppb oxygen. In addition, many vacuum towers will also require use of an antifoam, particularly when deaerating seawater.

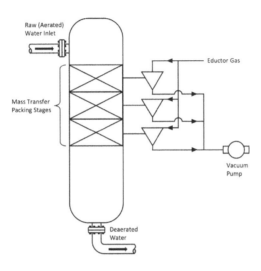

Figure 9.9 Schematic of a vacuum stripper tower.

9.6.3 Chemical Scavenging

The other method of oxygen removal is the use of chemical scavengers. These scavengers are usually one of the sulfite compounds.

9.6.3.1 Commonly Used Scavengers

The following are the four types of oxygen scavengers used most often by many operators:

- Ammonium bisulfite is used because of its higher solubility in brines and seawater than sodium bisulfite and sodium sulfite.
- Sodium sulfite is used in fresh water.
- Sodium bisulfite (or sodium metasulfite) is used in brines and seawater.
- Sulfur dioxide gas is used in fresh water and brine in areas where a packaged unit is provided.

Table 9.1 lists the reactions that occur when sulfites combine with oxygen. As with all chemicals, handling precautions are required, such as wearing protective clothing, safety glasses or a facemask, and providing adequate ventilation.

Table 9.1 Reactions of commonly used chemical scavengers with oxygen.

Scavenger (Sc)	Reactants	Reaction Products	Weight Ratio Sc:O_2
Ammonium bisulfite	$2NH_4HSO_3 + O_2$ →	$(NH_4)_2SO_4 + H_2SO_4$	6.2:1
Sodium bisulfite	$2NaHSO_3 + O_2$ →	$Na_2SO_4 + H_2SO_4$	6.5:1
Sodium sulfite	$2Na_2SO_3 + O_2$ →	$2Na_2SO_4$	8:1
Sulfur dioxide (gas)	$2SO_2 + 2H_2O$ → $2H_2SO_3 + O_2$ →	$2H_2SO_3$ $2H_2SO_4$	4:1

The weight ratio is the weight quantity of scavenger (e.g., milligrams per liter [mg/L]) required to scavenge a unit weight quantity of oxygen (in mg/L). The lower the number, the more efficient is the scavenger.

A discussion of each of the chemical scavengers follows:

- Ammonium bisulfite (NH_4HSO_3) is the most popular oxygen scavenger used in oil field systems. It is available as either a powder or a concentrated liquid (up to 70%), and most operators prefer delivery of the concentrated solution, ready to be pumped into their systems. Its advantages include its availability in highly concentrated solutions and its lower price. It does not react with atmospheric air and it can have low temperature storage capability, to −40°C (−40°F). Its disadvantages include its toxic and corrosive properties (because of its acidic pH) and a relatively slow reaction rate. Ammonium bisulfite can be catalyzed with cobalt or nickel metal salts to increase the reaction rate.
- Sodium bisulfite ($NaHSO_3$) is available as a powder with or without added catalyst or as a 20–35% catalyzed liquid solution. This is usually the fastest reacting scavenger for use in brines, particularly when used with a metal catalyst. Its advantages are its fast reaction rates and low-weight ratio (i.e., 6.5:1) (see Table 9.1). Its disadvantages are its corrosivity (because of its low pH) and its lower solubility in water than ammonium bisulfite. Solutions of sodium bisulfite cannot be winterized because of low solubility at low temperature.
- Sodium sulfite (Na_2SO_3) is usually obtained as a catalyzed powder and mixed onsite with water. It typically gives the fastest reaction rates of the sulfite scavengers in fresh water. It is a relatively low-cost material and can be shipped as 100% active material. Its main disadvantages are that it is inefficient (high-weight ratio of 8:1), it tends to cake as a solid, it requires mixing onsite, and the toxic and corrosive nature of its solution and fumes. It also has limited solubility in brine. Because sodium sulfite reacts with atmospheric air, storage tanks should be sealed from atmospheric exposure or have a floating cover.
- Sulfur dioxide (SO_2) is supplied as a liquefied gas under pressure. A liquid catalyst, such as a solution of $CoCl_2$ dissolved in water, is added upstream from the SO_2 injection if an increase in the reaction rate is desired. Because of the acidity of aqueous SO_2, special equipment is used for pumping and handling the SO_2. Its main advantages are cost-effectiveness and the relatively low quantities of gas used. One of its chief disadvantages is extensive personnel training is required in the hazards associated with its handling. Sulfur dioxide and its water mixture are hazardous and corrosive. Its low weight ratio (4:1) is offset by hazards in handling. Therefore, it is not commonly used as an oil field oxygen scavenger.

9.6.3.2 Interference with Oil Field Biocides

Scavengers based on sulfite and bisulfite interfere with common oil field biocides such as glutaraldehyde, 22-dibromo-3-nitrilopropionamide (DBNPA), and tetrakis(hydroxymethyl) phosphonium sulfate (THPS). For example, 2 ppm of ammonium bisulfite will remove 1 ppm of THPS. Interference with oil field biocides—reported for ammonium bisulfite and all sulfite-based scavengers—require the scavenger to be injected upstream of the biocide injection point with sufficient reaction time allowed.

Scavengers that are based on sulfite and bisulfite are reducing agents, so they react with sodium hypochlorite, which is an oxidizing agent used as a biocide in seawater and waterflood systems. Sulfite and bisulfite should be added first to remove dissolved oxygen, then the hypochlorite added far enough downstream to allow enough time for the oxygen scavenger to react.

9.6.3.3 Calculating Dosage of an Oxygen Scavenger

To calculate the dosage required for any of the oxygen scavengers described, the "dosage ratio" will be the above-weight ratio divided by the percent purity of the commercial product being used. The final dosage ratio (the actual feed ratio) should include a small excess to provide for fluctuations in oxygen content and to provide a residual that can be measured with easy-to-use field measurement kits. For example, sodium sulfite is typically used at a 10:1 ratio (that is, 10 mg/L of sodium sulfite for each 1 mg/L [1000 ppb] of oxygen to be removed). When very small amounts of oxygen (100 ppb or less) are to be removed, several times the theoretical amount of scavenger may be required to force the reaction to occur in a reasonable amount of time.

9.6.3.4 Injection Point Corrosion

It is important to know that all forms of sulfite and bisulfite oxygen scavengers are corrosive to carbon steel, and therefore the injection equipment should be made of stainless steel (type 316, UNS S31600). Injection nozzles or quills should ensure that the scavenger is introduced into the fluid flow and not allowed to impinge on the opposite wall of carbon steel piping to avoid injection point corrosion. Besides stainless steel, polyethylene and polypropylene are also acceptable construction materials for tanks.

9.6.3.5 Reaction (Holding) Time and Need for Catalyst

When using a chemical scavenger, remember that the scavenger reaction is not instantaneous. There must be sufficient holding time (or reaction time) designed into the system to allow the reaction to take place. The time required for the reaction is a function of the specific chemical, the dosage, the catalyst level, the water composition, and the temperature. For example, water at 60°C (140°F) requires almost 2 minutes to react with a scavenger that is not catalyzed. Oxygen scavenger reaction rate studies can be performed to select the scavenger, if required.[9] Reaction time in a dynamic system may be less than calculated.

Because it is very important that the chemical purchased is catalyzed, a catalyst may be required to start the scavenger reaction. Cobalt and nickel salts (typically chlorides) are the most common catalysts used. Catalysts can be supplied with the scavenger solution or injected separately directly upstream of the scavenger injection point. The metal catalyst ion (Co^{+2}) concentration in the water ranges from 1–100 ppb to be effective, although in practice a concentration of up to 5 ppm ensures that adequate amounts are present to overcome possible catalyst poisons. The need for a catalyst diminishes as water temperature increases and oxygen concentration increases. Catalysts may be required to deaerate raw seawater at 10–25°C (50–75°F), and even with a catalyst, a reaction time between 1–4 minutes is necessary (the shorter time corresponds to 40 ppb Co^{+2} concentration at a seawater temperature of 10°C [50°F]).

The use of catalyzed sulfite and bisulfite solutions in sour systems must be monitored carefully, because there is a risk that the metal catalyst will react with the H_2S to form an insoluble metal sulfide. (See the sidebar "Holding Time—Is It Really There?")

HOLDING TIME—Is It Really There?

Do not be fooled by believing there is sufficient holding time in your system without checking it. There was an instance where a small water flood used a supply water that contained some oxygen. Sulfur dioxide and catalyst were injected into the water source line to the supply tank as the oxygen scavenger. The scavenger was fed only when water was going into the tank. Sulfite residual determinations and some spot checks of dissolved oxygen content indicated that the water was successfully scavenged. However, corrosion coupons located at the plant discharge and failures in the manifold indicated otherwise. Investigation revealed that because of channeling in the tank, water was flowing directly from the inlet to the outlet while it was filling. There was insufficient reaction time (essentially zero) for proper scavenging. When the tank was full, and the supply line shut, there was sufficient time to complete the scavenging as the tank was emptying. During that time, the plant was handling specification water.

Because this was a cyclic condition, the spot checks of scavenger residual and oxygen content had indicated good scavenging, but the system was being corroded.

The problem was corrected by putting baffles in the tank so that water was held up long enough for the scavenger to do its job. Here is an instance when calculating holding time based on tank capacity and daily throughput did not work.

When scavengers are used, it is quite common to run residual sulfite tests as an indication that oxygen has been removed. Do not depend on these tests alone. Oxygen and sulfite can coexist in a system if the reaction has not reached completion.

9.6.3.6 Calcium Sulfate Precipitation

In water with high concentrations of calcium ion, a sulfite or bisulfite scavenger may add sufficient sulfate ion after reacting with oxygen to form calcium sulfate scale. Calcium sulfate is a hard, difficult to remove scale, Normally, only a small amount of oxygen is present, so there should be a small amount of sulfate ion present. However, the corrosion engineer should confirm with the chemical supplier if there is a potential for scaling. Gypsum ($CaSO_4 \cdot 2H_2O$) is the usual form of calcium sulfate scale encountered. Scale formation can be prevented using scale inhibitors.

9.6.3.7 Alternative Scavengers

Alternative scavengers are readily available, although they are principally used to remove oxygen from boiler feedwater systems. Because they were developed for higher temperature water systems, many require higher water temperature than might be commonly encountered in oil field operations to be effective. For example, hydrazine requires temperatures above 52°C (125°F) for the reaction. These scavengers include

- Hydrazine N_2H_4
- Carbohydrazide (N,N'-Diaminourea)
- N,N-Diethylhydroxylamine
- Hydroquinone
- D-Erythorbic Acid (D-Isoascorbic Acid)
- Methyl Ethyl Ketoxime

9.7 Dehydration of Gas

Corrosion in oil-water mixtures can be reduced or eliminated by dewatering the crude. Similarly, in gas systems, corrosion can be eliminated by dehydrating the gas. Because corrosion in gas systems is dependent on having moisture present, dehydration is the most practical way to control corrosion in these systems (gas transmission, gas lift, gas injection, and sometimes in gas gathering). Properly operated well site dehydration will often eliminate many field problems, particularly when handling gas with high CO_2 and/or H_2S levels. In fact, if it is dry enough, relatively pure CO_2 or H_2S does not create corrosion problems.

9.7.1 Water Dew Point

The water dew point of the gas is the measure of the degree of dryness. The water dew point is the temperature to which the gas must be cooled to reach water saturation. Any pipe metal temperature that reaches, or is colder than, the water dew point will become wetted. Dry gases have water dew points below the freezing point of water, even to a temperature as low as −40°C (−40°F).

The usual estimation made to ensure corrosion control is to "dehydrate the gas to a water dew point at least 5–6°C (10°F) below the coldest metal temperature to be encountered; however, in practice, a larger differential is maintained to account for fluctuations of gas moisture content, temperature (gas and atmospheric), pressure, and because many of the solids found in gas lines can be hygroscopic and will pick up water even from gas that is above its dew point temperature. Corrosion cells can be set up between the metal and these moist solids. Severe localized corrosion attack can result. However, when the gas's water content is less than 50% relative humidity, the gas will dry the solids and corrosion should not occur.

When determining the required water content, be sure to use the lowest wall temperature expected—*not* the lowest gas temperature. A cold area on the wall can be moist from condensation at that location even though the bulk gas temperature is not lowered. Joule-Thomson cooling[9] of the gas may drop the metal temperature locally. For bare pipelines located above ground, wind and cold atmospheric temperature can drop metal temperature.

9.7.2 Dehydration Versus Dehumidification

Maintaining humidity control (dehumidification) has been used in some cases; however, dehydration is a more reliable approach in most gas systems. Dehydration may be designed to only meet hydrate control conditions and, with just a bit more dehydration, corrosion control can also be achieved.

Dehumidification is not as thorough as dehydration in removing water in gas processing. However, dehumidification of small, enclosed spaces has worked well in controlling atmospheric corrosion where air exchange is limited. Relative humidity below 60% is achieved using air driers. An example where dehumidification has been used is the void space in offshore floating production storage and offloading vessels. Dehumidification is also used to prepare steel surfaces to be coated if the metal temperature is colder than the atmospheric water dew point.

9.7.3 Hydrate Control Is Not Corrosion Control

If methane hydrate deposition is to be controlled by direct methanol or glycol injection, do not expect corrosion control from that approach. CO_2 is soluble in methanol-water and glycol-water mixtures and will lower the pH just as in water alone.

9.7.4 Gas Dehydration Using Glycol

Glycol dehydration using a contactor (trayed or packed absorber column) is the most common method of field dehydration used for natural gas. Water is removed from the gas stream into the glycol solvent, typically triethylene glycol (TEG), by a process called absorption. In this absorption process, water in the gas stream is dissolved into a relatively pure TEG solvent stream as contact is made in the multistaged contactor (absorber) tower. The glycol dehydration system also includes a reverse process known as stripping (regeneration), where the water in the glycol solvent is transferred into the gas phase using heat and, in some cases, stripping gas. The terms "regeneration," "reconcentration," and "reclaiming" are also used to describe the stripping or purification process because the solvent is recovered for reuse in the absorption step.

9.7.4.1 pH Control

When using glycol dehydration for gases with a high CO_2 and/or H_2S content, do not overlook the need for corrosion control in the glycol circulation and regeneration systems. The usual method of corrosion control is to control the pH and stream quality of the circulating glycol. CO_2 and H_2S tend

---

9. In thermodynamics, the Joule–Thomson effect describes the temperature change of a real gas or liquid when it is forced through a valve while keeping it insulated so that no heat is exchanged with the environment. This procedure is called a throttling process or Joule–Thomson process. (Definition courtesy of Wikipedia <http://www.wikipedia.com>.)

to accelerate glycol degradation and can lead to creation of organic acids and solid particles in the system. The pH is usually maintained at around 7.0–7.5; a lower pH cannot control corrosion and accelerates the decomposition of glycol, and a pH >8.0–8.5 tends to make the glycol foam and emulsify. Control of pH is achieved by batch treating alkaline neutralizers such as borax or ethanolamines (such as monoethanolamine [MEA], diethanolamine [DEA], or triethanolamine [TEA]). Usually, 3 grams TEA per 10 liters glycol (0.25 pound of TEA per 100 gallons of glycol) will adequately raise the pH. Monthly analysis of the recirculating glycol for thermal decomposition products, chlorides, and pH is used to decide when pH adjustment is needed, or if the glycol needs to be replaced.

9.7.4.2 Filtration

"Dirty" glycol (i.e., glycol carrying solids) can cause system plugging and foaming as well as corrosion problems. Solids that fall out and collect in the line can lead to localized corrosion (under-deposit corrosion). Filtration is the usual method for solids removal. Many glycol systems use side stream filtration. Full stream filtration is costly but is a surer way to remove solids. Historically, partial filtration will not satisfactorily keep the system clean, particularly a very dirty stream.

9.7.4.3 Reflux Tower Corrosion

The glycol regeneration package is also susceptible to corrosion, particularly in the reflux tower (which sits above the reboiler) and the tower overhead system. The reflux tower overhead system is exposed to water, CO_2 and H_2S from the gas, and any short chain organic acids from glycol degradation. As the water condenses, this stream can be quite corrosive. Corrosion-resistant materials, such as stainless steels, are often specified for the reflux tower and overhead systems when dehydrating gas with high acid gas content.

9.7.5 Gas Dehydration Using Molecular Sieve

Another form of dehydration uses packed towers filled with molecular sieve ("mol sieve") pellets of aluminum silicate. Mol sieve units can achieve lower dew points than glycol dehydration and are used to prepare gas for cryogenic operations. Wet gas conditions exist before the inlet to the mol sieve vessels. Internal corrosion may occur in the wet gas inlet section to the mol sieve units, depending on gas temperature and composition. Elsewhere, the mol sieve unit is not wet and therefore does not experience internal corrosion. Regeneration of the mol sieve media requires heating the pellets to 175–300°C (350–570°F). Thermal fatigue cracking has been reported at the bottom of the mol sieve vessel due to the frequent temperature cycling and constraint of the inlet piping entering the bottom of the unit.

9.8 Control of Deposits (Cleanliness)

Corrosion problems in many systems can be avoided, or at least minimized, by keeping the surface clean. Examples were given in earlier chapters when scale, deposits, and debris can cause accelerated corrosion in many systems by trapping water against the metal surface, creating differential concentration cells, providing growth sites for sulfate reducing or acid producing bacteria, or by causing "hot spots" on heated surfaces and thus differential temperature cells. For example, refer to Chapter 3, Section 3.5.4. Operating and maintenance procedures should be set up to keep the equipment clean (for example, routine pigging of gas lines to remove water and deposits (often with corrosion inhibitor added), routine pigging of water injection lines and flushing of vessel bottoms, and periodic cleaning of tanks and vessels).

Because of the effect on operations caused by removing equipment from service for cleaning, routine vessel entry for cleaning is rarely practiced unless required for statutory internal inspection or because the quantity of deposits has adversely affected the operability of the equipment. Actual cleanout of vessels (that is, getting in and removing the solids and sludge) is required in some systems. Robotic tools have been developed to remove sludge from tanks and drums.

Oil-water separators that build up substantial amounts of sand are candidates for installing sand jetting equipment to allow for online cleaning.

9.8.1 Pigging

The reasons to pig a pipeline are to

- Remove sand, paraffin, sludge, scale, sessile bacteria colonies, water, and elemental sulfur deposits (corrosion and flow assurance)
- Remove the above mentioned deposits in preparation for inline- (or "smart" pig) inspection
- Reduce friction and pressure buildup in the pipeline (flow assurance)
- Control the quantity of methane hydrates (flow assurance)
- Batch treat corrosion inhibitors and biocides
- Improve the performance of the Chemical Management Program (see Chapter 8, Section 8.15)

Other reasons for pigging (not discussed in this book) include flushing after hydrotesting to remove water and debris, preparing a pipeline for abandonment or hot work repair, separation of two products in finished product pipelines, and applying an in situ coating.

9.8.1.1 Production Pipelines Benefiting Most from Maintenance Pigging

The following pipeline systems have the greatest potential benefit from maintenance pigging:

- Three-phase oil (full well stream) production pipelines
- Wet oil pipelines (two-phase oil and water) with substantial BS&W[10]
- Produced water pipelines
- Export crude (processed oil pipelines), with BS&W specification
- Gas pipelines with wet gas (containing equilibrium water) and dewpoint-controlled (dehydrated) dry gas
- Seawater pipelines

---

10. BS&W is Basic Sediment and Water. BS&W is measured from a liquid sample of the production stream. It includes free water, sediment, and emulsion, and is measured as a volume percentage of the production stream. BS&W is used to specify the maximum allowable sediment, water, and emulsion allowed in a crude sales contract. Most contracts specify BS&W below 1% or 0.5%.

9.8.1.2 Maintenance Pig Selection

There are two main considerations for pig selection:

1. What type of pig or pig sequencing is needed?
2. What is the best frequency to pig a pipeline?

Selecting the right pig is an art, although Table 9.2 provides guidelines. There are many exceptions to Table 9.2. In addition, rarely is only one type of pig used, and this is described later in this chapter in Section 9.8.1.5.

Table 9.2 Guidelines for selecting a cleaning pig, where
E = Excellent, G = Good, F = Fair, P = Poor, and NR = Not Recommended.

Pig Type	Sand/Sludge	Scale	Sessile Bacteria	Paraffin	Water
Polyurethane (PU) foam/swab	F	P	P	P	F
PU-external wrapped [a]	G	P	F	P	F
PU cast flexible	G	F	F	P	G
PU cup on steel mandrel ("super pig")	E	F	F	P	E
Bidirectional scraper discs	G	G	G	F	E
Scrapers (PU plow and blades)	P	F	F	E	P
Wire and pencil brush	P	G	E	P	P
Gel	G	P	P	P	E
Spheres	P	P	F	P	F
Studs	NR	E	P	NR [d]	NR
Spray pig [b,c]	G	G	E	G	NR

Notes:
a. External wrap of harder PU than pig core.
b. Specialty pig that directs high-pressure bypass flow onto the pipe wall.
c. Some are proprietary pig designs that apply chemical under pressure to the top of the pipeline.
d. Effective at breaking hard wax.

Pigs come in assorted sizes and styles. Generally, operators do not depend on spheres to remove solids—they are better for dewatering. Probably the most common pigs for solids removal are some version of the foam pigs. Multiple pigs can be used for cleaning, and foam pigs tend not to get hung up in the line.

Different pig designs are available to apply chemical treatment. Pigs that are most favored for applying biocides and corrosion inhibitors are cast polyurethane, polyurethane cups on a steel mandrel, and bidirectional discs. Wire and pencil brush pigs can be used in pitted pipelines to ensure that the biocide penetrates into the pits.

9.8.1.3 Progressive Pigging

A heavily fouled line requires progressive cleaning runs that begin with the most flexible, least aggressive pig (a polyurethane foam or swab), followed by progressively more aggressive pig runs. Pigging continues until there is only a small quantity of a dirty water envelope in front of the last pig as it nears the pig receiver, and few to no deposits remain on that pig after the pig trap is open. Soft urethane foam pigs are used initially to ensure there are no inline obstructions, large volumes of sediment, etc., especially in a pipeline that has not been cleaned recently or has an unknown but potentially high amount of deposits or flow obstructions. Soft pigs often get destroyed when they encounter obstructions in the pipeline.

After the foam pig, more aggressive pigs apply greater force perpendicular to the pipeline wall and are equipped with a combination of cups, disks, and brushes. Leaf-spring brush pigs or studded pigs are used to scour or hammer the pipe wall and remove hard deposits or scale. Brush pigs are used to clean corrosion pits prior to applying a corrosion inhibitor or biocide batch. These aggressive pigs should not be used at the beginning of the cleaning campaign because they risk sticking the pig in the pipeline.

9.8.1.4 Chemical Batching Pigs

A batching pig can be a cup pig, a bidirectional disc pig, or a combination of a disc and cup arrangement (see Chapter 8, Figure 8.4). There are proprietary pigs that spray chemical to the top of the pipeline, as might be considered to combat top-of-the-line corrosion problems.

9.8.1.5 Maintenance Pigging

Routine pigging is often performed using single pigs or "pig trains" that consist of two or more steel mandrels that contains a combination of cup, discs, and wire brush elements in one pig run. If pigging is performed frequently enough, the quantity of solids in the line is generally known, and there is little concern for sticking a pig if the same type of pig train is used. If there is a tight bend radius in the pipeline, a solid polyurethane core can be used instead of a steel mandrel.

Below are two examples of the construction of routine maintenance pig trains (the pipe (|) symbol is used to separate the components for readability).

Bidirectional (Bi-Di) cleaning pig construction:

gauge plate | bi-di disc #1 | bi-di disc #2 | bi-di disc #3 | steel mandrel | bi-di disc #4 | bi-di disc #5 | bi-di disc #6 | gauge plate

Cleaning Pig with wire brush:

bi-di disc #1 | bi-di disc #2 | bi-di disc #3 | steel mandrel | wire brush | bi-di disc #4 | bi-di disc #5 | bi-di disc #6

Figure 9.10 shows a similar design with two bi-di discs in front of the wire brush and two behind it. Prior to each inhibitor application, a brush pig could be run to prepare the pipe surface for bonding of the new inhibitor. To prevent damage to the inhibitor film a wire brush pig should not be used for batch treating.

Placement of the different pig train components is also an art. Some practitioners believe that a wire brush should not be the first component because it will get gummed up with waxy paraffinic deposits and be rendered less effective. However, vendors do sell arrangements with the brush component first, followed by discs or cups, with another brush behind. Just about every arrangement can be assembled on a mandrel, and there are no hard and fast rules about the most optimum arrangement to recommend. An all-cup pig is adequate for liquid removal in a gas system. In systems with solids, a cup-disc pig combination will be more effective than cups alone in moving debris.

Figure 9.10 Example of a cleaning pig with a two yellow bidirectional discs, a steel mandrel with wire brush, and two additional yellow bi-di discs.

Operators usually pig pipelines between monthly and quarterly, although pigging frequency varies with the system and service. Ideally, pigging is done often enough so that the line stays essentially solids-free. In other words, only a small quantity of solids accumulates between each routine pig run. There is an envelope of dirty water ahead of the pig's arrival. This dirty water persists for a while, and then the water clears up after the pig is in the trap. The sequence is "clean water, followed by dirty water for some time, followed by clean water." The pigging frequency can be optimized by measuring the length of time the dirty water envelope is in front of the pig when it nears the pig receiver. The longer the time duration of the dirty water envelope, the less effective pigging may be. (The pig could be the wrong type—too much bypassing—or the interval between pig runs is too long.)"

Another way to optimize pigging frequency is to measure the quantity of deposits remaining on the pig or in the pig trap after the pig trap is open and the pig is removed. The higher the amount of deposits, the less effective pigging may be. To increase accuracy of measurement, the pig trap should not be flushed until after the amount of deposit has been determined. This technique cannot be used if safety procedures require flushing the trap if the pig run was used to distribute a biocide through the pipeline. Residual biocide in the pig trap may be hazardous to personnel.

9.8.2 Flushing

Flushing is sometimes used to clean out piping and vessel bottoms and this will work well if the solids are loose (i.e., if they are not compacted or stuck to the walls).

When a piping segment or vessel is flushed by circulating water through it, the circulating water flow rate should be surged. That is, the valve should not just be opened for a constant flow rate. The velocity has to be varied by opening and closing to keep the solids moving. A reverse-flow valve manifold allows the flow to be reversed to dislodge solids that are up against obstructions.

The solids along the bottom are at equilibrium with the normal flow rate. Each time the rate is increased, some more of the solids will be moved. But in many cases, they will just move a bit farther along the bottom. By changing rates (and thus velocities) as rapidly as possible, a surging action is created that will keep the solids moving.

In many cases, flushing may not be sufficient. Manual clean out may be required in vessels. Lines may have to be pigged.

9.8.3 Filtration

Filtration is required in some systems to remove suspended solids from the fluids. The use of filters to maintain glycol quality in glycol regenerative systems was mentioned in Section 9.7.4.2. It is very important to remove the solid products of glycol degradation. In sour gas dehydration, large quantities of suspended iron sulfide deposits form and require filter maintenance.

Filtration is also used in amine sweetening systems, and the same comments apply here as in Section 9.7.4.2: use full stream filtration, not partial side stream filtration, for both amine and glycol streams.

Water injection systems are where filters are usually used. Most surface water supplies (seawater, lakes, rivers) will require filtration, but most subsurface water supplies do not. Filtration and the design of filter systems can be very involved, and a thorough discussion of it is beyond the scope of this book. Many texts, however, have information on filtration such as what is found in Reference 10.

Filters are also used offshore in seawater injection systems. Raw seawater is pumped to the topside facility where it is filtered and treated with biocide (usually continuous sodium hypochlorite and batch treatment of nonoxidizing biocides). In water systems, filters accumulate bacterial growth and require treatment to avoid becoming sources of MIC. In produced water systems, oil carryover tends to plug filters. At best, this requires very frequent backwash. At the worst, the filters will channel (that is, flow will be uneven with most of the water flowing through channels in the filter media and the solids pass right through).

There are many types, styles, and configurations of filters. The ones most common in oil field applications are either variations on the so-called "rapid sand" filters, both down-flow and up-flow styles, or one of the types of cartridge (or "sock") filters. The filters, using many materials other than sand as the filter media, are usually installed at a central water plant. Activated carbon is used as a filter medium to remove organic matter. Cartridge filters may be installed at the water plant, but they are often installed at the wellhead of water injection wells. Wellhead filters should be used as "emergency filters" (that is, they serve to protect the water injection well in case of upsets at the water plant).

When filtration is needed, sizing of the filter is very important. Generally, it is good practice to install filters rated at greater throughput than is required for the volumes of fluid to be filtered. Many manufacturers rate their filters based on tests using clean water with new media. In those cases, the throughput of dirty water will be much less than the rated capacity.

Although the installation of filters is intended to aid corrosion control by removing materials that can create harmful deposits, if not carefully designed and conscientiously operated, filters can contribute to corrosion. The filter media in "sand" filters present an ideal location for microorganism growth, so extra precaution should be exercised to avoid contamination. It is easy to decide to "clean-up" some wastewater through a filter in a water flood. If the wastewater is from a surface pit or is transported in a truck that has carried water from a pit, the odds are the water is infected with at least sulfate reducing bacteria (SRBs) and possibly other microorganisms. It can be very costly, difficult, and time consuming to sterilize a large filter system.

Any filter, and particularly cartridge filters, should be equipped with vents. When filter cartridges are changed, and the filter shell is closed, it is full of air. If this air is not vented as the filter is filled, a sizable volume of air (oxygen) can be introduced into the injection system, potentially resulting in a fire and/or explosion problem in the presence of hydrocarbon gas.

9.9 Control of Microbiologically Influenced Corrosion (MIC)

The next method of controlling the corrosive environment is control of MIC (i.e., the control of microorganisms, or "bacteria," often referred to in the field simply as "bugs"). Chapter 2, Section 2.5 addresses the basics of MIC, and Chapter 3, Section 3.5.4 addresses the types of localized corrosion caused by MIC.

As mentioned in Chapter 2, the most common microorganisms that create problems in oil field operations are sulfate-reducing bacteria (SRB) and acid-producing bacteria (APB). The SRBs are anaerobic bacteria (they thrive where there is no air) that use sulfates in their metabolism—the byproduct of which is H_2S. Although sulfate reducers are anaerobic, they are not necessarily killed by exposure to air. They merely become dormant in a hostile environment and can become active when placed in a friendly environment.

Bacteria can be found almost anywhere—a strain could probably be cultured off the floor; this is why the mere presence of SRBs and APBs does not necessarily mean control is required. There are two important considerations with MIC:

- How active are they?
- What problems, if any, are they causing?

Activity is often measured as population. There are two types of population: free-floating bacteria and attached bacteria. MIC is associated with attached bacteria, referred to as "sessile" bacteria. Sampling of free-floating ("planktonic") bacteria taken from water samples does not provide the corrosion engineer with an accurate understanding of sessile bacteria population. The detection and evaluation of SRBs and APBs are discussed in Chapter 10, Section 10.3.4.

Various problems may be caused by sulfate reducers:

- Souring an otherwise sweet system (production reservoir zone, produced fluids, and surface facilities), thus creating corrosion problems, possible well plugging due to solids (iron sulfide), and potential safety concerns due to the presence of H_2S gas formation. (Refer to Chapter 2, Section 2.4.3.2 for a discussion of sweet systems and sweet corrosion.)
- Plugging of injection wells due to iron sulfide corrosion byproducts.
- Souring of the injection reservoir (common when using surface waters).
- Failure of susceptible metallic components by sulfide stress cracking (SSC) and degradation of susceptible polymers by exposure to H_2S.
- Localized corrosion in separators, pipelines, storage tanks, and other equipment containing water, and operating under low velocity.

As far as producing wells are concerned, routine control (use of biocides) downhole is not usually required until proven to be necessary when one or more problems appears. MIC is found mostly in equipment handling produced water, starting with separators and following into the water injection systems. It also occurs in seawater injection systems. Biocide injection is prevalent in these systems. It is also good practice to control SRB in packer fluids, because they can add H_2S to an otherwise sweet fluid. In most cases, the corrosion inhibitor dosages used in clear brine packer fluids will effectively control the SRBs, especially if the inhibitor used is a water-soluble, quaternary amine, which has biocidal, as well as corrosion-inhibitive, properties. Batch biocide injection combined with maintenance pigging is a common preventative measure in pipelines with a water phase.

9.9.1 Water Injection Systems

As far as water injection systems are concerned, most systems using surface water as a water supply will require microbiological control. Most systems with underground source water do not have SRBs—unless the water has been contaminated on the surface by water handling practices.

Generally, even where sulfate reducers are native to a water source, design and operating practices can avoid problems, and thus, routine biocide treatment is not required. Below are a number of examples of this:

- Design systems so the water keeps moving. Because SRBs like dead areas, it is vital that water moves at velocities fast enough to prevent various problems. A liquid velocity of 1–1.5 m/s helps to control deposition given that deposition occurs mostly at a velocity of <1 m/s. However, a velocity of at least 3 m/s (10 ft/s) is needed to prevent bacteria from receiving the necessary nutrients and provides the best control of deposition.
- Avoid piping with no flow (dead lines, dead legs, etc.).
- Keep systems clean by flushing, pigging, etc., or with periodic shutdowns and clean outs.
- Do not inject wastewater from pits or tank bottoms back into a water injection system (use separate disposal).

Most times, these steps are all that is required. In some cases, however, occasional use or routine treatment with a biocide is required to control bacteria in a contaminated system. When biocides are required, biocide selection is very important, and the method of application is also important and should be matched to the specific system being treated and its problems.

Biocide treatment is expensive; therefore, the problem must first be defined to determine if biocide treatment is required. If it is, it has to be done correctly. The proper physical, mechanical, and chemical procedures should be initiated, and then a monitoring program to follow the progress should be created and initiated.

Seawater injection systems require microbiological control. Many operators inject sodium hypochlorite continuously into the seawater at a low free-chlorine residual of about 0.2 ppm. Operators also supplement hypochlorite with periodic (approximately monthly) injection of a nonoxidizing biocide and alternate this biocide with a different one to ensure that bacteria do not build resistance. Seawater filters are a breeding ground for bacteria, and they should be exposed to hypochlorite and biocide injection.

Because SRBs convert sulfate (SO_4^{2-}) to sulfide (S^{2-}) ions, some operators of seawater injection systems have installed ion exchange units to remove sulfate ions and prevent reservoir souring. No one really knows how little sulfate ion is required (the lesser, the better), and eventually, even systems with sulfate removal units require biocide treatment.

There are things to remember about biocide treatment:

- Halfway measures are expensive and ineffective.
- Chemical suppliers' data on biocide effectiveness may be optimistic and therefore may not be representative of actual field conditions.
- Most test data represent the performance of a biocide against a specific species of microorganism. This is because bacteria are cultured in media that promote the growth of a specific species.
- Bacteria samples should be taken from deposits, or scraped from corrosion coupons or monitoring devices, because these represent sessile populations, the kind that promote MIC.
- Better biocide performance data is becoming available. DNA and other modern molecular biology approaches are providing advanced understanding of the bacteria population and biocide performance (refer to Chapter 2, Section 2.5.5).

9.9.2 Reservoir Souring

Reservoir souring is defined as the phenomena of H_2S being produced in a reservoir that did not initially contain it. Reservoir souring is generally associated with seawater injection offshore. It results from sulfate reduction caused by microorganisms that may be indigenous to the reservoir or introduced through injected water. Reservoir souring affects

- Safety
- Cost of hydrocarbon processing
- Materials selection and corrosion management
- Meeting H_2S sales specification limits for produced hydrocarbons

There are no industry standards that predict the quantity of H_2S that could form because of reservoir souring. The science behind model-based predictive techniques is not well understood and approaches vary. Many factors affect the extent of souring, such as reservoir temperature around the injector and producing wells, the total dissolved solids and nutrient content of the injection water, the presence of competing microorganisms, the type of reservoir (carbonate or sandstone), and the ability of the reservoir to scavenge H_2S.

In the initial design of a new offshore project that involves seawater injection, the operating company must consider the potential for reservoir souring and assess the need to control it, and/or to specify higher-cost lower-strength and SSC-resistant materials, even though they may never be required. The addition of sulfate removal units and nitrate injection (to feed competing microbes in the reservoir so they dominate sulfate reducing bacteria) have been considered.[11] However, there is insufficient data to assess the level of effectiveness of these process-based mitigation schemes. The operating company is left to make its best assessment of whether or not to specify higher-cost materials of construction that are resistant to potential hydrogen-induced damage (SSC, HIC, etc.; refer to Chapter 3, Sections 3.10–3.13, inclusive).

If a field is already in operation and was constructed with carbon-steel casing, tubing, risers, pipelines, and pressure vessel equipment with the assumption that the reservoir is sweet, and over time the reservoir sours, the operator is confronted with determining the risks associated with failures of major equipment. Normally, a steadily increasing concentration of H_2S gas is first observed in the vapor space of water-handling equipment. Flexible risers used in deep water contain high strength steel reinforcing wires that are very susceptible to hydrogen damage (if they were not designed for H_2S), and failure of flexible pipe due to reservoir souring has occurred. High strength steel casing grades P-110 and Q125 are also susceptible.

Before reservoir souring causes sufficient H_2S to be generated and reaches the point where equipment is at risk, measures can be taken to scavenge H_2S. Below are some examples of scavenger types.

- Triazine is a saturated amine solution (trimethyltriazine or ethanolaminetriazine).[12]
- THPS is a biocide enhanced with nonfoaming polymers helps to penetrate biomass.[13]
- Acrolein has also been used to scavenge H_2S and remove iron sulfide plugging.[14]

The H_2S scavenger injection point depends on field test results at various locations. Locations that have been reported include the injection water and the topsides high-pressure gas-lift gas compression system. All scavengers require following unique safety procedures that are available from their manufacturers or suppliers.

9.9.3 Hydrotest Waters

Hydrostatic pressure testing (known as "hydrotesting") is the process of placing a noncompressible fluid (typically water) into a new or repaired pipeline or pressure vessel and then pressurizing the equipment above the intended operating pressure to check for leaks and equipment integrity.

The quality of hydrotest water varies from a municipal supply to open seawater. Water quality may differ in the concentration of dissolved oxygen, chloride concentration and total dissolved solids, in the types and concentrations of bacteria, and whether or not the water is filtered, which affects the quantity of suspended particles. The most desirable sources of water are demineralized water, high purity steam condensate, and potable water.[15] The next desirable source is seawater (filtered and taken more than 15 m (50 ft) above the seabed and 15 m (50 ft) below the sea surface. The least desirable source is taken from rivers, lakes, and finally brackish water. While fresh water is normally preferred, it is often not available or too expensive, particularly in offshore operations where seawater is readily available.

Hydrotest water can result in internal localized corrosion of the pipeline or equipment that was hydrotested. The mechanism most responsible for corrosion is MIC from SRB and/or acid-producing bacteria (APB); oxygen corrosion may also play a significant role. The length of time the water remains stagnant in the equipment has a large effect on the corrosion that may occur. Most hydrotests are completed within one week, often within 72 hours. In some cases, the water can reside within the pipeline for a month. Commissioning schedule delays and other factors can result in longer exposure; refer to Section 9.10.1, below, for more information.

The need for corrosion protection during hydrotesting has been recognized, and the most common corrosion control method is to use one or more chemicals added to the hydrotest water.[15-16]

9.9.3.1 Chemical Treatment of Hydrotest Water

A chemical triad is necessary consisting of water-soluble (in the type of water used for hydrotesting) oxygen scavenger, biocide, and organic corrosion inhibitor, at concentrations of approximately 100 ppm, 200–300 ppm, and 100 ppm, respectively (concentrations are expressed as active ingredient and are shown for seawater). Although the hydrotest water is aerated, the pipeline or equipment to be tested is closed, so additional sources of oxygen are absent. When aerated water contacts the steel surface, corrosion reaction begins. Over time, corrosion will consume dissolved oxygen and the corrosion rate will drop. However, damage of the new pipeline has been initiated. Damage does not occur uniformly, but in the form of pits because of the formation of oxygen concentration cells. "Oxygen levels fall to concentrations in the low parts per billion level (<50 ppb) within 18–48 hours so pit depth is expected to be shallow if it occurs."[16] Because oxygen concentration corrosion is self-limiting as the oxygen concentration depletes as corrosion progresses, some operators choose to omit adding oxygen scavenger.

Compatibility of Biocide and Oxygen Scavenger

If a scavenger is used, compatibility with the biocide, as described in this Section 9.6.3.2, must be attained. The two most commonly selected biocides are glutaraldehyde and THPS, and both react with sulfite and bisulfite oxygen scavengers. When hydrotesting subsea pipelines, there normally is no way to mix chemicals in the pipeline, so they are added to the hydrotest water during filling. The oxygen scavenger should be added before the biocide in a way that allows for reaction time. This may not be practical, and a compromise is necessary: omit the scavenger, or allow the scavenger and biocide to react and add enough of both to compensate.

Need for a Corrosion Inhibitor

Because hydrotest waters typically have varying salinities and may contain compounds such as CO_2, organic acids, and sulfur compounds, the use of corrosion inhibitors should be considered to prevent damage to the line. Inhibitors are most needed if the water is to remain in the line for more than several days and the line is not pigged and cleaned prior to being placed in service. Corrosion inhibitors used for this function are typically water-soluble, oil-dispersible amine-based inhibitors that evenly disperse through the hydrotest brine and coat the inner wall of the equipment. Corrosion inhibitors are not effective in preventing corrosion by oxygen. In many systems where bacterial attack is not a consideration, a combination treatment of corrosion inhibitor and oxygen scavenger has been effective. Because oxygen and bacteria are a threat in most instances when hydrotesting with seawater, some operators elect to add a compatible biocide and oxygen scavenger without a corrosion inhibitor for shorter duration hydrotesting. In rare circumstances where water discharge is not a consideration, a water-soluble quaternary amine can be used as both an inhibitor and a biocide.

Chloride Content of Hydrotest Water

The chemicals added must be compatible with the type of water selected. The chloride content of the water plays a significant role in this consideration because some chemicals are either not sufficiently soluble in seawater or sufficiently soluble in fresh water.

Regulatory Requirements on Discharge

Any water that contains chemical additives may be subject to regulatory control prior to being discharged to an open body of water. Biocides are particularly addressed by regulations. The ideal hydrotest biocide is one that is fast acting and effective at moderate use concentrations but has a short enough half-life such that the hydrotest water disposal will not require biocide deactivation prior to or during discharge. All the chemicals selected to treat he hydrotest water must be compatible with environmental discharge requirements. Meeting regulatory requirements has the largest impact on chemical selection, even more than chemical effectiveness.

Gluteraldehyde. Some biocides are permitted to be discharged into surface water within regulated concentration limits. The most common of these is glutaraldehyde[17]; regulations vary, but a discharge limit of 10 mg/L active ingredient is typically permitted for glutaraldehyde. It has been shown to be biodegradable as defined by different and widely accepted biodegradation protocols. Glutaraldehyde has been shown to leave significant portions (>50%) of aquatic microbe populations viable after they were exposed to 30 mg/L dosages. Because glutaraldehyde is typically dosed at 150–200 mg/L, and its reaction with bacteria is relatively fast; the discharge dosage of glutaraldehyde after hydrotesting could result in a significantly toxic situation in most cases. Therefore, the operator must ensure maintaining the discharge concentration within the regulation.

THPS. Toxicity testing indicates that water containing less than 4 ppm THPS will generally pass the 48-hour toxicity requirements for the Gulf of Mexico. Permissible discharge limits change with revisions to regulations, and the user should check the regulations that set rules for the affected body of water. The chemical supplier or manufacturer can provide toxicological data and predict biodegradability for different systems.

If the concentration of THPS could exceed the discharge permit, hydrogen peroxide can be used to neutralize THPS prior to discharge (approximately 17 ppm hydrogen peroxide is required per 100 ppm THPS [active]).

No Chemical Additive. Because of regulations, many operators choose not to add any chemicals to hydrotest water for short duration exposure. Some operators use a 7-day guideline, others a 30-day guideline. There is no evidence that corrosion damage requires 7 days, or 30 days, or more to initiate, so opting out of injecting chemicals is a risk that some operators believe they can manage. The shortest exposure time is less prone to initiating corrosion. A sounder approach is to select a compatible biocide and oxygen scavenger that meets discharge requirements.

Draining and Drying After Hydrotesting. There is no way to tell how many times leaks that occurred within a year or so after hydrotest could have been prevented by killing the bacteria at the time of the test. After hydrotesting, a carbon steel pipeline can be dried by first pigging, and then flowing a dry gas through the line. Warm, dry gas or methanol can be used to dry relatively small-volume equipment and can be considered for stainless steel heat exchanger tubes, piping, tanks, and pressure vessels. Pitting failures of austenitic, ferritic, and martensitic grades of stainless steel equipment have been reported many times soon after hydrotesting (refer to Chapter 3, Figure 3.10). In welded austenitic stainless steel, corrosion pits initiate at welds because of the action of chloride ions and MIC. These pits have a pinhole opening at the surface of the pit below which a large cavern is formed inside the metal. The cavity is generated by MIC and the build-up of acidic components such as ferric chloride inside the cavern (for an example, refer to Chapter 3, Figure 3.38). Draining and drying immediately after hydrotesting is critical with stainless steel equipment. Low points must be drained.

9.10 Other Methods and Environments

This section addresses protecting equipment that will be out of service for an extended period from corrosion. It also addresses controlling corrosion in wet acid gas systems and wet gas pipelines using pH control and pH stabilization. Controlling pH to control corrosion in water cooling and heating systems is also described.

9.10.1 Mothballing, Layup, and Decommissioning

There are different methods used to preserve equipment during short or extended outages. They are, in order from longest to shortest length of time out of service are decommissioning, mothballing, dry layup, and wet layup.

9.10.1.1 Decommissioning

Equipment that is decommissioned means it is no longer in service and will remain that way. Decommissioned equipment should be physically removed from service. Often, if left in place and unused, decommissioned equipment creates dead legs in the piping system, which can result in internal corrosion. A good example of this is a bank of heat exchangers that are no longer all required. (Dead legs are described in Chapter 1, Section 1.5.5.3.)

9.10.1.2 Mothballing

The intent of mothballing is to preserve the equipment that will be removed from service for future use. Mothballing of piping, vessels, and equipment is important if they will be out of service for a long time, especially if the equipment is exposed to a coastal or a marine atmospheric environment. Often, equipment is shut down when not needed, as during layoffs and other times of reduced staffing, and no preservation steps are taken. Severe internal and external corrosion can result, leaving the equipment inoperable when it is needed in the future. The details of equipment preservation vary depending on the item, the materials of construction, and the fluids it has been handling. For example, equipment that was in sour service may contain pyrophoric iron sulfide scale that could ignite if it contacts air. Equipment that was in full well stream service might contain corrosion products, sand, and other deposits that attract moisture and lead to under-deposit corrosion.

Most manufacturers of compressors, pumps, and engines have specific instructions for mothballing their equipment. Refer to References 18–19 for guidelines on mothballing equipment.

9.10.1.3 Dry Layup

Dry layup involves cleaning, drying, and sealing the equipment with a positive pressure of an inert atmosphere, usually nitrogen (nitrogen can leak from the enclosed space if not properly sealed). Some inhibitors are specifically designed for mothballing and long-term preservation including vapor phase inhibitors (Chapter 8, Section 8.13.2); these volatilize and fill the sealed equipment with corrosion inhibitor transported by the product's high vapor pressure. Other forms of wet and dry mothballing include filling the equipment with clean, rust preventative oil, or connecting the equipment to dehumidification equipment.[19] Refer to Section 9.7.2 for information on dehumidification.

9.10.1.4 Wet Layup

Wet layup is a form of mothballing used when the equipment will be out of service for an intermediate time. It is like hydrotesting, but the amount of time water resides in the equipment is substantially longer: six months to two or more years. The chemical formulations used in wet layup must account for the half-life of the chemical, that is, the time required to reduce the concentration of the chemical by half because the chemical decomposes or hydrolyzes. Biocides have half-life considerations, and the initial dosage must account for reduction in concentration. THPS degradation increases with an increase in temperature or in pH.[20] Data on glutaraldehyde and THPS half-life are provided in the references and are also available from the chemical supplier.[16-17, 20-21] Section 9.9.3.1 also provides data. Sampling valves should be installed on the equipment so that chemical and bacterial analysis of the water can be taken. Disposal of chemically treated water should meet all relevant regulations.

9.10.2 pH Control

Controlling the pH (neutralization) is used in some cases for corrosion control. The use of pH control for dehydration glycol is described in Section 9.7.4.1.

9.10.2.1 pH Control in Wet Acid Gas

There are other cases where pH control can be used (e.g., wet gas streams with high CO_2 and/or H_2S). When water is being condensed as the gas passes through coolers, it is very difficult (if not impossible) to get good distribution of conventional film-forming liquid inhibitors in each tube because pressure drop is unequal in so many parallel flow passes. There are some vapor phase amines (ammonia is the cheapest) that are carried in the gas and will dissolve in condensed water to neutralize the acid gases that have dissolved in the water. The main problem is dosage and monitoring. About the only choice is to monitor the pH at the water draw off and adjust the neutralizer dosage accordingly. It is because of the dosage adjustment and monitoring problems that it is usually much more cost effective to use corrosion-resistant materials for coolers, and to coat the vapor space in vessels.

9.10.2.2 pH Stabilization in Wet Acid Gas Pipelines

The pH stabilization method is being used as an alternative to film-forming organic corrosion inhibitors to inhibit CO_2 corrosion in wet gas pipelines.[22-23] The method has been applied in sweet systems with mono ethylene glycol (MEG) as the hydrate inhibitor. An alkali is added to the bulk MEG–water phase that promotes the formation of a protective corrosion film. Alkalis considered include NaOH, KOH, $NaHCO_3$, $KHCO_3$, Na_2CO_3, K_2CO_3, and methyl diethanolamine (MDEA). The pH stabilization method has been applied in systems where there is no risk of calcium carbonate scale formation, i.e., no risk of formation water breakthrough occurring. The pH stabilization method is effective in controlling corrosion in the MEG–water phase in a wet gas pipeline.

Controlling Top-of-the-Line Corrosion in pH Stabilized Pipelines

In sweet, wet gas pipelines that go subsea, top-of-the-line corrosion is a concern because the CO_2 condenses when pipeline metal temperature drops. Top-of-the-line corrosion is a mechanism that occurs in the first or hot part of the pipeline where the temperature drops below the water dew point. After condensation is complete, corrosion at the bottom of the line predominates. To inhibit top-of-the-line corrosion, the MEG with pH stabilizer must be injected a distance upstream of the pipeline inlet, and the MEG–water liquid must be transported to the top of the pipeline as droplets. This can be achieved through an understanding of flow regime using transient flow analysis or by pigging. Slug flow, annular, annular-mist, or bubble flow brings liquid to the top of a horizontal pipeline (see Chapter 10, Figure 10.3). Pigs can be used when the flow regime itself will not transport liquid droplets to the top of the pipeline.

9.10.2.3 pH Control in Open and Closed Cooling or Heating Water Systems

In recirculating cooling water systems (open or closed), pH control is critical to the success of the chemical treatment program. In open water systems (cooling towers), the chemical treatment depends on maintaining a target pH range that both helps to form protective corrosion films on steel equipment, as well as to control scaling. Open cooling water pH is often maintained in the scaling range for calcium carbonate; scaling is prevented by adding scale-control chemicals. As the cooling water recirculates, picks up heat, becomes hotter, and partially evaporates out of the cooling tower stack, its pH increases due to an increase in bicarbonate ion concentration. Depending on the cooling water chemistry (calcium, bicarbonate, sulfate concentration, etc.), pH is controlled by introducing fresh makeup water. An acid such as sulfuric or hydrochloric, may be required to keep the cooling water pH from becoming too alkaline and heavily scale-forming. Because of the open, aerated nature of the water, bacteria is free to enter and biocide addition (including chlorine, chlorine substitutes, and nonoxidizing biocides) is routine, and can be continuous and/or batched. Corrosion control is maintained using inorganic or organic phosphorous compounds. Refer to Chapter 8, Section 8.10.2.6 for additional discussion of corrosion inhibitors and biocides for open water systems.

In closed water systems, the water is typically selected to contain lower concentrations of calcium and other scaling ions than in open water systems because closed systems rarely refresh the water, making scaling more likely. Water can be distilled, boiler feedwater, or steam condensate. The pH of closed cooling water is maintained much higher (e.g., pH 8–10.5) than open cooling water. Alkalis such as sodium borate are added. The water is not aerated, so oxidizing corrosion inhibitors such as sodium nitrite, molybdate, or a combination of both, are used (refer to Chapter 8, Section 8.10.2.7 for additional discussion). Biocides are only added infrequently, usually batch treated as required.

References

1. API RP 14E (latest edition), *Design and Installation of Offshore Production Platform Piping Systems*, 5th ed., Section 2.3 "Sizing Criteria for Liquid Lines" (Washington, DC: American Petroleum Institute, 2013).
2. Salama, M.M., "Influence of Sand Production on Design and Operations of Piping Systems," CORROSION 2000, paper no. 00080 (Houston, TX: NACE International, 2000).
3. Russel, Y.R., H. Nguyen, and K. Sun, "Choosing Better API RP 14E C Factors for Practical Oil Field Implementation," CORROSION 2011, paper no. 11248 (Houston, TX: NACE International, 2011).
4. University of Tulsa Erosion/Corrosion Research Center (Tulsa, Oklahoma: University of Tulsa, 2019) http://www.ecrc.utulsa.edu/ (Oct. 5, 2019).
5. NACE SP0198 (latest edition), "Control of Corrosion Under Thermal Insulation and Fireproofing Materials—A Systems Approach" (Houston, TX: NACE International, 2016).
6. Peabody, A.W., "Peabody's Control of Pipeline Corrosion," 2nd ed., Ronald Bianchetti, Ed. (Houston, TX: NACE International, 2001): pp. 249–254.
7. NACE SP0286 (latest edition), "Electrical Isolation of Cathodically Protected Pipelines" (Houston, TX: NACE International, 2007).
8. Byars, H.G., and B.R. Gallop, "Injection Water + Oxygen = Corrosion and/or Well Plugging Solids," *Materials Performance* 13, 12 (Houston, TX: NACE International, 1974) p. 33.

9. Mateer, M.W., "Selection of Oxygen Scavengers for Oil Field Waters," CORROSION '86, paper no. 178 (Houston, TX: NACE International, 1986).
10. Davies, M., and P.J.B. Scott, *Oil Field Water Technology* (Houston, TX: NACE International, 2006): pp. 440–466.
11. Jordan, L.C., A. Simsek, C.L. Bargas, J.T. Bracey, and F.M. Erdal, "Dealing with Uncertainty–Impact of Scaling Prediction on Concept Selection for Deepwater Production Systems," CORROSION 2011, paper no. 11346 (Houston, TX: NACE International, 2011).
12. Kumar, R., L. Zea, and P. Jepson, "Role of Pressure and Reaction Time on Corrosion Control of H_2S Scavenger," paper no. SPE-114175-MS (Aberdeen, Scotland: Society of Petroleum Engineers, 2008).
13. Larsen, J., P.F. Sanders, and R.E. Talbot, "Experience with the Use of Tetrakishydroxymethylphosphonium Sulfate (THPS) for the Control of Downhole Hydrogen Sulfide," CORROSION 2000, paper no. 00123 (Houston, TX: NACE International, 2000).
14. Salma, T., "Cost Effective Removal of Iron Sulfide and Hydrogen Sulfide from Water Using Acrolein," SPE Paper 59708 (Richardson, TX: Society of Petroleum Engineers, 2000).
15. Darwin, A., K. Annadorai, and K. Heidersbach, "Prevention of Corrosion in Carbon Steel Pipelines Containing Hydrotest Water–An Overview," CORROSION 2010, paper no. 10401 (Houston, TX: NACE International, 2010).
16. Pensalka, J.E., J. Fichter, and S. Ramachandran, "Protection Against Microbiologically Influenced Corrosion by Effective Treatment and Monitoring During Hydrotest Shut-In," CORROSION 2010, paper no. 10404 (Houston, TX: NACE International, 2010).
17. McGinley, H.R., G. Hancock, M. Miksztal, M. Enzien, and S. Gonsier, "Gluteraldehyde: An Understanding of Its Ecotoxicity Profile and Environmental Chemistry," CORROSION 2009, paper no. 09405 (Houston, TX: NACE International, 2009).
18. Twigg, R.J., "Guidelines for the Mothballing of Process Plants," MTI Publication No. 34 (out of print) (St. Louis, MO: Materials Technology Institute, 1989).
19. Lyublinski, E., Y. Vaks, T. Natale, M. Posner, W. Rohland, S. Woessner, R. Singh, and H. de Souza Siriaco, "Corrosion Protection of Mothballed Equipment," CORROSION 2014, paper no. 4335 (Houston, TX: NACE International, 2014).
20. Wilmon, J., "THPS Degradation in the Long-Term Preservation of Sub-sea Flow-lines and Risers," CORROSION 2010, paper no. 10402 (Houston, TX: NACE International, 2010).
21. Zhou, J., K., Wen, T. Gu, and A. Kopliku, "Mechanistic Modeling of Anaerobic THPS Degradation in Seawater Under Various Conditions," CORROSION 2008, paper no. 08512 (Houston, TX: NACE International, 2008).
22. Olsen, S., "Corrosion Control by Inhibition, Environmental Aspects, and pH Control. Part II: Corrosion Control by pH Stabilization," CORROSION 2006, paper no. 06683 (Houston, TX: NACE International, 2006).
23. Olsen, S., and A.M.K. Halvorsen, "Corrosion Control by pH Stabilization," CORROSION 2015, paper no. 5733 (Houston, TX: NACE International, 2015).

Bibliography

Craig, B.D., "Predicting Critical Erosion-Corrosion Limits of Alloys for Oil and Gas Production," *Materials Performance* 37, 9 (1998): p. 59.

Davies, M., and P.J.B. Scott, *Oil Field Water Technology* (Houston, TX: NACE International, 2006): pp. 481–499.

DNVGL-RP-O501 (latest edition), "Managing Sand Production and Erosion" (Oslo (Bærum), Norway: DNV GL, 2015).

Miksic, B.A., *Preservation, Lay-Up and Mothballing Handbook,* 3rd ed. (Houston, TX: NACE International, 2013).

NACE Publication 3A194 (out of print), "Oxygen Scavengers in Steam Generating Systems and in Oil Production" (NACE International: Houston, TX, 1994).

Salama, M.M., "Erosion Velocity Limits for Water Injection Systems," *Materials Performance* 32, 7 (1993): p. 44.

Shirazi, S. A., B. S. McLaury, J. R. Shadley, and E.F. Rybicki, "Generalization of the API RP14E Guideline for Erosive Services," *Journal of Petroleum Technology* 47, 8 (1995): p. 693.

Appendix 9.A: Checklist for Troubleshooting Oxygen-Free Water Injection Systems

The following items should be checked to determine if water injection systems are set up and operating oxygen-free:

Gas Blankets on Tanks and Vessels

1. Check source of blanket gas. It must be oxygen-free.
2. Check for individual gas blanket regulators on each tank. One regulator cannot properly serve two tanks. Regulators should be mounted at the top of the tank with a gas line diameter of 25 mm (1 in) or larger.
3. Check gas blanket regulators by opening the thief hatch when the tank level is dropping (not when it is filling). A rush of gas out of the hatch should be heard and the regulator should open to supply more gas to the tank. To confirm this observation, use an oxygen meter, place the oxygen probe in the vapor space of the tank, shut the hatch, and take a reading.

Water Source Wells

4. Check gas regulator on the annulus of a water supply well. The regulator should maintain a positive pressure when the well first comes on.
5. Check water supply wells without a foot valve (check valve) for leaks at fittings. Although it is best to have foot valves, they are not used in cold climates. That way water will drain from the source well wellhead fittings to prevent freezing when the well is shut in. When the well pump has just shut down, check all fittings and valves on the wellhead for vacuum leaks. Fittings and valves can leak under vacuum even when they do not leak under pressure. Check for oxygen at the wellhead when the pump first comes on.

Pumps

6. Check all transfer pump check valves for proper operation. When the pump is shut off, it should not run backward. Transfer pumps are usually centrifugal pumps used to move water from a battery location to the main injection plant.
7. Check how much water is bypassed from the high-pressure pumps back into the tanks. Pumps should be sized to minimize bypass.
8. Check for leaks on centrifugal pump shafts (air can leak in while water leaks out).

Tanks

9. Check produced and supply water tanks for a downcomer[11] pipe on the inlet line. The pipe should extend to below the lowest operating level. (Water should not free-fall in the tank.)
10. Check level controls on all water tanks. Controls should be set to minimize the level fluctuations.
11. Check diverter valves for leaks. These valves are used when two incompatible waters are alternately injected to avoid water mixing. Leaks can be detected by collecting water samples from each source water and from downstream of the diverter valve. Use temperature and/or chemical analysis (chlorides, conductivity) to judge if the diverter valve is leaking.

---

11. A pipe that connects liquids downward in a vessel or tank.

Water Injection Wells

12. Check injection wells that use a vacuum to inject water into the injection well. The master valve should be pinched back (partially closed) so that a positive pressure is maintained on all wellhead fittings.
13. Check for air leaks where tubing is packed off around the wellhead (the rubber seal); pour water around the seal. If there is a leak, the water will disappear into the seal.
14. Check casing (annular space) with a compound gage (a pressure gauge that measures both positive pressure and vacuum). A tubing leak could be indicated if a vacuum is observed.

Operating Procedures

15. Check the procedure used to change the filter cartridge (if installed in the system). The procedure to change a filter should include bleeding air from the filter before it is put back into service.
16. Check with the plant operator about the disposal of any water picked up by vacuum trucks after a leak. No oxygenated water (water that has been in contact with air) should be pumped into the injection system.
17. Check to see if any water that is discharged to a pit on occasion is pumped into the system. It should not be pumped back into the injection system.

Corrosion Monitoring and Inspection

Timothy Bieri

10.1 Introduction

Broadly speaking, corrosion monitoring provides a representative corrosion rate for a given piece of equipment exposed to a corrosive environment. Monitoring can be divided in any number of ways. NACE 3T199[1] provides a useful framework for the categorization of monitoring techniques, Figure 10.1. This chapter will generally follow that framework in describing surface and downhole methods of detecting internal corrosion and monitoring corrosion control programs.

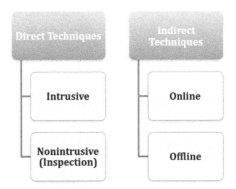

Figure 10.1 Corrosion monitoring technique definitions.

Corrosion monitoring and inspection are key components for effective corrosion management. Corrosion monitoring activities typically require equipment offering high sensitivity and relatively short response times. Inspection activities typically require equipment with high degrees of accuracy, but relatively long response times. However, these requirements overlap and certain technologies (e.g., ultrasonic measurement) can be used for both corrosion monitoring and inspection activities. The purpose of corrosion monitoring is to provide leading indicators of corrosion damage to equipment *before* significant damage occurs and allow mitigation methods to be employed.[2] The purpose of inspection is to provide assurance that equipment is fit for continued service. Inspection can be considered a lagging indicator of corrosion because it measures how much corrosion has occurred.

Experience has shown when the corrosion monitoring device shows increasing corrosion, actual corrosion rates in the system have increased accordingly. Likewise, when a corrosion control program is implemented (or modified) and monitoring indicates that corrosion is under control, it actually is under control. This phenomenon holds true whether the monitoring is done with corroding specimens (e.g., coupons), probes (electrical resistance or linear polarization), chemical analysis (e.g., analysis of dissolved iron in the fluids), or other monitoring techniques. There are instances where increased corrosion is occurring without being detected, and, conversely, where a monitoring device indicates corrosion is increasing, but no changes in metal loss are found by inspection. These circumstances are addressed later in the chapter. They highlight the need to use a systems approach to corrosion monitoring, which uses two or more independent monitoring techniques whose results can be synthesized into making sound corrosion management and inspection decisions.

All monitoring data must be interpreted properly with the knowledge of the monitoring location, the system's history and the degree of reliability of the test procedure in the specific environment. An understanding of the monitoring location and primary corrosion threats is critical to proper interpretation and utilization of the results. For example, changes in corrosion rates at wellsite surface monitoring locations generally correlate with changes in wellbore corrosion rates. The person evaluating the results needs to be aware of such situations, and how the difference in temperature at the wellhead and at downhole points, and the partial pressure of acidic gases such as CO_2 and H_2S, can play a large role in determining corrosion rates. The analyst must learn to rule out spurious results when common sense indicates that a system is more corrosive or less corrosive than the measurements indicate. In many cases, where corrosion control is critical, or the system is unusually corrosive, and/or risks are high, multiple monitoring techniques are used to evaluate the system and its control program.

The selection of appropriate monitoring methods, device types, locations, and orientations is critical for successful corrosion monitoring. Selection of inappropriate methods and locations results in wasted effort and expense. Worse still, inappropriate monitoring methods or equipment can generate misleading information and provide an unjustified degree of confidence concerning the corrosivity of a process stream, which is not always questioned. "False positives," "false negatives," and indications related to process variable interferences can seriously affect the credibility and usefulness of corrosion monitoring activities.

10.2 Location

Corrosion monitoring techniques, other than actual physical inspection of equipment, only provide indicators of corrosion rates and are not necessarily the actual corrosion rates in the system. Regardless of which monitoring technique or device deployed, their location is going to be important. Essentially, the results of a monitoring device or activity will reflect the corrosion rates at that location, under the conditions at that location. Consequently, corrosion monitoring should be placed at locations that allow optimal monitoring of the corrosion threats. Monitoring location considerations include

- Where the location is along the length of a process stream
- Changes to the process stream itself
- Where it is within the fluid stream (i.e., which fluid phase)
- Effects from the flow rate and flow regime

While physical accessibility should not dictate the location, it is an important consideration (refer to Chapter 9, Section 9.2.4). Often there is a balance between the optimum corrosion location and the ability to safely access and maintain the location. Some additional location considerations are presented in the following sections.

10.3 Indirect Techniques

Indirect techniques measure parameters that either influence or are influenced by corrosion.[1] These techniques can be separated into "online" and "offline" (Figure 10.2).

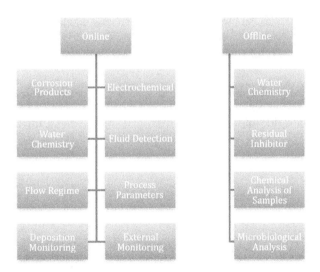

Figure 10.2 Indirect measurement techniques.

10.3.1 Online Techniques

Online techniques allow for collecting data in real time, or near-real time, without the need to remove a device from the system.

10.3.1.1 Process Parameters

There are several process parameters that have an effect on corrosion rate. These parameters can be measured online in real time.

- Pressure: System pressure affects the composition of the fluids. This is important when dealing with acid gases and their respective partial pressures.
- Temperature: Higher temperatures can increase the rates of chemical reactions. In gas systems, lower temperatures can lead to condensation of water.
- Dew Point: Dew point is related to temperature and moisture content of a gas. Below the dew point, condensation can be expected.
- Water Cut: Water cut, in multiphase systems, is the proportion, or percentage, of water to the total fluids.
- Production rates: Production rates, or throughput, can affect corrosion and is discussed in the next section.
- Base Sediment and Water (BS&W): BS&W is a measure of impurities in a finished crude oil. The effect on corrosion is closely related to fluid velocity. Low velocity in pipelines can allow BS&W to settle in low points.

Many of these parameters will have a normal operating limit. Excursions outside of these limits are important to understand because they may trigger a corrective action.

10.3.1.2 Fluid Detection

In single-phase flow, the flow regime is a straightforward calculation and can be either laminar or turbulent. Important factors for this calculation include fluid density and viscosity. In multiphase systems, there can be a variety of flow regimes depending on the liquid and gas rates (Figure 10.3).

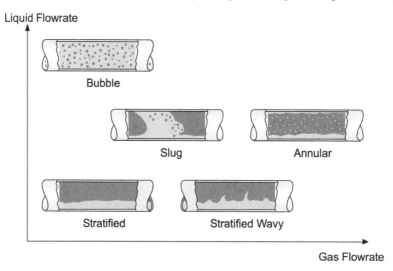

Figure 10.3 Multiphase flow regime.

In both cases, there are predictive models that can be used to help identify locations where corrosion is likely.[3]

10.3.2 Offline Techniques

Offline techniques are used when a sample is collected for later analysis.

10.3.2.1 Chemical Methods

Chemical analyses come into play at various times in monitoring programs. Sometimes samples are taken when troubleshooting problems, or evaluating procedures (e.g., those collected during injection line pigging). Other times, the need is for routine monitoring of liquid streams (e.g., dissolved iron in gas well water samples, dissolved oxygen in injection waters, or even glycol samples to determine pH, inhibition levels, and glycol quality).

Analysis of the H_2S and CO_2 content of gases will indicate changes in the systems and can give an indication of corrosiveness in the system. There are now oxygen meters available that will detect oxygen in gases in the parts per billion (ppb) range. These meters have proved particularly valuable in troubleshooting oxygen entry into gas lift systems.

The analysis of liquid samples can cover both composition of the liquid stream and composition of contaminants. Sampling techniques become extremely important if the analysis is to be representative of the conditions in the system.

Sample Connections

Many of the troubleshooting and monitoring procedures involve analysis of the gases, liquids, and solids from a system. As a guideline, sampling valves should be installed at each point in a facility where a change can occur (i.e., at producing, supply, and injection wells, at manifolds, and before and after each vessel or piece of equipment). It is a good practice to "double valve" sampling connections, particularly in high-pressure applications. Typically, sample valves are needle valves. A large pressure drop across valves can "cut" the valve out in a short timeframe—needle valves can be quite easily plugged by any debris that is present in the system. In either case, the sample location is unusable until the system can be shut in for valve repair, replacement, or cleanout. Figure 10.4 illustrates the close-coupled double-valve sample connection that simplifies valve maintenance. As noted, the nipple connecting the valves to the system should be as short as practical to ensure that representative samples can be collected while minimizing the amount of fluids required to flush through the sample point. The "inner" valve is preferably a ball or other "full opening" valve—it is always opened to its fullest extent for sampling. The "outer" valve is a needle valve and is used to control the flow while sampling. By having double valves, the outer valve can be changed whenever it cuts out. This valve arrangement may be used for all types of sampling (oil, gas, water, or mixtures).

Figure 10.4 Double valve sampling connection.

Gas condensate wells often make small amounts of water or produce water in heads (or slugs). Therefore, a sample accumulator (sample pot or sample receiver) at the well is needed. Figure 10.5 presents a sketch of a sample receiver for a gas well.[4]

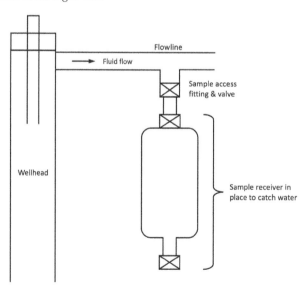

Figure 10.5 Typical double-ended sample receiver, and connection on the bottom of a flowline.

Dissolved Iron

One method of attempting to detect corrosion and to evaluate control effectiveness is to measure the iron content of the fluid. In the field, the iron content analysis is, more often than not, referred to as the "iron count." An iron content study can be broken down into three basic steps: sampling, analysis, and evaluation or interpretation. Monitoring of dissolved iron is covered in more detail in NACE SP0192.[4]

Sampling

Because the usefulness of the study depends entirely on the validity of the data, the most important step of the three is the collection of representative, uncontaminated samples. In some studies, only the produced water is collected and analyzed. However, many companies measure iron in the oil or iron in the total fluid. The technique used when collecting the sample is of utmost importance. The following should always be considered when collecting samples for iron content:

- Cleanliness of equipment is very important. The sample jar must be clean and free from iron. Glass bottles or jars are the preferred sample container because they are impermeable to air (oxygen). Plastics such as polyethylene and polypropylene are permeable to oxygen. Oxygen in the water will precipitate the dissolved iron. Many times, unless special steps are taken to avoid precipitation (such as acid in the sample jar), the precipitated iron will be difficult to remove or to put back into solution, and a lower-than-actual iron content will be reported. Standard glass sample bottles are satisfactory containers. Normally glass bottle lids will be heavy plastic. However, jar lids are usually metal, and a nonmetallic insert should be used to prevent contact between the well fluids and the metal lid. This insert can act as the sealing gasket.
- The sample valve must be free of rust and scale. Any nipples or fittings downstream of the sample valve, which are open to the air, should be removed if possible. One technique found, as an effective aid to sampling, is to use brass or polyvinyl chloride (PVC) plastic sample fittings screwed into the sample valve. These fittings serve to minimize splash and replace steel fittings, which may have rusted in the air. It is sometimes advisable to wash out the sampling valve. Diluted hydrochloric acid in a plastic wash bottle can be used to rinse iron oxide particles from a valve before it is flushed for sampling.
- The sample valve and upstream nipples should be flushed out completely. Because it is usually desired that the sample represent the flowing fluid, the fittings must be flushed to ensure that fresh fluid is being sampled, not the fluids that have been standing in the dead space in a nipple or bull plug.
- Splash and overflow should be avoided when collecting samples for iron analysis. Iron in solution may oxidize while it is being collected, or particles of corrosion product may be carried in the fluid. Splash and overflow of the jar can tend to concentrate these solids and can result in misleading results.
- Above all, the sample must represent the fluid being sampled, and must not be contaminated by rust, scale, etc., from around the sample valve.

Cleanliness and flush out have been emphasized because it is very easy to contaminate an iron count sample. Even in the worst conditions, only a few hundred parts of iron are found in a million parts of water, and very often in the range of around 20 ppm. A piece of iron oxide (rust) about the size of a large pinhead dissolved in a pint (0.48 L) of fluid is equivalent to about 10 ppm of iron. It is very easy, therefore, to contaminate a sample by carelessly letting rust or scale particles get into it.

Quite often, the water in a sample turns red or rust colored as it is collected, or just after collection. This usually means that iron that was dissolved in the sample has reacted with air to form iron oxide. If it turns black, iron sulfide is probably being formed (although some oxides of iron are also black; refer to Chapter 2, Table 2.2). To keep the iron oxide or sulfide from forming, the sample may be collected with a few milliliters of acid, such as diluted HCl, in the jar. If acid is used to maintain iron solubility, these samples should be dedicated to iron count analysis, and never used for water analysis because pH and chloride (from the acid) affect the results.

Analysis

There are several methods of actually analyzing the amount of iron in a fluid.[5-7] Some methods can be used in the field, while other techniques are used only in the laboratory. When starting iron content surveys in a new area, it is necessary to run some checks on the analytical method with the new fluid, because other materials dissolved in the fluids can affect the iron analysis and give erroneous results.

Evaluation and Reporting

Finally, evaluation of the data must be done. In some gas wells, where production is essentially the same from day to day and where only small quantities of water are produced, it may be satisfactory to use the iron content as is—that is, to compare the iron contents themselves from sample-to-sample or well-to-well. However, high iron counts in wells with low water production do not necessarily indicate severe corrosion; similarly, low iron counts in wells with high water production are not necessarily indicative of low corrosion. Much more important is the amount of iron produced in the system on a daily basis.

Therefore, in most wells and systems, for the best comparison, the iron count should be converted to "mass of iron per day." Pounds-iron-per-day calculations take into consideration the volumes of the produced fluids and, in effect, gives a value to "iron production." A nomograph to simplify this conversion is presented as Figure 10.6. Although the nomograph is handy, actually the calculation [Equation ([10.1)] is fairly straightforward:

$$\frac{\text{lb of Iron}}{\text{day}} = 0.00035 \frac{\text{mg}}{\text{L}} \cdot x \frac{\text{bbl}}{\text{day}} \tag{10.1a}$$

$$\frac{\text{kg of Iron}}{\text{day}} = \frac{y \frac{\text{mg}}{\text{L}} \cdot x \frac{\text{m}^3}{\text{day}}}{1000} \tag{10.1b}$$

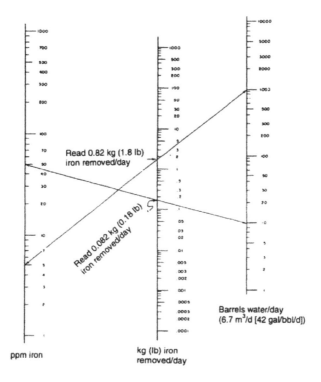

Figure 10.6 Nomograph is useful in calculating pounds of iron lost per day. Diagonal lines represent two calculations with different parts per million and barrels of water per day.

Fluid volumes produced on a daily basis can increase over a period of time due to increases in water production or changes of lift conditions. Also, daily production volumes usually vary from well to well in the same field. When evaluating iron content results, other factors should be considered:

- How long has the well been producing prior to sampling? If it is sampled once just after it has started producing and again after it has produced for several days, the iron content may be very different.
- Has the well recently been worked on? If a well has had equipment changes within a few days before sampling, the iron count may be higher than normal—even on successfully inhibited wells—due to the dislodging of scale or corrosion products, etc., when running the equipment in and out of the hole.
- Does the water have the same composition of minerals each time it is sampled? This can be particularly important in gas well work. Most water in the early life of a gas well is so-called "condensed water." This water was a vapor in the reservoir and at the bottom of the hole. As the vapor comes to the surface, temperature and pressure changes allow the vapor to condense. When the vapors condense, acid gases dissolve in the liquid and corrosion can take place. This is the water that is sampled and the iron found is from corrosion. In many gas wells, however, formation water may be produced along with the condensed water. Formation water is usually salty and contains many minerals. It may even contain dissolved iron from the formation. Un-

fortunately, the amount of formation water in the total water produced is usually not constant. One sample from a well may analyze as a very pure condensed water, and the next day, a sample might be mostly formation water. As can be seen, if the formation water contains dissolved iron from the formation, this iron will be analyzed along with the iron from corrosion, and the iron count may be higher than usual. Spent acid water—produced after an acid job on a well—may also contain large amounts of iron, which the acid dissolved from the formation. One simple way to document the type water is to collect two samples at the same time—one with acid for iron count analysis, the other without acid for analysis of the chloride content. The chloride results can be compared to a previous analysis of the formation water to determine if the current sample is condensed water, formation water, or a mixture.

- Where in the system were the samples collected? Were all samples from the same point? As might be imagined, this can be important when the water is very corrosive, because the iron content may vary depending on the amount of steel that it contacts. When studying corrosion of wells, it is always best to obtain iron content samples as close to the wellhead as possible.

There are various types of sample arrangements used at the wellhead. On oil wells, it is usually sufficient to have a bleed valve. On most gas wells, as mentioned earlier in the chapter, a sampling pot of some type is required.

While iron content is a useful monitoring tool, it does need to be validated for each system and should be combined with other monitoring techniques. An example application of this would be evaluating the effectiveness of a corrosion inhibitor program where the reduction of iron content from the pretreatment level indicates the success of the control. Figure 10.7 presents such a story.

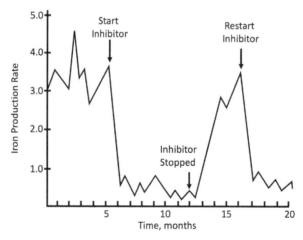

Figure 10.7 Graphical presentation of iron production rate vs. time with pertinent operating information.

Corrosion Products Analysis

Chemical analysis of corrosion product samples and deposits from a system can be another important part of a monitoring program. Samples may be taken directly from the equipment, coupons, or test nipples. Knowing the composition of such deposits helps to evaluate the type of problem and to detect changes in the system. For accuracy, analyses should be performed in the laboratory. Chapter 2, Section 2.7 provides laboratory methods and qualitative field analysis techniques.

Again, sample collection and handling are important for proper interpretation of the results. It is important the person submitting the sample provides full details on the conditions and location of the sample.

Samples of corrosion products can change chemically once they are removed from a system. For example, when iron sulfide comes in contact with air, it will start to oxidize and convert to iron oxide. A sample that was black from iron sulfide when collected may be red-brown from iron (ferric) oxide when received in the laboratory. Thus, the color of the sample at the time of collection becomes important information.

Dissolved Gas Analysis

The gases most important to corrosion include carbon dioxide (CO_2), hydrogen sulfide (H_2S), and oxygen. The degree of importance, and thus the monitoring required, will vary with systems. In water systems, the dissolved gases become very important and monitoring of dissolved H_2S and oxygen is a critical part of troubleshooting and control.

Acid Gases

The most important dissolved acid gases (that is, the gases that ionize in solution to form acids) are H_2S and CO_2. The H_2S may be native to the formation or may be generated by sulfate reducing bacteria (SRB). The CO_2 is native to the formation except in some enhanced oil recovery (EOR) projects. The CO_2 injection, flue gas injection, and in situ combustion (fire flood) all add CO_2 to the formation fluids.

While laboratory test procedures usually are the most accurate, the problems of collecting and transporting representative samples from the field to the laboratory make field tests very attractive. Field tests are readily available to check dissolved H_2S and CO_2 contents. The results of such tests are usually quite acceptable. In many cases, the presence or absence of the acid gas is what is the most important. Sometimes when troubleshooting a problem, the increase or decrease of the acid gas content through the system is the item of primary interest (refer to Section 10.3.3).

Dissolved Oxygen

As mentioned in almost every chapter—oxygen is the greatest accelerator of corrosion in all oil field systems. Therefore, the monitoring of dissolved oxygen is a very important part of any corrosion control program. Because oxygen is the principal corrodent in most oil field water systems, dissolved oxygen monitoring is particularly important in those systems. Oxygen can be measured with fair accuracy on location in most systems—down into the parts per billion (ppb) range. As with other specialized tools, experience is very important in obtaining data and interpreting the results.

Chemical tests that are relatively easy, rapid, inexpensive, and accurate are also available for field use. Like acid gases, oxygen does not lend itself to sampling and transportation to a laboratory so field tests are ideal. Most of these tests are colorimetric determinations, so the eyes and experience of the tester are very important—and the sampling technique must avoid oxygen contamination. When considering their use in a specific water system, the test procedure and reagents should be checked for interference from ions native to the water. As a generality, seawater is less likely to interfere than produced brines.

Inhibitor Presence or Concentration Tests

The determination of the presence and/or concentration of a corrosion inhibitor in the fluid can provide useful information when evaluating an inhibition program. Analytical procedures are available to determine the amount of inhibitor in a system. For some materials, laboratory analysis is required; for others, a simple field test can be used to determine the presence and concentration of an inhibitor in the liquid stream. Many of these tests measure the total amine concentration, and it is important to remember there are other components within many inhibitors.

The so-called "copper ion" test can be used for some inhibitors to determine the presence of an inhibitor in well fluids. In this case, a coupon dipped in or exposed to the liquids is immersed in a copper solution, typically copper sulfate. If or where an inhibitor film is present, no copper will deposit. On the other hand, copper will deposit on any bare steel on the coupon. Thus, an empirical correlation can be made as to an inhibitor's presence or effectiveness. This test is not valid for most water soluble or highly dispersible chemistries.

pH

Because pH measures the acidity or alkalinity of fluids, pH determination is a part of any corrosion study. There are two pH measurements that can be important. The pH in the system at system temperatures and pressures can be quite different than the pH determined on a sample collected from the system. When the pH is primarily acidic due to dissolved acid gases, these gases will escape when the sample is collected and the pH of the sample will be higher (less acid) than the pH in the system. Pressure pH probes are available, but the usual approach is for the analyst to calculate the system pH based on the acid gas concentrations, the system temperature, the chemical analysis of the water (anions and cations), and the acid gas solubilities at system temperatures and pressures.

When the pH is the result of components in the water, such as spent acid from well acidization or materials native to the water, the pH of a sample is meaningful. A pH meter should be used (rather than pH paper) whenever accurate readings are required. As with any analytical instrument, the calibration and maintenance of the pH probes are critical for accurate, reproducible data.

The pH of fluids besides water also can be determined. For example, the pH of dehydration glycol, heating/cooling water solutions, and gas sweetening solutions indicates the "health" of those systems. Some solutions require special procedures to determine the pH. The pH of a glycol is usually determined as an "apparent pH." The glycol is usually diluted with distilled water (often 50/50), and the pH is measured on that mixture and reported as an "apparent pH."

10.3.3 Gas Analysis

In gas well or gas handling applications, determination of CO_2 and H_2S is fairly routine when they are present in large quantities. The same is true, along with oxygen, in plant applications. Trace quantities of H_2S and oxygen are harder to detect, but can be of extreme importance. For example, under the right conditions, traces of H_2S can cause cracking of high-strength steels. In oil wells, the measurements may be more difficult, and the important factor is usually the presence or absence of the gas. H_2S in gas should be measured at the well. Laboratory analysis of H_2S will invariably be lower than actual. A report of "nil" H_2S basically says "there was none present in the sample as analyzed," or the quantity was below the detection limit of the measurement technique. There are several methods for measuring H_2S on location, including several brands of "gas tubes."

The determination of oxygen in the lift gas is often quite meaningful in evaluating corrosion possibilities in gas lift systems—particularly in rotative systems and those using return gas from a plant with vacuum gathering systems.

10.3.4 Microbiological Detection and Monitoring

As mentioned in Chapter 2, Section 2.5, bacteria, particularly the SRB, can cause a number of problems, particularly in water handling systems. Much has been written and said concerning microorganisms and the importance of their role in corrosion reactions. Progress in understanding the exact or complete role played by various microorganisms in corrosion reactions is continuing. Microorganisms are quite numerous and exist in many systems, however, the mere presence of bacteria is not necessarily bad. The important thing is how active are they, and do they create or contribute to a particular problem.

Coverage of this subject is quite involved.[8] The important point is that there are monitoring techniques to determine the activity of microorganisms and these should be included in the monitoring program where applicable.[9-10] Techniques for both planktonic (those in the water stream) and sessile (those attached to surfaces) bacteria are covered.

Because the collection of water samples to determine planktonic bacteria is fairly easy as long as sterile containers and techniques are used, they are the most common tests run. However, the results can be quite variable from moment to moment depending on the number of bacteria caught in any sample. In many cases, the main value of planktonic tests is to indicate that there are active, viable bacteria upstream of the sample location. Sessile bacteria are those colonies that are actively growing, are involved in the corrosion process, and are contributing bacteria to the planktonic count. Cultures of sessile bacteria may be made from scraping equipment surfaces during inspections, from special "bacteria devices" exposed in the system, and from corrosion coupons when they are removed from a system.

10.4 Direct Techniques

Direct techniques provide a measurement data that are directly influenced by corrosion. Direct techniques can be further divided into intrusive and nonintrusive.

10.4.1 Intrusive Techniques

Intrusive monitoring is defined as monitoring that requires penetration through the pipe or vessel wall to gain access to the interior of the equipment.[1] Intrusive does not mean the monitoring device needs to 'intrude' into the production stream; there are many examples of monitoring devices that are installed flush with the equipment wall.

Intrusive monitoring methods can be further divided according to Figure 10.8, and intrusive monitoring is historically the most common method for the detection and monitoring of internal corrosion. Essentially, this method consists of the physical exposure of test specimens to the corrosive environment—pieces of a material are placed into the environment to corrode and then be analyzed; these includes coupons, test nipples, spool pieces, and special devices using "corrosion probes." The use of intrusive monitoring does carry the risk of exposing personnel to process fluids which may be under pressure. Hence intrusive monitoring activities require appropriate procedures, tools and training to execute safely.

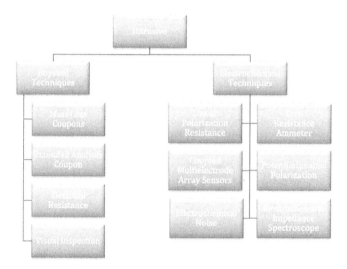

Figure 10.8 Direct-intrusive monitoring techniques.

10.4.1.1 Access Fittings

There are several ways to accomplish intrusive monitoring, including the use of specially designed flanges and piping sections. However, the most common method involves the use of an access fitting, which is welded or bolted onto the equipment. These fittings provide an opening into the fluids through which a monitoring device can be inserted. A typical access fitting, known as a "2-inch access fitting," has a 50.8 mm (2 in) opening through it, and can be purchased to contain pressures as high as 41.3 MPa (6,000 psi). Figure 10.9 shows a typical access fitting "kit," which includes the fitting (prepared for welding onto a pipe), a solid plug (which is installed until a monitoring device is installed), a pressure cap, and a pressure gauge (to indicate potential leaks). Lower pressure fittings are also available with a 25.4 mm (1 in) diameter opening.[2]

Figure 10.9 Access fitting: kit and example installation.

The fittings can be attached onto the equipment wherever there is a suitable space. For piping, common locations are at the 12 and 6 o'clock locations. The physical accessibility needs to be considered at the access fitting location. Tools that retrieve monitoring devices while the system is under pressure require physical space, Figure 10.10. Typical clearances are at least 2 m (6.5 ft) vertical and 1 m (3.3 ft) horizontal. Additionally, permanent platforms may be required for access fittings located more than 2 m (6.5 ft) above the working height.

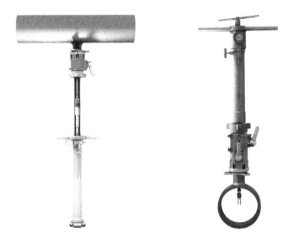

Figure 10.10 Example retriever tools for installing and removing intrusive monitoring under pressure. The tool is connected to the access fitting via a high-pressure service valve (single or double isolation).
Images provided courtesy of Cosasco.

10.4.1.2 Mass Loss Coupons

A mass loss coupon is a small piece of metal that is inserted in the system and allowed to corrode for a period of time. It is common in production to "install a coupon" when corrosion in a system is to be evaluated. Many pipelines and process industries also monitor their internal corrosion with coupons. NACE SP0775[11] provides guidance on the preparation, installation, and analysis of coupons. Coupons are carefully cleaned and weighed before and after they are exposed in a system. From the mass, or weight, loss of the coupon, the corrosion rate can be determined. Pit depths can be measured and pitting rates determined. Coupons also can be used to evaluate erosion and collect samples of materials that precipitate, deposit, or grow on the surface. As with other tools, coupons are not infallible because there are many variables that can affect the results.

Coupon Types and Mountings

Coupons come in several shapes, sizes, and configurations. The size and configuration will depend on the system, the type holder being used, the line size, and the orientation. The most common shapes are flat strips, round rods, and disks. Each type has advantages and drawbacks. The flat strips are very popular because they have a larger surface-to-weight ratio, they have flat surfaces for determining pit depths, and the surface area is relatively constant even at high corrosion rates. The round rods are simpler to install and remove because they simply screw into the insulator, and bolts, nuts, and wrenches are not required. Furthermore, depending on the holder design, multiple coupons can be mounted on the same fitting. Disk shape coupons can be installed at locations where being flush with the pipe wall is important, for example in a pipeline that has pigging operations. Numerous other shapes and sizes are available for special purposes. For example, ring type coupons are made to be run in drill pipe tool joints. Most coupons run in the field are made of low-carbon steel (such as grade 1020 [UNS G10200]). However, coupons of API grade steels or corrosion-resistant alloys (CRAs) are used in special studies. There are many types of coupon holders (mounting devices). The holder assembly is basically a means of supporting the coupon in the fluid stream while electrically isolating it from other metal (Figure 10.11). Figure 10.12 shows example installations. The configuration and placement of coupons depends on the operating conditions and where in the flow corrosion rates are desired.

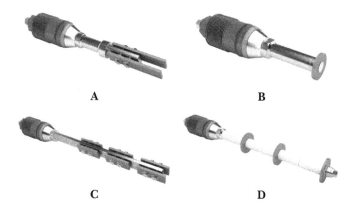

Figure 10.11 Typical strip and flush mass loss coupon assemblies: (A) typical strip coupon assembly, (B) typical flush mount disk coupon assembly, (C) ladder strip coupon assembly, and (D) multiple disk coupon assembly.
Images provided courtesy of Cosasco.

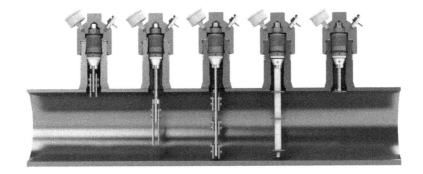

Figure 10.12 Intrusive and flush coupon schematic.
Image provided courtesy of Cosasco.

There are other configurations of coupon holders that are made to be installed and removed when a system is depressurized.

Location and Position

The coupon's location in a system can greatly affect the results because corrosion does not always take place uniformly throughout the system. Multiple coupon locations are required for most surface facilities to adequately monitor the various environments, and because corrosive conditions often will be different in the various parts of a system. Basically, a monitoring location should be installed at the beginning and end of flow, gathering or injection lines, and before and after each vessel, tank, and piece of equipment. In other words, to have a comprehensive monitoring system, locations are required each place the environment changes (or could change).

Figure 10.13 shows coupon locations in a gas condensate well facility where corrosion problems are expected and a corrosion inhibitor is being used for control. Coupons exposed at the well are used to monitor downhole inhibition; those ahead of the choke in the heater, monitor inhibition of the high-pressure flowline, and those after the heater reflect the increased liquids and the new pressure and temperature. Coupons exposed at the separator inlet reflect the environment in the heater-to-separator flowline; the locations on the condensate, gas, and water lines off the separator monitor each of those streams, and the sales gas location monitors the effectiveness of the dehydration tower in providing dry gas to sales. The two locations in the glycol system monitor the dry and wet glycol. By changing coupons in all these locations on the same schedule, a corrosion profile, or history, is developed. Profiling the system becomes quite useful for troubleshooting when changes in corrosion rates and problems occur.

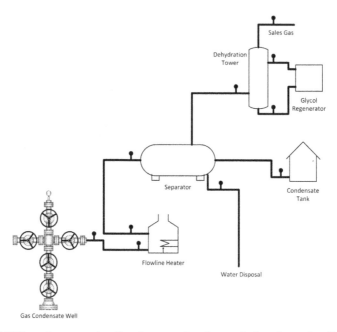

Figure 10.13 Corrosion coupon locations for comprehensive monitoring of a gas handling system.

Multiple coupon locations are particularly important in water flood systems. A typical water injection system will require coupons in each of the water supplies, after tanks and pumps (possible points of air entry), at intermediate headers, and at key injection wells scattered throughout the system. Locations only at the water plant or at the injection wells will not tell the full story. NACE SP0499[12] contains additional guidance for monitoring seawater injection systems.

Coupons in multiphase systems must be oriented so that they will be exposed to the water present, or they will not reflect corrosion in water wet areas. Figure 10.14 shows three different coupon positions in a portion of wet gas piping under two different operating conditions. The coupon results at each position will be different. In fact, under the conditions illustrated, coupons exposed in position "A" would indicate corrosion rates of "little-to-none," because they would not normally be wetted with water. Positions "B" and "C" should show corrosion because of water buildup and carryover. However, the numerical values of the corrosion rates may not be the same—it will depend on how much of the time position "B" is exposed to water and water wet conditions have been maintained. Fortunately, experience has shown that changes that affect corrosion rates, such as inhibitor addition, will affect both coupons.

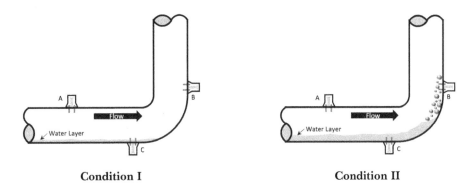

Figure 10.14 Corrosion coupon positions (A, B, C) in wet gas piping under different conditions: (A) lower flow rate with lower water volume; (B) higher flow rate with higher water volume.

As can be seen, "C" is the ideal position for wet gas piping. Even under condition (I), the submerged portion of the coupon will reflect the corroding conditions (general attack, pitting, attack at the water-gas-interface, etc.), although the overall corrosion rate may have to be recalculated based on the actual water wetted or corroding area. When "C" is not possible because of piping configuration and insufficient clearance, "B" is the preferred alternate position. As a generality, "A" should only be used for water-packed piping where submergence is assured. If "A" is the only choice in multiphase piping, it should be located in areas where the turbulence will be the highest. Thus, the chance of water wetting the coupon will be increased, and if reproducible corrosion rates occur on coupons in position "A," it can be used in that system. NACE SP0775[11] provides additional considerations for installation and interpretation of coupons.

Environment

The fluids to which the coupons are exposed are also important. As can be imagined, a blob of paraffin or an oil film coating part of a coupon can give inconsistent results. Paraffin or oil cannot be expected to coat the same area on every coupon put in the system. Consequently, coupons are most often used in gas wells, water systems, and other situations where such problems are kept at a minimum. However, the use of coupons can be helpful in any environment if factors such as water dropout, velocity, multiphase flow, etc., that may affect the results are taken into consideration. The visual descriptors of the type of attack (discussed under "Results, Evaluation, and Reporting" below) can be very important in evaluation of coupon exposures from variable environments.

Care and Handling

The handling of the coupon during installation and removal can affect the results. A drop of sweat or sweaty handprints can increase the rate of corrosion at that point. A greasy thumbprint can protect or partially protect an area of the coupon. The coupon must be free from corrosion when it is installed in the system and must be prevented from corroding further while it is being shipped back to the laboratory. One way of protecting coupons is to use inhibited paper envelopes. When installing coupons, they should be carefully handled by their edges and with clean rags or gloves. To avoid further corrosion after removal from the system, coupons should be immediately returned to their protective envelopes or wrappings (most coupon suppliers provide vapor phase inhibited paper envelopes or wrappers).

Exposure Time

Exposure time (that is, the time the coupon is in the system) is also important. Figure 10.15 illustrates the concept of exposure time. There is no rule to determine the exposure time—the time required to reach the system corrosion rate depends on the system corrosivity. Short-term exposures give a quick answer, but these can be misleading. For example, if the coupon's surface was prepared by abrasive blasting, a short (only a few days) exposure may indicate higher than actual corrosion rates due to the removal of the surface roughness. On the other hand, if the coupon has a polished surface, a short exposure may be misleadingly low because it takes time for corroding sites to begin. If pitting is a problem in a system, the pits may take several weeks to develop to the point where they may be detected and measured. Therefore, if an exposure time of a month or less indicates that a system is under control, the exposure time should be increased and the results compared with the shorter-term data.

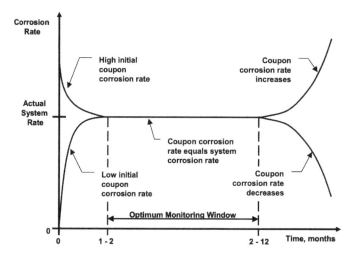

Figure 10.15 Example of coupon corrosion rate vs. exposure time for CO_2 corrosion.[2]

Results, Evaluation, and Reporting

There are other variables such as surface preparation, cleaning, etc., which can affect the results of coupon studies. Thus, to compare results from one exposure period to another and one coupon location to another, coupons should all be supplied by the same vendor. Each supplier has a somewhat different procedure for coupon preparation and processing; therefore, the corrosion rates determined from one supplier will not agree with those from a different supplier. Fortunately, experience has shown that even though the numbers are not the same, different vendors' coupons will track each other. That is, when corrosion rates vary, the results of the different vendors will likewise vary. As a generality, vendors are consistent with their coupon procedures, and results can be compared from one exposure to the next.

When analyzing the results of a coupon exposure, it is important to get the complete story. Often there are two levels of reporting; field grading by personnel that remove the coupons, and laboratory data. Field grading is a qualitative assessment and can provide an early indication of the results. Figure 10.16 shows an example of a visual comparator that field crews can use in grading coupons removed from the system.

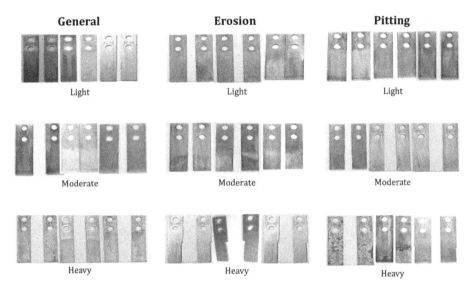

Figure 10.16 Visual comparator for coupon field grading.

Laboratory data is quantitative and will provide specific data about the coupons. The following types of information are determined from coupon exposures:

- Overall (or mass loss) corrosion rate
- Pitting details (pit depths, pitting frequency, and or pitting rate equivalent)
- Erosion
- Visual appearance of the type of attack
- Analysis of deposits or microbiological culture

Overall corrosion rate is based on the coupon's mass loss, its surface area, and the exposure time. The corrosion rate is determined using Equation (10.2) and Table 10.1:

$$\text{Corrosion Rate} = \frac{(\text{Initial Weight} - \text{Final Weight}) \times \text{Unit Factor}}{\text{Area} \times \text{Density} \times \text{Exposure Period}} \quad (10.2)$$

Table 10.1 Common oil field corrosion rate calculation parameters.

Variable	SI Units	US Units
Corrosion Rate	millimeters per year (mm/y)	mils per year (mpy)
Mass	grams (g)	
Density	grams/cubic centimeter (g/cm^3)	
Exposure period	days (d)	
Area	centimeters squared (cm^2)	inches squared (in^2)
Unit Factor	3,650	22,300

The overall corrosion rate values, determined from the weight loss, assumes the corrosion was uniform over the coupon's entire surface. Of course, this is not usually true. Therefore, a complete report should include a visual inspection of the coupon to determine the type of attack. Figure 10.17 shows example general corrosion field grades.

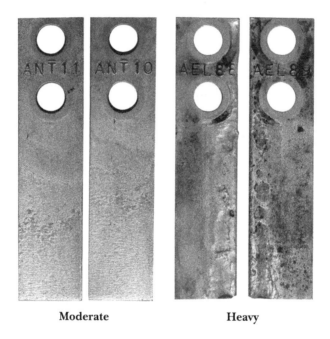

Moderate   Heavy

Figure 10.17 Examples of general corrosion field grades.

Pitting is often the most important corrosion characteristic determined by coupon exposures. As discussed in Chapter 3, Section 3.4, pitting (and similar forms of localized attack) is the usual corrosion mechanism in oil field systems. Figure 10.18 shows examples of pitting corrosion field grades. Additional guidance on pitting evaluation can be found in ASTM G46.[13] Longer-term (one month or greater) coupon exposure is the one monitoring method that allows for an evaluation of pitting:

- Visual examination can describe the shape and form of pits (broad and flat, small and sharp, round, or irregular, etc.). Changes in pitting shape can flag changes in the corrosion environment.
- The number of pits or pitting frequency (per coupon or per unit area) can be determined.
- Pit depths may be measured, and the "pitting rate equivalent" can be determined by Equation (10.3). The pitting rate can be misleading as the result indicates a linear rate. However, for carbon steel pit growth rate typically follows a power law where the time exponent is between 0.5 and 3.

$$\text{Pitting Rate} = \frac{\text{Depth of Deepest Pit} \times 365}{\text{Exposure Time (days)}} \tag{10.3}$$

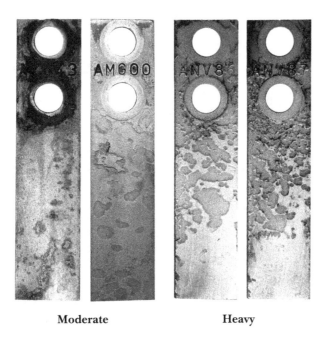

Moderate   Heavy

Figure 10.18 Examples of pitting corrosion field grades.

Erosion is another type of attack that can be determined by coupons. Figure 10.19 shows damage (notch) to the leading edge of a standard mass loss coupon due to erosion. It is difficult to calculate an erosion rate on the basis of coupon data alone, but the qualitative results can be used to inform the overall integrity program.

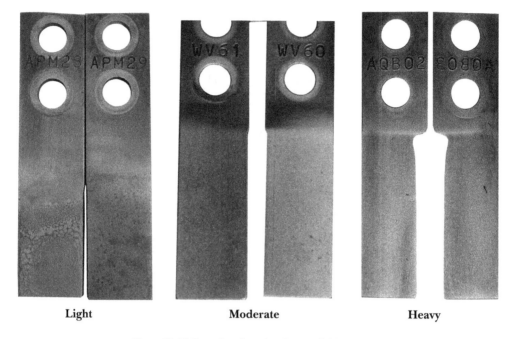

Figure 10.19 Examples of erosion damage field grades.

High-resolution surface profilometry can also be used to scan the surface of coupons and provide even more quantifiable data (Figure 10.20). Additional details such as maximum pit depth, pit density, average pit depth, number of pits, volume/depth of individual pits, and total volume, can also be determined from this method. There are limitations to this method, for example how to handle the edges and mounting holes or the fact that larger coupons may require multiple scans that need to be combined into a single file.

Figure 10.20 Laser profilometry scan of coupon surface.

"Type of attack description" is a method of evaluating the corrosion on a coupon. The descriptor can include various terminology (also refer to Chapter 3, Sections 3.2–3.4):

- "Etched" would indicate a more or less uniform roughening of the coupon.
- "Overall attack" would be obvious corrosion over the entire coupon surface.
- "Areas of attack" would indicate nonuniform corrosion over irregular areas, as opposed to distinct pitting. Some of the surface would show little if any corrosion while other areas could be severely corroded.
- "Localized pits" would describe a few pits scattered over the coupon (refer to Chapter 3).
- "Pitting" would imply pitting over most of the coupon's surface (refer to Chapter 3).
- "Sharp pit," "broad pit," or "round pit" would identify the shape and appearance of pits (refer to Chapter 3).
- "Erosion" would indicate leading edge damage that is not corrosion.

These descriptions can aid in the interpretation and evaluation of coupon results as long as they are well defined and consistently applied. Changes in the type of attack from previous exposures can be an indication of a change in the corrosion environment. The descriptors also can help explain differences in results in systems where paraffin or deposits may cover parts of the coupon's surface. In some cases, such deposits may shield the surface from attack; in others, the opposite may be true.

Analysis of deposits collected from the coupon when it's removed from the system can be used to distinguish between scale, precipitates, and corrosion products, as well as determine the type of corrosion product. Samples of deposits or films on a coupon may be placed in prepared culture media for microbiological analysis.

Special Studies

Although, as stated earlier, most coupons used in routine monitoring are made of low-carbon steel, often grade 1020 (UNS G10200). In some cases, the heat treatment of the coupon should match that of the equipment, especially for higher strength grades of pipeline and downhole tubing. Coupons can also be fabricated from other materials for special studies. Alloy coupons, including some of the stainless steels and the other CRAs discussed in Chapter 4, are exposed in field systems to evaluate their performance in actual producing situations. NACE RP0497[14] and ASTM G4[15] provide additional guidance on different types of coupon testing. Figure 10.21 illustrates how coupons can be used in corrosion inhibitor performance evaluations. New coupons should be installed for each chemical change when performing evaluations of this type. While laboratory studies can give very good data, it is virtually impossible to duplicate all system conditions in the laboratory. Thus, just as mentioned in other chapters, field studies and tests are necessary.

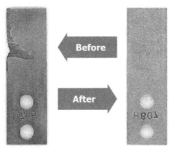

Figure 10.21 Coupons used for corrosion inhibitor evaluations.

10.4.1.3 Extended-Analysis (EA) Coupon

Extended-analysis coupons are similar in size and shape to mass loss coupons. The main differences are the purpose of the evaluation, the level of care in handling, the length of exposure period, and the extent of the evaluation performed.[1,9] EA coupons are capable of providing more insight as to the corrosion mechanisms versus standard mass loss coupons.[16] Figure 10.22 shows scanning electron microscope images of EA coupons.

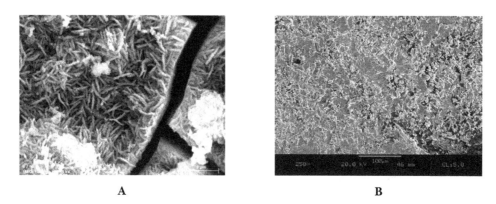

Figure 10.22 Scanning electron microscope (SEM) scans of EA coupons: (A) dried biofilm (mag 10,000×), and (B) pit initiation on EA coupon surface.

10.4.1.4 Electrical Resistance Probes

The electrical resistance (ER) technique determines metal loss by measuring the increase in resistance of a metal specimen as its cross-sectional area is reduced by corrosion.[1,17] The ER probe is similar to a coupon, except the corrosion rate is measured as a function of the change of resistance with time. There are standard and high-resolution ER probes. Corrosion rate is calculated by a simple formula [Equation (10.4)] where "K" is the probe factor supplied with each probe:

$$\text{Corrosion Rate} = \frac{(\text{change in dial reading}) \times (\text{probe factor})}{\text{time between readings}} = \frac{\Delta d \times K}{\Delta t} \quad (10.4)$$

Several styles of standard ER probes are available to fit different situations. The measuring element may be in the form of a wire, thin wall tube, or flush surface element. Probes are available with different wire diameters and tube wall thicknesses to provide a range of sensitivities. Figure 10.23 shows examples of several different probe types.

Figure 10.23 Typical ER probes.

Like coupons, ER probes can be used in many situations. The same comments on location and orientation apply to ER probes, as were discussed for coupons. In fact, in many cases, coupons and ER probes are used together, as seen in Figure 10.24. The ER probe is used to closely follow changing rates and/or to flag sudden changes (e.g., the lack of inhibition), and the coupon results are used to determine overall rates and pitting effects over longer periods of time. ER probes do not measure pitting rates. In fact, severely pitting situations may dictate the use of the tubular (cylindrical) element probes rather than the wire element probes (the cross section [resistance] of a tubular probe is less affected by pits). Another consideration of wire versus tubular probe elements is when iron sulfide (FeS) is present in the system. Iron sulfide is electrically conductive and can bridge across the wire element and result in erroneous measurements. Tubular elements can be used to avoid this problem.

Figure 10.24 ER probe installed in an access fitting and connected by cable to an automatic data-logging device. This example is from an offshore platform, which has space constraints.

The ER probe is installed in the system and the change of resistance is measured with a meter. These probes can be read periodically using a handheld logger, connected to an automatic data logger, which is downloaded periodically; or it is continuously monitored through a hardwired or wireless connection to the data acquisition system (Figure 10.25).

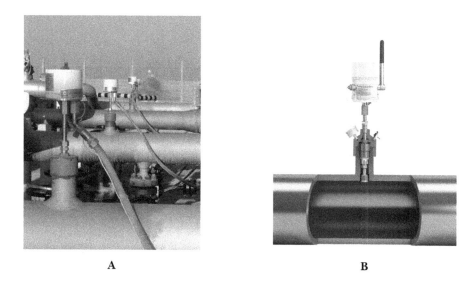

Figure 10.25 High-resolution ER probes mounted in parallel lines and hardwired to the data acquisition system: (A) hardwired, and (B) wireless.
Photographs provided courtesy of Cosasco.

A baseline reading is made after the probe is installed and a suitable time elapses for it to come to equilibrium with the environment (temperature, surface conditioning). The equilibrium time will vary with the environment; the least will be a few hours, but more commonly will require 1–2 days.

The time required to measure a given change in corrosion rate, known as "response time," is a function of several variables.[18] The response times for a standard ER probe with a 5 mil (0.13 mm) element, at various corrosion rates, is shown in Figure 10.26. Within the range of most commonly experienced corrosion rates (0.03–0.76 mmpy [1–30 mpy]) response times will range from 6–175 hours. These numbers are somewhat optimistic, in the sense that several data points are required to establish a corrosion rate. These estimates should be doubled to give practical response times ranging from 12–350 hours. High-resolution ER probes can reduce the response time significantly allowing for real-time monitoring.

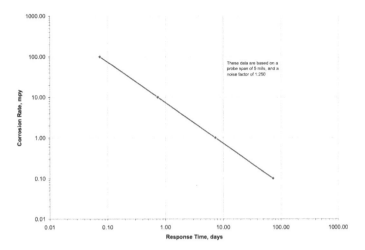

Figure 10.26 Standard ER probe response time.[18]

Data may be plotted graphically to obtain corrosion rates. Typical metal loss data obtained from high resolution ER probe is shown in Figure 10.27. This example shows the probe response to changing corrosion inhibitor concentrations. The points indicate metal loss (dial) readings and the solid lines show the slope of the different segments. These slopes are the corrosion rate for each segment. The steeper the slope, the higher the corrosion rate. For each step increase in corrosion inhibitor there was a corresponding decrease in the corrosion rate.

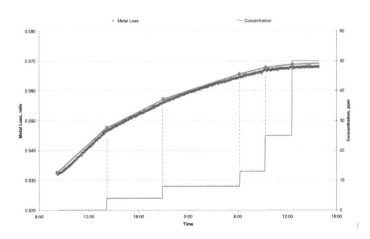

Figure 10.27 High resolution ER probe response (top line) due to increasing corrosion inhibitor concentration (bottom stepped line). The red line represents the slope of the metal loss data at a given inhibitor concentration.[18]

CHAPTER 10: Corrosion Monitoring and Inspection

Interpretations of the data obtained with an ER probe are subject to the same general limitations as the data from coupons. For example, deposits on the probe surface can shield the element from the corrosive environment. In some cases, iron sulfide scale laid down on the element is electrically conductive, causing an apparent increase in the cross section of the exposed element. This results in a lower corrosion rate measurement, zero apparent corrosion, or a negative reading (as an apparent increase in metal rather than a decrease through corrosive attack). Probes are more fragile than coupons and, even with a perforated shield around the element, their use may be limited in high velocity situations. Shields can also create an artificial "dead" (no flow) area.

As implied, ER probes are the most useful in systems or situations where corrosion rates may change rapidly and may be varying. Probes have been used to troubleshoot corrosion problems and to evaluate treating schedules and chemical changes in corrosion inhibition programs. Corrosion rates determined from ER probes are typically not used to determine inspection frequency.

ER probes can also be used for erosion rate monitoring. Essentially, the carbon steel element is replaced with a corrosion resistant alloy. In doing so, change in metal loss over time would be due to erosion. There are only a few probe configurations that lend themselves to erosion rate monitoring. Figure 10.28 shows an erosion ER probe with the element angled at 45° to create an impingement area for solids in the fluid stream.

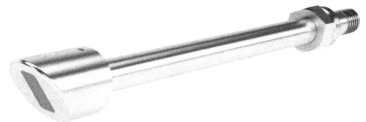

Figure 10.28 ER based erosion probe.
Photograph provided courtesy of Cosasco.

10.4.1.5 Test Spools

Specially prepared nipples or spools can be installed in a system to serve as large coupons. Such test spools are often used in connection with other monitoring techniques and may be exposed for several months (e.g., 6–24 months). The results of their exposure reflect the long-term effects of corrosion. They also represent a 360° sampling of the interior pipe surface and give an accurate sampling of dynamic effects. Test spools may be weighed, and corrosion rates determined as with coupons. Furthermore, the ends of the spool can be capped and sealed as soon as it is pulled to preserve any deposits and corrosion products for laboratory examination and analysis. However, their biggest contribution is usually their physical appearance. The test spools are sacrificial in nature and once removed can be cut open, inspected, and pit depths measured. For new processes, isolated test spools made from different construction materials (e.g., carbon steel and 316SS) and exposed to the same fluids can provide useful performance data.

10.4.1.6 Electrochemical Methods

Because corrosion is fundamentally an electrochemical process, electrochemistry can be used for monitoring. The main advantage of electrochemical techniques is they provide the corrosion rate as the corrosion occurs.[19] The primary electrochemical methods include linear polarization resistance (LPR) probes that measure corrosion rate, galvanic probes that are primarily used as oxygen or inhibitor film detectors, and hydrogen probes that detect corrosion by detecting hydrogen atoms passing through metal. Each technique has its advantages and limitations, but all can provide useful information for evaluating corrosion control programs.

Linear Polarization Resistance Probes

This technique measures instantaneous corrosion rates by utilizing an electrochemical phenomenon known as linear polarization resistance (LPR). LPR involves measuring the current required to change the electrical potential of a specimen corroding in a conductive fluid by a few millivolts. It has been shown that as long as the change in potential is no more than 10–20 mV, the rate and nature of the corrosion reactions are undisturbed, and the amount of applied current necessary to effect the potential change is proportional to the corrosion current. Thus, the instrument can be calibrated to give a direct readout of corrosion rate.[1,17]

LPR instruments are available that use either two or three electrodes for corrosion rate measurements. Several electrode configurations are also available. Some are made with standard pipe plugs. Others are made for insertion and removal through access fittings (Figure 10.29).

Figure 10.29 Two-electrode linear polarization probe.
Photograph provided courtesy of Cosasco.

Use of the LPR technique for corrosion rate measurements is limited to electrically conductive solutions because current must flow from one electrode to another through the solution. Measurements cannot be made in gas or oil; reliable measurements are difficult to make in intermittent gas, oil, and water flow. However, the corrosiveness of oil-water mixtures can be measured if water is in the continuous phase. Probe elements can become "shorted" by conductive media to produce erroneous results. Once installed in a system, sufficient time must be allowed for the electrodes to reach equilibrium with the environment before a valid measurement is possible. LPR probes should be installed in such a manner that they can be easily removed for cleaning, inspection, or replacement.

Galvanic Probe

As the name "galvanic probe" implies, it is a bimetallic couple (usually brass and steel) so constructed that the current output can be measured on a microamp meter.[20] When a galvanic probe is placed in a water system, the current output will stabilize at some value that reflects the conditions in the system. Any change in the system that will change the current output will be reflected on the meter. While the probe is affected by many conditions (flow rate, temperature, inhibitor films, etc.), it is very sensitive to oxygen. In most oxygen-free systems, the probe will polarize and the output will approach zero. Even small amounts of oxygen will depolarize the probe, and the current output will be increased. Anyone who has not used a galvanic probe will be surprised at its sensitivity and response to minute quantities of oxygen. An example of a galvanic probe application is described in Chapter 9 Section 9.5.2.

Because of these properties, the galvanic probe can be an extremely valuable asset for routine monitoring and troubleshooting problems. The galvanic probe is a rugged, inline instrument that requires a minimum of care. It provides a "telltale sign" that the operator can watch. It will "flag" air leaks and malfunctions in water systems so that corrective measures may be taken before severe damage has occurred. The body of a two-electrode LPR probe designed for installation in access fittings can be equipped with dissimilar metal elements and converted to a galvanic probe.

Like most tools, the galvanic probe has its limitations. Unlike the previous probes that have meters connected to their probes only when taking readings, the galvanic probe and meter must be connected all the time so the probe can remain polarized continuously. In a few produced-water systems, it fouls quickly and may require frequent cleaning. In many systems, such as seawater, only infrequent cleaning is required. Most systems can get by with only quarterly maintenance. A limitation of galvanic probes has been found to occur in sour water systems. The two dissimilar metal probe elements are selected so that one is always anodic to the other. The probe is calibrated assuming that carbon steel is the anode and a copper or copper alloy is the cathode. Exposure to sour liquid can result in copper becoming anodic to carbon steel. In this case, a CRA element should be substituted for copper, but the current readout and its conversion to percentage of oxygen saturation will not be accurate.

Although the galvanic probe oxygen detector is not a true oxygen meter (that is, it does not give a readout in parts per million of oxygen), it does indicate very small changes in low levels of dissolved oxygen. It cannot be used at high oxygen levels, because once it is fully depolarized, the reading levels out. Fortunately for the purposes elaborated here, it is the most sensitive at zero to about 200 ppb (parts per billion). Best practice would be to calibrate the probe response to oxygen measurement from another technique.

Hydrogen Probe

The simplest form of a hydrogen probe consists of a hollow steel tube sealed on one end and equipped with a pressure gage on the other end. Such probes are primarily used in systems containing H_2S and take advantage of the fact that, as discussed in Chapter 1, Section 1.2 and Chapter 3, Section 3.11, sulfides retard hydrogen atoms combining to form hydrogen gas molecules. Similar to the phenomena of hydrogen blistering, a portion of the atomic hydrogen generated by the corrosion reaction will diffuse through the tube wall. Once inside the void space in the tube, the hydrogen atoms will form hydrogen gas molecules, which are too large to diffuse back through the tube wall.

Thus, the pressure in the tube will rise in proportion to the amount of hydrogen in the tube, which is a function of the amount of hydrogen generated by the corrosion reaction. In other words, the higher the rate of hydrogen generation, the higher the corrosion rate.

In addition to the simple probe, other configurations have been developed that do not require entry into the corrosive media.[21] By various means, these devices collect and measure the hydrogen that diffuses through the pipe or vessel wall. Sometimes referred to as "strap-on" probes, the trapped hydrogen is measured electrochemically in some devices and by reducing a vacuum in others. All seem to work well as long as the device is sealed at the wall of the pipe.

The hydrogen probe is a qualitative or semiquantitative tool that can be used with other corrosion rate measurement techniques. It has been most commonly used in sour systems, but has found application in some sweet systems. However, in the absence of sulfide, the sensitivity is much lower. If oxygen is present, the hydrogen probe will not work because the oxygen will combine with the hydrogen atoms to form water and there will be very few, if any, atoms to penetrate the probe.

10.4.2 Nonintrusive Techniques (Inspection)

Nonintrusive measurement techniques do not require access to the interior of the equipment. Consequently, these techniques avoid the risks associated with intrusive monitoring activities. These techniques are commonly referred to as "inspection" or "nondestructive examination" (NDE) and they serve as a vital part of any corrosion monitoring program; as such, they become a vital part of any corrosion control program. Many of these techniques can be labor intensive and require competent personnel to perform the activities and interpret the results.[22]

Various inspection techniques are available: a simple "look in it," inspection scopes, feeler and/or depth gauges, electromagnetic, ultrasonic and radiographic methods. Tools and methods for inspection of downhole, as well as surface facilities are available. Like the other parts of the monitoring program, the types and extent of monitoring will vary with the project, its location, stakes, and risks.

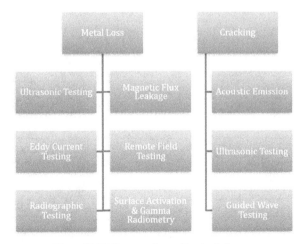

Figure 10.30 Common inspection techniques.

10.4.2.1 Visual Testing (VT)

Of all the methods available for detecting corrosion, visual inspection is the most reliable. The capability of visual inspection is usually dependent on the available access, the surface condition of the equipment or component, and the level of lighting. It may be difficult to arrange in many cases and impossible in others. However, every opportunity for visual inspection should be utilized. In fact, careful visual inspection should be a part of maintenance procedures any time a line is open, a pump is down, or a tank is being cleaned. At a minimum, the procedure should include taking notes, making sketches, and measuring details that might be relevant. Use a camera where and when possible. Also, probe a sample area for pitting, particularly if significant corrosion product is present. Pits can be obvious, but often they will be hidden beneath scale or deposits. Pit depth gages are available (refer to Chapter 3, Figure 3.21).

Visual inspection may include the use of equipment such as borescopes (including those using fiber optics), and video equipment to record the observations. Subsea equipment is inspected by divers or by using remote operated vehicles equipped with cameras. Rope access (abseiling) may be necessary to gain access to difficult-to-reach areas such as stacks and below the cellar deck of offshore platforms. However, unmanned aerial vehicles (UAV, or drone) are being used more frequently and they can be fitted with a variety of high powered optics. Meaningful visual inspections can be accomplished with as little equipment as a tape measure, mirror, safety flashlight, and a pocketknife. The key words are "look, see, and document." Figure 10.31 shows different examples of visual testing.

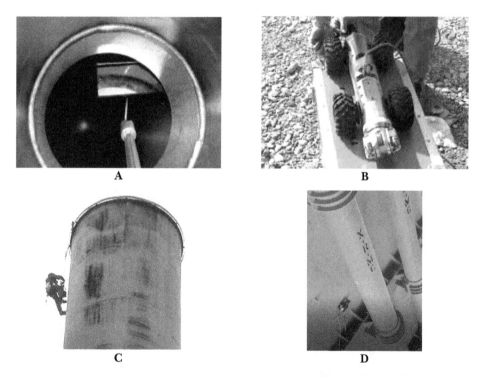

Figure 10.31 Visual inspection examples: (A) visual inspection using a mirror, (B) robotic crawler used for visual inspection of pipeline, (C) visual inspection of a stack using rope access, and (D) magnetic crawler performing visual inspection offshore.

Another area of visual inspection that is usually not even thought of as inspection is the external inspection that can be made whenever a person is around wells and facilities. Keeping on the lookout for changes or other signs that affect corrosion control can "flag" problems before they occur. Just as visual inspection is the basic inspection, the human eye is the basic inspection tool. However, like any tool, it needs to be properly used. A great thinker once said, "We see with our eyes. We observe with our minds."

It is very important when personnel are around any equipment, that they use their eye/mind tools to visually inspect for signs that indicate corrosion may be occurring, or that control programs have gone awry. Watch for anything that might indicate a problem developing. For example, an open thief hatch on a water tank that would allow air into an otherwise air-free system, or seals on a centrifugal charge pump leaking, or damage to the covering over thermal insulation that could allow water entry, or a shorted isolation (insulating) flange.

10.4.2.2 Ultrasonic Testing (UT)

The determination of component thickness by UT is probably the most common application of NDE after visual inspection. UT inspection techniques use ultrasonic energy to measure the thickness of a metal object and to locate defects or flaws in the metal. In ultrasonic inspection, an ultrasonic wave, generated by a piezoelectric transducer, is transmitted through a liquid couplant to the metal surface. Ultrasonic waves will travel through the metal until they encounter a change in material property, interface, discontinuity, or the opposite surface of the wall.[23-24] At that point, the waves are reflected to a receiver and transformed into an electric impulse by the transducer (Figure 10.32). The time between the initial pulse and the first reflection is relative to the distance traveled by the sound wave with respect to the material's velocity properties. This distance is equal to twice the thickness of the metal or twice the distance to a discontinuity. UT is also known as the "pulse-echo" technique and is commonly used for both thickness measurements and flaw detection.

In ultrasonic flaw detection, a small probe is used to scan the component material for defects, whether planar (crack-like) or volumetric. By scanning the probe and observing the response on the flaw detector screen, the location of defects can be determined and their size estimated. By suitably designing the probe, ultrasonic beams can be introduced into the component material at almost any angle.

While thickness gauging can provide an accurate measurement of component condition, erroneous results can be reported. For example, spot measurements will more than likely miss pitting corrosion or other localized forms of damage. Also, the presence of paint and similar coatings on the component surface can introduce significant errors, adding several times their thickness to the total ultrasonic reading. While coating thickness can be compensated for by the inspection technician, some users remove the coating under the transducer. Thickness gauge scanning of selected areas of equipment with relatively simple geometry, e.g., pressure vessels and pipelines, can be automated (AUT).

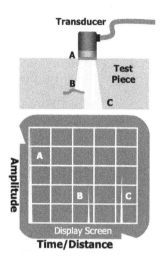

Figure 10.32 Pulse echo display.[2]

Ultrasonic testing (UT) is usually a manual operation, using a small ultrasonic probe connected to a handheld gauge. The main use of thickness gauging is to determine the remaining wall thickness, particularly in component areas where internal corrosion/erosion is suspected. For assessing component condition, thickness surveys are made by making a number of "spot" measurements in a grid pattern covering the component's surface or the local area of concern. Selections of appropriate thickness monitoring locations, and the procedure for calibration and collecting readings, are important factors. Figure 10.33 shows examples of manual and automated UT.

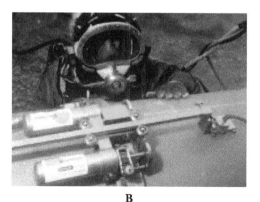

A   B

Figure 10.33 Ultrasonic inspection examples: (A) UT of a weld on a pressure vessel nozzle, and (B) AUT inspection of a subsea pipeline.

UT thickness measurements are referred to as "scans." There are three types of scans designated as "A-scans," "B-scans," and "C-scans." The terms have to do with the patterns used to take the readings:

- "A-scan" is a single point thickness reading. A-scan display is a front view showing UT "pips" from LCD display. Pips are the vertical high points in the signal on the UT screen (cathode ray tube, etc.). The pips occur where the sound wave enters the component and where it reflects off the back wall (e.g., corroded surface or interior, imbedded flaw). Figure 10.32 shows an example A-scan.
- "B-scan" is a number of thickness readings in a line. The line may be along the bottom of a pipe or vessel, along the back of an elbow, circumferentially around a pipe or vessel, vertically across the fluid level in a vessel or tank, etc. B-scan results are often presented as numerical thickness readings, and graphically as a thickness profile (cross section).
- "C-scan" is a large group of thickness readings in a grid pattern. The images provide a top (plan) view, where a color indicates flaw depth relative to a depth chart. C-scans represent an area of the pipe or vessel wall, and the C-scan results are most often processed by computer and presented as thickness maps, or as computer generated 3-dimensional graphics with numerical notations for areas of critical metal loss. Automated C-scans use probes that are mechanically held in place and moved across the metal surface with computer guidance. The readings are recorded, processed, and displayed by a computer. The automated C-scan (AUT) can take and process thousands of readings per square foot and thus present very detailed thickness maps or other output. Where appropriate, the computer can construct B-scan type cross sections from the stored data and report specific points of interest as if they were A-scans. Figure 10.34 provides an example of how UT can be displayed.

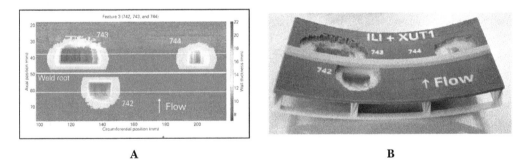

Figure 10.34 Visualization of UT data: (A) C-Scan showing pipeline anomalies adjacent to a weld, and (B) 3D-printed model of the pipeline anomalies.[25]

Sometimes, when less detailed information is required (such as verifying that a wall has sufficient thickness for welding or hot-tapping), a technique known as "scrubbing" may be used. As the name implies, the probe is moved back and forth over an area looking for the lowest wall thickness. The lowest thickness will often be the only one recorded.

Permanently attached transducers can improve the accuracy and repeatability of measurements by both removing errors associated with the placement of transducers and the interpretation of the measurements by different personnel. Figure 10.35 shows two different examples of permanently installed monitoring devices: an array of UT transducers, and a guided wave sensor ring. These techniques are capable of being monitored online.

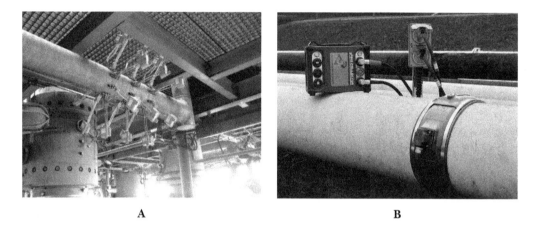

Figure 10.35 Examples of permanently installed monitoring: (A) UT sensor array installed on piping, and (B) guided wave sensor on a pipeline.

UT is one of the most powerful NDE methods available. The capability of this method to accurately determine defect size, in particular the height or depth from the measuring surface, makes it an integral part of fitness-for-service assessment. A high level of operator skill is required, although this is partly mitigated by semi- and fully automatic techniques. Thin materials and welds are difficult to test, as are coarse-grained structures, such as castings and austenitic stainless steels.

10.4.2.3 Radiographic Inspection (RT)

In petroleum production, radiography is used primarily to inspect welds at construction sites and to inspect piping and production equipment for corrosion and erosion. In conventional radiographic testing, a source of ionizing radiation ("X" or gamma) is used to produce an image of the component on film by placing the radiation source on one side of the component and the film on the other side.[26-27] Following exposure to radiation, the film is processed and viewed on an illuminated screen for visual interpretation of the image. Radiography gives a permanent record (the exposed film), which is a major advantage of the method, and is widely used to detect volumetric flaws (surface and internal).

Thin areas or voids in material are indicated by darker areas because more energy is able to penetrate the piece during the exposure period while thicker or denser areas are indicated by lighter areas because less energy was able to penetrate the piece during the exposure. The images are viewed as contrast in shades of gray and appear much like shadows. Figure 10.36 shows an example of a radiographic image of a corrosion coupon inserted into a "T" piece section of piping.[2] Chapter 3, Figure 3.23 shows a radiographic image of sand buildup in a pipe with resulting under-deposit corrosion.

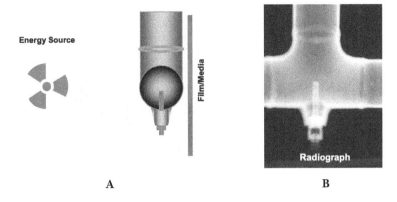

Figure 10.36 The radiographic technique where (A) is an artist's rendering of how a radiograph is produced, and (B) is the actual radiograph produced.

X-ray equipment ranges from about 20 kV to 20 MV (the higher the voltage, the greater the penetrating power of the radiation and the greater the thickness of component that can be tested). Gamma radiography is carried out using radioactive isotope sources (e.g., cobalt-60, iridium-192), although its sensitivity is generally less than that achievable by X-ray radiography. It is widely used for fieldwork because of its greater portability. More recently developed techniques include real time and digital/computed radiography in which the image is produced electronically, offering numerous advantages in terms of speed, storage, and image interpretation, Figure 10.37.

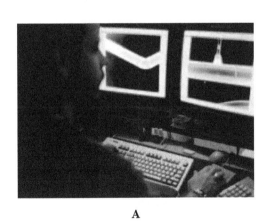
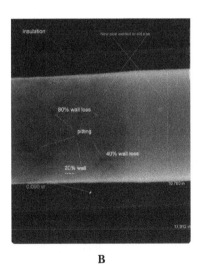

Figure 10.37 Digital radiography interpretation and advanced software allowing detailed measurements: (A) a radiography interpreter reviewing images, and (B) an annotated radiograph with example measurements.

CHAPTER 10: Corrosion Monitoring and Inspection

Radiography can be used to determine the wall thickness of pipe, as well as to detect pitting or other localized corrosion damage. Metal defects such as porosity or inclusions can be detected, and weld quality can be determined. Figure 10.38 has representative photographs of a pipeline and inspection radiographs. It is easy to see the agreement between RT images and the physical examples.

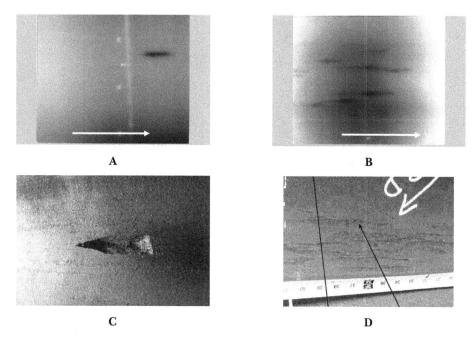

Figure 10.38 Representative samples: (A–B) RT images, and (C–D) physical samples.

10.4.2.4 Magnetic Particle Inspection (MPI)

Magnetic particle testing is an NDT technique used to locate surface and slight subsurface discontinuities in materials that have high magnetic permeability.[28-29] Lines of magnetic flux must cross the flaw, and not be parallel to it. These discontinuities are associated with crack-like or planar flaws. MPI is not used to measure wall thickness. In MPI, the component is magnetized either locally or overall. If the component is sound, the magnetic flux is predominantly inside the material. If, however, there is a surface-breaking defect, the magnetic field is distorted, causing local flux leakage. The flux leakage is detected and displayed by covering the surface with fine iron particles, usually in dry powder form but sometimes suspended in a liquid. The particles accumulate at the regions of flux leakage revealing the defect as a line of iron particles on the component surface (Figure 10.39).

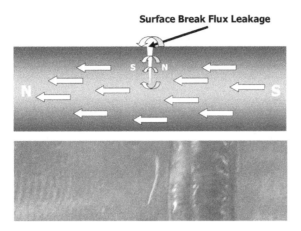

Figure 10.39 Principle of MPI: (Top) artist's rendering showing the magnetic flux direction, and (Bottom) image of the fluorescent particles under black light.[2]

MPI is applicable to metals that can be magnetized. It is important to ensure that the direction of the magnetic flux is appropriate for the defects expected. A variety of manually operated equipment is available, the most common method of magnetization being the application of a permanent or electromagnet (AC yoke) to the component surface. On hot surfaces, dry powders may be, and on dark surfaces a thin layer of white paint increases the contrast between the background and the black magnetic particles. The most sensitive technique uses fluorescent particles viewed under UV (black) light.

Figure 10.40 Flaw detection using MPI: (A) wet fluorescent MPI showing multiple linear defects in cold rolled tubing, and (B) MPI of weld for crack-like flaws.

MPI practices consist of dry and wet, visible and fluorescent methods. Wet visible and wet fluorescent methods use magnetic particles suspended in liquid. The visible method uses white light, while the fluorescent method requires ultraviolet light for evaluation. Because of the simplicity of application, the dry powder visible method is the most common practice employed for field inspection because the technique does not require carrier liquids or special lighting.

10.4.2.5 Penetrant Testing (PT)

Penetrant testing is a relatively fast, low-cost method to inspect for finding breaks in continuity at the surface of specimens. In PT, liquid penetrant is drawn into surface-breaking defects by capillary action; application of a developer draws out the penetrant in the defect producing an indication on the component surface.[30-31] Penetrant testing can be applied to any nonporous clean material, metals or nonmetals, but is unsuitable for dirty or very rough component surfaces.

PT is a chemical test, which uses a dye-carrying liquid that is applied to the surface of a component to make visible surface-related discontinuities such as cracks, porosity, and forging laps (Figure 10.41). While there are several options available, red-dye penetrant is the most commonly used; fluorescent penetrants are used when maximum flaw sensitivity is required. Penetrant testing can be automated, but in the field, the method is manually applied. The common application for field inspections is kits containing aerosol cans of dye penetrant, cleaner/remover, and developer.

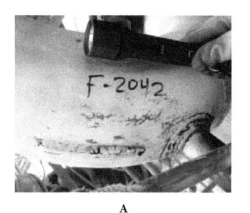

A   B

Figure 10.41 Dye penetrant testing: (A) external chloride stress corrosion cracking (SCC) of stainless steel piping, and (B) dye penetrant inspection of stainless steel socket welded fittings.

10.4.2.6 Eddy Current Testing (ET)

In eddy current testing, a coil carrying an AC current is placed close to the component surface. The current in the coil generates circulating eddy currents in the component close to the surface; these in turn affect the current in the coil by mutual induction. Surface-breaking defects and material variations in the component affect the strength of the eddy currents. Therefore, by measuring the resultant electrical changes in the coil, defects can be detected.

Basic test equipment consists of an alternating current source, a connected coil of wire (referred to as a probe), and a voltmeter to measure the voltage change across the coil. The examination process involves moving and rotating the probe or coil over and through the test piece to measure the voltage changes across the coil. Figure 10.42 shows an inspector using ET to examine a bank of boiler tubes.

Eddy current testing is applicable to all electrically conducting materials. Conventional ET is essentially a near-surface technique, especially for steels because of their relatively high magnetic permeability. It is widely used in plant inspection for nonferrous materials, where eddy current penetration is deeper, especially for corrosion detection in heat exchanger tubes.

Advantages of ET- over UT flaw detection are faster scanning speeds and no requirement for a couplant. It can be used through surface coatings up to a few millimeters thick, whereas ultrasonics can only be used through relatively thin layers. The inspection of welds in ferritic steels can be problematic because the response is dominated by changes in the magnetic permeability across the weld.

Figure 10.42 Tube set inspection using ET.[2]

10.5 Intrusive Inspection

While the inspection techniques outlined earlier are typically deployed in a nonintrusive manner for surface equipment, there are several circumstances where the techniques are also deployed in an intrusive manner. This section describes the deployment of inspection techniques into a few different equipment types and locations.

10.5.1.1 Downhole Inspection

All downhole inspection techniques are intrusive and require specific well intervention equipment, as well as requiring the well to be out of service throughout the inspection. The two main types of inspection tools available are multifinger caliper and electromagnetic.

Multifinger Caliper

Downhole caliper survey tools have been in use for many years. The caliper measures the internal diameter of tubing or casing and indicates general corrosion or pitting attack. Essentially, these tools use feelers that ride the inside wall of the pipe as the tool is pulled out of the well. These feelers draw a profile of the wall as they extend into pits or collars, or pick up other variations in diameter. The number of feelers vary with the size of the tubing being calipered and with the service companies' proprietary design of the tool. The idea is for the feelers to cover as much of the inside wall of the pipe as is practical.

Initial caliper technology would only record the feeler that moves the farthest and, thus, report the deepest penetration at a specific location in the tubing. The accuracy of this tool is dependent on centralization of the tool. The current generation of caliper tools are capable of recording the movement of all the feelers, as well as the tool inclination and relative bearing of each finger. This allows a better understanding of corrosion mechanisms by providing more accurate mapping of the corrosive damage recorded and any correlation to well bore trajectory (Figures 10.43 and 10.44).

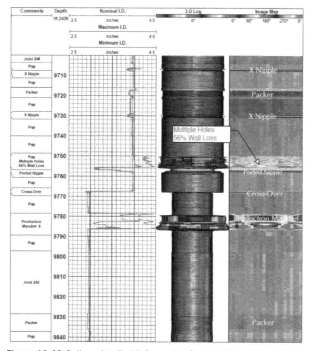

Figure 10.43 Caliper detailed information for a section with wall loss.

Joint No.	Jt. Depth (Ft.)	Pen. Upset (in.)	Pen. Body (in.)	Pen. %	Metal Loss %	Min. I.D. (in.)	Comments	Damage Profile (% wall) 0 50 100
224	8767	0	0.01	5	2	3.92		
225	8803	0	0.02	6	4	3.92		
226	8842	0	0.01	5	3	3.93		
227	8880	0	0.02	6	2	3.91		
228	8918	0	0.02	6	3	3.92		
229	8957	0	0.01	4	1	3.90		
230	8995	0	0.01	5	1	3.91		
231	9032	0	0.02	6	2	3.90		
232	9069	0	0.02	6	2	3.92		
233	9107	0	0.02	7	3	3.93		
234	9145	0	0.01	5	3	3.92		
235	9183	0	0.03	10	4	3.92		
236	9221	0	0.01	4	2	3.92		
237	9258	0	0.02	7	1	3.92		
238	9294	0	0.02	6	2	3.91		
239	9331	0	0.02	8	3	3.92		
240	9369	0	0.02	8	3	3.91		
241	9407	0	0.02	6	4	3.94		
242	9447	0	0.01	5	1	3.94		
243	9484	0	0.03	11	7	3.91		
244	9521	0	0.01	4	1	3.93		
245	9559	0	0.02	6	2	3.91		
246	9596	0	0.02	6	2	3.91		
247	9633	0	0.03	10	7	3.92		
248	9670	0	0.02	8	5	3.92		
248.1	9704	0	0.01	3	1	3.96	Pup	
248.2	9706	0	0	0	0	3.80	4.5" OTIS X NIPPLE	
248.3	9708	0	0.02	7	1	3.94	Pup	
248.4	9716	0	0	0	0	3.86	7" X 4.5" BAKER SABL-3 PACKER	
248.5	9721	0	0.01	3	0	3.89	Pup	
248.6	9729	0	0	0	0	3.82	4.5" OTIS X NIPPLE	
248.7	9731	0	0.02	6	2	3.93	Pup	
248.8	9740	0	0.02	6	1	3.94	Pup	
248.9	9750	0.26	0.35	100	56	3.97	Pup Probable Hole in Upset! Multiple Holes!	
249.1	9756	0	0	0	0	3.07	4.5" OTIS PORTED NIPPLE	
249.2	9758	0	0.01	4	1	3.95	Pup	
249.3	9767	0	0	0	0	2.88	4.5" X 3.5" CROSS-OVER	
249.4	9768	0	0.01	5	0	2.92	Pup	
249.5	9778	0	0	0	0	2.91	PRODUCTION MANDREL 8 (LATCH @ 5:00)	
249.6	9789	0	0.01	4	1	2.94	Pup	
250	9797	0	0.02	8	2	2.95		
251.1	9829	0	0	0	0	2.93	7" X 3.5" BAKER FB-1 PACKER	
251.2	9837	0	0.02	7	3	2.92	Pup	

Figure 10.44 Excerpt from a typical caliper log.

Caliper surveys are of great value when used on a comparative basis to determine (1) how effective a corrosion inhibition program is, and/or (2) if there is a change in well conditions. A background profile should be run before starting the program, and this should be followed by subsequent caliper surveys after a suitable time has elapsed. Within the limitations of the tool, such surveys are a direct measurement of the progress of corrosion in subsurface equipment.

Like other tools, profile calipers are not the perfect monitoring tools because they have limitations:

- Scale and corrosion products can mask the pits. When possible, the well should be acidized before running a caliper survey if a known scale problem exists in the well.
- Some pits may be missed because of spacing of the feelers; however, the general condition of the tubing will be evaluated.
- Plastic lined tubing could be damaged by a caliper survey. While this was a potential issue due to the finger pressure in early tools, the current tools have a low finger pressure (~2 lbs [1 kg]). An

operator should always check with the lining applicator and the caliper company before running a survey in lined tubing to determine if the tool can be run safely.
- Caliper feelers can remove protective scale or an inhibitor film from the tubing during a survey. If the well is put back on production without immediate retreatment with a corrosion inhibitor, a phenomenon known as "caliper track corrosion" can occur. Note that the wireline "cut" is not a smooth wear surface, but is pitted because of wear-corrosion (discussed in Chapter 3, Section 3.7). The wireline-removed corrosion product and the bare area became anodic. Because the bare area was much smaller than the rest of the surface (the cathodic area) corrosion was greatly accelerated. This should not be a frequent or serious problem today because operators have learned the need to inhibit any well after wireline or survey work.

Electromagnetic Inspection Tools

Electromagnetic casing caliper tools differ from the profile caliper in that they use an electromagnetic tool to measure the inside diameter of casing without using feelers to contact the metal surface. Current imposed on a coil generates a magnetic flux in the casing, which is evaluated by the tool. Electromagnetic wall-thickness surveys measure casing wall thickness and provide a log of corroded areas. The tool is precalibrated for the casing in the well. Changes in wall thickness (due to internal or external loss of metal) of the casing within the sensitivity range of the instrument are detected. The tool can be run in the absence or presence of fluids and is not affected by scale deposits or cement.

The inside diameter is measured over a length of 1–2 in, and the average diameter is determined. The caliper is particularly sensitive to vertical cracks or pits, but will also detect holes or pits that have a diameter of about 1 in. Limitations of this tool are as follows:

- The diameter of holes or pits less than 1 in diameter may not be detected with the tool.
- A background log should be run on a well for reference purposes before corrosion takes place.

Other Techniques

Surface inspection of downhole equipment includes electromagnetic as well as visual inspection. The use of electromagnetic type inspection of tubing out of the hole is a viable method for sorting tubing. It also can be used to evaluate the extent and location of corrosion in the string if the joints are numbered as they are pulled. The joint number can then be correlated with depth to determine at what depth corrosion occurs.

"Downhole camera" is a term that has been applied to a variety of inspection tools. The fiber optics video cameras can provide excellent views of corroded tubing when the well conditions permit viewing. Others have used ultrasonics or laser profilometry to scan the inside of the pipe; computer processing of the signals present a "picture" of the inside of the pipe.

10.5.1.2 Pipeline Inline Inspection

Inline Inspection (ILI) tools are available, which can be sent through a pipeline and will provide a continuous record of pipe wall conditions. Sometimes referred to as a "smart pig," "intelligent pig," or "electronic pig," ILI tools are self-contained units. They are based on either the electromagnetic flux (MFL) leakage technique or ultrasonic technique to detect both internal and external defects.[32-34] Figure 10.45 shows an MFL tool at a pig receiver.

Figure 10.45 ILI of an oil pipeline with a MFL tool.

ILI tools are available for most pipeline diameters. The smaller diameters have become more readily available as the electronics, power, and recording packages have been further miniaturized. The use of ILI tools continues to increase, and their use has become routine for many companies. Even lines built without pigging facilities can sometimes be modified to accept temporary pig launchers and receivers.

Not only do the tools show linear location and depth of corrosion, but the orientation is also recorded. Thus, the circumferential location is identified. The technology continues to evolve with continuing improvements in accuracy, reliability, and interpretation. There are also limitations with ILI technologies including velocity, temperature, and pressure limitations. Tables 10.2–10.3 provide the typical probability of detection (POD) and sizing capabilities for different anomaly types.

Table 10.2 Typical detection and sizing specifications for ultrasonic metal loss tools.[32]

	General Metal Loss	Pitting	Axial Grooving	Circumferential Grooving
Base material				
Depth at POD = 90%	1.0 mm (0.04 in)	1.0 mm (0.04 in)	1.0 mm (0.04 in)	1.0 mm (0.04 in)
Depth sizing accuracy at 80% certainty	±0.4 mm (0.016 in)	±0.4 mm (0.016 in) diam.: >10 mm (0.40 in)	±0.4 mm (0.016 in) width: >10 mm (0.40 in)	±0.4 mm (0.016 in) Length: >10 mm (0.40 in)
Width sizing accuracy at 80% certainty	±12 mm (0.47 in)	±12 mm (0.47 in)	±12 mm (0.47 in)	±12 mm (0.47 in)
Length sizing accuracy at 80% certainty	±6 mm (0.24 in)	±6 mm (0.24 in)	±6 mm (0.24 in)	±6 mm (0.24 in)

Table 10.3. Typical detection and sizing specifications for MFL tools (t = wall thickness).[32]

	General Metal Loss	Pitting	Axial Grooving	Axial Slotting
Base material				
Depth at POD = 90%	0.15-0.20t	0.15-0.20t	0.10-0.15t	0.20-0.25t
Depth sizing accuracy at 80% certainty	±0.15-0.20t	±0.20-0.25t	±0.20-0.25t	±0.15-0.20t
Width sizing accuracy at 80% certainty	±15-20 mm (0.60-0.80 in)	±15-20 mm (0.60-0.80 in)	±15-20 mm (0.60-0.80 in)	±15-20 mm (0.60-0.80 in)
Length sizing accuracy at 80% certainty	±15-20 mm (0.60-0.80 in)	±10-15 mm (0.40-0.60 in)	±15-20 mm (0.60-0.80 in)	±15-20 mm (0.60-0.80 in)

While an ILI program can be important element in the overall corrosion and integrity management program, it should be considered like any other inspection or monitoring technique: as a tool to be applied where it delivers the most value.

10.5.1.3 Tank Bottom Inspection

Tank bottom inspection is paralleling the development of new and better pipeline inspection equipment. Spurred by leak prevention regulations for inspection of aboveground storage tanks, several "floor scanning" devices have been developed. Both magnetic flux and ultrasonic technology are being used to determine the condition of tank floors. While visual inspection can evaluate the interior side, the soil side is often the most severely corroded. The new scanners allow for a more accurate determination of damage. The computer-processed data from the scanners can create "contour maps" of the corroded areas. Because most of the scanners are manually guided over the tank floor, the tank must be clean, purged, and cleared for safe entry. For proper operation, most of the machines require a relatively clean floor for the scan. Figure 10.46 shows a tank floor scanner in use. Additionally, acoustic emission testing has been used successfully to not only prioritize the tanks for inspection, but to test the tank bottoms for active corrosion. Comparisons of MFL tank floor scanners and AE testing have shown good correlation.[35]

Figure 10.46 Storage tank floor scanning.

10.6 Records and Failure Reports

Many people do not think of records as monitoring, but the importance of keeping complete records of corrosion control programs and equipment failures cannot be overemphasized. The maintenance of good records is vital for troubleshooting problems, evaluating the results of other types of monitoring, and in optimizing corrosion control programs. A record of performance is the ultimate proof of the effectiveness of any program and provides a factual basis for future decisions.

The broad category of records not only includes corrosion control program records, but also failure and/or inspection records and very importantly production and process records. If good records are not available, it is only a guess as to what may have changed, or why the system performed as it did.

10.7 Summary

Below is a summary of monitoring issues:

- Keep in mind there are a number of techniques available.
- Plan the monitoring portion *along with* the rest of the planning.
- Install the necessary equipment during the initial installation.
- Use the appropriate monitoring techniques throughout the project's life.

Table 10.4 presents a high level summary of the various monitoring techniques covered in this chapter. Table 10.5 maps many of the monitoring techniques against common oil and gas production systems. Typically, multiple techniques are integrated in the development of an effective corrosion monitoring program.

Table 10.4 Summary of techniques.

Method	Most Common Usage and Remarks
CORRODING SPECIMENS	
Corrosion Coupons	Widely used for routine monitoring to determine overall corrosion rates, pitting rates and type of attack in water injection systems, gas well and gas handling systems. May be used in any environment. (A comparative technique to evaluate corrosion and corrosion control in various parts of a system and changes with time.)
Test Nipples (Spools)	Used in water and gas lines for long-term exposures. (May be used for periodic inspection or may be processed as coupons.)
Electrical Resistance Type (ER)	May be used in any environment—usually used in plant type applications.
Linear Polarization Type (LP)	Requires sufficient water to provide current path between electrodes.
CHEMICAL TESTS	
Routine Water Analysis	Used to identify type of water and water sources.
Dissolved Iron	Determined on samples collected specifically for iron analysis. Used to monitor inhibition programs in gas wells and oil wells. Iron in water used for gas wells. Iron in total fluid used for oil wells.

Table 10.4 (cont.) Summary of techniques.

Method	Most Common Usage and Remarks
CHEMICAL TESTS (cont.)	
Sulfite Residual	Used to determine treatment dosages where sulfite or sulfur dioxide is used as an oxygen scavenger.
Chlorine Residual	Is used to determine chlorine dosages in fresh water where chlorine or hypochlorite is used for microorganism control.
Inhibitor Presence	May be used in some water systems with certain inhibitors to determine the amount of inhibitor carry through.
pH Determination	Used to monitor and adjust pH where control is used as corrosion control, such as glycol systems, drilling fluids, and plant applications. A pH meter should be used (rather then pH paper) whenever accurate readings are required.
OTHER ANALYSIS	
Gas Analysis	Carbon dioxide and hydrogen sulfide contents are determined to establish the environment. (For accuracy, measurements should be made on location.)
Scale and Deposits	Samples of scales and corrosion products are analyzed to identify material and determine type of problems.
OXYGEN MEASUREMENT AND DETECTION	
Oxygen Meters	Used to spot check for content of dissolved oxygen in water and oxygen in gas. (Continuous measurement possible but not usually practical.)
Galvanic Probe	Most practical approach to "flag" oxygen entry in water systems. Permanently mounted in system, read on meter or recorder. May be used with recorder to spot "cyclic" air entry.
MICROORGANISMS	
Bacterial Activity	Used to determine activity of bacteria, usually in water injection systems. Preferred method of detection.
Bacteria Count	Occasionally used in study of microorganisms. Not normally recommended. (Many consider the activity is more important than the number.)
Microscopic	Laboratory tests often used to characterize or identify organisms.
Specific Property Tests	Tests are continually being developed to detect the presence of specific strains or specific properties of bacteria.
INSPECTION	
Visual Inspection	Scheduled and occasional, a most improtant detection and monitoring technique applicable to all situations. Records of equipment condition should be kept.
Optical (Borescope)	Used to inspect inside tubular goods and exchanger tubes.
Magnetic Particle Inspection	Used to locate cracks in steel materials, particularly thread areas and equipment parts.
Ultrasonic Thickness	Used to measure wall thickness of vessels and pipe. May be run while equipment is on stream.
Radiographic	Used for inspection of wellheads, valves, manifolds, and lines in critical applications. May be run while equipment is on stream.
In-Line Inspection	Used to determine remaining wall in pipelines. Instrumented pig pumped through the line.
Downhole Corrosion Calipers	Used to determine conditions of downhole pipe. Tubing calipers usually run in gas wells and flowing oil wells. Casing calipers used in all types of wells. Pipe should be free of deposits. Usually run to verify results of surface monitoring.

Table 10.5 Summary of monitoring techniques applicability.

| | DIRECT MONITORING METHODS | | | | | | INDIRECT MONITORING METHODS | | | | |
| | Intrusive | | | Nonintrusive | | | Online | | | | Offline |
	Mass Loss Coupons	ER Probes	LPR Probes	Galvanic Probes	Permanently Installed Ultrasound Devices	NDE Using Mobile Equipment e.g., Ultrasound, Radiography	Hydrogen Probes and Patches	Acoustic Solid Particle Detectors	Online Process Analysis e.g., pH, Chloride Concentration	Online Process Monitoring e.g., Temperature, Dew Point	Offline Chemical and/or Solids Sampling and Analysis	Microbiological Sampling and Analysis
Seawater injection and cooling systems	✓	✓(1)	✓(2)	✓	✓	✓	✗	✓	✓	✓	✓	✓
Produced water treatment and injection systems	✓	✓	✓	✓	✓	✓	✓	✓	✓	✓	✓	✓
Aquifer water	✓	✓	✓(3)	✗	✓	✓	✗	✓	✓	✓	✓	✓
Multiphase flow (gas, oil, and water)	✓	✓	✓(4)	✗	✓	✓	✓	✓	✓	✓	✓	✓
Unstabilized crude oil	✓	✓	✗	✗	✓	✓	✓	✗	✗	✓	✓	✗
Produced gas	✓	✓	✗	✗	✓	✓	✓	✓	✗	✓	✓	✗
Vacuum units and pipework	✓	✓	✗	✗	✓	✓	✗	✗	✓	✓	✓	✗
Storage vessels and tanks with separated water bottoms	✓	✓	✗	✗	✓	✓	✗	✗	✓	✓	✓	✓

Key: ✓ Possible application   ✗ Not applicable

Notes:
1. Flush-mounted ER probes provide misleading data if there is any biofilming tendency. Nonflush probes projecting into the process stream are more effective.
2. Applicability depends on water quality. LPR probes provide misleading data if there is any biofilming tendency.
3. Applicability depends on water quality. LPR probes are unsuitable if there is a low ion content, or there is a strong scaling tendency, or other forms of electrode contamination are likely.
4. Applicable only if flow is stratified and water cut is above approximately 20%.

References

1. NACE Report 3T199-2012, "Techniques for Monitoring Corrosion and Related Parameters in Field Applications" (Houston, TX: NACE International, 2012).
2. Hedges, B., T. Bieri, K. Sprague, and H. Chen, "A Review of Monitoring and Inspection Techniques for CO_2 & H_2S Corrosion in Oil & Gas Production Facilities: Location, Location, Location!" paper no. 06120, CORROSION 2006 (Houston, TX: NACE International, 2006).
3. NACE Publication 21413, "Prediction of Internal Corrosion in Oil Field Systems from System Conditions" (Houston, TX: NACE International, 2016).
4. NACE SP0192-2012, "Monitoring Corrosion in Oil and Gas Production with Iron Counts" (Houston, TX: NACE International, 2012).
5. Rydell, R.G., and W.H. Rodewald, "Iron in Oil Technique as a Corrosion Control Criterion" *Corrosion* 12, 6 (1956): p. 271.
6. API RP 45, "Recommended Practice for Analysis of Oil Field Waters" (Washington, DC: American Petroleum Institute, 2014).
7. ASTM D 1068-15, "Standard Test Methods for Iron in Water" (West Conshohocken, PA: ASTM International, 2015).
8. Eckert, R.B., *CorrCompilations: Introduction to Corrosion Management of Microbiologically Influenced Corrosion* (Houston: TX: NACE International, 2015).
9. NACE TM0212-2012, "Detection, Testing, and Evaluation of Microbiologically Influenced Corrosion in Internal Surfaces of Pipelines" (Houston, TX: NACE International, 2012).
10. NACE TM0194-2014, "Field Monitoring of Bacterial Growth in Oil and Gas Systems" (Houston, TX: NACE International, 2014).
11. NACE SP0775-2013 (formerly RP0775), "Preparation, Installation, Analysis and Interpretation of Corrosion Coupons in Oil Field Operations" (Houston, TX: NACE International, 2013).
12. NACE SP0499-2012 (formerly TM0299), "Corrosion Control and Monitoring in Seawater Injection Systems" (Houston, TX: NACE International, 2012).
13. ASTM G46-94 (2013), "Examination and Evaluation of Pitting Corrosion" (West Conshohocken, PA: ASTM International, 2013).
14. NACE RP0497-2004, "Field Corrosion Evaluation Using Metallic Test Specimens" (Houston, TX: NACE International, 2004).
15. ASTM G1-01 (2014), "Conducting Corrosion Tests in Field Applications" (West Conshohocken, PA: ASTM International, 2014).
16. Eckert, R.B., *Field Guide to Internal Corrosion Mitigation and Monitoring for Pipelines* (Houston, TX: NACE International, 2016).
17. ASTM G96-90 (2013), "Online Monitoring of Corrosion in Plant Equipment (Electrical and Electrochemical Methods)" (West Conshohocken, PA: ASTM International, 2013).
18. Bieri, T., M. Reading, D. Horsup, and R. Woollam, "Corrosion Inhibitor Screening Using Rapid Response Corrosion Monitoring," paper no. 06692, CORROSION 2006 (Houston, TX: NACE International, 2006).
19. NACE Publication 310104, "Field Monitoring of Corrosion Rates in Oil and Gas Production Environments Using Electrochemical Techniques" (Houston, TX: NACE International, 2014).
20. NACE Publication 1C187-2005, "Use of Galvanic Probe Corrosion Monitors in Oil and Gas Drilling and Production Operations" (Houston, TX: NACE International, 2005).
21. NACE Publication 1C184-2008, "Hydrogen Permeation Measurement and Monitoring Technology" (Houston, TX: NACE International, 2008).
22. ASNT Recommended Practice No. SNT-TC-1A-2016, "Personnel Qualification and Certification in Nondestructive Testing" (Columbus, OH: American Society for Nondestructive Testing, 2016).

23. ISO Standard 9934 (latest revision), "Non-Destructive Testing–Ultrasonic Testing–Reference Blocks and Test Procedures for the Characterization of Contact Probe Sound Beams" (Geneva, Switzerland: International Standards Organization).
24. ASTM E213, "Standard Practice for Ultrasonic Testing of Metal Pipe and Tubing" (West Conshohocken, PA: ASTM, 2014).
25. Hedges, B., K. Sprague, T. Knox, and S. Papavinasam, "Monitoring and Inspection Techniques for Corrosion in Oil & Gas Production," paper no. 5503, NACE 2015 (Houston, TX: NACE, 2015).
26. ASTM E1742, "Standard Practice for Radiographic Examination" (West Conshohocken, PA: ASTM International, 2012).
27. ISO Standard 5579:2013 (latest revision), "Non-Destructive Testing–Radiographic Testing of Metallic Materials Using Film and X- or Gamma Rays–Basic Rules" (Geneva, Switzerland: International Standards Organization, 2013).
28. ISO Standard 9934 (latest revision), "Non-Destructive Testing–Magnetic Particle Testing" (Geneva, Switzerland: International Standards Organization, 1998).
29. ASTM E709, "Standard Guide for Magnetic Particle Testing" (West Conshohocken, PA: ASTM International, 2015).
30. ISO Standard 3452 (latest revision), "Non-Destructive Testing–Penetrant Testing" (Geneva, Switzerland: International Standards Organization, 1998).
31. ASTM E165, "Standard Practice for Liquid Penetrant Examination for General Industry" (West Conshohocken, PA: ASTM International, 2012).
32. NACE Publication 35100, "In-Line Inspection of Pipelines" (Houston, TX: NACE International, 2017).
33. NACE SP0102-2017, "In-Line Inspection of Pipelines" (Houston, TX: NACE International, 2017).
34. API Standard 1163, "In-line Inspection Systems Qualification" (Washington, DC: American Petroleum Institute, 2013).
35. Papasalouros, D., K. Bollas, D. Kourousis, and A. Anastasopoulos, "Acoustic Emission Tank Floor Testing: A Study on the Data-Base of Tests and Follow-Up Inspections," 31st Conference of the European Working Group on Acoustic Emission (EWGAE 2014) (Dresden, Germany: EWGAI, 2014).

Corrosion Management

Timothy Bieri

11.1 Introduction

Many major incidents could have been prevented by using information and learned knowledge that was available. Failure to recognize a threat, integrate all available data, or respond quickly with corrective action can result in serious consequences. Until oil and gas processing equipment can be manufactured using corrosion-proof materials (also known as "unobtainium"), consideration of corrosion is required. When corrosion is addressed from conception to abandonment, or over the equipment lifecycle, it is called "corrosion management." As mentioned in previous chapters, corrosion is time based and Figure 11.1 illustrates how wall thickness changes with time.

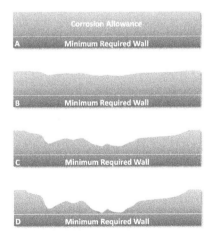

Figure 11.1 Wall thickness changes with time: options decrease while risk increases.

When equipment is new (A) the full wall thickness is available; typically this would include the required wall thickness for pressure containment plus some additional wall thickness for corrosion allowance. As corrosion initiates (B), corrosion barriers can be used to mitigate the corrosion. Poor corrosion control (C) begins to limit the available options and requires more vigilance in the deployment of mitigation measures and assurance of their performance. Once the corrosion allowance is gone (D), the equipment will need a fitness-for-service assessment.[1] If the equipment is no longer fit-for-service, there are only a handful of options that are available: repair, rerate, replace, or shut-in. As time passes, the options for intervention and remaining life decrease and risk increases. This chapter will present a high-level discussion on risk, corrosion management programs, and the associated economics.

11.2 Risk

Risk is the combination of the probability, or likelihood, of an event occurring and the consequence of the event having occurred. One popular method for illustrating this concept is using a 'bow-tie' (Figure 11.2).

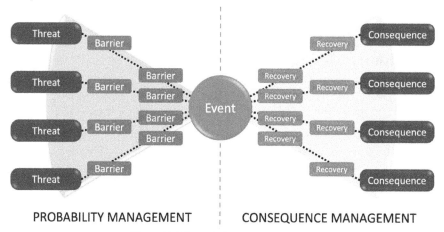

Figure 11.2 Example bow-tie risk diagram.

Threats can have barriers to minimize the probability of an event; consequences can have recovery barriers to minimize the consequence. The bow-tie approach is generic because it can be applied to any number of different events in multiple industries.

11.2.1 Risk Assessment

Reviewing all the corrosion threats, barriers, and consequences is a process called "risk assessment." Risk assessments (RAs) can be qualitative, quantitative, or a hybrid.

- Qualitative: Relies more on institutional knowledge and engineering judgment in review of the data. Qualitative assessments can be performed quickly (relative to Quantitative), but are depen-

dent on having people with the right expertise involved in the process. The results are often presented in terms such as Higher, Medium, or Lower risk.
- Quantitative: More rules based and requires a high level of detailed information. Frequently, there is an algorithm that calculates a risk number based on the detailed input information. An example output is mean time to failure (in hours).
- Hybrid, or Semiquantitative: Tries to leverage the best qualities from each. It has the speed of qualitative with the systematic benefits of quantitative.

There are several risk assessment references that can provide a more thorough discussion of the process.[2-5]

11.2.1.1 Setting the Stage

Regardless of which type of RA process is run, a minimum level of data is required. Additionally, a cross discipline team will be required to perform the assessment. Table 11.1 lists the typical disciplines and data required.

Table 11.1. Typical disciplines and data requirements.

Typical Disciplines	Typical Data Requirements
Corrosion & Materials	Design Data
Inspection or Integrity	Process and Operational data
Operations	Sampling Data
Process & Process Safety	Mitigation Data
–	Inspection Data
–	Monitoring Data

11.2.1.2 Segments or Circuits

The next step is to take the system being assessed and divide it into segments or circuits that have common threats, and common barriers and mitigation to those threats. A change can affect either the internal or external conditions; the following are two examples:

- The transition of a pipeline or piping from below grade to above grade, changes the external threats and barriers.
- Gas compression would have different internal threats and barriers upstream versus downstream of the compressor.

The level of granularity in creating segments depends on the type of RA that is being performed.

11.2.1.3 Threat Assessment

Much of the first five chapters covered the wide variety of corrosion threats, which are summarized in Table 11.2. However, not all threats are credible for all materials or systems. A credible threat is defined as one that can be a reasonably expected result in corrosion. For example, CO_2 corrosion would not be a credible threat in a seawater injection system.

The credible threats for each segment need to be identified along with the expected unmitigated corrosion rate or cracking susceptibility. Corrosion rates can be based on corrosion models, industry guidance, or engineering judgment. In addition to corrosion rate, corrosion morphology, e.g., general corrosion, localized corrosion is also useful to determine at this stage because morphology affects the mode of failure (and leak size) and plays a role in determining inspection strategy.

Table 11.2 Common internal and external threats.

Damage Type	Internal Threats	External Threats
Wastage	Amine corrosion	Atmospheric corrosion
	Chloride pitting corrosion	Chloride pitting corrosion
	CO_2 corrosion	Corrosion under insulation (CUI)/Corrosion under fireproofing (CUF)
	Corrosion by chemicals (including well workover fluids)	Crevice corrosion (including flange faces)
	Crevice corrosion (including flange faces)	Galvanic corrosion
	Erosion, erosion-corrosion, and solids-free flow-induced damage	Mechanical damage (fretting)
	Galvanic corrosion	
	H_2S (localized) corrosion	Soil corrosion (including MIC)
	High-temperature degradation	Stray current corrosion
	Mechanical damage (abrasion, wear, galling, and fretting)	–
	Microbiologically influenced corrosion (MIC)	–
	Other acidic fluids (e.g., glycol, well stimulation)	–
	Oxygen corrosion	–
	Preferential weld corrosion (PWC)	–
	Stray current corrosion	–
	Under-deposit corrosion	–
Cracking	Alkaline stress corrosion cracking	Alkaline stress corrosion cracking
	Chloride stress corrosion cracking (SCC)	Carbonate stress corrosion cracking
	Fatigue (corrosion)	Chloride stress corrosion cracking (SCC)
	Hydrogen embrittlement	Fatigue (corrosion and mechanical)
	Hydrogen induced cracking (HIC) and stress-oriented HIC (SOHIC)	Hydrogen embrittlement
	Sulfide stress cracking (SSC)	–
	Liquid metal embrittlement (Hg)	–

11.2.1.4 Barriers

Barriers can be a design or operational activity designed to reduce or eliminate a corrosion threat. Previous chapters have presented many different methods for controlling corrosion in oil and gas production operations. They generally have been handled on an individual basis. The various barriers are summarized in Table 11.3, which also indicates the most common use of each barrier. Many times, more than one barrier is valid for a specific threat. The challenge in developing a corrosion control program is to select the best barrier(s) to address the complexity of corrosion threats.

Table 11.3 Corrosion barrier summary.

Method	Most Common Usage
Environmental (or Process) Control	
Design	All Systems: · Consideration of dead legs (Chapters 1, 9) · Velocity (erosion) (Chapters 1, 2, 3, 9) · Coatability (Chapter 6) · Bimetallic couples, etc. (Chapter 3)
Oxygen Exclusion	All Systems—Prime method of control in many systems. (Chapter 9)
Dehydration and/or Humidity Control	Gas gathering, pipeline, injection; gas-lift systems; oil pipelines may involve water removal or maintenance of elevated temperature of gas above the dew point (Chapter 9)
Neutralization	Concerning Acids and/or Acid Gas—Special applications such as wet gas streams (with H_2S and/or CO_2) in processing plants, or pH control of glycol. (Chapter 9)
Sweetening	Removal of acid gases (such as H_2S and CO_2) from gas streams and injection waters. (Chapters 1, 8)
Line Pigging	Flowlines, Gathering Systems, Injection Lines (Chapter 9): · Gas—To remove water and solids—often in connection with an inhibition program. · Oil—To remove water from low places, particularly when water percentage is low. · Water—To remove settled solids to avoid microorganism and/or pitting sites.
Cleanouts (Equipment)	Surface Vessels—Oil and water handling: for periodic removal of accumulated solids that can accelerate corrosion. (Chapter 1)
Operating Procedures	Sometimes operating procedures can be modified to minimize effects of corrosion or erosion-corrosion. For example, flushing dead areas; controlling pressure and velocity. (Chapters 1, 2, 3, 9)
Microbiological Control	Cooling Water Systems—Water injection systems; sometimes producing wells. Design and cleanouts may be all that is required; biocides and combinations of corrosion-inhibitor-biocides are sometimes justified. (Chapters 2, 8, 9)
Chemical Inhibition (Chapters 8, 9)	
Film Forming Corrosion Inhibitors	Downhole in oil and gas wells. Flowlines and gathering lines. Water injection systems (if required, when oxygen-free operation does not completely control).
Neutralizing Inhibitors	Special wet gas applications (see also "Neutralization").

Table 11.3 (cont.) Corrosion barrier summary.

Method	Most Common Usage
Chemical Inhibition (Chapters 8, 9), (cont.)	
Other Corrosion Inhibitors	Cooling waters, heat-cooling mediums.
Scale Inhibitors	May sometimes be used when scale deposits cause concentration cells and thus pitting.
Protective Coatings and Linings (Chapter 6)	
Internal Tubing (Plastic Coatings)	Gas wells, gas-lift wells, critical oil wells. Water supply and injection wells may be supplemented with inhibitor in critical wells.
External Pipe	"Pipeline" Coatings—For buried and submerged lines. May be supplemented with cathodic protection.
Internal Cement Linings	Water gathering and injection lines. May be used when line is welded.
Vessel Internal Coatings	Water sections and vapor space of production and water injection vessels and tanks. May be supplemented with cathodic protection in submerged service. Vessels must be "designed" to be coated.
Fiberglass Epoxy (or Polyester) Linings	Repair of vessels or piping handling water and crude storage tank bottoms.
Atmospheric Coatings	Exterior of vessels, pipes, structures · Onshore—Standard paints · Offshore, Marine, or Severe Environments—Plastic coatings
Metallic Coatings or Linings	Internals of valves and wellheads in severely corrosive service.
Cathodic Protection (Chapter 7)	
External	Buried and submerged pipelines, flowlines, gathering systems often with pipeline coatings. Well casing, tank bottoms, and offshore structures (submerged zone).
Internal	In water handling portions of separation equipment and tankage. Water handling vessels.
Metallics (Chapter 4)	
Crack-Resistant Steels	All equipment and tubulars where H_2S is handled.
Low-Alloy Steels	Valves and wellheads in severely corrosive service (may use for trim or for complete assembly). Tubulars in highly corrosive service downhole equipment. Special service piping, exchanger tubes, pumps, and chemical tanks.
High Alloys	Tubulars, valves, wellheads, and special equipment in extremely corrosive service.
Copper Alloys	Exchanger tubes, pumps, etc., in water service.
Nonmetallics (Chapter 5)	
Plastic Pipe (Thermoplastic and Fiberglass Reinforced)	Flowlines, gathering systems for produced fluids and water (usually low pressure).
Fiberglass Reinforced Tanks	Special services such as chemical tanks (must match to service) or produced water tanks.

Corrosion allowance is not listed as a barrier because it does not alter the corrosion rate. Corrosion allowance can be useful if corrosion is uniform. However, corrosion allowance may not be dependable for pitting and highly localized attack. Corrosion allowance also carries a cost in terms of both additional material *and* weight.

Assurance

Most barriers require some level of assurance that they have the intended effect. Barrier assurance can also be considered monitoring and inspection, which is covered in Chapter 10. The type and level of assurance may vary during the life of the project as corrosion control methods change. Data from assurance activities can be used in determining corrosion rates and equipment integrity.

11.2.1.5 Probability of Failure (POF)

The probability of failure is an attempt to predict the frequency of equipment failure. The corrosion threats, corrosion morphology, and barrier effectiveness are inputs into the POF calculation. POF can also be based on generic failure frequencies based on company or industry data. For time-based threats like corrosion, POF can be considered in determining the "remaining life" of the equipment. A short remaining life equates to a high POF. If both the corrosion rate and wall thickness are known, then remaining life is straightforward to calculate. Corrosion-related failure modes that involve environmental cracking (e.g., SCC, SSC, etc.) are more condition-based than time-based; hence, POF is not a useful determinant of remaining life.

11.2.1.6 Consequence of Failure (COF)

The determination of the consequence of failure is complex and corrosion practitioners rarely have all of the information required to make the assessment. Typically, a multidisciplinary team is needed to make the assessment. COF is a function of several factors including, but not limited to,

- Failure mode and resulting leak size (e.g., pinhole or rupture)
- Composition of the process fluid (e.g., toxicity, flammability, dispersion, etc.)
- Estimated volume released
- Recovery barriers such as fire and gas detection, emergency shutdown valves, fire suppression equipment, and secondary containment
- People that may be affected—including people onsite and the public
- Costs associated with equipment repairs, equipment downtime (i.e., production deferral) and response

11.2.1.7 Risk Matrix

Once the POF and COF have been determined, they can be plotted on a risk matrix similar to Figure 11.3. Typically, each column or row in the matrix represents an order of magnitude difference from the one that comes before or follows it. Figure 11.3 is an illustrative example. Each operating company maintains its unique risk matrix which has been thoroughly reviewed by its technical, managerial and legal specialists. Different situations (such as inherent or unmitigated risk and mitigated risk) can be plotted on the risk matrix. For example, if points 2 and 3 on the matrix are compared, it is evident that the consequence of event 2 is significantly less than event 3, whereas the probability remains the same.

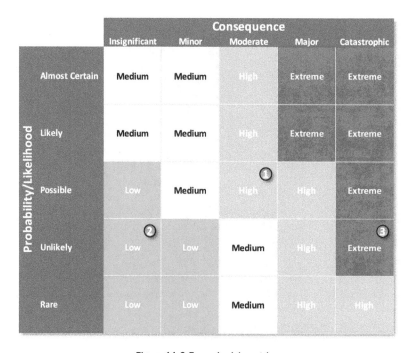

Figure 11.3 Example risk matrix.

A corrosion management program can reduce the probability of an event, and when coupled with appropriate recovery measures to reduce the consequence, the overall risk is reduced even further. There is no such thing as "zero risk" so the remaining amount is often called the "residual risk." For example, the event might be a leak or loss of containment. On the probability side, an example threat is internal corrosion due to CO_2, and a barrier could be corrosion inhibition. The barrier reduces the probability of the event. On the consequence side, a recovery example would be fire and gas detection, which would alarm and/or shut-in the system. The recovery reduces the consequence of the event. In this scenario, the consequence may be limited to lost production and repair costs for the associated equipment. Figure 11.4 illustrates this concept of inherent and residual risk.

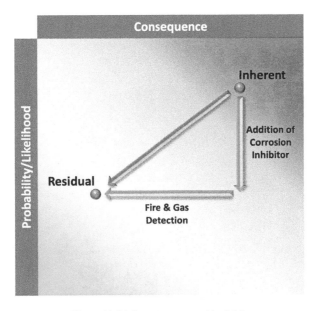

Figure 11.4 Inherent versus residual risk.

This exercise allows for a comparison of the relative risks across the spectrum of equipment, facility, or operating entities. Different risk matrices can be developed for safety, environmental, or business risk.

11.3 Corrosion Control Programs

The overarching goal of corrosion control is to manage the equipment such that it is fit-for-service over its lifecycle. In doing so, corrosion related failures and unplanned activities would be minimized. That does not mean there is *zero* corrosion or *zero* maintenance required. A successful program is influenced by the items covered in the previous chapters:

- Identification of credible corrosion threats
- Corrosion mitigation (barriers) to manage the credible threats, which include corrosion mitigation changes and operational changes
- Corrosion monitoring to assess corrosivity provides assurance that the corrosion mitigation is working, and/or can assess potential changes within a system
- Inspection programs to develop the required understanding of both damage assessment and barrier assurance
- Operational activities such as well work or changing production profile (e.g., water cut) that can affect the corrosion rates of the equipment

11.3.1 Unsustainable Cycle

Figure 11.5 presents a corrosion management cycle[6] that, while not uncommon, is unsustainable.

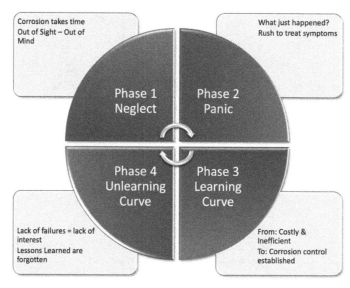

Figure 11.5 Unsustainable corrosion management.

Unfortunately, for some field operations, there is little if any planning for corrosion. Corrosion becomes a concern only after the first failure. At that time, the pressure is to get back on production and the most cost-effective control method may not be selected. Justification for the selection is quite often "Justification by Panic" rather than a sound engineering-operations-management decision. For example, as mentioned in Chapter 1 Section 1.5.1.2, most oil wells do not have corrosion problems until appreciable water is produced, therefore corrosion issues are often ignored until a failure occurs. Deciding the appropriate time to treat a well is discussed in Chapter 9 in the sidebar entitled "Control Corrosion Now or Later?"

Another example is where the initiation of treating programs are delayed at facility startup because the programs do not directly affect startup operations. More than one seawater injection system has regretted operating on raw untreated water for several weeks before starting their biocide program. Those systems were so badly infected that it took many months, innumerable cleaning pig runs, and large volumes of biocide to get the system under reasonable control.

When corrosion has been under control for several years and there are few if any failures, it is very easy to forget the program. This has happened often, when personnel have changed since the program was started or since a failure. Vessels with internal coating could have attachments welded to the outside that destroy the coating. Isolation flanges become shorted, inhibitor treatments dropped, gas blankets not maintained, and similar other occurrences. At least one field superintendent even questioned the need for cathodic protection of buried flowlines because there were no failures! Eventually, the cycle begins anew.

11.3.2 Sustainable Cycle: Plan, Do, Check, Act

A sustainable corrosion management program is one that is integrated and follows a continuous improvement philosophy. Figure 11.6 presents one example of the simple quality management continuous improvement approach of Plan, Do, Check, Act.

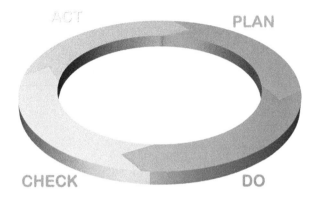

Figure 11.6 Continuous improvement cycle: Plan, Do, Check, Act.

11.3.2.1 Plan

Plan is where the purpose or objective is set and typically includes the specific target that performance will be measured against. The first key step to economically successful corrosion control is planning. The most cost-effective time to consider corrosion is as early as possible. Recalling Figure 11.1, options decrease and risks increase with the passage of time.

At the planning stage, the minimum corrosion considerations should provide for corrosion detection, and monitoring facilities should address the design issues in Chapter 9, Section 9.2 to avoid or minimize building in corrosion problems. Contingency plans to handle future problems can be developed for many facilities. For example, if inhibitor injection will be required for later water production, fittings and valves dedicated as "future chemical injection points" can be installed during initial construction at a fraction of the cost of later shutdown and retrofit. When corrosion control issues are reviewed and control methods are selected, it is very important to consider changing environments or extraneous conditions that can affect the control program.

11.3.2.2 Do

Do is where the Plan is executed. Installation of corrosion controls include not only the more obvious construction items like coatings and corrosion resistant materials, it also includes the initiation of other methods such as chemical treatment, cathodic protection, and oxygen-free operation. When chemical treatment is to be used on new installations, it is important that the treating program begins on startup. Whether it is corrosion inhibitor in a well flowline or biocide in a seawater injection, the treatment should start while the metal is still clean (no corrosion product or microbiological growth). An important part of the Do step is having the proper procedures defined. Procedures are required to operate and maintain the corrosion control efforts in a safe and consistent manner.

11.3.2.3 Check

Check is where the performance is measured against the target. Essentially, did execution (Do) of the Plan have the desired results.

Because the monitoring program is an integral part of any corrosion control program, monitoring should be in use from a project's startup until its abandonment. The type and level of monitoring may vary during the life of the project as corrosion control methods change. Today's computerization of data has made it much simpler to integrate corrosion monitoring and inspection results with equipment performance and production data and information. Records set up to "flag" changes and excursions will alert personnel of anomalies.

11.3.2.4 Act

Act, also sometimes referred to as *Adjust*, is where the learning takes place. If the Check deviates from the target, then corrective actions can be Planned. If the performance matches the target, then the new standard is established.

Periodic reviews are another step in a successful program. Because production changes with time, it is important to determine if the corrosion control program is still adequate, or in some cases, if it is excessive. Both external and internal programs should be reviewed.

- Pipeline coatings can deteriorate: *ask* "Is the cathodic protection adequate?"
- Offshore platform coatings weather and are damaged: *ask* "Is it time for inspection and maintenance painting?"
- Rod pumping well water cuts and volumes increase: *ask* "Does the inhibitor dosage or treatment frequency need to be changed?"
- A gas well flowing pressure has decreased and a compressor has been installed: ask "Are the acid gas partial pressures low enough to discontinue the well's inhibition?"

These and similar changes should be on the review schedule. Ideally, engineering and operating personnel, as well as those directly responsible for corrosion control, should get together for the review. The objective of the reviews should be to maintain two-way communication and to determine what, if any, program changes are to be made.

Making the necessary changes and maintaining an updated program will ensure a cost-effective program.

11.3.3 Program Integration

One way to achieve cost-effective programs is to have an "Integrated Corrosion Control Program" or an "Overall Corrosion Control Program" for a project—that is, a program that uses several methods of corrosion control in the same project.

An integrated corrosion control program involves the use of several methods listed in Table 11.3 in the same system, project, or equipment. It uses the best combination of technical and economic choices throughout the system—not just piecemeal. This approach is easiest to put into practice on a new project—one in the planning and design stage. At that point, it is easy to look at the whole project: the proposed flow diagram, the process variables, and the various corrosion environments. Developing an integrated program as part of the design allows the selection of the most cost-effective combination of corrosion control methods. Large installations and offshore facilities present excellent opportunities to apply the integrated approach.

The integrated approach is not limited to large installations. It also can be developed for individual well projects. For example, the corrosion control program for a corrosive gas well could include internally plastic-coated tubing supplemented with a tubing displacement corrosion inhibitor program; corrosion-resistant alloy downhole equipment, and wellhead; monitoring facilities at the well (coupon and water sample [iron content]); continuous corrosion inhibitor injection into the flowline; and the buried flowline's external coating supplemented with cathodic protection.

There are many aspects to corrosion management that are *people dependent*. Thus, training of operating and maintenance personnel is critical. Training should include the "why something is important," not just "how and what to do." No matter how good a program has been devised, if the operations and maintenance personnel have not "bought in" on it, the program will not be a success.

11.3.3.1 Ask Yourself

Consider the following questions when considering the corrosion management program:

- What are the credible corrosion threats?
- What are the barriers to the credible corrosion threats?
- How is barrier performance assessed?
- How is barrier performance assured?
- What are the corrective actions?

By asking these leading questions, the strength (or weakness) of the corrosion program can be readily assessed.

11.3.3.2 Consider the Unexpected—Look at the Big Picture

Sometimes corrosion control programs do not give the expected results because a key issue in the corrosive environment was overlooked. Consider the following examples:

A refinery was being built at a seacoast location. A large tanker berth was included. Corrosion control and environmental concerns were addressed from the planning stage through commissioning. For example, the lines from the onshore tank farm to and from the tanker loading arms were coated with a coating system designed to last for many years in the marine atmosphere. However, within months the coating started to fail. Poor surface preparation? No! Poor application? No! What was missed? Seagulls like to roost on overwater lines. Roosting seagulls unload their waste. Yes, the coating system that would withstand the rugged marine environment was not resistant to the chemicals in seagull droppings.

An onshore production facility and tank farm to handle offshore production was constructed on an island in the Middle East. The production platforms were many miles from the island. A subsea pipeline brought the crude to the island. A coastal atmospheric coating system was selected for the overhead piping and receiving tanks at the facility. The coating did not hold up well. With high-humidity coastal air, atmospheric temperatures in the 120+°F (50°C) range, and crude that was cooled as it was pipelined along the sea bottom to less than 80°F (27°C), there was continuous condensation on the lines and tanks. In fact, the report was that walking under a pipe rack was like walking in the rain. Under the actual circumstances, rather than the atmospheric coating system chosen, an immersion coating would have been a better choice.

A West Texas water flood was started using deaerated Ogallala fresh water for supply water. The designers installed type 410 stainless steel (UNS S41000) valves at the well (at that time, this was the cheapest corrosion-resistant alloy valve). Type 410 stainless steel had worked well with the Ogallala water in other installations. Overlooked was the fact that as produced water increased, Ogallala injection wells were switched to the San Andres produced water—a sour brine. Type 410 stainless steel failed because of chloride pitting.

In all three cases, the primary environment (or what was thought to be the primary environment) was considered—but the look to the future or the look for other influences was omitted.

11.4 Economics of Corrosion Management

The most recent study on the cost of corrosion, the IMPACT study,[6] sets the global impact of corrosion at US$2.5 trillion annually. Furthermore, the study estimates the savings using currently available corrosion control practices to be US$375–875 billion. Split the difference and there is a 25% savings to be gained by applying known corrosion management technologies. While it is not clear what percentage of the total is attributed to the oil and gas sector, the percentage reduction does not seem unreasonable.

It should be clear that beyond the obvious safety and environmental reasons for corrosion management, there are business drivers. The goal is to be safe and competitive. The question always is: "What's the most cost-effective way to handle this corrosion problem?" Sometimes the choice is obvious, but other times, a detailed economic analysis is required. Corrosion engineers should examine the economic analysis performed by others for assumptions, such as the expected life of carbon steel with inhibition and check these assumptions against knowledge gained in this, or other, fields. Seeking investment for corrosion management means competing against several priorities. Project engineers are under pressure to reduce capital investment, and spending on corrosion control measures (e.g., valves and fittings) required for future inhibition needs will be scrutinized in a tight economic environment. Corrosion practitioners need to become skilled at understanding the business aspects as well as the safety aspect of their solutions.

11.4.1 Consider It an Investment

An investment is usually required to make a profit. Oil and gas companies often think in terms of investments—investments in property, in drilling wells, and in facilities. While corrosion control is seldom thought of as an investment, a properly planned, operated, and maintained corrosion control program will generate profit, thus it should be thought of as an investment.[8]

The typical appraisal method is geared to look at corrosion control as an added cost. However, in many cases the cost is an investment and will pay dividends in the future. It is a fact that corrosion control itself does not generate revenue, reserves, or assets. Rather, it is a method of "profit improvement"—achieving maximum profit from existing income and assets. The profit improvement is the result of decreasing costs due to failure and reduced business risk. Failure costs include costs of repair or replacement and the costs of lost or deferred production.

One explanation of the difficulty to evaluate or understand corrosion control costs is that management is trained to evaluate costs that can be easily quantified and audited. There are two challenges: corrosion is time based, and you cannot add metal back; and the costs due to corrosion are only estimates until after the fact—and then it is too late.

11.4.2 Comparative Economic Evaluations

The evaluation of which corrosion control method to use in a given situation, is based on both technical and economic considerations. Sometimes there is only one technical solution; however, often there are several solutions to choose from—and like all engineering and operating decisions—that is where economics comes into play. Again, the economics may be obvious in some cases; in other instances, a comparison between methods (including the "do nothing but repair and replace" option) come into play. When considering the costs, all associated costs should be considered:

- Capital costs
- Repair and replacement costs
- Operating and maintenance costs (including personnel requirements)
- Risk factors (of safety, damage, pollution, etc., should a failure occur)
- Revenue loss from downtime and, in some cases, lost production

Not only are the economics of various control methods compared, but they should be compared with the cost of doing nothing. That is, letting the equipment corrode, then replacing it when it is no longer fit for service. When the severity of corrosion on specific equipment has not been established and the control is very expensive, the best decision may be to let the equipment corrode to determine its life. At that time, the decision can be made to continue as is or to take corrosion control measures. For example, a design calls for a low-pressure water transfer pump. A corrosion-resistant pump will cost four times the cost of a standard carbon steel pump. The pump is to be needed for 10 years. The decision could be to initially install the carbon steel pump; if it needs replacing in three years or more, the expense of an alloy pump has been deferred. At the time of replacement, the economics should be reevaluated to determine which pump is more cost-effective at that time.

The details of the economic decisions will depend on the way a company looks at money—its tax position, its rate of return, its policies on what costs are expensed and what costs are capitalized, and whether it considers the salvage value of equipment. As far as salvage value is concerned, many producing companies do not include equipment salvage value in their economics. When an asset reaches its economic limit for the owner company, it is sold. Traditionally, the sale price is based on the value of the remaining reserves—not on the value of any equipment.

In simple cases, the only comparison may be between the cost of one technical solution and the cost of doing nothing but repair and replace. For example, when a sucker rod pumping well starts having rod and tubing failures, the most cost-effective approach to corrosion control is usually inhibition. The question becomes which is more cost effective—an inhibition program or continuing to pull and replace rods. In most cases, the cost of repair will easily exceed the cost of an inhibition program. In other cases, the cost of inhibition could be comparatively high and the answer might not be as clear cut. In those wells, more detailed estimates may be required. Oftentimes, the reduction of costs other than repair or replace may be the controlling factors. In many cases, the elimination of a failure will increase the profit from a well by eliminating downtime and lost or deferred production. This can be especially true in marginal wells that "water out" whenever they are shut-in. Such wells may take several days after the repair to return to normal production rates—the lost production is not just for the day or two it took to get a pulling unit and replace the rod, but could amount to the equivalent of 4–7 days income.

Because each company has its own appraisal methods, and most calculations are handled by computer programs, it is difficult to present the variations. However, economics texts cover the calculations for time-based investment procedures. Other references on corrosion also show examples of corrosion and materials related economics.[6,9]

11.4.3 Another Way: The Break-Even Figure

Sometimes when evaluating corrosion control, particularly on new projects, there is no detailed failure history. In such cases, a slightly different approach to economics is required to justify the initial costs of corrosion control. When the complete picture is not available because all the facts are not known, it may be necessary to show a payout based on the number of failures that will have to be avoided.

Offshore Gulf of Mexico gas condensate fields often have multiple pay zones. Some of the zones have caused serious corrosion problems—others are much less aggressive. Typically, a given well may be equipped to produce from several zones—sequentially or concurrently. Consider the case where a 30-well project was being developed. Several wells had been completed and were producing. It was established that at least one zone's gas was sufficiently corrosive to justify corrosion control. The obvious choices at that time were (1) monitoring each well to establish which wells were problem wells and then start an inhibition program, hoping that not too much damage had occurred before the problems were identified; (2) inhibit all wells; or (2) run internal plastic coated (IPC) tubing when the wells were completed, replacing the existing bare tubing when it failed with IPC tubing.

Because the producing formations were extremely sensitive to back pressure, all forms of inhibition had been ruled out. That left IPC tubing as the choice. Management wanted to know which wells would become future problems so that the IPC tubing would be run only in "corrosive" wells. In this case, it would be virtually impossible to determine which wells would be corrosion problems until failure history was established. Thus, a stalemate developed.

The stalemate was broken when Management was shown that the expense of coating *all* wells in the project would be less than the expense of repairing and restoring *only one* well whose tubing had failed due to corrosion. Based on the experience to date, there was no doubt (and no disagreement), that if left uncontrolled, there would be several corrosion failures. The cost basis for running IPC tubing in all wells included the cost of properly handling and running the IPC pipe, as well as the cost of the coating. The repair estimates for a failure included not only the cost of replacing the tubing (rig up, pull tubing, run new IPC tubing, and rig down), but also the costs of restoring these wells to production (clean-out or stimulation to overcome the formation damage from the downhole work).

11.5 Summary

Corrosion control is an important and major part of safe and reliable operation. Corrosion control programs are methods of profit improvement. Understanding the relative risk is an important concept in corrosion management, because it helps to prioritize available resources as well as potential gaps. Integrated corrosion control programs use several methods to provide the most cost-effective control for a asset. Following a continuous improvement process is part of an integrated program and will help to minimize unplanned activity related to corrosion. The answers to the five simple questions under the section "Ask Yourself" tests the strength of the program. Lastly, corrosion control activities have to compete for investment dollars and corrosion engineers need to be skilled at communicating the business aspects—as well as the safety aspects—of solutions.

References

1. API 579 (latest revision), "Fitness-for-Service" (Washington, DC: American Petroleum Institute).
2. API 580 (latest revision), "Risk Based Inspection" (Washington, DC: American Petroleum Institute).
3. API 581 (latest revision), "Risk-based Inspection Methodology" (Washington, DC: American Petroleum Institute).
4. DNVGL-RP-G101, "Risk Based Inspection of Offshore Topsides Static Mechanical Equipment" (Amhem, Netherlands: DNV-GL, 2017).
5. DNVGL-RP-F107, "Risk Assessment of Pipeline Protection" (Amhem, Netherlands: DNV-GL, 2017).
6. Bowles, J.C., "Some Benefits of Long-term Thinking," *Pipeline & Gas Journal* 224, 3 (1997): p. 20.
7. Koch, G., J. Varney, N. Thompson, O. Moghissi, M. Gould, and J. Payer, "International Measures of Prevention, Application and Economics of Corrosion Technologies Study" (Houston, TX: NACE International, 2016).
8. Byars, H.G., "Corrosion Control Programs Improve Profits, Part 1—How to Approach the Problem" *Petroleum Engineer International* 57, 11 (1985): p. 62.
9. Verink, E.D., Jr., "Economics of Corrosion," Chapter 3, *Uhlig's Corrosion Handbook* (3rd Edition), R. W. Revie, ed. (New York, NY: J. Wiley and Sons, 2011).

Index

A

acetic acid, 54, 60, 306, 340, 390
acetylenic alcohols, 340
acid gases, 8
   in coal seam gas, 30
   corrosion in, 427
   corrosion control in, 301, 385
   corrosion monitoring in, 429, 430
   in gas condensate well, 14
   internal coatings and, 255
   internal corrosion caused by, 16, 18, 19
   neutralizing, 335
   partial pressure of, 18, 308, 385, 422, 484
   pH control in wet, 413
   pH stabilization in wet, 413
   reflux tower corrosion and, 399
   removal of, in gas sweetening systems, 19, 335
   in separators, 18
   solubility of, 342, 430
   in water systems, 27, 29
   wet, corrosion control in, 411
acid inhibitors, 340–341
acid producing bacteria (APB), 59, 337, 399, 405, 409
acid solutions, 340
   corrosion reaction in, 45
   pH of, 46–48

acid(s), 8, 190
   acetic, 54, 60, 306, 340, 390
   carbonic, 50, 131
   carboxylic, 306
   compositions, 165
   corrosion, 29–30, 76
   D-Erythorbic, 397
   formic, 60, 340, 390
   galvanic corrosion by, 98
   hydrochloric, 62, 64, 65t, 96, 113, 138, 340, 414, 425
   hydrofluoric, 113, 174, 198, 210, 340
   lactic, 60
   mill scale and, 100
   mineral, corrosion in, 77, 96
   mud, 340
   neutralizing, 335
   nitric, 29, 113
   organic, 52, 54, 92, 306, 340, 343, 399, 410
   oxalic, 113, 390
   pickling, 115
   polythionic, 113
   producers, 73
   propionic, 54
   solutions. *See* acid solutions
   sulfuric, 29, 64, 113, 414
   volatile organic, 54

weak, 113
well stimulation, 340
acrolein, 408
additives, 230, 232, 301, 305, 306, 342
   antifoam chemicals required, 336
   corrosion inhibition and, 307
   foaming process and, 315
   regulatory requirements, 410
age hardening, 151t, 173
aging, 151t, 157
air dried, for atmospheric coatings, 233, 238
alkaline amines, 130, 133
alkaline neutralizers, 399. *See also specific alkaline neutralizers*
alkaline SCC, 130–131
alkaline solutions, 45–47, 110f
alkanolamines, 335–336
alloy(s)
   aluminum, 79, 83, 113, 137, 156, 176, 335
   austenitic, 133–136, 164, 172
   cobalt-based, 174
   copper-based, 136, 175–176
   corrosion resistant, 95–97, 157
   defined, 149
   ferrous, 97, 156
   heat-treated, 157
   heterogeneous, 156
   homogeneous, 156
   kinds, 156
   metallurgy and, 149–150
   nickel-based, 171–174
   nonferrous, 156
   SSC-resistant precipitation, 167
   titanium, 174–175
   tungsten, 96, 176
aluminum alloys, 79, 83, 113, 137, 156, 176, 335
aluminum anode galvanic system, 296
aluminum anodes, 270, 291
amine SCC, 130
amine sweetening equipment corrosion, 19
ammonium bisulfite (NH_4HSO_3), 393–395, 394t
amorphous thermoplastics, 184
anchor patterns, 243, 243t
anisotropic polycrystalline metals, 146–147, 151t
annealing, 151t, 153, 166
anodic inorganic inhibitors, 303
API 579-1/ASME FFS-1, 74–75, 79, 116
API RP 14E, 21–22, 378

area principle, 98–99
aromatic compounds, 215
arsenic ions, 3
artificial aging, 157
artificial lift wells, 12–14, 16, 385
atmosphere, severity of, 5t
atmospheric corrosion, 5, 5t, 372
   coatings are barrier to, 30
   cathodic protection and, 95
   dehumidification controls, 398
   low-alloy steel bolts and, 25
atmospheric protection, 207, 250–254, 258
atmospheric zone, 32, 33, 251
   coating system used in, 252, 253t
atomic hydrogen, 3, 115, 132
   as H_2S corrosion product, 55
   generated by corrosion reaction, 116, 450
attached bacteria, 406
austenite, 150f, 163, 168, 168f
austenitic stainless steels, 95–96, 96f, 133–136, 151f, 163–165, 172
   corrosion in marine atmospheres, 164
   crevice corrosion, 165
   CSCC of, 133–136
   free machining grades, 136, 160, 163
   heat-tinted welds, 164–165
   highly alloyed, 170–171
   intergranular corrosion in, 113
   oxygen, 134–135
   pH, 133–134
   precipitation hardening, 166–167
   PWC of, 103–105, 104f–105f
   SCC, 165
   sensitization of, 113–114, 114f
   sensitized microstructure, 135–136
   strain hardening, 135
   temperature, 133
   thermally insulated, 135
automated C-scan (AUT), 453–455
automated (semibatch) treatment, 327, 328
auxiliary equipment, 8, 15, 372

B

baked coatings, 233–234
barrier coating, 234f, 236
batch and circulate treatment, 327, 328, 349, 352. *See also yo-yo treatments*

circulation time required for rod pumped well, 369–370
batch and fall method, 321–322, 322t, 350, 352–354
batch and flush, 349
　circulate treatment, 327, 328, 367–368
　circulation time required for, 369–370
batch treating frequency
　for gas wells, 323
　and inhibitor volume, 368t
　for open-annulus oil wells, 328
BCC. *See* body centered cubic (BCC)
bi-directional (Bi-Di) pig, 331, 402–403
bimetallic couple, 97
binder, 176, 231, 233. *See also* resin
biocide, 16
　batched into treating lines, 331
　in chemical treatment, 375, 401
　compatibility of, 410, 411
　distribution through pipeline, 404
　gluteraldehyde, 395, 410–411, 412
　microbiological corrosion, 59, 301, 339, 407
　for microorganism control, 339, 391
　nonoxidizing, 30, 337, 405, 407, 414
　in seawater injection, 483
　scavengers interfere with oil field, 395
　THPS, 408
　treatment, 407
　use of, 406
blistering, hydrogen, 20, 115–117, 115f, 450
body centered cubic (BCC), 145, 145f, 150, 152, 156
body coat. *See* intermediate coats
bolted equipment, 25–27, 26f
bolt failures, 114f
　due to CSCC, 135
　due to exposure to H_2S, 26
　stainless steel, 27
　subsea, 27
borax, 399
brine packer fluid, 310, 406
butt fusion, 193, 197, 223
butyl, 204

C

calcium sulfate precipitation, 396
capital costs (CAPEX), 143–144
　vs. OPEX, 229f, 345

carbide distribution, 154
carbohydrazide (N,N'-Diaminourea), 397
carbon and low-alloy steels, 43, 58, 151f, 158
　alkaline SCC of, 130–131
　behavior at hot and cold temperature, 159
　CO–CO_2 combinations, 132
　CP for, 280
　cracking by ammonium, calcium, and sodium nitrates, 132
　erosion-corrosion in, 106
　external SCC of, 131–132, 132t
　heat treatment, 160
　localized corrosion of, 94, 94f
　MIC of, 92f–93f
　PWC of, 103
carbon capture and sequestration, 29–30
Carbon Dioxide (CO_2):
　Analysis in acid gas, 423, 429–430, 431, 468t
　CO_2 corrosion appearances, 51, 80, 81ff, 83f, 87f, 94f, 101–103
　effect on corrosion, 7, 48f, 50-53, 78, 318
　　in amine systems, 41, 335
　　in steam systems, 42
　　at the top of pipelines, 413
　effect of CO_2/H_2S ratio on corrosion, 57
　effect on corrosion fatigue life, 128, 129t
　effect on elastomers and thermoplastics, 215, 215t
　effect of oxygen on CO_2 corrosion, 385
　effect on pH, 13, 48f
　effect on SCC of steel, 130–131
　effect of chromium concentration in steel on CO_2 corrosion, 159
　effect of dehydration on CO_2 corrosion, 334, 339, 397
　effect of turbulent flow on CO_2 corrosion, 22
　effect of steel microstructure on CO_2 corrosion, 154, 324
　effect of volatile organic acids on CO_2 corrosion, 54
　in carbon capture and sequestration, 29-30
　in coal seam gas, 30
　in deep, hot acid gas wells, 14
　in dissolved acid gases, 429
　in horizontal wells, 343
　in insitu combustion, 29
　in stripping gas, 391
　in threat assessments, 475, 476t, 480

in water-alternating gas (WAG) injection, 28-29, 210
iron carbonate corrosion product, 45, 52-53, 55, 57, 61, 64, 65t, 53, 94, 106, 111, 154
Testing elastomers and thermoplastics for CO_2 service, 220, 224
case hardening, 151t
casing corrosion, 15-17
  external, 15-16, 289
  internal, 16
  packer fluids, 16-17
  surface, 16
cast iron, 171
cathodic disbondment, 238, 298
cathodic inorganic inhibitors, 303-304
cathodic polarization, 3, 279, 280
cathodic protection (CP), 20, 97, 118, 262
  application. *See* cathodic protection application
  circuit, 268f
  corrosion fatigue and, 73, 129
  corrosion mitigation by, 86, 88, 92f, 95
  criteria. *See* cathodic protection criteria
  defined, 267
  installing, 144, 182
  interference, 276-278, 277f
  overview, 267
  principles, 267-268
  role of protective coatings, 298
  in submerged zone, 167, 295-298
  survey and test methods. *See* cathodic protection survey and test methods
  systems. *See* cathodic protection systems
cathodic protection application, 128, 287-298
  CP design, 287-288
  flowlines, gathering lines, distribution lines, injection lines, and pipelines, 290
  galvanic anodes, 296-297
  offshore structures, 295-298
  storage tanks, 293-295
  surface equipment and vessels, 290-293
  well casings, 289
cathodic protection criteria, 278-280
  criteria for determining, 280
  important terms used in, 279-280, 280f
  reference electrodes, 278-279, 278f, 279t
cathodic protection survey and test methods
  close-interval potential survey, 282
  current requirements for, 286-287
  IR drop measurement of current flow, 283-285, 284f
  soil or water resistivity measurements, 285-286, 286f
  structure-to-electrolyte potential measurements, 281-283, 281t, 282f, 283f
  surface potential survey, 284-285, 285f
  two-electrode survey method, 285
cathodic protection systems
  galvanic anode, 269-271
  impressed current, 271-278
caustic embrittlement, 130
cavitation, 112, 112f
cement and concrete, 181-182, 182f, 341
  linings, 259, 339
chemical analysis inside cracks, 64
chemical batching pigs, 331-332, 331f, 402
  commercial pig, 332
  typical batching pig, 332
chemical composition, 2-3
  coatings categorized by, 232, 233
  control of, 114-115
  CRAs depend on, 157, 177
  weld line corrosion and, 155
chemical degradation, 215
chemical delivery and reliability, 344-345
chemical injection
  controlling, 344
  equipment, 164
  facilities, 314, 375
  future points, 483
  lines, 95
  strings, 197
  tanks and pumps for, 31, 355
  valves, 313, 326, 349, 355, 363-365
chemical methods in corrosion monitoring, 423-430
chemical reactions
  of chemical scavengers with oxygen, 394, 394t
  corrosion, 44-46, 61
  curing methods and, 233-234
  temperature and, 422
chemical resistance
  of elastomers and thermoplastic seals, 215-216
  as factor of nonmetallic piping, 191-192
  protective coatings and, 227, 231

chemical scavengers
  calcium sulfate precipitation, 396
  commonly used scavengers, 393-394, 394t
  dosage ratio of, 395
  injection point corrosion, 395
  interference with oil field biocides, 394
  in oxygen removal, 390, 393-397
  reaction time and need for catalyst, 395-396
chemical treatment
  for corrosion control, 375
  in corrosion management program, 483-484
  dead legs and, 377-378
  of hydrotest water, 409-411
  pH control in, 414
  pig selection in, 401
chloride stress corrosion cracking
  (CSCC), 74, 334
  austenitic alloys, 133-136, 134f, 165, 460f
    free machining grades, 136
    oxygen, 134-135
    pH, 133-134, 134f
    repair system for, 211-212
    sensitized microstructure, 135-136
    strain hardening, 135
    temperature, 133
    thermal insulation, 135
  failures, 27
  as form of environmental cracking, 74, 118
chlorinated polyvinyl chloride (CPVC), 184,
  185t, 186f, 187f, 190
Christmas tree, 11, 17, 173
chromium-containing steels, 158-159
chromium-molybdenum steels, 159
close-interval potential survey, 282
coal seam gas, 30
coating(s). *See also* protective coatings
  application. *See* coating application
  as barrier to corrosion, 30
  based on chemical composition, 233
  design for, 373
  drying and curing requirements, 233-234
  external, 17, 259-263
  formulation. *See* coating formulation
  internal, 20, 28, 116
  of production facilities and
    equipment, 250-264
  reinforced, 234
  protective, 227-264, 298

  selection. *See* coating selection
  spray, 244-245
  system. *See* coating systems
  thickness, 232
  in tubular goods, 255-258, 256f-258f
  types, 232-234
  in vessels and tanks, 254-255
coating application, 241-250
  inspection, 245-250
  potential problems in, 244t-245t
  structural, 244-245
  surface preparation, 241-243
coating formulation, 230-232
  additives, 232
  pigment, 231
  resin or binder, 231
  vehicle, 231-232
coating selection, 237-240
  application considerations, 238-240
  functional requirements, 238
  service environment, 237-238
coating systems, 234-237
  intermediate coats in, 236, 236t
  primers in, 235-236, 235t
  topcoat in, 236-237, 236t
cobalt-based alloys, 174
cold weld compounds, 212
composites, 187
concentration cell corrosion, 19, 23, 33, 73,
  85-97, 85f-86f
  crevice corrosion, 86-87, 87f
  differential aeration cells, 88-89, 89f
  oxygen tubercles, 88, 88f
  scale and deposits, 90-93
concrete materials, 181-182, 182f
conductivity, electrical, 46
connections
  eliminating, 195-196
  flange, 195, 195f
  metallic, 214
  out-of-service branch, 333
  sample, 423-424, 424f
  submerged, 271
  threaded, 106, 194-195, 201
  welded, 106, 201
continuous treatment, 323, 325, 326, 328
controlling corrosion, 1, 4, 4t, 298, 389
  CP in, 295

in oil and gas production, 477
protective coatings and, 227, 255, 298
in wet acid gas systems, 411–414
control of deposits (cleanliness), 399–405
  filtration, 404–405
  flushing, 404
  pigging, 400–404
cooling water systems, 61
  corrosion inhibition in, 302, 336–337
  recirculating, 338, 414
  types, 336
copper-based alloys, 136, 175, 251
  impingement corrosion in, 109, 110f
  SCC in, 136
corrosion
  atmospheric, 5, 5t
  casing, 15–17. *See also* casing corrosion
  classifications, 72–76
  concentration cell, 73, 85–97, 85f–86f
  control methods, 4, 4t. *See also* corrosion control
  on coupon, 443
  crevice, 86–87, 87f
  defined, 1, 44
  electrochemical, 101f
  erosion-, 106–107, 106f–107f
  external, 19
  fatigue failure, 125–129, 126f–128f, 129t
  galvanic, 97–105, 97f, 165
  general, 76–79, 76f–78f
  impingement, 109, 110f
  injection point, 309, 395
  intergranular, 113–115
  internal, 19
  of iron and steel, 43–44
  localized, 79–85
  mechanisms involving wall-loss damage, 73
  mechanisms not involving wall-loss damage, 74
  microbiologically influenced, 73, 92–93, 165
  offshore zones, 32–33. *See also specific zones*
  physical variables, 61–62
  preferential weld, 102–105
  rebar, 182
  reflux tower, 399
  ringworm, 101–102, 102f, 155
  soil, 5–6, 6t
  sour, 50
  in storage tanks, 19–20, 20f
  submerged, 6, 6t
  sweet, 50
  types/forms of, 71–72, 72t
  under-deposit, 90, 90f, 91f
  wear-, 108–109, 108f–109f
  weld line, 155
corrosion cell, 304, 397
  basic, 1–4
  electrochemical, 267, 268f
  important points about, 3
  parts of, 45
  schematic of, 2f
corrosion control, 31
  control of deposits (cleanliness), 399–405
  control of MIC, 405–411
  decommissioning, 412
  dehydration of gas, 397–399
  designing for, 372–376
  dry layup, 412
  electrical isolation, 380–384
  external, 31–33, 31f
    atmospheric zone, 33
    splash zone, 32–33
    submerged zone, 32
  internal, 37–42
  mothballing, 412
  overview of, 371
  oxygen exclusion, 385–390
  oxygen removal in, 390–397
  pH control, 413–414
  piping design and, 376–380
  pre-FEED stage, 372
  process layout, 371
  wet layup, 412–413
corrosion control design, 372–376
  for chemical treatment, 375
  for coating, 373
  for inspection, 374
  mixing waters from various sources, 376
  for monitoring, 375
  for pigging, 375–376
  of piping systems, 376–380
corrosion control options, corrosion inhibitor vs., 345–346
corrosion control programs, 481–486
  integration, 485–486
  sustainable cycle, 483–484
  unsustainable cycle, 482

corrosion coupon location and position, 435–437, 436f, 437f
corrosion damage
 alternative definitions of, 74–75
 applications of definitions, 75–76
 on metal surface, 94
 in splash zone, 32
corrosion fatigue, 28, 74, 125–129, 126f–128f, 129t
corrosion inhibition
 in cooling water systems, 336–337
 in ESP wells 325
 in gas compression systems, 334–335
 in gas handling systems, 333–334
 in gas sweetening systems, 335–336
 in gas wells 320-323, 325–326, 349, 351, 352, 354–355, 357, 359, 360, 362, 363-366
 in gas-lift oil wells 323–324, 354, 365–366
 in hydraulic pump oil wells 324–325
 in heating and cooling media, 337–338
 in piping systems, 333
 in rod pump wells 327–328, 357, 359, 360, 367-368
 in shale gas wells, 344
 in shale oil wells, 343–344
 summary of, 346
 timing, 373
 in vessels and tanks, 332–333
corrosion inhibitor(s)
 applications, 319
 chemical delivery and reliability, 344–345
 classification, 310–311
 classification and use, 302t
 compatibility of, 312–313
  with downstream process, 315
 components of, 305f
 concentration of, 430
 vs. corrosion control options, 345–346
 defined, 301
 effectiveness, 308, 310, 316, 324, 368
 emulsification properties, 313
 environmental concerns, 315–316
 filming efficiency, 309
 filming time, 309–310
 film persistency, 309
 foaming properties, 315
 formulations, 305–307
 freeze point, 313
 freeze-thaw stability, 313
 fundamentals of, 302
 in gas injection wells, 339
 in horizontal wells, 341–344
 injection fittings and, 345
 inorganic, 303–304
 laboratory inhibitor testing, 317–319
 materials compatibility, 314
 mobility, 314
 need for, in hydrotest water, 410
 organic, 304, 304f
 partitioning between oil and water, 311–312
 pour point, 313
 properties, 307–316
 selecting, 316–319
 solubility or dispersibility, 310–311
 solvent, 305, 305f, 306–307, 311, 314, 323, 357, 365
 sticks treatment, 328, 329, 349, 362–363
 of surface facilities, 329–338
 thermal stability, 313–314
 treatment methods. *See* corrosion inhibitor treatment methods
 volatile corrosion inhibitors, 341
 in water injection systems, 338–339
 well acidizing inhibitors, 340–341
 in well or system, 308–309
corrosion inhibitor treatment methods, 319–329
 batch and fall method, 321–322, 322t
 batching frequency for gas wells, 323
 continuous downhole treatment, 326
 ESP oil wells, 325
 formation squeeze treatment, 325–326
 gas-lift oil wells, 323–324
 hydraulic pump oil wells, 324–325
 nitrogen squeeze and nitrogen displacement treatments, 321
 overview of, 319–320
 partial tubing displacement, 322
 treating string approach, 323
 tubing displacement treatment, 320–321
 wells without packers, 326–329
 yo-yo method, 322
corrosion inhibitive coating, 234f
corrosion management
 corrosion control programs, 481–486
 economics of, 486–489
 overview, 473–474

risks in, 474–481
corrosion monitoring and inspection
   direct techniques, 431–461
   indirect techniques, 421–431
   intrusive inspection, 461–466
   location, 420–421
   offline techniques, 423–430
   online techniques, 421–423
   overview of, 419–420
   records and failure reports, 467
   summary of, 467t–469t
corrosion probes, 375, 432
corrosion product and scale analysis, 62–64
   chemical analysis inside cracks, 64
   elemental analysis, 63
   field test reference guide to, 65t
   original manufacturing defect, 63
   quick field analysis methods, 64
   x-ray diffraction (XRD), 63
corrosion reactions, 44–46, 55. *See also* chemical reactions
corrosion resistance
   alloys and, 174–176
   carbide distribution and, 154
   in CO_2, 54, 154
   of FRP, 29, 198
   effect of heat treatment on, 155
   of polymers, 192, 215
   of stainless steels, 95, 160, 163, 165, 167
corrosion-resistant alloys (CRAs), 58, 79, 95–97, 157, 434
corrosion under insulation (CUI), 5, 33–35, 34f–35f, 254, 380
corrosion variables, 46–59
   conductivity, 46
   dissolved gases, 47–59
   pH, 46–47
corrosive environments, 8, 22
   external, 5–6
      atmospheric corrosion, 5
      CUI, 5
      soil corrosion, 5–6
      submerged corrosion, 6
   internal, 6–10
CPVC. *See* chlorinated polyvinyl chloride (CPVC)
crack-like flaws, 74, 123
crater, 73
crevice corrosion, 73, 85, 86–87, 161, 165, 213
   of flanges, 87f
crude dehydration, 18
crude stabilization, 18

D

DBNPA. *See* 22-dibromo-3-nitrilopropionamide (DBNPA)
dead legs, 333, 407
   corrosion inhibition and, 316
   locations, 23–25, 24f
   in piping design, 377–378
   stagnant zones and, 23–25
   types of
      mothballed equipment, 23
      operational, 23
      permanent, 23
      physical, 23
dealloying, 74, 175
dehumidification, dehydration *vs.*, 398
dehydration of gas, 397–399
   vs. dehumidification, 398
   using glycol, 398–399
   using molecular sieve, 399
   water dew point of, 397–398
D-Erythorbic Acid (D-Isoascorbic Acid), 397
desulfovibrio desulfuricans, 60
22-dibromo-3-nitrilopropionamide (DBNPA), 395
differential aeration cells, 48, 85, 88–89, 89f
directional drilling, 137, 341, 342
direct techniques, in corrosion monitoring, 431–461
   intrusive techniques, 432–451
   nonintrusive techniques, 451–461
disk shape coupons, 434
dissolved carbon dioxide (CO_2), 50–53, 50f, 51f, 53f
   factors affecting corrosion by, 54
   partial pressure guidelines, 52
   solubility of, 53f
dissolved gases, 47–59. *See also specific types*
dissolved hydrogen sulfide (H_2S), 55–59
   conditions leading to corrosion, 56–57
   effect of CO_2, 55
   forms of corrosion product, 55
   injection well plugging, 59
   mitigation of plugging, 59

need for corrosion inhibition, 55-56, 56f
producing well plugging, 59
sulfide stress cracking (SSC), 57-59, 58f
dissolved oxygen (DO), 48-50, 50f
monitoring of, 429-430
sources of, 49
dosage ratio of oxygen scavenger, 395
downhole corrosion inhibitor application techniques
for oil and gas wells, 349-368
batch and fall treatment, 352-354
batch and flush/circulate treatment down annulus, 367-368
chemical injector valve, 363-365
dump bailer treatment, 359-360
formation squeeze, 355-357
gas-lift gas treatment, 365-366
inhibitor sticks treatment, 362-363
nitrogen formation squeeze, 351-352
treating string, 354-355
tubing displacement, 349-351
wash bailer treatment, 360-361
weighted liquids, 357-359
yo-yo treatments, 352-354
downhole inspection, 462-464
downhole treating methods, infrequently used, 328-329. drilling operations, 28
dump bailer treatment, 359-360
duplex stainless steel, 83, 95, 96, 151t, 199, 378
microstructure of, 167-171, 168f
pitting corrosion of, 84f
subsea structures and, 297

E

economics
of corrosion management, 486-489
as factor for nonmetallic piping, 192-193
eddy current testing, 460-461, 461f
elastomers, 181, 212-216
chemical resistance of, 215-216
classification of, 214-215
commonly used seal materials, 214-215
polymers and, 183
seals and packing of, 212-214, 216t
flange isolation, 213-214
gaskets, 213
O-rings, 213
other seals, 213

electrical isolation, 213
avoiding galvanic couples, 383-384, 384f
insulating flanges, 380-383, 382f-383f
monolithic isolating joints, 380-383
electrical resistance probes, 444-448
in access fittings, 445f
installation of, 446f
response time, 447f
uses, 448f
electrical submersible pump (ESP) oil wells, 13, 325
electrochemical corrosion, 101f, 267, 268f
electromagnetic inspection tools, 464
elemental analysis, 62, 63
enhanced oil recovery (EOR) projects, 28-29
environmental cracking, 74, 118, 168, 343, 346, 479
EPDM. *See* ethylene propylene diene terpolymer (EPDM)
equipment
coating of production facilities and, 250-264
flanged/bolted, 25-27, 26f
piping, 20-27
processing, 18-19
gas dehydration, 19
gas sweetening, 19
separators, 18
storage tanks, 19-20
surface, 17
erosion, 73, 111-112
corrosion damage and, 342
ER probe, 448, 448f
internal, 17
mass loss coupons and, 434, 441
problems in slurry pipeline liners, 203
sand, 111f, 379
soil, 205, 378-379
of valve plug, 112f
erosional velocity, 21, 378, 379
erosion-corrosion, 73, 106-107, 106f
at elbows, 107f
in flow lines, 17
impingement corrosion, 109, 110f
wear-corrosion, 108-109, 108f-109f
ethanolamines, 399
ethylene propylene diene terpolymer (EPDM), 204, 214, 216
explosive decompression, 215, 255

extended-analysis coupons, 444, 444f
extended batch treatment, 327
external casing corrosion, 15–16, 289
external corrosion, 19
   control of, 31–33, 31f
      atmospheric zone, 33
      splash zone, 32–33
      submerged zone, 32
   CUI, 5, 77, 380
   of drill pipes, 28
   in flow lines, 17, 258, 290
   mitigation, 86
   in packer fluids, 16
   protective coatings and, 371
   in storage tanks, 19, 255
   of well casing, 289
external thermal insulation, 33. *See also* lagging

F

face centered cubic (FCC), 145, 145f
fatigue failure, 28
   corrosion, 125–129, 126f–128f, 129t
   of metals in air, 123–125, 125f
   in plants, 124, 125f
FCC. *See* face centered cubic (FCC)
FEPM. *See* tetrafluoroethylene propylene (FEPM)
ferrite, 150t, 151t
ferritic stainless steel, 151t
ferrous alloys, 95, 97, 156
   carbon and alloy steels, 158
   cast irons, 171
   chromium-containing steels, 158–159
   chromium-molybdenum steels, 159
   nickel-containing steels, 159–160
   stainless steels, 160–171
FFKM. *See* perfluoroelastomers (FFKM)
fiber reinforced plastic (FRP), 181. *See also* FRP pipe; FRP sucker rods; FRP tanks
   applications of nonmetallic pipe and, 198–199
   disadvantages unique to, 191
   offshore firewater pipe, 210–211
   piping, 188f–189f, 193
   production and injection tubing, 209–210
   repairing corroded steel, 211–212
   thermosetting, 187, 194, 198–199
   for tubular goods, 188f–189f

field grades of test coupons
   erosion damage, 442f
   general corrosion, 440, 440f
   pitting corrosion, 440, 441f
filming efficiency, 309
filming time, 309–310, 361
film persistency, 307, 309, 316, 325
filtration, 399, 404–405
fire floods. *See* in situ combustion projects
fire performance
   as factor for nonmetallic piping, 192
   of plastics, 197
FKM. *See* fluorocarbon (FKM)
flange connections, 195, 195f
flanged joints, 25–27
flange isolation, 213–214
flexible pipe, 199–200, 408
flowing wells, 14–15
   auxiliary equipment, 15
   gas wells, 14–15
   oil wells, 14
flowlines, 162, 201
   cathodic protection of, 290, 482
   FRP pipe for, 206–207
   and gathering systems, 329–332
   internal corrosion of, 343
   and protective coatings, 258–259
   subsea, 263, 297
   as surface equipment, 17
flow velocity, 21–22, 329, 377
fluid detection in online measurement techniques, 422–423, 422f
fluids, packer, 16–17, 99, 406
fluorocarbon (FKM), 215, 216, 216t
   fluorocarbon polytetrafluoroethylene, 25, 26f, 183, 213–216
fluoroelastomers, 215, 314
flushing, 404, 407
force curing, 233
formation squeeze treatment, 321, 324, 325–326, 355–357
formic acid, 60, 340, 390
free-floating bacteria, 406
free machining grades, 136, 160, 163
FRP pipe
   designing, 205–206
   failures of, 205
   for flowlines, well lines, manifolds, 206–207

manufacturing, 204–205
offshore firewater, 210–211
oil and gas production, 204–207
portability of, 211–212
FRP sucker rods, 209
FRP tanks, 207–209, 208f–209f
fusion-bonded coatings, 234
future inhibitor injection skid, 373

G

galvanic anode in CP systems, 269–271
    advantages and disadvantages of, 270t
    definition of, 269
    design and installation of, 271
    installed on offshore structures, 296–297, 296f, 297f
    performance analysis, 269t
galvanic corrosion, 2, 73, 97–105, 97f, 165
    area principle, 98–99
    electric isolation to prevent, 213, 384
    electrochemical corrosion, 101f
    failure mechanism of, 165
    galvanic series. *See* galvanic series
    mill scale, 100–101, 101f
    oxygen and depolarizers, 100
    polarity reversal, 100
    preferential weld corrosion (PWC), 102–105
    ringworm corrosion, 101–102, 102f
galvanic couples, avoiding, 383–384
galvanic probe, 449, 450
galvanic series, 99, 100, 269, 281t, 383
gas blanket
    checking, 387
    design of, 386–388
    oil blanketing, 388
    in power oil tank, 324
    self-blanketing, 387
    on tanks, 19, 417
gas compression systems, corrosion inhibition in, 334–335
gas dehydration, 19
    filtration in, 404
    using glycol, 398–399
    using molecular sieve, 399
gas handling system
    corrosion coupon location for, 435–437, 436f, 437f
    corrosion inhibition in, 333–334

gaskets, 163
    fluoropolymer elastomeric, 175
    heat-resistant, 259
    insulating, 214, 381
    nonmetallic, 213
gas-lift gas, 323, 365–366
gas-lift oil wells, 323–324
gas stripping towers, 391–392, 392f
gas sweetening systems, 19
    corrosion inhibition in, 335–336
gas wells, 14–15, 108, 325. *See also* oil and gas wells
    batch treating frequency for, 323
    downhole corrosion inhibitor application techniques for, 349–368
    treating string approach in, 323
general corrosion, 73, 76–79, 76f–78f
general metal loss, 74, 465t, 466t
glass-reinforced epoxy (GRE), 198, 214
glass-reinforced plastic (GRP), 198
glutaraldehyde, 395, 410–411, 412
glycol
    Filtration, 399
    pH control, 398–399
    reflux tower corrosion, 399
gouge, 74
grain boundaries, 104, 113, 114, 135, 146, 173
grain size effects, 146, 147–149, 147f–149f, 153, 159
groove, 75, 77f, 103, 213
gypsum ($CaSO_4 \cdot 2H_2O$), 396

H

hardenability, 151t, 158
HDPE, *See* polyethylene
heating and cooling media
    corrosion inhibition in, 337–338
    molybdate inhibitors, 338
    nitrite inhibitors, 337
    silicate inhibitors, 338
    water quality, 338
heat tint (heat-tinted welds), 96, 104, 105f, 164–165
heat-treated alloys, 157
heat treatment, 129
    alloys, 157, 171
    defined, 151t
    localized, 155

postweld, 131, 214, 381
preventing intergranular corrosion using, 114
SSC, 58
of steels, 152–155, 161, 166
heat welding, 193
heterogeneous alloys, 156
highly alloyed austenitic stainless steels, 95, 163, 170–171
high temperature attack, 122
HNBR. *See* hydrogenated acrylonitrile butadiene (HNBR)
homogeneous alloys, 156, 157
horizontal drilling, 341. *See also* directional drilling
horizontal well(s)
corrosion inhibition in, 341–344
definition of, 342
materials and corrosion concerns, 342–343
shale oils, unique properties of, 342
stimulation with mud acid, 340
hybrid risk assessment, 475
hydraulic pumped wells, 13, 326
hydraulic pump oil wells, 324–325
hydrazine (N_2H_4), 397
hydrogen blistering, 20, 115–117, 115f, 450
hydrogen embrittlement, 27, 32, 118–119, 125, 297
CP result in, 129, 173
cracking of high-strength steels by, 132
SSC by, 162
in wet H_2S, 137
hydrogen-induced failures, 115
hydrogen induced cracking (HIC), 74, 115–116, 116f
hydrogen induced stress cracking (HISC), 167
hydrogen probes, 449, 450–451
intrusive, 116, 117f
patch/nonintrusive, 116–117, 117f
Hydrogen Sulfide (H_2S)
chemical scavenging, *See* reservoir souring
corrosion caused by 6, 14, 19-20, 47-48, 55-57, 80, 82, 94
corrosion product analysis, 64-65
effect on corrosion fatigue 128-129
effect on elastomers and thermoplastics, 216t
effect on galvanic corrosion, 99-100
fugacity, 318
gas analysis, 431

in horizontal wells, 343
hydrogen blistering, 115-116
hydrogen embrittlement, 27, 118-119, 136
hydrogen induced cracking (HIC), 75, 115-116
reaction with catalyzed O_2 scavenger, 396
removal by gas sweetening, 109, 110f, 335-336
reservoir souring, 30, 408-409
service envelopes for 13Cr, 161
service envelopes for 17-4 PH stainless steel, 166
service envelopes for duplex stainless steel, 168
service envelopes for precipitation hardened nickel-based alloys, 173-174
service envelopes for solid solution nickel-based alloys, 171-173
sulfate reducing bacteria (SRB), 59-60, 92-93, 405-406
sulfidation (high temperature), 77, 159
sulfide stress cracking (SSC) 26-27, 57-59, 74, 95, 119-122
hydrogenated acrylonitrile butadiene (HNBR), 215, 216, 216t
hydroquinone, 397
hydrotest water
chemical treatment of, 409–411
chloride content of, 410
compatibility of biocide and oxygen scavenger, 410
draining and drying after hydrotesting, 411
glutaraldehyde, 410–411
in MIC, 409–411
need for corrosion inhibitor, 410
no chemical additive to, 411
regulatory control on discharge, 410
THPS, 411

I

imidazolines, 306, 306f
impingement corrosion, 73, 107, 109, 110f
impressed current anode, 273–275
impressed current CP system, 271–278
advantages and disadvantages of, 273t
in CP application, 297
impressed current anode, 273–275, 274f, 275f
installation of, 271–272, 272f
power sources, 276

indirect measurement techniques, 421–431
  gas analysis, 431
  microbiological detection and monitoring, 431
  offline techniques, 423–430
  online techniques, 421–423
injection fittings, 345
injection point corrosion, 309, 335, 395
inorganic corrosion inhibitors, 302, 337
  anodic, 303
  cathodic, 303–304
in situ combustion projects, 29
inspection. *See also* corrosion monitoring and inspection
  in coating application, 245–250
  coating, check points, 246t
  crack, 131
  equipment, 144
  periodic, 25
  radiographic, 199
  timing, 246
  tools, 246–250, 247f–250f
  visual, 25, 79, 126, 452–453
insulation, 380–384. *See also* electrical isolation
  thermal, 5, 33, 135, 263, 453
integrated corrosion control program, 485
intergranular corrosion, 74, 104, 113–115
  in austenitic stainless steels, 113–114, 114f
  preventing, 114–115
intermediate coats, 236, 236t
internal casing corrosion, 16
internal corrosion, 8, 419
  control of, 373, 375
  corrosion inhibitor as barrier to, 342–343, 480
  in dead legs, 23–25
  of flow lines, 17, 342, 343
  galvanic couple prevention, 380
  monitoring with coupons, 434
  in production system, 10
  in storage tanks, 19
  of wells, facilities, and equipment, 37–42
  in wet gas, 399
internal polymer liners, 200–204
  for pipelines, 203
  polyethylene liner, 200–203
  tank and slurry, 203–204
intrusive hydrogen probes, 116, 117f

intrusive inspection
  downhole inspection, 462–464
  pipeline inline inspection, 464–466, 465f, 465t–466t
  tank bottom inspection, 466, 466f
intrusive monitoring methods, 432–451
  access fittings, 432–433, 433f
  defined, 432
  electrical resistance probes, 444–448
  electrochemical methods, 449–451
  extended-analysis coupons, 444, 444f
  mass loss coupons, 434–443
  test spools, 448
iron corrosion, 43–44, 336
isotropic polycrystalline metals, 146–147, 152t

K
knife-line attack, 103, 105

L
laboratory inhibitor testing, 317–319
lagging, 33
laser profilometry scan of coupon surface, 442f
lifecycle cost, 144, 229
linear polarization resistance (LPR) probes, 449
Liquid Metal Embrittlement, 138
localized corrosion, 5, 56, 73, 79–85, 81f–83f
  in carbon and low-alloy steel, 30, 94, 94f
  concentration cell corrosion, 85–97, 85f–86f
  conditions for H_2S, 56–57
  corrosion resistant alloys, 95–97
  in dissolved CO_2, 50, 50f, 51f, 92
  general *vs.*, 77
  oxygen tubercles, 88, 88f
  pitting corrosion, 79–85, 83f–85f
  pitting of metals and, 94–97
local thin area, 75
low-alloy steel, 151t. *See also* carbon and low-alloy steel
lower critical temperatures, 152t

M
mackinawite, 55, 56, 57, 69
magnesium anodes, 269–270, 292
magnetic particle inspection, 458–460, 459f
maintenance coating, 251–252
martensite, 150, 150t, 154

martensitic stainless steels, 95, 96, 109, 151t, 160–162, 165
  cold temperature service, 162
  effect of oxygen, 162
  precipitation hardening, 166
  service envelopes, 161
  weldability, 162
mass loss coupons, 434–443
  care and handling, 437–438
  coupon types and mountings, 434–435
  exposure from environments, 437
  exposure time, 438
  location and position, 435–437, 436f
  results, evaluation and reporting, 438–443, 443f
Material Safety Data Sheet (MSDS), 315, 316
materials selection. *See also* metallic material selection; nonmetallic material selection
  metallic, 143–176
  nonmetallic, 181–220
  standards and papers on, 144
  strategy, 142
  types of costs with, 143–144
mechanical properties, 152t
  as factor for nonmetallic piping, 192
mesa corrosion, 51f, 80
metallic material selection
  alloys, 149–150, 156–157
  crystalline structure, 145–149, 145f
  ferrous alloys, 158–171
  grain size effects, 147–149, 147f–149f
  heat treatment of steels, 152–155
  nonferrous alloys, 171–176
  polycrystalline metals, 146–147
  specifications, standards and codes, 176–177
  steels, 150–152
metal (iron and manganese) oxidizing bacteria (MOBs), 59, 60
methyl ethyl ketoxime, 397
microbiologically influenced corrosion (MIC), 30, 59–61, 62, 73, 92–93, 165
  APBs, 60
  with attached bacteria, 406
  of carbon steel, 92f–93f
  control of, 336–337, 343, 405–411
  field monitoring, 60–61
  hydrotest water, 409, 411
  MOBs, 60
  MMMs, 61
  pitting attack of alloys with, 80
  problems by sulfate reducers, 406
  PWC caused by, 105, 105f
  reservoir souring, 408–409
  SRBs, 59–60
  water injection systems, 405, 406–407
mild carbon steel, 151t, 152
mill scale, 100–101, 101f
molecular microbiological methods (MMMs), 61
molybdate inhibitors, 338
Monel K-500 (N05500), 136, 138, 173
morpholine, 335
mothballed equipment, 23
multifinger caliper
  as inspection tool, 462–464, 462f–463f
  limitations, 463–464

N

NACE MR0175/ISO 15156, 26, 43, 58, 119, 122, 123, 135, 155, 157, 161, 162, 166-174, 177
NBR, *See* nitrile butadiene rubber (NBR) nitrile
neoprene (polychloroprene), 204, 216
nickel-containing steels, 159–160
nitrile butadiene rubber (NBR) nitrile, 204, 214, 215, 216, 216t
nitrite inhibitors, 337
nitrogen displacement treatments, 321
nitrogen formation squeeze treatment, 321, 351–352
N,N-Diethylhydroxylamine, 397
nonferrous alloys, 156
  aluminum and aluminum alloys, 176
  cobalt-based alloys, 174
  copper-based alloys, 175–176
  nickel-based alloys
    precipitation-hardened, 173–174
    solid-solution, 171–173
  solid-solution nickel-based alloys, 171–173
  titanium alloys, 174–175
  tungsten alloys, 176
nonintrusive hydrogen probes, 116–117, 117f. *See also* patch probes
nonintrusive techniques
  eddy current testing, 460–461, 461f
  magnetic particle inspection, 458–460, 459f
  penetrant testing, 460, 460f
  radiographic inspection, 456–458, 457f–458f

ultrasonic testing, 453-456, 454f-456f
visual testing, 452-453, 452f
nonmetallic material selection
  applications of, 197-200
  cement and concrete, 181-182, 182f
  elastomers, 212-216
  factors to consider for, 191-196
  FRP offshore firewater pipe, 210-211
  FRP pipe for oil and gas production, 204-207
  FRP production and injection tubing, 209-210
  FRP sucker rods, 209
  FRP tanks, 207-209, 208f-209f
  nonmetallic pipe, 190-204
  offshore riser splash zone protection, 216-217
  polymers, 183
  repair systems for corroded steel, 211-212, 211f
  standards for, 220-225
  thermoplastics, 183-187, 185t, 186f-187f
  thermosetting, 187-189, 188f-189f
nonmetallic pipe, 190-204
  advantages of, 190
  applications in oil and gas production, 197-200
  disadvantages of, 190-191
  factors to consider before selecting, 191-196
  internal polymer liners for, 200-204, 202f-203f
  joining methods for, 193
normalizing, 102, 152t, 153, 154, 155
Nylon 6, 185t

O

offline techniques, in corrosion monitoring and inspection, 423-430
  acid gases, 429
  analysis, 426
  chemical methods, 423-430
  corrosion product analysis, 428-429
  dissolved iron, 425f
  dissolved oxygen, 429-430
  evaluation and reporting, 426-428, 427f-428f
  inhibitor presence or concentration tests, 430
  pH, 430
  sample connections, 423-424, 424f
  sampling, 425-426
offshore operations, 30-33, 409
  corrosion control, 31-33
  onshore vs., 30
  reservoir souring, 30
offshore riser splash zone protection, 216-217
oil and gas wells, 14
  downhole corrosion inhibitor application techniques for, 349-368
  sources of oxygen in, 49, 385
oil blanketing, 388
oil soluble-water dispersible inhibitors, 310
oil soluble-water insoluble inhibitors, 310
online techniques
  in corrosion monitoring and inspection, 421-423
  fluid detection in, 422-423, 422f
  process parameters in, 422
operating costs (OPEX), 144, 345, 390
  CAPEX vs., in coatings, 229f
operational dead legs, 23
organic corrosion inhibitors, 302, 304, 304f, 309, 330, 413
  other forms of, 306
original construction coating, 232, 251
original manufacturing defect, 63
O-rings, 204, 212, 213, 220
overall corrosion control program, 485
oxalic acid, 113, 390
oxygen
  for CSCC, 134-135
  and depolarizers, 100
  entry in water systems, 388-389, 388f
  sources of, in gas plants, 389-390
oxygen exclusion, corrosion control and, 385-390
oxygen removal
  chemical scavengers, 393-397
  in corrosion control, 390-397
  in gas stripping towers, 391-392, 392f
  in vacuum stripping towers, 392-393, 393f
oxygen tubercles, 49, 88, 88f

P

packer fluids, 14, 16-17, 99
  brine, 310, 406
  SRB in, 406
partial pressure, 26, 51-52
passive oxide film, 79, 93, 160, 164, 165, 181
patch probes, 116-117, 117f

PE, *See* polyethylene
Peabody, A.W., 281
pearlite, 150*t*
PEEK. *See* polyetheretherketone (PEEK)
penetrant testing (PT), 460, 460*f*
perfluoroelastomers (FFKM), 215–216, 216*t*, 314
permanent dead legs, 23
PE-RT, *See* polyethylene
PEX, *See* polyethylene
pH, 46–47, 47*f*, 48*f*, 57
   in corrosion monitoring, 430
   for CSCC, 133–134, 134*f*
   of solution, 122
pH control, 398–399
   in cooling water systems, 414
   in heating water systems, 414
   in wet acid gas, 413
physical dead legs, 23
physical variables, 46, 61–62
   pressure, 62
   temperature, 61–62
   velocity, 62
pigment, 231, 233
pig/pigging, 400–404
   chemical batching pigs, 402
   corrosion control design for, 375–376
   maintenance, 402–404
   maintenance, pipelines benefiting from, 400
   maintenance pig selection, 401
   routine, 402
   train, 402–403
   types of, 332
pipeline
   benefiting from maintenance pigging, 400
   inline inspection, 464–466, 465*f*, 465*t*–466*t*
   pH stabilization in wet acid gas, 413
   top-of-the-line corrosion, 330
pipe/piping, 20–27
   dead legs, 23–25, 24*f*
   design. *See* piping design
   external coatings in, 259–263
   flexible, 199–200
   flow velocity, 21–22
   FRP, 188*f*–189*f*. *See also* FRP pipe
   internal coatings in, 258–259
   internal polymer liners for, 200–204
   joining methods for, 193
   polyethylene liner, 200–203

   spoolable, 199
   stagnant zones, 23–25
   subsea, 263
   turbulence areas, 22
pipe shoes, 379
piping design, 376–380
   dead legs in, 377–378
   sand effects in, 378–379
   support design, 379–380
   thermally insulated piping, 380
   velocity effects in, 378
   water traps in, 377, 377*f*
pit, 28, 73, 182, 409, 411
pitting, 75
pitting corrosion, 24, 73, 79–85, 81*f*–85*f*, 94–97, 94*f*, 96*f*
pitting resistance equivalent number (PREN), 167, 169
planktonic bacteria, 406, 431
platform coating, 251
poisons. *See* arsenic ions; sulfide ions
polarity reversal, 100
polarization, 3, 100, 279, 288, 449–450
polyamide (PA), 185*t*, 199, 203
polycrystalline metals, 152*t*, 156
   anisotropic, 146–147
   isotropic, 146–147
polyetheretherketone (PEEK), 183, 213, 215
polyethylene (PE, HDPE, PE-RT, PEX), 184, 185*t*, 190, 199–200
polyethylene liner, 200–203, 202*f*–203*f*
polyphenylene sulfide (PPS), 183–184, 215, 216
polypropylene (PP), 185*t*, 190
polytetrafluoroethylene (PTFE), 25, 26*f*, 183, 213–216
polyvinylidene fluoride (PVDF), 184, 199, 215
population, types of, 406
potential of the metal, defined, 44
PPS. *See* polyphenylene sulfide (PPS)
precipitation hardening, 152*t*, 165–167
   austenitic, 166–167
   high strength, SSC-resistant, 167
   martensitic, 166
preferential weld corrosion (PWC), 90, 102–105
   of austenitic stainless steel, 103–105, 103*f*–105*f*
   caused by MIC, 105, 105*f*
   of steel, 102–103, 103*f*

preferred orientation, 146
pressure, in corrosion reaction, 62
primers, 233, 235*t*
   barrier, 235
   inhibitive, 235-236
   sacrificial, 236
processing equipment, 18-19
   gas dehydration, 19
   gas sweetening, 19
   separators, 18, 90*f*, 290-291, 332, 334, 390
progressive cavity pumped wells, 13-14
propionic acid, 54
protective coatings
   CAPEX *vs.* OPEX, 229*f*
   chemical composition, 233
   corrosion under insulation (CUI), 254
   curing methods, 233-234
   formulation, 230-232
   misconceptions in, 264
   overview, 227-230
   performance considerations, 229*f*
   production facilities and equipment, 250-264
   reinforced coatings, 234
   role in cathodic protection, 298
   selection dependencies, 237-240
   systems, 234-237
   thickness, 232
   types, 232
   typical liquid-applied, 230*t*
PTFE. *See* polytetrafluoroethylene (PTFE)
pumps, 417
   as source of oxygen entry, 388, 388*f*
PVDF. *See* polyvinylidene fluoride (PVDF)
pyrite, 55
pyrrhotite, 55, 56, 57

Q

qualitative risk assessment, 474-475
quantitative risk assessment, 475
quenching, 152*t*, 153-154
quick field analysis methods, 64, 65*t*

R

radiographic inspection, 199, 456-458, 457*f*-458*f*
rebar corrosion, 182
reference electrodes, 284, 285*f*
   in CP application, 295-296
   in CP criteria, 278-279, 278*f*, 279*t*
reflux tower corrosion, 399
reinforced coatings, 234
reservoir souring, 30, 372, 407
   defined, 408
   in MIC, 408-409
resin, 90, 231
resin rich layer, 205
ring type coupons, 434
ringworm corrosion, 101-102, 102*f*, 155
risk assessment, 474-479
   barriers, 477-479, 477*t*-478*t*
   data requirements in, 475, 475*t*
   level of assurance in, 479
   segments or circuits in, 475
   setting the stage, , 475, 475*t*
   threats in, 475-476, 476*t*
risks
   assessment, 474-479
   consequence of failure, 479
   in corrosion management, 474-481
   matrix, 479-481, 480*f*, 481*f*
   probability of failure, 479
routine pigging, 400, 402

S

sacrificial anode systems
   design and installation of, 271, 271*f*
   use of, 298
sacrificial coating, 233, 234*f*
sacrificial primers, 236, 251
sand effects in piping design, 378-379
sand erosion, 57, 378, 379
   mitigation of, 111, 111*f*
scale and deposits, 90-93
   microbiologically influenced corrosion (MIC), 92-93, 92*f*-93*f*
   under-deposit corrosion (UDC), 90, 90*f*, 91*f*
self-blanketing, 387
semicrystalline thermoplastics, 184
semiquantitative risk assessment, 475
sensitization, 104, 113-114
sensitized microstructure, 133, 135-136
separators, as processing equipment, 18, 390, 406
sequestration, carbon capture and, 29-30
sessile bacteria, 400, 406, 431
shale oils, 342

silicates, 338
sodium bisulfite (NaHSO$_3$), 393, 394, 394t, 395
sodium hydroxide (NaOH), 130, 198
sodium sulfite (Na$_2$SO$_3$), 393, 394, 394t, 395
soil
    components in galvanic anode installation in, 271
    corrosion, 5–6, 6t, 15, 16
    resistivity measurements, 285–286, 286f, 289
solid-solution nickel-based alloys, 171–173
solubility, 51, 310–311
    in carrier fluid, 311
    of CO$_2$, 53, 53f
    of oxygen in water, 68f
    at system temperatures, 311
solution annealing, 152t, 166, 168, 173
solution heat treatment. *See* solution annealing
solvent
    coatings, 231, 241, 242, 244–245
    corrosion inhibitor, 305, 305f, 306–307, 311, 314, 323, 357, 365
    welding, 193, 194, 194f
sources of oxygen
    in gas plants, 389–390
    in producing wells, 385
    pumps as, 388, 388f
    water injection wells as, 389
    water supply wells as, 389
    in water systems, 385–389
sour corrosion, 50, 318
spheroidizing, 153
splash zone, 32–33, 251
    protection systems, 216–217
spoolable pipe, 193, 196, 198, 199
stagnant zones, 23–25
stainless steel, 151t, 160–171. *See also specific types*
    austenitic, 95–96, 96f, 163–165
        PWC of, 103–105, 104f–105f
        sensitization of, 113–114, 114f
    bolt failures, 27
    definitions of, 151t
    duplex, 96, 167–171
    ferritic, 162–163
    vs. FRP, during designing, 205–206
    heat treatment of, 152–155
    high strength, 132–133
    martensitic, 95, 160–162
    metallurgy and, 150–152
    microstructure, 54–55
    names for, 151
    precipitation hardening, 165–167
    structure of, 150, 150t
    UDC pitting of, 93f
stepwise cracking (SWC), 59, 74–75, 115, 115f
storage tank, 19–20
    corrosive zones in, 20f
    in CP application, 293–295
    external protection, 294–295
    floor scanning, 466f
    FRP, 208f
    internal protection, 293
strain hardening, 27, 133, 135
stress corrosion cracking (SCC), 74
    aluminum alloys, 137
    austenitic alloys, 133–136
    austenitic stainless steels, 164–165
    carbon and low-alloy steels, 130–132
    copper-based alloys, 136
    high strength steels, 132–133
    resistant precipitation hardened alloys, 167
    titanium alloys, 137
structure-to-electrolyte potential measurements, 281–283, 281t, 282f, 283f
submerged corrosion, 5, 6, 6t
submerged zone, 32, 251, 295
subsea, 132, 144
    applications, 173, 348
    bolt failures, 27
    pipelines, 263, 263f
    structures, 297, 298f
sucker rod pumping wells, 13, 326
sucker rods, FRP. *See* FRP sucker rods
sulfate reducing bacteria (SRB), 30, 59–60, 92, 290, 337, 405, 408, 429
sulfide ions, 3, 60, 115
sulfide stress cracking (SSC), 57–59, 58f, 74, 119–122
    conditions required for, 121
    factors affecting susceptibility, 122
    failure, 26, 30, 55
    typically intergranular, 120–121, 121f
    susceptibility and steel strength, 119–120, 119f–120f
sulfur dioxide (SO$_2$), 5, 9, 393, 394, 394t, 396

surface casing corrosion, 16
surface equipment, 17, 461
   bolted equipment, 25–27
   external cathodic protection, 294-295
   internal cathodic protection, 290–293, 291f, 292f
   flanged joints as, 25–27
   flow lines, 17
   wellhead and Christmas tree, 17
surface facilities, 49, 94, 95, 168, 274, 451
   chemical batch treatment, 331–332
   flowlines and gathering systems, 329–332
   inhibition of, 329–338
   production and plant facilities, 332–338
surface potential method, 284, 285f
surface preparation standards, 243t
sustainable corrosion management, 483–484, 483f
sweet corrosion, 50, 391, 406

T

tank bottom inspection, 207, 271, 278, 466, 466f
tank(s). *See also* storage tanks
   batteries, 381
   coatings in, 254–255
   externally coated, 228f
   FRP. *See* FRP tanks
   newly coated, 228f
   shell, 207
   and slurry pipeline liners, 203–204
temperature, 46
   bolt, 134
   in corrosion reaction, 61–62
   for CSCC, 133
   maximum and minimum service, 216
tempering, 152t, 153–154
tensile strength, 123, 152t, 220
tetrafluoroethylene propylene (FEPM), 215
tetrakis(hydroxymethyl) phosphonium sulfate (THPS), 395, 408, 410, 411, 412
thermal insulation, 5, 33, 135, 263, 453
thermally insulated piping, 254f, 374, 380
thermogalvanic corrosion, 73
thermoplastics, 183–187
   amorphous, 184
   applications of nonmetallic pipe, 197–198
   maximum and minimum service temperature for, 216
   semicrystalline, 184
   service considerations for, 184–185, 185t
thermoplastic seals, 181, 215–216, 216t
thermosetting, 187–189, 190, 210
   FRP, 194, 198–199
   materials, 183
THPS. *See* tetrakis(hydroxymethyl) phosphonium sulfate (THPS)
threaded connections, 86, 124-125, 189f, 194–195
titanium alloys, 79, 137–138, 174–175
topcoat in coating systems, 236–237, 236t
top-of-the-line corrosion, 330
   controlling pH stabilized pipelines, 413
toughness, 145, 152t, 159, 162, 164, 168, 231
transformation ranges, 152t
Transport Canada TP14612, 184, 192, 224
treating string approach, 323, 354–355
triazine, 408
tubercles, defined, 49, 50f
tuberculation, 49, 50f
tubing displacement treatment, 320–321, 349–351
tubular goods, coatings in, 255–258, 256f–258f
tungsten alloys, 96, 176
turbulence areas, 20, 22
two-electrode survey method, 285

U

ultrasonic testing, 374, 453–456, 454f–456f
under-deposit corrosion (UDC), 18, 90, 90f, 91f, 165, 338, 412, 456
Unified Numbering System (UNS), 145
uniform corrosion, 72, 73, 75, 77
United States Coast Guard (USCG), 184, 192, 194, 225
unsustainable corrosion management, 482, 482f
upper critical temperature, 152t, 153

V

vacuum stripping towers, 392–393, 393f
vehicle, 231–232
velocity effects, in piping design, 378
velocity, 378
   in corrosion reaction, 62
   erosional, 21

as factor affecting corrosion rates in
    production environments, 7t
as factor affecting corrosivity of waters, 6t
flow, 21–22
vessels and tanks
    corrosion inhibition in, 332–333
    internal cathodic protection, 290-293, 291f,
        292f
    protective coatings, 254–255
visual inspection, 25, 79, 126, 295, 374, 452,
    452f, 453, 464
visual testing, 452–453, 452f
volatile corrosion inhibitors (VCIs), 339, 341
volatile organic acids, 54

W

wall-loss damage
    mechanisms involving, 73
    mechanisms not involving, 74
wash bailer treatment, 329, 360–361
water alternating gas systems (WAG)
    injection, 28–29
water corrosivity, 6, 6t
water dew point, 132, 333, 397–398
water injection systems, 377
    checklist for troubleshooting oxygen-free,
        417–418
    control of MIC in, 406–407
water injection wells, as source of oxygen entry,
    389
water quality, 338, 409
water resistivity measurements, 285–286, 286f
water soluble-oil dispersible inhibitors, 310
water soluble-oil insoluble inhibitors, 310
water supply wells, as source of oxygen entry,
    389
water systems, 27–28
water traps, in piping design, 377, 377f, 379
wear-corrosion, 73, 108–109, 108f–109f, 464
weighted liquids treatment, 357–359
welding, 103, 123
    heat, 193
    solvent, 194

weld line corrosion, 103, 104, 104f, 155
well acidizing inhibitors, 340–341
well casings, 289, 298
wellhead equipment, 157, 161
    defined, 17
wells, 11–17, 11f–12f. See also oil and gas wells
    artificial lift, 12–14
    ESP pumped, 13
    flowing, 14–15
    gas, 14–15
    hydraulic pumped, 13
    oil, 14
    progressive cavity pumped, 13–14
    sucker rod pumping, 13
wells without packers treatment, 326–329
    automated (semibatch) treatment, 328
    batch and flush treatment, 327
    batch treating frequency, 328
    continuous treatment, 328
    extended batch treatment, 327
    fluid level, 326–327
    infrequently used downhole treatment
        methods, 328–329
wet acid gas
    pH control in, 411, 413
    pH stabilization in, 411, 413
wireline/wireline damage, 29, 57, 108, 108f, 255,
    257, 258f, 263, 324, 329, 346, 359, 360, 366, 464

X

X-ray diffraction (XRD), 63, 64, 145

Y

yield strength, 26, 118, 132, 147, 152t, 173
yo-yo treatments, 322, 352–354

Z

zero corrosion, 481
zinc anodes, 270, 292
zinc-rich primers, 236

CPSIA information can be obtained
at www.ICGtesting.com
Printed in the USA
JSHW062339260723
45462JS00002B/9